Natural Product Chemistry

Natural Product Chemistry

A mechanistic and biosynthetic approach to secondary metabolism

Kurt B. G. Torssell

Kemisk Institut
Aarhus Universitet

JOHN WILEY & SONS LIMITED
Chichester · New York · Brisbane · Toronto · Singapore

Library of Congress Cataloging in Publication Data:
Torssell, Kurt B. G., 1926–
Natural product chemistry.

Includes bibliographies and indexes.
1. Metabolism, Secondary, 2. Natural products.
I. Title.
QP171.T67 574.1′33 82-2045

ISBN 0 471 10378 0 (cloth) AACR2
ISBN 0 471 10379 9 (paper)

British Library Cataloguing in Publication Data:
Torssell, Kurt B. G.
Natural product chemistry.
1. Biosynthesis 2. Natural products
I. Title
574.19′29 QH345

ISBN 0 471 10378 0 (cloth)
ISBN 0 471 10379 9 (paper)

Typeset by Bath Typesetting Ltd., Bath
and printed in Great Britain by Page Bros. (Norwich) Ltd.

To Magda

Contents

Preface

The chemistry of natural products has made great progress during the last decades as a result of our better comprehension of enzymatic processes, and the development of biogenetic and biosynthetic theories which logically classify and link together an immense variety of compounds. The endless compilation of compounds could eventually be substituted by biosynthetic schemes easy to survey. Natural product chemistry gradually changes from being traditionally descriptive in nature to being mechanistic and predictable. An enzymatic reaction is no longer just a transformation of reactant into product with unknown transition, but can be or should be understood in the same way as any other analogous reaction carried out non-enzymatically in the test tube. In the present text I have tried to present the formation and reactions of natural products using the vocabulary of ordinary organic reaction mechanisms. It is important to understand the technique nature uses to bring about its multitude of astonishlingly elaborate compounds. The underlying biosynthetic principles and enzyme mechanisms have had a great impact on the imagination of many organic chemists, and imitation of bio-organic processes in the laboratory has led to important advances in synthetic methods. It is my belief that the further development of natural product chemistry will become more biologically oriented. Secondary metabolites shall no longer be regarded as waste products, we just do not know yet how they function. We are at present in a transition of re-evaluation. They certainly play some role necessary for the existence of the organism in its environment. Fruitful research on biochemical interactions in ecosystems is ahead of us which must involve cooperation between chemists and biologists. Recent developments in the area of chemical communication among insects is a stunning example of the opening of new frontiers in science and of the importance of multidisciplinary teamwork.

I have not restricted myself to present established facts only in this text. A textbook ought to be at the frontier. I have tried to demonstrate the dynamic state of mechanistic biosynthetic research by topically following the current development up to the end of 1981 and occasionally I have preferred to give my own mechanistic interpretation of the events. The critical reader will certainly be able to differentiate the speculative from the accepted. There is often a rich flora of mechanistic proposals to choose among.

I have grouped together the nitrogen containing compounds—the amino acids, alkaloids and *N*-heteroaromatics—in Chapters 6, 7 and 8. However, it is recommended to read Chapter 6, particularly the pyridoxal catalysed reactions (section 6.4), after the carbohydrates (Chapter 2) because they provide the basis for some of the reactions discussed in the other chapters.

I wish to express may thanks to Professors R. H. Thomson, Aberdeen, and A. Kjaer, Lyngby, for reading the whole manuscript and for their valuable criticism. Not the least are my thanks due to Mrs. Ella Larsen for her indefatigable typing work and for drawing all the figures. I appreciate any comments on errors and suggestions for improvements of the text.

KURT TORSSELL
Aarhus 1981

Chapter 1

Introduction and general considerations

1.1 The literature

In addition to current journals several comprehensive series are issued containing reviews on various topics of natural product chemistry. Some of these are issued annually or regularly at longer intervals. A great many monographs on special topics are published. The following condensed bibliographic list represents the main sources consulted by the author for the present textbook and is recommended for further studies. Every chapter is supplied with a list of references documenting recent developments. Regrettably, some references must be chosen by chance and consequently they are not always fully representative.

Textbooks

Organic Chemistry

March, J. *Advanced Organic Chemistry. Reactions, Mechanisms and Structure* 2nd Edn. McGraw-Hill, New York, 1977.

Noller, C. R. *Chemistry of Organic Compounds* 3rd Edn. Saunders, Philadelphia, 1965.

Biochemistry

Lehninger, A. L. *Biochemistry*, 2nd Edn. Worth, New York, 1978.

Stryer, L. *Biochemistry*, 2nd Edn. Freeman, San Francisco, 1981.

Walsh, C. *Enzymatic Reaction Mechanism*, Freeman, San Francisco, 1979.

Natural Product Chemistry

Geissman, T. A. and Crout, D. H. G. *Organic Chemistry of Secondary Plant Metabolism*, Freeman, San Francisco, 1969.

Manitto, P. *Biosynthesis of Natural Products*, Ellis Horwood, Chichester, 1981.

Tedder, J. M., Nechvatal, A., Murray, A. W. and Carnduff, J. *Basic Organic Chemistry. Part 4. Natural Products*, John Wiley, London, 1972.

Comprehensive series, monographs and review articles

Biogenesis of Natural Compounds, Bernfeld, P. (Ed.) 2nd Edn. Pergamon, Oxford, 1967.

Bioorganic Chemistry **1–4** (1977–1978). van Tamelen, E. (Ed.), Academic Press, New York

Biosynthesis. Specialist Periodical Reports, **1–6** (1972–1980), Geissman, T. A. and Bu'Lock, J. D. (Eds.), The Chemical Society, London.

Comprehensive Organic Chemistry, Vol. 5, *Biological Compounds*, Haslam, E. (Ed.), Pergamon, Oxford, 1979.

Devon, T. K. and Scott, A. I. *Handbook of Naturally Occurring Compounds* **1–2** (1972–75). Academic Press, New York

Hegnauer, R., *Chemotaxonomie der Pflanzen*, **1–6** (1962–1973). Birkhäuser, Basel

Karrer, W. *Konstitution und Vorkommen der organischen Pflanzenstoffe*, (1957), Karrer, W., Cherbuliez, E. and Eugster, C. H. *Ergänzungsband 1* (1977). Birkhäuser, Basel.

Marine Natural Products, **1–4** (1978–1981). Scheuer, P. J. (Ed.), Academic Press, New York

Miller, M. W. *The Pfizer Handbook of Microbiological Metabolites*, McGraw-Hill, New York, 1961.

Phytochemistry **1–3** (1973). Miller, L. P. (Ed.), van Nostrand Reinhold, New York

Progress in the Chemistry of Organic Natural Products, **1–39** (1938–1980). Founded by Zechmeister, L., Springer, Vienna.

Advances in Carbohydrate Chemistry and Biochemistry **1–36** (1945–1979), Academic Press, New York.

Carbohydrate Chemistry, Specialist Periodical Reports, **1–11** (1968–1979), Guithrie, R. D. and Brimacombe, J. S. (Eds.), The Chemical Society, London.

Ganem, B. *From glucose to aromatics. Recent developments in natural products of the shikimic acid pathway*. Tetrahedron Report 59, *Tetrahedron* **34** (1978) 3353.

Haslam, E. *The Shikimate Pathway*, Butterworths, London, 1975.

Biosynthesis of Aromatic Compounds, Billek, G. (Ed.), Pergamon, Oxford, 1966.

Packter, N. M. *Biosynthesis of Acetate-Derived Compounds*, J. Wiley, London, 1973.

Crabbé, P. *Prostaglandin Research*, Academic Press, New York, 1977

Thomson, R. H. *Naturally Occurring Quinones*, Academic Press, London, 1971.

Fieser, L. F. and Fieser, M. *Steroids*, Reinhold Publ. Corp., New York, 1959.

Terpenoids and Steroids. Specialist Periodical Reports **1–9** (1971–1979), Overton, K. H. and Hanson, J. R. (Eds.), The Chemical Society, London.

Aberhart, D. J. *Biosynthesis of penicillins and cephalosporins*, Tetrahedron Report 30, *Tetrahedron* **33** (1977) 1545.

Amino Acids, Peptides and Proteins, Specialist Periodical Reports **1–10** (1969–1979), Young, G. T. and Sheppard, R. C. (Eds.), The Chemical Society, London.

Recent Advances in Phytochemistry, **10**, *Biochemical Interaction between Plants and Insects*. Wallace, J. W. and Mansell, R. L. (Eds.), Plenum Press, New York, 1975.

Chemical Ecology, Sondheimer, E. and Simeone, J. B. (Eds.), Academic Press, New York, 1970.

The Alkaloids, I–XVIII (1950–1981), Manske, R. H. F., Holmes, H. L. and Rodrigo, R. G. A. (Eds.), Academic Press, New York.

The Alkaloids, Specialist Periodical Reports **1–9** (1971–1979), Saxton, J. E. and Grundon, M. F. (Eds.), The Chemical Society, London.

Dalton, D. R. *The Alkaloids*, M. Dekker. New York, 1979.

The Porphyrins, I–VII (1978–1979). Dolphin, D. (Ed.). Academic Press, New York.

1.2 Background

Natural product chemistry is in its different aspects an ancient science. The preparation of foodstuffs, colouring matters, medicinals or stimulants are examples of activities as old as mankind. When chemists in the late eighteenth century took the final jump from the world of myths into modern science, the true properties of extracts obtained from nature aroused great curiosity amongst scientists. They began to separate, purify, and finally analyse the compounds produced in living cells. Separation methods were developed and without doubt natural product chemistry has brought great stimuli to the development of the refined techniques we have today, such as the various analytical and preparative chromatographic methods: column chromatography, GC, TLC, HPLC, paper chromatography, electrophoresis, ion exchange, etc. These methods have made it possible to isolate compounds present in extremely small quantities. Structural elucidation was typically carried out by degradation to smaller fragments of known structure combined with investigations of the reactivity pattern and elementary analysis of the compounds. These earlier works led in many instances to discoveries of new valuable reactions and rearrangments. However, without spectroscopy in its service, natural product chemistry would never have attained the status and refinement it has today. UV, IR, NMR, MS, ESR, CD and ORD changed working methods and habits considerably through the years. A large number of spectroscopic data, correlating spectral properties with structure, was collected. These data give us sometimes so much information about the structure that pure chemical transformations can be reduced to a minimum. This enables us now to solve structures with much less material than was needed earlier, but as a consequence of these developments, no doubt, valuable information about the chemistry is lost. When the amount of sample available is very limited and the structure too complicated, X-ray crystallography is an ultimate resource. Programmes are now available that make the structural elucidation almost a routine job for the specialist. The classical cumbersome, degradative work is almost turned into a separatory and X-ray crystallographic problem—provided we can obtain suitable crystals.

1.3 Synthesis and biosynthesis

Two other important aspects of structural determination are synthesis and biosynthesis. A structure was earlier not considered to be rigorously proved unless the compound was also synthesized. The structure of fragments was often confirmed by an unambiguous synthesis. The motivation nowadays for total synthesis is not so much the question of confirmation as the challenge inherent in synthesizing intriguing structures in much the same way as the mountaineer is challenged by the precipices of the mountain ridge. Practical and commercial interests often require alternative preparative procedures. These efforts have brought considerable elegance, creativity and efficiency to the art of synthesis. New methods have been designed and worked out specifically for the purpose so that natural product chemistry has stimulated the development of organic synthesis.

The recognition of biosynthetic principles is the most significant development in natural product chemistry. During the last century a great number of new structures was determined. In the earlier phase organic chemists were just content to solve the structure of natural products and group them according to origin, pharmacological activity or structure, but rather soon the mass of information suggested the need for a more coherent view of biogenesis. The ingenuity and intuition of a few chemists led the painstaking search for the hidden biosynthetic pathways in the right direction. It became evident rather early to Wallach and Ruzicka that the terpenes had a common building block, the isoprene unit, the true origin of which was concealed for a long time, and Winterstein and Trier suggested on good grounds that alkaloids were formed from α-amino acids, ideas which were further developed by Schöpf and Robinson. A peep-hole into the fascinating synthetic workshop of the living cell was gradually opened. The biosynthetic routes of the various classes of compounds were mapped out and precursors and intermediates identified. The enzyme catalysed reactions in the cell have their *in vitro* counterparts and their mechanisms can be correlated with known organic reaction mechanisms.

Biomimetic studies have led to novel and elegant synthetic procedures and biosynthetic principles are constantly applied in the process of structural determination. The substituent pattern is often defined by such considerations and consequently many 'unnatural' structures violating these principles can be discarded as incorrect.

1.4 Primary and secondary metabolism

It is customary to distinguish between primary and secondary metabolism. The former refers to the photosynthetic processes producing simple and widely distributed low molecular weight carboxylic acids of the Krebs cycle, α-amino acids, carbohydrates, fats, and proteins. These compounds are commonly regarded as the domain of biochemists. We will follow this classification and discuss them in their capacity as starting materials—the precursors—of the secondary metabolites. They are more characteristic for the particular biological group, such as family or genus, and apparently the synthetic machinery involved here is related to the mechanism of evolution of species (*cf.* section 1.5). The specific pattern of constituents in species has, in fact, been used for systematic determination. This field of research, called biochemical systematics or chemotaxonomy, has gained increasing interest during the last decades. Natural product chemistry as defined today concerns mainly the formation, structure and properties of the secondary metabolites. There is no sharp division line between the primary 'biochemical' metabolites and the secondary metabolites. The common sugars glucose, fructose, mannose, the function and chemistry of which has been intensively studied by biochemists, are ranged in the first group, whereas closely related rare sugars, such as chalcose, streptose, mycaminose, discovered as constituents of antibiotics and investigated by organic chemists, are ranged as secondary metabolites. The essential amino acid proline is regarded as a primary metabolite but the likewise widely distributed 6-membered analogue pipecolic acid is classified as a secondary metabolite or an alkaloid (Fig. 1).

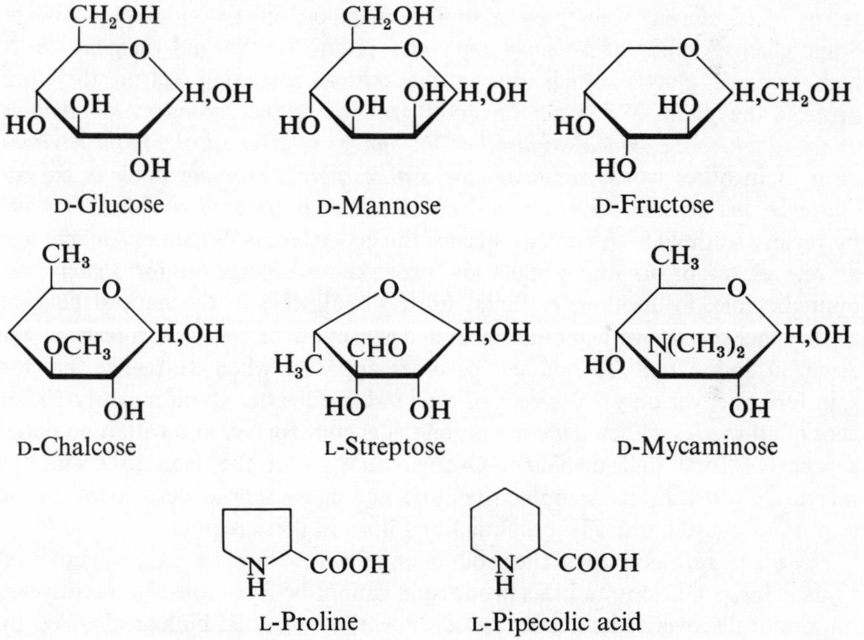

D-Glucose D-Mannose D-Fructose

D-Chalcose L-Streptose D-Mycaminose

L-Proline L-Pipecolic acid

Fig. 1 Primary metabolites and structurally closely related secondary metabolites

1.5 Chemical ecology

A characteristic feature of secondary metabolites is that their function is principally unknown. They are definitely not just waste products; we know too little about them yet. As evident from grafting experiments plants flourish remarkably well both in the absence of several of their normal, characteristic metabolites and in the presence of many extraneous, in somewhat the same way as a human being functions excellently without his blind gut. We will never know, though, what effect a certain metabolite could have had at one stage of the evolution. It is hard to believe that the organism should allocate so large a proportion of metabolic resources for purposes void of sense. The production of secondary metabolites is connected with several external factors, such as replicatory growth, flowering, season, temperature, habitat, length of daylight, etc. For example, young oak leaves contain very little tannin but the concentration increases during the summer to reach a maximum in the autumn. On the other hand, the solanine concentration in potato leaves is reduced during the growing season. For what purpose? It is suggested that an apparent plant like the oak tree forming large populations should be correspondingly exposed to the attack of different species of animals. An increasing concentration of astringent, indigestible tannin, acting as growth inhibitor of larvae, combined with the toughness of mature leaves serve as protection.[1] Tannins are, in fact, a common constituent in the leaves of woody plants. The large amounts of distasteful resins in coniferous trees have a similar effect on most non-adapted insects. Some glucoalkaloids of *Solanum* spp. repel potato beetles and their larvae. A high level of glucoalkaloids during the critical sprouting season therefore protects the plant. We could on good grounds define *secondary metabolites as non-nutritional chemicals controlling the biology of other species in the environment* or in other words *secondary metabolites play a prominent role in the coexistence and coevolution of species*, an idea that is phrased in different ways by various authors.[2,3] As a consequence the ecosystem is dynamic. Any change at one elevel or in one population provokes a change, unfortunately yet unpredictable, in another. A threat on a global scale to the natural balance is the unchecked overexploitation and destruction of the rain forests in the Amazon and Africa. The danger becomes apparent when we realize that the rain forests cover only 7 per cent of the land surface but contain nearly 50 per cent of all species. When a species is gone, it is gone forever and with it potential sources of food and medicine. Overexploitation of the mangrove swamps increase the risk of catastrophic floodings and cause serious destruction to the important coastal and offshore fauna and flora in these regions.

The organism has adapted the production of metabolites, i.e. enzyme activities, to its living conditions and this production cannot be fortuitous. In recent years important discoveries have been made showing unexpected biological effects by seemingly uninteresting compounds. Under the heading chemical ecology we can gather together a number of chemically guided behaviours in animals and plants. Biologists realized long ago that scents induce specific behaviour in animals but it was not until 1959, when the first sex-attractant bombycol (Fig. 2), was isolated in the female silk moth, *Bombyx mori*,[4] that real interdisciplinary research started on pheromones, the chemical compounds acting as attractants. The signal power of some sexual attractants is enormous. The attractants released by a single female silk moth under favourable wind conditions can

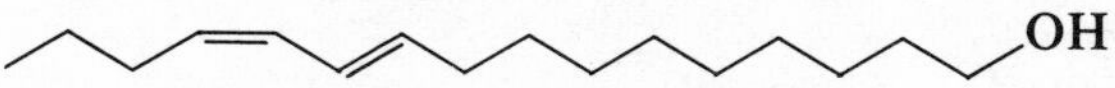

Bombycol (*Bombyx mori*)

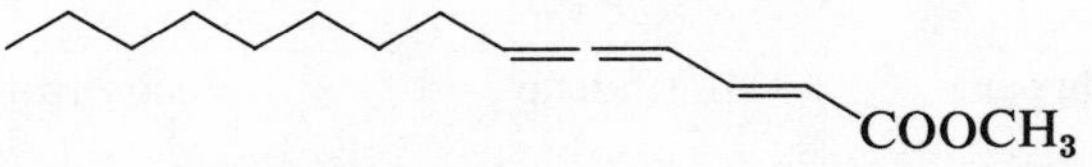

$COOCH_3$

Methyl (*E*)-2,4,5-tetradecatrienoate[5] (*Acanthoscelides obtectus*)

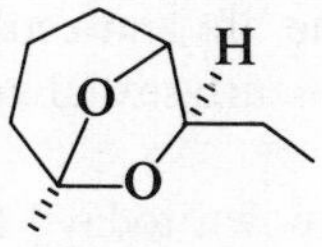

exo-Brevicomin[6]
(*Dendroctonus brevicomis*)

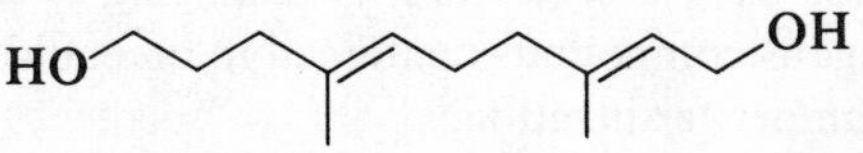

(*E*,*E*)-3,7-Dimethyl-2,6-decadien-1,10-diol[7]
(*Danaus gilippus*)

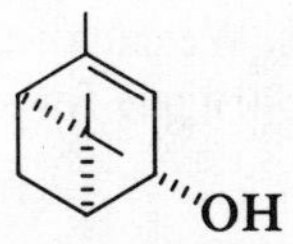

cis-Verbenol
(*Ips typographus*)

Fig. 2 Sex attractants isolated from insects

affect a male at a downwind distance of several kilometres. The male begins to react at concentrations as low as *ca.* 100 molecules per cm^3. We can imagine what great confusion these signals will create in a community containing thousands of species if they were not highly specific.

Scent from flowers of *Ophrys* orchids attracts insects and induces copulatory behaviour in specialized bees.[8] During these movements on the labellum pollination is accomplished. Some of the compounds responsible for this chemical signal are simple sesquiterpenes. It is sometimes observed that the individual component has a comparatively low activity, whereas the composite is highly active, a phenomenon termed synergism.

The compounds could be used by man for the control of insect populations. *Ips typographus* is a beetle that causes great damage in the Scandinavian pine forests. It is attracted by rather simple monoterpenes, e.g. *cis*-verbenol[9] and field tests with traps charged with pheromones are promising.

Female gametes of the brown algae *Ectocarpus siliculosus*, *Cutleria multifida*, and *Fucus serratus* produce simple low molecular unsaturated hydrocarbons that attract and effect the locomotory behaviour of the male gametes[10] (Fig. 3). The true mechanism of these chemical signals now observed in plants is still unknown. The findings raise serious questions and problems concerning oil

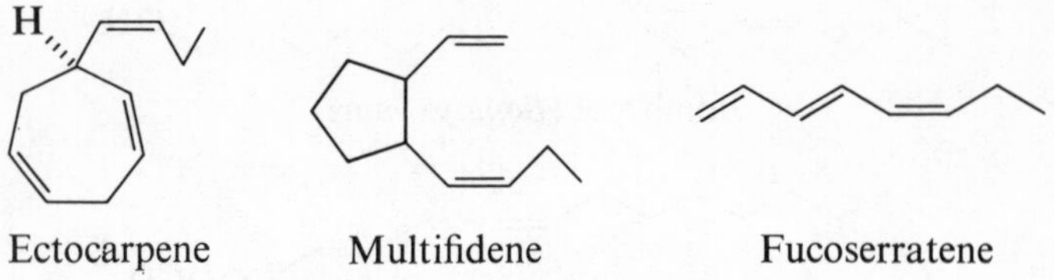

Fig. 3 Sex attractants from brown alga

pollution in sea water and its damaging effect on marine life and ecosystems. Analogous compounds contained in oil could very well disturb sexual approach and hamper fertilization.

The structures of several specific sex attractants are known today[11] (Fig. 2). They have generally a polyketide or isoprenoid structure, but still our knowledge is in an embryonic state.

Secondary metabolites produced in plants or animals have thus a strong behaviour control on other species. We can distinguish several areas of chemical control applied in the Darwinian struggle for survival:

(a) sexual attractants (male–female);
(b) feedants, antifeedants, repellants and toxins (animal–plant, animals);
(c) defence and alarm (animals);
(d) development, e.g. sterilization by queen substance (bees), metamorphosis, egg laying, growth suppressors (animal–plant, animals, plants);
(e) social behaviour, e.g. building behaviour (termites), territorial claims (bees), track indicators (ants), accumulation (locusts), etc.

These regulators show large variation of structure (Fig. 4). Every class of secondary plant constituent is represented as a particular attractant, repellant, etc. The cabbage butterfly *Pieris brassicae* has developed a taste for glycosinoates, e.g. sinigrin (section 6.8), which is a feeding attractant for this particular insect but simultaneously it acts as repellant or is poisonous for most other insects. Insects adapted to these glycosinoates and derived isothiocyanates have the advantage that they do not need to compete for their food with other species. But is such a situation of advantage to the plant? In many cases the insect becomes dependant on the attractant. It can only be fooled to feed on other plants by infiltrating it with the attractant and it prefers starvation to death to consuming other food. The general nutritional value of most plants varies comparatively little. Consequently the insect population is also controlled by the host plant. Oviposition is guided by the presence of a certain attractant for oligophagous and monophagous insects. The larvae give up feeding in the absence of the stimulant. Warburganal[12] isolated from the East African tree *Warburgia stuhlmannii* is an antifeedant against larvae of the army worm but it does not have any repellant effect on locusts. On the other hand the African neem tree *Melia azadirachta* is never attacked by desert locusts. It contains a complex triterpene as active principle.[13]

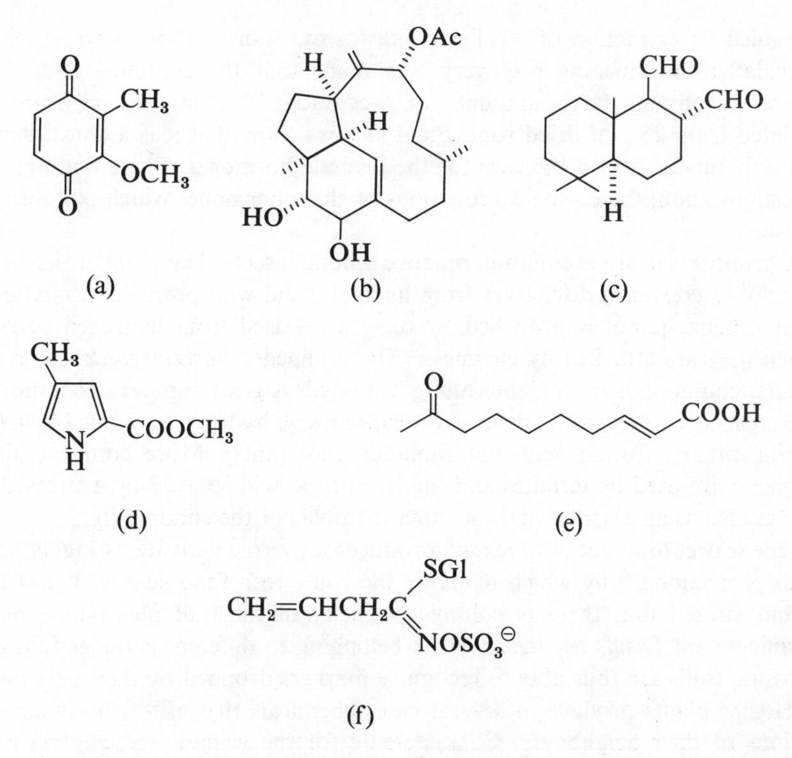

Fig. 4 (a) Quinone exuded in defence by a millipede *Narceus gordanus*. (b) Compound isolated from the defensive secretions of the termite *Trinervitermes gratiosus*. (c) Antifeedant from *Warburgia stuhlmannii* against the African army worm. (d) Trail pheromone from *Atta texana*. (e) Queen substance. (f) Sinigrin, stimulant for feeding and oviposition for the cabbage butterfly, *Pieris brassicae*

Several insects are adapted to toxins in their food. They can even accumulate them, thereby making themselves distasteful for predators, e.g. birds. *Danaus* monarch butterflies feed as larvae on milkweeds, *Asclepiadaceae*, rich in toxic cardenolides (section 5.9). The toxins are accumulated by the caterpillar and it is established that even one adult contains sufficient cardenolide to cause a predatory bird to vomit.[14] Wise from the experience it will hesitate to catch another one. No plants are found in nature without insects thriving on them. They have apparently found a way to deal with noxious metabolites either by an effective detoxification mechanism or by developing impermeable membranes.

The queen substance, 9-oxo-2-decenoic acid, and closely related compounds produced by the honeybee queen have complex effects on the bee community. It is a sexual pheromone which stabilizes worker swarms and controls queen cell building.

The steroid ecdysone[15] (section 5.9) stimulates metamorphosis, i.e. an effect opposite to juvenile hormone[16] activity (section 5.4). It is required in extremely low concentrations for normal insect metamorphosis. 25 mg of ecdysone was obtained by extraction of 500 kg of silkworms. Shortly after its structure was elucidated the amazing discovery was made that the common yew, *Taxus baccata*, contains large amounts of β-ecdysone.[17] 25 mg of ecdysone was isolated from 25 g of dried root. Plant surveys show that it is a constituent of many plants as is also the case for the juvenile hormones. These findings pose questions about the ecological functions of these hormones which still await an answer.

Chemical warfare is common practice among insects. The bombardier beetle, *Brachynus crepitans*, discharges from its anal gland with precision a mixture of simple benzoquinones propelled by oxygen released from hydrogen peroxide when they are attacked by enemies.[12] The millipede, *Narceus gordanus*, exudes similar compounds from glands along its body.[19] A great number of compounds are claimed to be used as defensive weapons, e.g. hydrogen cyanide (*Apheloria corrugata*), or formic acid and monoterpenes (ants). More complex higher terpenes are used by termites and toads. Formic acid sprayed by a stressed ant serves also as an alarm signal for other members of the community.

The leaf cutting ant *Atta texana* produces a pyrrole derivate as highly active trail pheromone,[20] by which it marks the route to a food source. It has been domonstrated that there is colony specificity in the trail pheromone of the formicine ant *Lasius neoniger*.[21] Ants belonging to different colonies following crossing trails are thus able to recognize markers dropped by their nest mates.

Higher plants produce in several cases chemicals that affect the living conditions of their neighbours. Characteristic for the walnut tree, *Juglans regia*, is the poor vegetation under the tree. It produces juglone, 5-hydroxynaphthoquinone, which is not tolerated by most plants. The shrub, *Salvia leucophylla*, in the Californian chaparral has a bare zone of soil surrounding each shrub. It produces 1,8-cineol and camphor as most active agents, which kill the neighbouring vegetation. During long wet periods the monoterpenes are washed out from the soil and the grassland moves closer to the shrubs.[22]

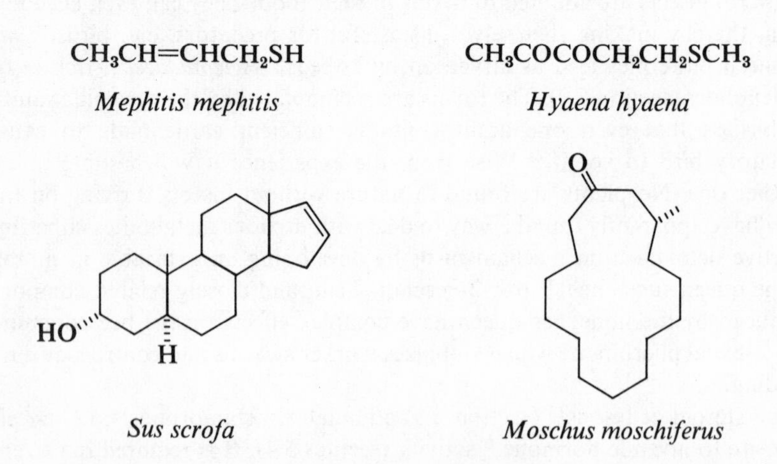

Fig. 5 Mammalian pheromones

Before leaving chemical ecology it ought to be pointed out that chemical signals are not restricted to insects;[23] they are common among mammals as well but our knowledge in this area is still primitive. The skunk, *Mephitis mephitis*, responds to danger by releasing evil smelling mercaptans from its anal glands. The smell is intense and rather persistent. I accidentally ran over a skunk with my car which became literally useless for any kind of human dating for one month. The hyena and the deer release anal exudates under stress. Well documented is the effectiveness of the musky boar odour in arousing sows. The active agent is a steroid, related to the male sex hormone androsterone (Fig. 5).

1.6 Biochemical reactions and organic reaction mechanisms

Biochemical processes are basically the same as other organic reactions, familiar to us from organic chemistry both as far as thermodynamic and mechanistic considerations are concerned. The stage for the enzyme catalysed reactions is the three-dimensional asymmetric surface of a protein. As a result of the chiral environment the products become enantiomeric. A special case is depicted in Fig. 6. The two methylene protons of ethanol are different with respect to their orientation towards the oxidant $NAD^{\oplus}$, and H_S is selectively removed. The two 'faces' of $NADH_S$ are different, since there is no free rotation around the N–R bond and from NADH one of the protons is specifically transferred in the reversed reduction. This has been demonstrated by specific labelling. The methylene carbon is said to be prochiral.

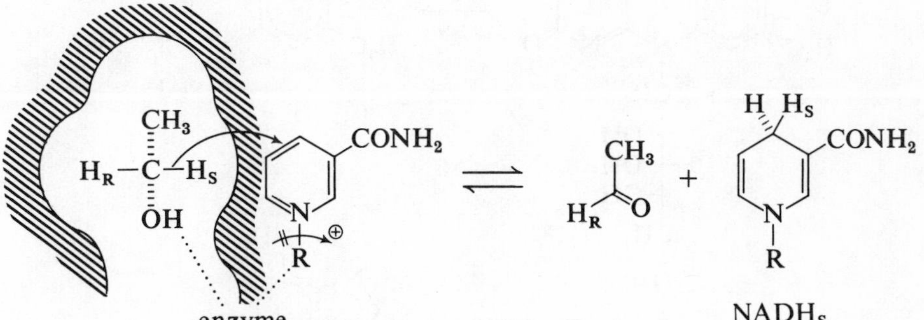

Fig. 6 Specific removal of one of the methylene hydrogens in enzymatic oxidation of ethanol

There are several electron transferring systems known:

1. pyridine based, $NAD^{\oplus}$:
2. flavin based, FAD;
3. iron-porphyrins, cytochromes;
4. ubiquinones or coenzymes Q (Fig. 7).

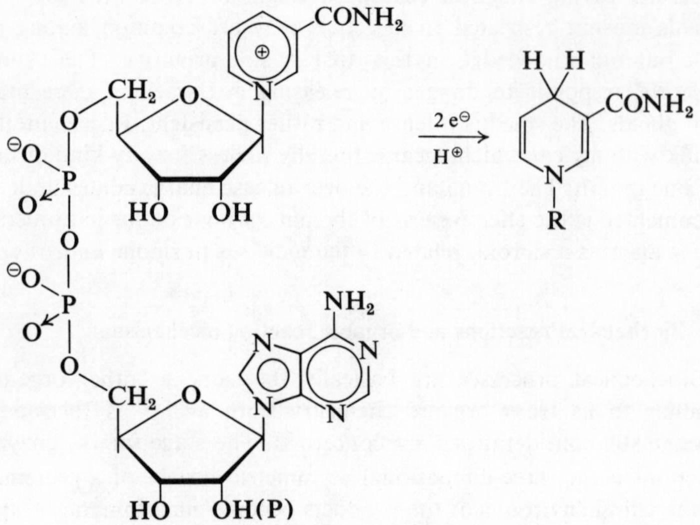

NAD $^+$, Nicotinamide adenine dinucleotide
NADP$^+$ = NAD $^+$-2-phosphate

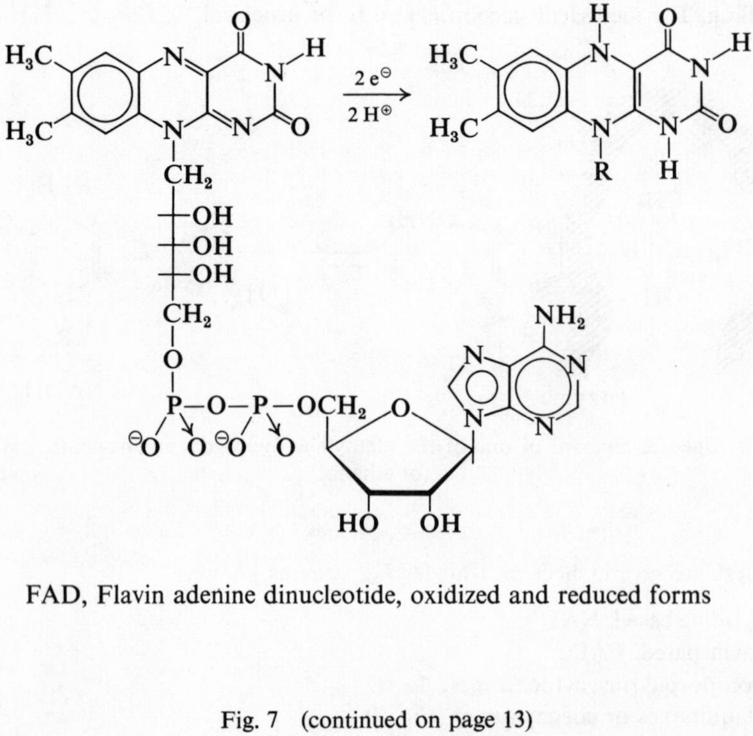

FAD, Flavin adenine dinucleotide, oxidized and reduced forms

Fig. 7 (continued on page 13)

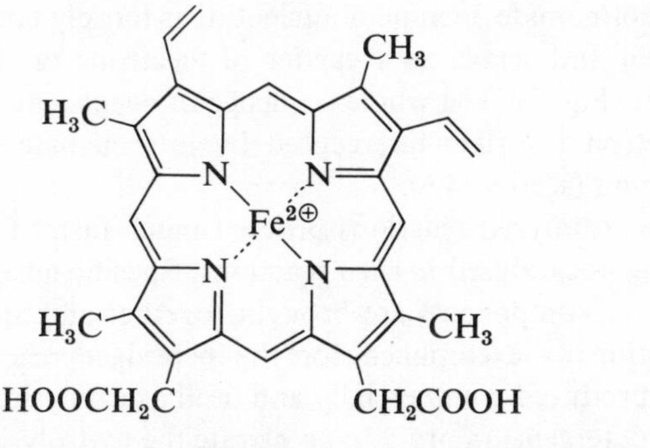

Haem, complex of protoporphyrin IX and ferrous ion

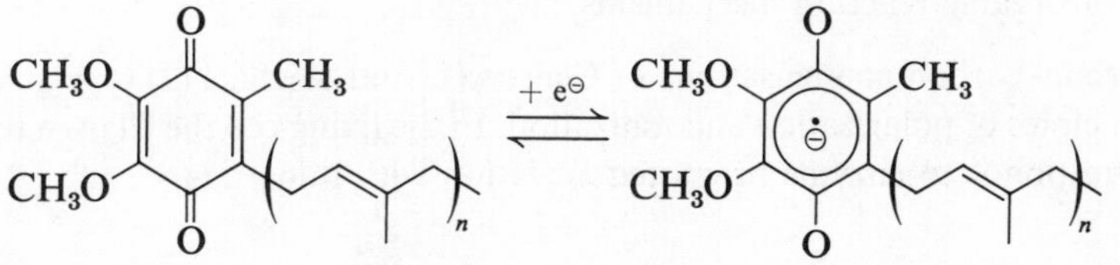

Ubiquinone, $n = 6$–12

Fig. 7 Structures of electron transferring cofactors

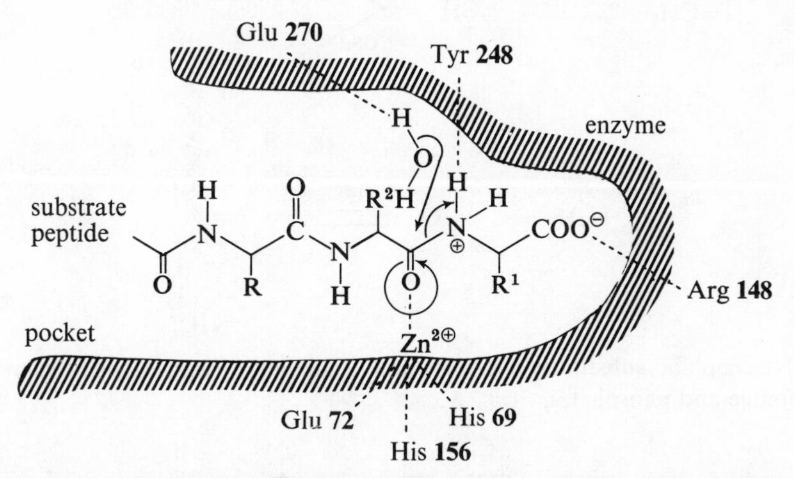

Fig. 8 Model of the hydrolysis of the terminal amino acid, R^1, of a peptide at the active site of carboxypeptidase A. The substrate is held in proper conformation in a pocket of the enzyme by hydrogen bonding. The carbonyl group is associated with the essential zinc ion of the enzyme and the terminal carboxyl and amino groups with arginine and tyrosine residues. Carboxypeptidase A contains a single polypeptide chain with 307 amino acids. The efficiency of the peptidase depends on its ability to carry out three steps simultaneously: (a) enhancement of the dipolar character of the carbonyl bond by its association as ligand to the zinc ion; (b) introduction of a molecule of water close to the carbonyl group; (c) protonation of the amino function thus facilitating the cleavage of the peptide bond

NAD$^{\oplus}$ or nicotinamide adenine dinucleotide is loosely bounded to the dehydrogenase protein and serves as a carrier of electrons in the form of a hydride transfer agent (Fig. 7). The whole sequence of events can also be formulated as two one-electron transfers intercepted by intermediate hydrogen abstraction and protonation (section 4.6).

The enzyme catalyzed reactions proceed much faster ($\leq 10^{10}$ times) than the corresponding uncatalyzed *in vitro* reactions. Specific adsorption on the protein ensures that the components are brought together and are oriented favourably for the reaction as exemplified for the peptidase reaction (Fig. 8). Several enzymes are produced commercially and used as catalysts, e.g. peptidase, which are added to detergents in order to accelerate the hydrolysis of proteins in stains.

We can distinguish several types of reactions, here systematized in the terminology of organic reaction mechanisms.

1. Carbon–carbon condensations of Claisen (1) and Michael (2) type guided by principles of polarization and ionization. In the living cell the Claisen reaction corresponds to enzyme promoted acylation with thioesters,

$$-\overset{|}{\underset{|}{C}}H_2 \underset{}{\overset{-H^{\oplus}}{\rightleftharpoons}} -\overset{\ominus}{\underset{|}{C}}H \begin{cases} \xrightarrow{\;\diagdown CO\;} -\underset{|}{C}H\overset{|}{\underset{|}{C}}-O^{\ominus} \\ \xrightarrow{\;-COSR\;} -\underset{|}{C}H\underset{|}{C}O + RS^{\ominus} \end{cases} \tag{1}$$

$$-\overset{\ominus}{\underset{|}{C}}H + \;=\!\!\diagup\!\overset{\diagdown}{C}O \overset{H^{\oplus}}{\rightleftharpoons} -\underset{|}{C}H-CH_2-CH_2-\overset{\diagdown}{C}O \tag{2}$$

2. Nucleophilic substitution. *C*-, *N*- and *O*-alkylation with *S*-adenosyl methionine and phosphates,

$$-\overset{\ominus}{\underset{|}{C}}H + RX \rightleftharpoons -\underset{|}{C}HR + X^{\ominus} \tag{3}$$

$$R^1OH + RX \rightleftharpoons R^1OR + HX \tag{4}$$

$$\rangle\!=\!\langle^{H} + RX \longrightarrow \rangle\!=\!\langle^{R} + HX \tag{5}$$

$$X = R_2S^{\oplus},\ H_2PO_4(PO)$$

3. Elimination. In biological systems phosphate and ammonia are the leaving groups par excellence and B is a nucleophilic group of an enzyme, -OH, $-NH_2$, or SH (hydroxy, amino, or mercapto group),

$$\text{B:}\ \text{H–C–C–OP} \rightleftharpoons \text{>C=C<} + BH^{\oplus} + PO^{\ominus} \tag{6}$$

4. Oxidation, reduction, dehydrogenation. The exact mechanism is uncertain. They can be formulated either as hydride transfer, as one-electron transfer followed by hydrogen abstraction, or as a two-electron transfer. They are promoted by $NAD^{\oplus}$, FAD enzyme systems.

$$\text{H–}\overset{|}{\underset{|}{C}}\text{–OH} \rightleftharpoons \text{>C=O} \tag{7a}$$

$$\text{H(>C=O)} \rightleftharpoons \text{–COOH} \tag{7b}$$

$$\text{–}\overset{H}{\underset{|}{C}}\text{–}\overset{H}{\underset{|}{C}}\text{–} \rightleftharpoons \text{>C=C<} \tag{7c}$$

$$\text{–}\overset{H}{\underset{|}{C}}\text{–}NH_2 \rightleftharpoons \text{>C=NH} \tag{7d}$$

$$\text{R–C}_6\text{H}_4\text{–OH} \underset{}{\overset{-e^{\ominus},\ H^{\oplus}}{\rightleftharpoons}} \text{R–C}_6\text{H}_4\text{–O}^{\bullet} \xrightarrow{\text{Dimerization}} \text{(HO)(R)C}_6\text{H}_3\text{–C}_6\text{H}_3\text{(R)(OH)} \text{ or } \text{(HO)(R)C}_6\text{H}_3\text{–O–C}_6\text{H}_4\text{(R)} \tag{7e}$$

A special case is hydroxylation, insertion of molecular oxygen into hydrocarbons to yield alcohols (7f, g), sometimes proceeding via epoxides as intermediates but not necessarily radical in nature.

$$\text{C}_6\text{H}_5\text{R} \xrightarrow{[O]} \text{HOC}_6\text{H}_4\text{R} \qquad (7f)$$

$$\text{AlkH} \xrightarrow{[O]} \text{AlkOH} \qquad (7g)$$

5. Wagner–Meerwein or carbonium ion rearrangements occur frequently, especially in terpene biosynthesis, and account for structures which formally do not seem to obey the 'isoprene rule'.

$$-\overset{\oplus}{\underset{|}{C}}-\underset{|}{\overset{R}{\overset{|}{C}}}- \rightleftharpoons \left[\;\overset{\overset{R}{\oplus}}{C\!-\!C}\;\right]_{\text{TS}} \rightleftharpoons -\underset{|}{\overset{R}{\overset{|}{C}}}-\overset{\oplus}{\underset{|}{C}}- \qquad (8)$$

R = H, alkyl, aryl

6. Carboxylations and decarboxylations.

$$-\overset{|}{\underset{|}{C}}^{\ominus} - CO_2 \rightleftharpoons -\overset{|}{\underset{|}{C}} - COO^{\ominus} \qquad (9)$$

This 'Grignard'-like reaction is an essential step, e.g. in fatty acid synthesis when acetate is converted into malonate. The reaction is reversible.

1.7 Principal pathways

The main streams of secondary metabolism is outlined in Fig. 9. One remarkable feature is that most metabolites originate from a very limited number of precursors. They are the link to primary metabolism in which they also play an important role. *Acetic acid* has a central position in the form of its thioester, acetyl CoA, or acetyl coenzyme A (Fig. 10). It is produced in the cell from pyruvic acid or fatty acids, or it may be directly formed from acetate and coenzyme A with ATP (Fig. 12) as mediator.

From acetic acid *mevalonic acid* is derived, from which, via 3,3-dimethylallyl pyrophosphate and the isomeric isopentenyl pyrophosphate—the isoprene unit—the terpenoids are formed. From carbohydrates *shikimic acid* is derived which is the key to a wealth of aromatics. Finally, it is worth pointing out the importance of amino acids as precursors of the great variety of nitrogen containing compounds.

Several groups of metabolites have mixed biogenesis, i.e. an intermediate or metabolite from one principal pathway acts as a substrate for another metabolite from a different pathway. Thus, flavonoids are derived from a polyketide (three acetate units) and a cinnamic acid (shikimic acid). The indole alkaloids come from shikimate and a monoterpene (loganin).

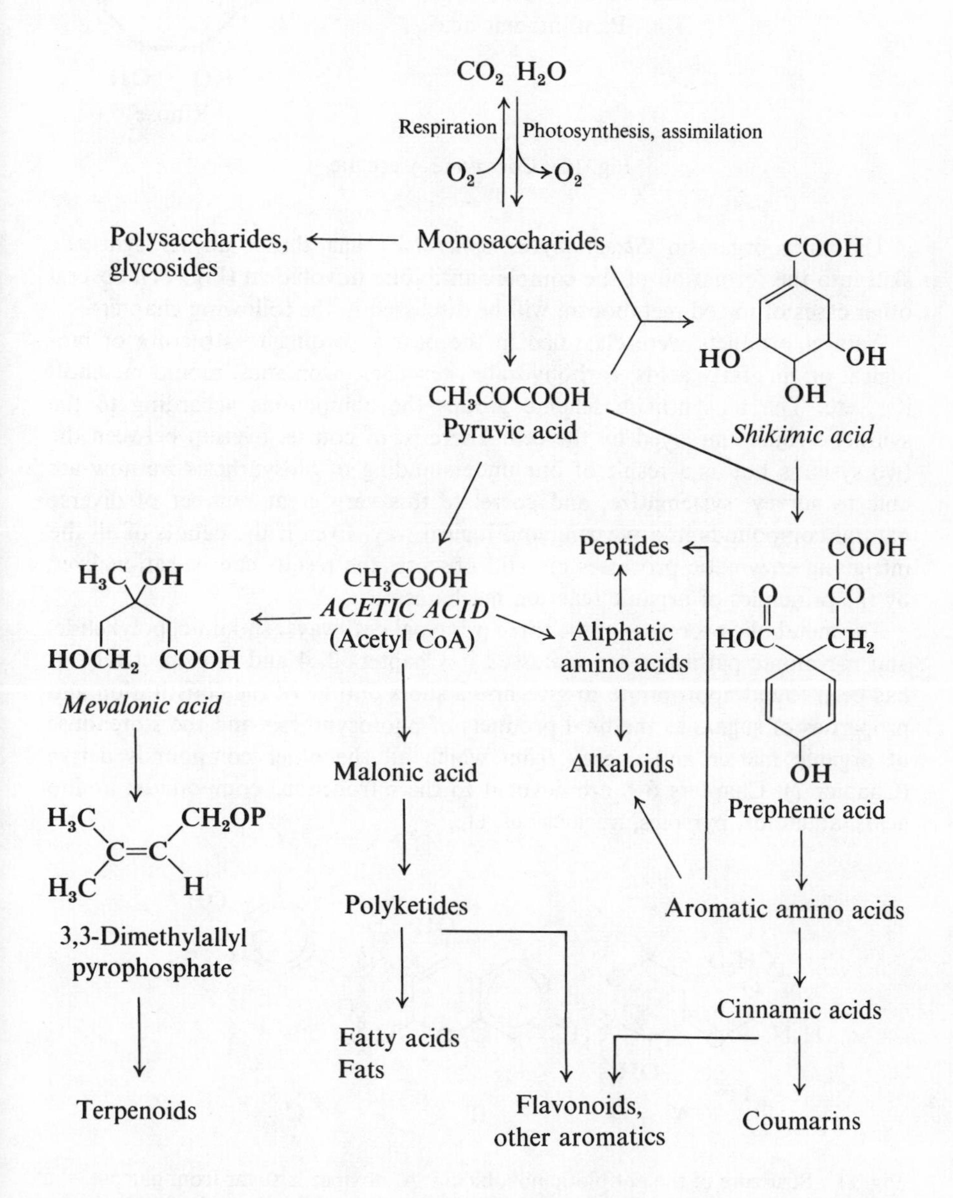

Fig. 9 Main streams of secondary metabolism

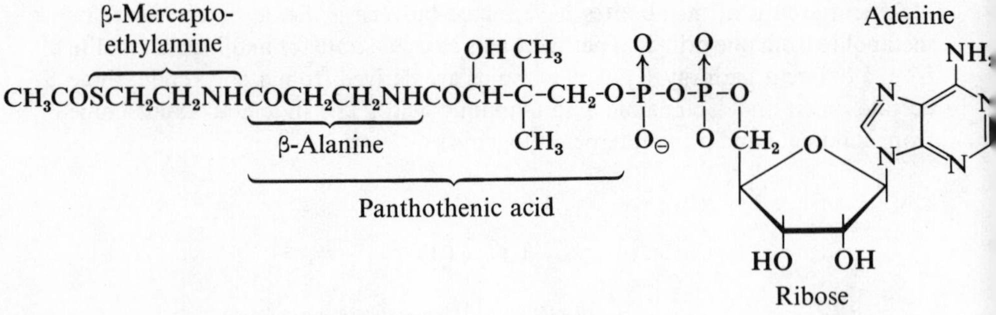

β-Mercapto-ethylamine

Adenine

$CH_3COSCH_2CH_2NHCOCH_2CH_2NHCOCH(OH)\text{-}C(CH_3)_2\text{-}CH_2\text{-}O\text{-}P(O)(O^{\ominus})\text{-}O\text{-}P(O)(O)\text{-}O\text{-}CH_2$

β-Alanine

Panthothenic acid

HO OH

Ribose

Fig. 10 Coenzyme A acetate

The microorganism *Streptomyces spheroides* has thrust all its synthetic skill into the formation of the complex antibiotic novobiocin (Fig. 11). Several other cases of mixed metabolism will be discussed in the following chapters.

Natural products were classified in the past according to structure or biological origin: fatty acids, carbohydrates, terpenes, aromatics, mould metabolites, etc. The biosynthetic scheme groups the compounds according to the synthetic route employed by the cell. There is, of course, overlap between the two systems but as a result of our understanding of biosynthesis we now are able to survey, systematize, and correlate this very great number of diverse natural compounds in a pleasing and logical way. Even if the details of all the intriguing enzymatic processes are still obscure, the results can be rationalized by the principles of organic reaction mechanisms.

The metabolic processes of the three principal pathways: shikimic, polyketide, and mevalonic pathways are discussed in Chapters 3, 4 and 5, respectively. It has been found appropriate to give first a short outline of the constitution and properties of sugars as the final products of photosynthesis and the storehouse of organic matter and energy from which all the other compounds derive (Chapter 2). Chapters 6–8 are devoted to the nitrogenous compounds: amino acids, alkaloids, pyrroles, nucleosides, etc.

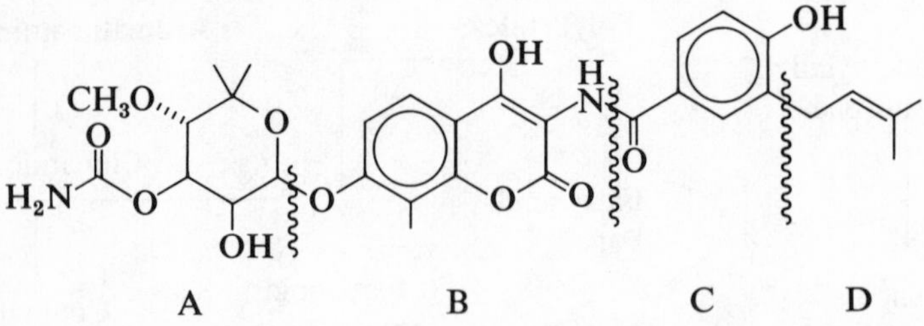

A B C D

Fig. 11 Structure of the antibiotic novobiocin. A, noviose, a sugar from glucose + a $CONH_2$ group + 2 CH_3; B, a coumarin derivative from shikimic acid; C, *p*-hydroxybenzoic acid from shikimic acid; D, an isopentenyl group from mevalonic acid

1.8 The one carbon fragment

C-, *N*- and *O*-methyl groups are frequently found in natural products, and they cannot be accounted for by any of the already mentioned pathways. Most biological methylation in animals, plants, and bacteria involves methionine, the methyl group of which is activated by *S*-adenosylation with ATP (Fig. 12;

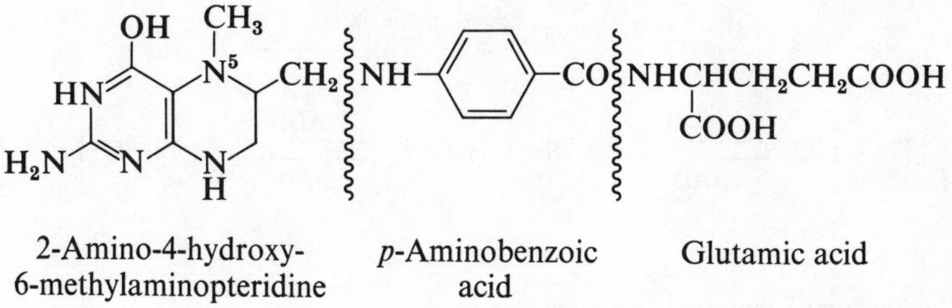

2-Amino-4-hydroxy-6-methylaminopteridine — *p*-Aminobenzoic acid — Glutamic acid (I)

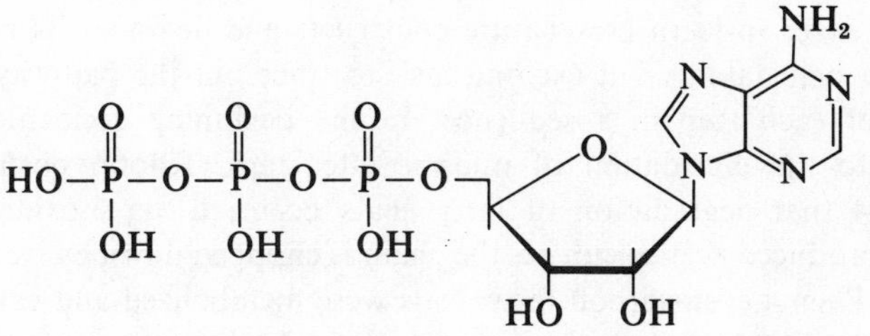

Fig. 12 N^5-Methyltetrahydrofolic acid, I, and adenosine triphosphate, ATP, II

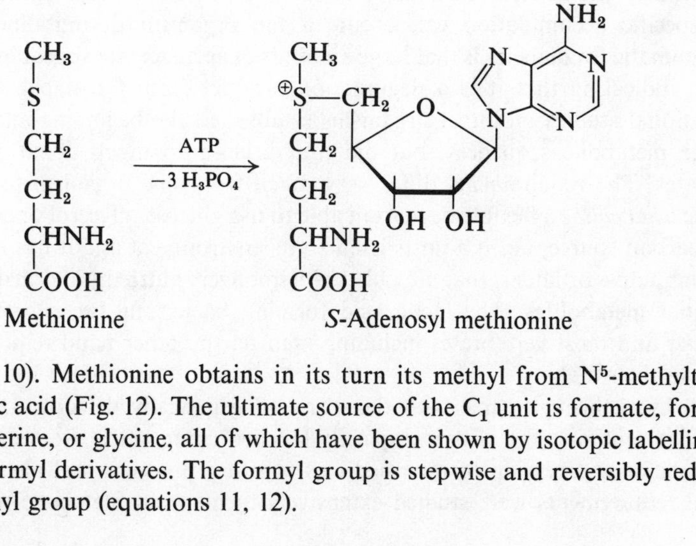

$$\text{Methionine} \xrightarrow[-3\,H_3PO_4]{ATP} S\text{-Adenosyl methionine} \qquad (10)$$

Methionine — *S*-Adenosyl methionine

equation 10). Methionine obtains in its turn its methyl from N^5-methyltetrahydrofolic acid (Fig. 12). The ultimate source of the C_1 unit is formate, formaldehyde, serine, or glycine, all of which have been shown by isotopic labelling to give *N*-formyl derivatives. The formyl group is stepwise and reversibly reduced to a methyl group (equations 11, 12).

$$\text{5,6,7,8-Tetrahydrofolic acid (N}^5\text{-CH}_3\text{)} + \text{Homocysteine} \xrightarrow{B_{12}} \text{Methionine} \quad (11)$$

5,6,7,8-Tetrahydrofolic acid

$$>N^5{-}CHO \underset{NAD^{\oplus}}{\overset{NADH}{\rightleftharpoons}} >N{-}CH_2OH \underset{NAD^{\oplus}}{\overset{NADH}{\rightleftharpoons}} >N{-}CH_3 \quad (12)$$

1.9 Elucidation of metabolic sequences

For millions of years nature has refined her synthetic skill; no wonder that chemists are eager to learn how nature constructs and degrades its molecules. It has been a painstaking, but exciting task to trace out the pathway and the mechanism of each step in a sequence. In the beginning accidental results contributed to the elucidation of intermediate steps. Knoop postulated as early as 1904 that degradation of fatty acids occurred via β-oxidation and presumably produced acetic acid, i.e. the chain is chopped down by two carbons at a time. ω-Phenyl-even-carbon fatty acids were metabolized and extracted in the urine of the test animal as phenylacetic acid, whereas ω-phenyl-odd-carbon fatty acids gave benzoic acid. Collie hypothesized at about the same time (1907) that the reversed reaction, the acetate condensation of the Claisen type, is the origin of many naturally occurring phenolics.

Normally the intermediates are present in a low steady state concentration but specific accumulation can occur in the organism during illness. Thus symptomatic for diabetes is that large amounts of acetoacetate appear in blood or urine, indicating that it is a degradation product from fatty acid oxidation. Nutritional studies on intact organisms established the beginning and the end of the metabolic sequences, but did not disclose so much about the intermediates. The requirements differ very much from one organism to another. *Escherichia coli* is a flexible bacterium able to use glucose, glycerol or acetates as sole carbon source, i.e. it adjusts itself to its environment and it has enzymatic systems active or latent, that are able to interconvert nutrients and manufacture essential metabolites. The lactic acid forming bacterium *Leuconostoc mesenteroides* and most vertebrates including man on the other hand require amino acids in their diet.

The breakthrough came with the work on genetically defective organisms, so-called mutants, and with isotopically labelled compounds. Accumulation of intermediates, diversion of intermediates to other products and altered nutritional requirements were studied extensively to map out metabolic sequences.

Mutants can arise spontaneously or can be produced by the actions of chemicals or irradiation (X-rays, UV). Provided that the damage is not lethal, it often happens that only one gene is damaged, i.e. the cell is unable to produce one specific enzyme. A damage in the primary metabolism results in inhibited growth. It is more difficult to trace a damage to the secondary metabolism since their products are not essential for the development of the organism. However, the underlying principles are the same. Suppose that we are studying the essential sequence $A \rightarrow E$ (Fig. 13). The normal organism can carry out all steps and A is required for growth. In mutant 1 $D \rightarrow E$ is blocked, D will accumulate and the growth can be sustained only by adding E to the nutrient. In mutant 2 C will accumulate and addition of D or E will restore normal activity of the organism. In mutant 3 $B \rightarrow C$ is blocked, but B will not necessarily accumulate, because B can be used as substrate for another enzyme and this special strain will produce a somewhat different spectrum of metabolites, F, G. The filtrate of 1 will restore the growth of 2 but 2 cannot support 1. The shikimic acid pathway was mainly elucidated by Davis using a number of mutants of *E. coli.*

Normal organism $A \longrightarrow B \longrightarrow C \longrightarrow D \longrightarrow E$

Mutant 1 $A \longrightarrow B \longrightarrow C \longrightarrow D \not\longrightarrow E$

Mutant 2 $A \longrightarrow B \longrightarrow C \not\longrightarrow D \longrightarrow E$

Mutant 3 $A \longrightarrow B \not\longrightarrow C \longrightarrow D \longrightarrow E$

$\qquad\qquad\qquad\qquad F \dashrightarrow G$ (from B)

Fig. 13 Mutants with defective enzymes at different points along a metabolic chain

The most powerful method of establishing a metabolic sequence involves the use of isotopes. Table 1 shows the properties of some isotopes used as tracers. The fundamental investigations were run with the radioactive isotopes, ^{3}H, ^{14}C, and ^{32}P, but in recent years with the advent of pulse Fourier-transform ^{13}C NMR spectroscopy biosynthetic studies witness a new explosive development. Radiotracing and mass spectrometry are the most sensitive methods.

Table 1 Properties of isotopes used as tracers

Isotope	Relative natural abundance (%)	Radiation	Half-life	Spin
^{2}H	0.015		Stable	1
^{3}H		β	12.1 years	$\frac{1}{2}$
^{13}C	1.1		Stable	$\frac{1}{2}$
^{14}C		β	5.700 years	
^{15}N	0.37		Stable	$\frac{1}{2}$
^{18}O	0.20		Stable	
^{32}P		β	14.3 days	

Radiotracing is preferably used in cases where quantitative data of incorporation are needed, but where information of location of label in the molecule is less important. The high sensitivity is of advantage if the efficiency of incorporation or conversion is low.

If the label can be excised from the molecule by some simple well defined degradative method, the radiotracer technique is the method of choice. Suppose that in a metabolic sequence A–E (Fig. 13) we know the compounds fairly well but we do not know with certainty their exact order of formation, we could apply the radiotracer technique in a primitive and simple way. If radioactive carbon dioxide is administered, it will first appear in A and then in B etc. The measurement of the relative activity of the isolated compounds after a certain period of time would provide us with the correct order. This has been done in the morphine series. Radioactivity appeared first in thebaine, then in codeine and last in morphine. Thus, the methylation is not, as one possibly would guess, the final step of this sequence (13).

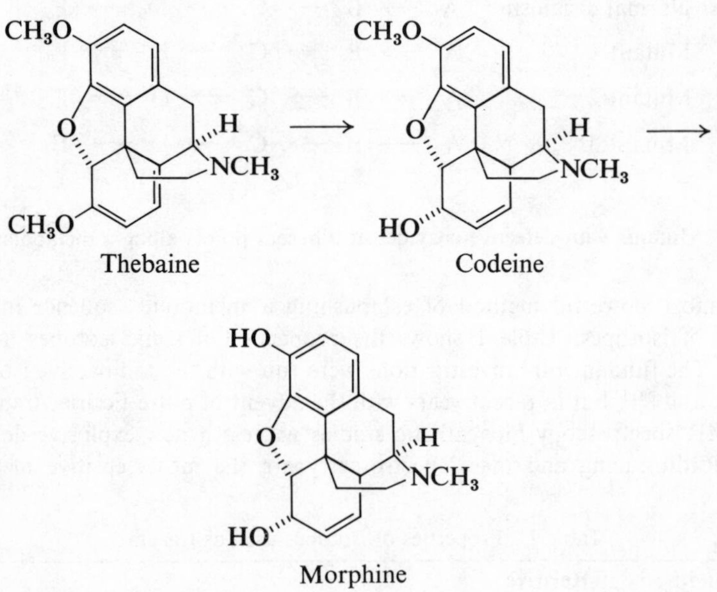

Thebaine → Codeine →

Morphine

(13)

In order to locate exactly the labelled atoms in a metabolite it has to be degraded in an unambiguous way. The activity of each fragment, isolated as CO_2, acetic acid or other well defined small organic molecule, is then measured. Controlled degradation of a big molecule is a very difficult and time-consuming undertaking. This methodology was skilfully demonstrated by Bloch, Lynen, Popjak and Cornforth in the biosynthesis of cholesterol as being ultimately derived from acetate (14).

$$CH_3-\overset{*}{C}OOH \longrightarrow \text{Cholesterol skeleton} \quad (14)$$

Acetic acid

Cholesterol skeleton

Calvin used radiotracers for the elucidation of the mechanism of carbon dioxide fixation in green plants, i.e. photosynthesis. The green alga *Scenedesmus obliquus* was found to incorporate $^{14}CO_2$ extremely rapidly. Within seconds the label was observed in glycine, alanine, aspartic acid, malic acid, citric acid, triose phosphates, hexose phosphates, glyceric acid-3-phosphate, etc. By reducing the time of the exposure glyceric acid-3-phosphate was recognized as the first stable intermediate in carbon dioxide fixation. This type of investigation called for a sensitive and rapid analytical method. The metabolism was quenched by hot alcohol and the metabolites were separated and identified by paper chromatography. The radioactive spots were localized by scanning the paper with Geiger-Müller counter or by autoradiography. In the latter method the chromatogram is brought into close contact with a photographic film, sensitive to the short-range β-irradiation of ^{14}C. The active spots develop as black areas on the film and can be identified by the use of reference compounds. This same method is also used for studies of uptake, transport and accumulation of radioactive metabolites in plants. The whole plant is rapidly pressed in frozen condition and applied to a photographic film.

A mass spectrometric determination requires only a few micrograms of the sample, but in order to be able to locate the exact position of the isotope, the fragmentation pattern of the compound must be well understood. Mass spectrometry has the advantage that no previous degradation is necessary; the instrument takes care of that part. ^{2}H, ^{13}C, ^{15}N, and ^{13}O isotopes have been used successfully in structural determinations but scrambling has to be considered when ^{2}H is used.[24,25]

^{13}C NMR spectrometry along with ^{1}H NMR is today the most efficient tool for structural determination of organic molecules that the chemist possesses. One can usually determine routinely the intensity, shift, and ^{13}C–^{1}H multiplicities of each carbon in most low molecular compounds. More complex molecules with a number of very similar carbons are exceptions, e.g. carotenes, long chain fatty acids. That means that a label will be recognized by a change in intensity and its location by its shift. Quantitative measurements are difficult because intensities of ^{13}C absorptions of different carbons vary considerably due to differences in relaxation times, NOE effects and instrumental fluctuations during recordings. For a well resolved spectrum we need approximately 10 mg of the metabolite and an enrichment of label of *ca.* 50 per cent i.e. for a 90 per cent enriched precursor one can accept a dilution to *ca.* 1: 200 in the metabolic process. These experimental requirements set the limit of the applicability of the method.[26] Very important is the use of doubly labelled 1,2-$^{13}C_2$-acetate enabling detailed investigations of bond formed and broken in the polyketide pathway. Multiple labelled ^{2}H–^{13}C compounds allow studies of the integrity of the H–C bond in a metabolic process.

^{2}H NMR spectroscopy has the advantages of being inexpensive and unlike ^{3}H it does not require special handling. Due to the short relaxation time of ^{2}H and the absence of NOE effect minimizing saturation, it integrates accurately. ^{2}H has a low natural abundance (0.016 per cent) and therefore incorporation of ^{2}H can be detected at a very high dilution; *ca.* 1: 10 000 for 50 per cent enrichment. Poor resolution and spectral crowding are the main disadvantages.[27]

^{3}H NMR spectroscopy has gained interest in recent time. It is a sensitive method. ^{3}H has the nuclear spin 1/2 and gives proton-like spectra with narrow line widths and small NOE effects. The spectra can therefore be integrated and as a result of its low natural abundance enrichment of label can be detected at extremely low level.[27]

Administration of labelled precursors to living organisms often presents several kinds of problems. The precursor added to the nutrient may have difficulty in diffusing through the cell wall of the microorganism. If higher plants are studied, the precursor may not be absorbed by the roots, or if absorbed it may not be transported or may be degraded before it reaches the tissue where the metabolism occurs. A similar fate may also be shared by precursors applied by injection, spreading on leaves or via a wick through the stem. Higher concentrations of a normal plant constituent may have undesirable toxic effects. Dilution presents a serious problem since the added precursor has to compete with the normal pool of metabolites in the cell during the experiment. The cellular reactions are reversible; a constant degradation and rebuilding occur that causes a slow dispersion of the original label. Several of these problems have been circumvented in cell-free preparations with intact enzymatic systems. In fact, the fundamental steps of the mevalonic acid pathway were elucidated by this technique.

Tissue cultures (undifferentiated callus cells) seem to offer several advantages provided that the cells have retained the synthetic power of the parent organism to produce secondary metabolites, actually a serious restriction. Translocation and permeability problems are reduced and there are no seasonal variations. Callus cells have a simpler organization and they are easier to reproduce. It is necessary to work under aseptic conditions but this ensures on the other hand that the metabolites originate from the plant and not from symbiotic bacteria. By proper choice of problem the utilization of this technique will undoubtedly be extended in the future.[28]

Bibliography

1. Feeny, P., *Ecology* **51** (1970) 565.
2. Erlich, P. R. and Roven, P. H., *Evolution* **18** (1965) 586.
3. Whittaker, R. H. and Feeny, P., *Science* **171** (1971) 757.
4. Butenandt, A., Beckmann, R. and Hecker, H., *Hoppe-Seyler's Z. physiol. Chem.* **324** (1961) 71.
5. Horler, D. F., *J. Chem. Soc.* (C) **1970**, 859.
6. Silverstein, R. M., Brownlee, R. G., Bellas, T. E., Wood, D. L. and Brownee, L. E., *Science* **159** (1968) 889.
7. Meinwald, J., Meinwald, Y. C. and Mazzocchi, P. H. *Science* **164** (1969) 1174.
8. Kullenberg, B. and Bergström, G., *Endeavour* **34** (1975) 59.
9. Krawielitski, S., Klimetzek, D., Bakke, A., Vite, J. P. and Mori, K., *Z. Angew. Entomol.* **83** (1977) 300.
10. Müller, D. G. in *Marine Natural Products Chemistry*. Falkner, D. J. and Fenical, W. H. (Eds.), Plenum Press, New York, 1977, p. 351.
11. Brand, J. H., Young, J. C. and Silverstein, R. M. in *Prog. Chem. Org. Nat. Prod.* **37** (1979) 1.
12. Kubo, I., Lee, Y. W., Pettei, M., Pilkiewicz, F. and Nakanishi, K. *J. Chem. Soc. Chem. Commun.* **1976,** 1013.
13. Nakanishi, K. *Rec. Adv. Phytochem.* **9** (1975) 283.
14. Roeske, C. N., Seiber, J. N., Brower, L. P. and Moffit, C. M. *Rec. Adv. Phytochem.* **10** (1976) 93.
15. Butenandt, A. and Karlson, P. *Z. Naturforsch.* **9b** (1954) 389.
16. Röller, H., Dahm, K. H., Sweeley, C. C. and Trost, B. M. *Angew. Chem.* **79** (1967) 190.
17. Takemoto, T., Ogawa, S., Nishimoto, N., Arihari, S. and Bue, K., *Yakugaku Zasshi* **87** (1967) 1414.
18. Schildtknecht, H. and Holoubek, K., *Angew. Chem.* **73** (1961) 1.
19. Monro, A., Chada, M. S., Meinwald, J. and Eisner, T., *Ann. Entomol. Soc. Amer.* **55** (1962) 261.
20. Tumlinson, J. H., Silverstein, R. M., Moser, J. C., Brownlee, R. G. and Ruth, J. M., *Nature* **234** (1971) 348.
21. Traniello, J. F. A., *Naturwissenschaften* **67** (1980) 361.
22. Muller, C. H., *Rec. Adv. Phytochem.* **3** (1970) 106; Schildknecht, H., *Angew. Chem.*, **93** (1981) 164.
23. Harborne, J. B., *Introduction to Ecological Biochemistry*, Academic Press, London, 1977, p. 162.
24. Budzikiewicz, H., Djerassi, C. and Williams, D. H., *Mass Spectrometry of Organic Compounds*, Holden-Day, Inc., San Francisco, 1967.
25. Grostic, M. F. and Rinehart, Jr., K. L. in *Mass Spectrometry: Techniques and Applications*, Milne, G. W. A. (Ed.), J. Wiley, New York, 1971, p. 217.
26. Simpson, T. J., *Chem. Soc. Revs.* **4** (1975) 497.
27. Garson, M. J. and Staunton, J. *Chem. Soc. Revs.* **8** (1979) 539.
28. Overton, K. H. and Picken, D. J. in *Prog. Chem. Org. Nat. Prod.* **34** (1977) 249.

Chapter 2

Carbohydrates and primary metabolites

2.1 Classification. Structure of glucose

Carbohydrates (sugars) is the general term for polyhydroxy compounds usually containing carbonyl functions. They are classified according to the numbers of carbons, C_7 heptoses, C_6 hexoses, C_5 pentoses, etc., or according to the number of units in the molecule, monosaccharides, disaccharides and polysaccharides. Sugars containing an aldehyde function are termed aldoses

$$
\begin{array}{c}
CO_2H \\ | \\ CHOH \\ | \\ CHOH \\ | \\ CHOH \\ | \\ HCOH \\ | \\ CO_2H \\
\text{Glucaric acid} \\ \text{Mannaric acid}
\end{array}
\xleftarrow{HNO_3}
\begin{array}{c}
\overset{1}{C}HO \\ |_2 \\ CHOH \\ |_3 \\ CHOH \\ |_4 \\ CHOH \\ |_5 \\ HCOH \\ |_6 \\ CH_2OH \\
(+)\ \text{Glucose} \\ (+)\ \text{Mannose}
\end{array}
\xrightarrow{PhNHNH_2}
\qquad (1)
$$

$$
\begin{array}{c}
CH{=}NNHPh \\ | \\ C{=}NNHPh \\ | \\ CHOH \\ | \\ CHOH \\ | \\ HCOH \\ | \\ CH_2OH \\
\text{Glucosazone}
\end{array}
\xleftarrow{PhNHNH_2}
\begin{array}{c}
CH_2OH \\ | \\ CO \\ | \\ CHOH \\ | \\ CHOH \\ | \\ HCOH \\ | \\ CH_2OH \\
(-)\ \text{Fructose}
\end{array}
$$

and if they contain a keto function, ketoses. They are present in all living material where they play a cental role in primary biosynthesis, production and storage of energy and matter. Hexoses are by far the commonest type of sugars and glucose is one of the most widely distributed and abundantly occurring compounds. The structural elucidation of glucose by Fischer in 1896 belongs to the classical era of stringent structural reasoning long before modern spectroscopy was available. No absolute configuration was known at the time and Fischer suggested that the configuration of C^5 was as written in (1). The gross structure of glucose, mannose and fructose was known, and it was found that they gave the same osazone, i.e. they have identical configuration at C^3–C^5 and fructose must have the carbonyl group at C^2. When glucose and mannose were oxidized by nitric acid, different optically active glucaric acids were obtained (Fig. 1). The symmetric structures I and II can therefore be eliminated and also III because the configuration of glucose and mannose differs only at C^2. A change of the configuration at C^2 of III leads to either I and II. Thus, the structure of mannaric and glucaric acids must be represented by IV–VI.

When (-)arabinose was reacted with hydrocyanic acid and hydrolysed, a mixture of gluconic and mannonic acids was obtained, and oxidation with nitric acid gave an optically active dibasic acid, VII (2). The configuration at C^2–C^4 in arabinose is thus identical to the configuration at C^3–C^5 in glucose

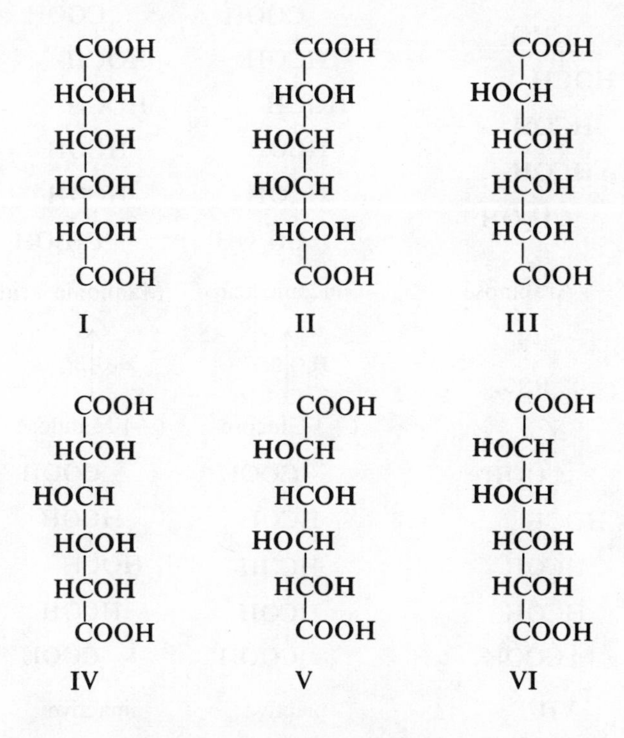

Fig. 1 Structure of glucaric acids

VIII	IX	X	XI
CHO	CHO	CH_2OH	CH_2OH
HCOH	HOCH	CO	HCOH
HOCH	HOCH	HOCH	HOCH
HCOH	HCOH	HCOH	HCOH
HCOH	HCOH	HCOH	HCOH
CH_2OH	CH_2OH	CH_2OH	CHO
(+) Glucose	(+) Mannose	(−) Fructose	(+) Gulose

Fig. 2 Structure of hexoses

and mannose and since the only optically active dicarboxylic acid from arabinose is represented by VII, it follows that the structure of (+)glucose and (+) mannose is either VIII or IX, the structure of (—)fructose is X (Fig. 2), and the structures of arabinose, gluconic and mannonic acids are as depicted in (2).

(−) Arabinose	Gluconic acid	Mannonic acid
CHO	COOH	COOH
HOCH	HCOH	HOCH
HCOH	HOCH	HOCH
HCOH	HCOH	HCOH
CH_2OH	HCOH	HCOH
	CH_2OH	CH_2OH

(−) Arabinose $\xrightarrow[2.\ H_2O]{1.\ HCN}$ Gluconic acid + Mannonic acid (2)

(−) Arabinose $\xrightarrow{HNO_3}$ VII; (+) Glucose $\xrightarrow{H_2O\ Br_2}$ Gluconic acid; (+) Mannose $\xrightarrow{H_2O\ Br_2}$ Mannonic acid

VII		
COOH	COOH	COOH
HOCH	HCOH	HCOH
HCOH	HCOH	HOCH
HCOH	HCOH	HCOH
COOH	COOH	COOH
VII optically active	inactive	inactive

A fourth sugar was available to Fischer, (+)gulose, which on oxidation with nitric acid gave the same glucaric acid as (+)glucose. This observation is only compatible with structure XI for (+)gulose, VIII for (+)glucose and IX for (+)mannose. An interchange of the C^1 and C^6 functions in IX leads to the same compound.

2.2 Conformation and stereoisomerism

The open form of hexoses exists in equilibrium with their hemiacetals or hemiketals (3a,b). Glucose is predominantly present in the pyranose form and both the α- and β-anomers have been isolated. By definition, α denotes the isomer which has C^1–OH and C^5–C^6 in the *trans* position. Cyclization creates a new asymmetric centre. The anomers differ in physical properties and in optical rotation. In solution the two forms approach an equilibrium, a reaction that can be followed by measuring the optical rotation. The change is called mutarotation. The six-membered rings exist in chair and boat forms (Fig. 3).

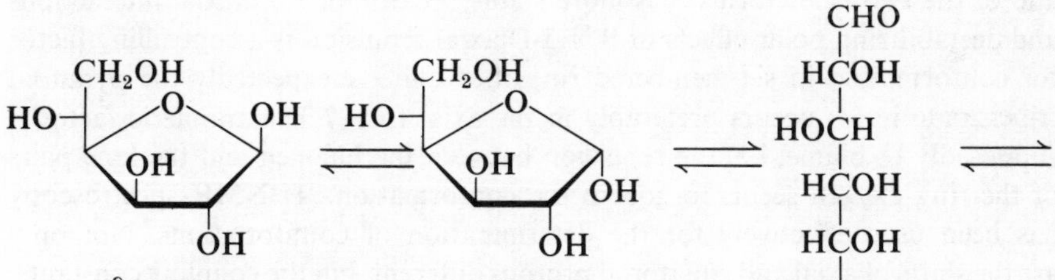

CH₂OH, HO, O, OH, OH, OH — β-anomer ⇌ CH₂OH, HO, O, OH, OH, OH — α-anomer ⇌ CHO / HCOH / HOCH / HCOH / HCOH / CH_2OH ⇌

Furanose form of glucose (3a)

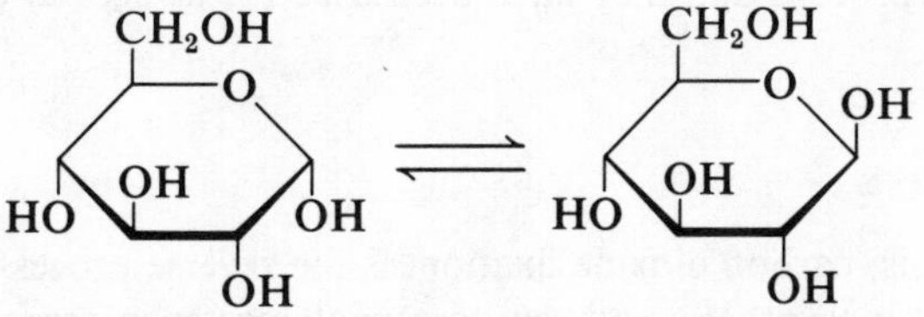

CH_2OH, O, OH, HO, OH, OH — α-anomer ⇌ CH_2OH, O, OH, OH, HO, OH — β-anomer

Pyranose form of glucose

$$RCHO + R'OH \rightleftharpoons RCH(OR')(OH) \underset{}{\overset{R'OH}{\rightleftharpoons}} RCH(OR')(OR') + H_2O \quad (3b)$$

Hemiacetal — Acetal

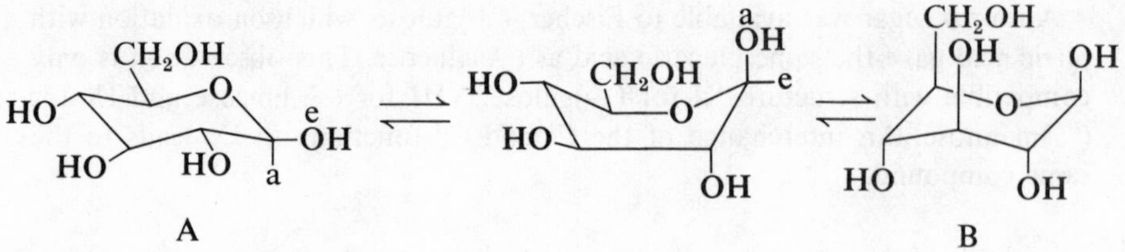

Fig. 3 Chair and boat forms of β-glucose. e denotes equatorial bond, a axial bond

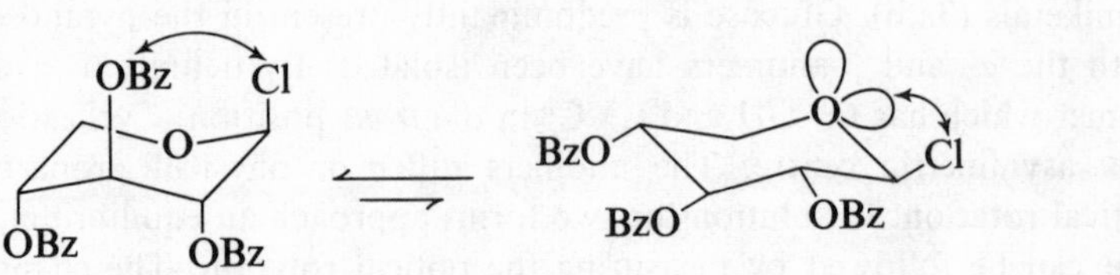

(4)

Because of reduced steric interaction, the chair form is normally more stable and of the two chair forms A is more stable because of 1,3-diaxial interactions and destabilizing polar effects in B. 1,3-Diaxial repulsion is a controlling factor for conformation in six-membered rings but quite unexpectedly the pyranoid tribenzoate in (4) occurs preferably in the axial form? The anomeric factor is supposedly to blame, i.e. the repulsion between the halogen and the lone pairs of the ring oxygen seems to govern the conformation. ^{1}H NMR spectroscopy has been used effectively for the determination of conformations. Not only are the shifts of axial and equatorial protons different, but the coupling constants, $^{3}J_{aa}$, of 1,2-diaxial protons are larger (*ca.* 9 Hz) than coupling constants of J_{ee} and J_{ae} of 1,2-diequatorial and 1,2-axial-equatorial protons (*ca.* 2–4 Hz). The coupling constants, $^{1}J_{^{13}C^{1}H}$, are also consistently different for axial and equatorial anomeric protons and they have been used for structural determination.[2]

2.3 Photosynthesis

Photosynthesis, carbon dioxide fixation, is the reverse process of respiration. It is the central biosynthetic pathway to carbohydrates in green plants. Energy is required for the process and it is provided by light that is absorbed by the chlorophylls in green plants. The details of the overall reaction (5) are extra-

$$nCO_2 + H_2O \underset{\text{Respiration}}{\overset{\text{Assimilation} \atop h\nu,\ \text{ATP, NADPH}}{\rightleftharpoons}} (CH_2O)_n + nO_2 \qquad (5)$$

ordinarily complicated, but are now known to a great extent. Life on earth depends on this reaction and it is of the utmost importance that man economizes with our large remaining areas of green vegetation in the temperate and tropical zones, e.g. in the Amazon region. They are part of our renewable energy sources. Animals use the energy in the respiration process and carbon dioxide is returned, thereby contributing to the oxygen–carbon dioxide balance.

The exceedingly large combustion of fossil fuels in the last decades seems to have increased the carbon dioxide content in the atmosphere slowly which in the long run could bring about climatic changes of unforeseen consequences.

Carbon dioxide uptake and the synthesis of sugars occur in a series of steps. In the first stage the energy of light is used for building up energy-rich ATP molecules (6) from which energy then is released to sustain the activities of the cell. The terms 'energy-rich' and 'release of energy' commonly used in biochemical literature in connection with ATP mean actually 'reactive'. It is not a question about transfer of a quantum of energy but simply of using ATP as a phosphorylating agent, thus making functions in other molecules (substrates) more reactive, e.g. –COOH, $-CONH_2$, –OH, are transformed to –COOP, –C(=NH)OP, –OP. The nucleophile reacts with one of the P–O–P bonds of ATP (section 1.8, Chapter 1 Fig. 12), and one or two phosphate units are transferred to the nucleophile.

$$H_2O + NADP^{\oplus} + P + ADP \xrightarrow{h\nu} \tfrac{1}{2}\,O_2 + NADPH + ATP + H^{\oplus} \qquad (6)$$

It was found by using oxygen-labelled water that the oxygen produced did come from water and not from carbon dioxide. Some sulphur bacteria use hydrogen sulphide as hydrogen donor, while other bacteria use certain organic substrates such as 2-propanol.

In the second phase ATP and NADPH are used for the carbon dioxide reduction. Calvin and coworkers found by using radioactive CO_2 that 3-phosphoglyceric acid was one of the first intermediates and the label was predominantly localized in the carbonyl group. They proposed the following sequence of reactions (7)–(9). Ribulose-1,5-diphosphate enolizes and the anion is carboxy-

CH_2OP–CO–HCOH–HCOH–CH_2OP (Ribulose-1,5-diphosphate) $\underset{}{\overset{-H^{\oplus}}{\rightleftharpoons}}$ CH_2OP–C(–O)=C(–O$^{\ominus}$)…H–HCOH–CH_2OP $\xrightarrow{^{*}CO_2}$ [CH_2OP–C(OH)($^{\ominus}$OO$\overset{*}{C}$)–C=O–HCOH–CH_2OP + H_2O] $\longrightarrow$ (7)

Ribulose-
1,5-diphosphate

$\overset{*}{C}OO^{\ominus}$–HCOH–$CH_2OP$ + $COO^{\ominus}$–HCOH–CH_2OP

3-Phosphoglycerate

lated at the 2-position giving an unstable intermediate which undergoes a cleavage to two moles of 3-phosphoglycerate (7). 3-Phosphoglycerate now enters the reverse glycolytic pathway (8), (9). ATP transfers 3-phosphoglycerate to the more reactive 3-phosphoglyceroyl phosphate. It is reduced by NADPH to 3-phosphoglyceraldehyde, one molecule of which rearranges to dihydroxyacetone phosphate and undergoes an aldol condensation with another molecule of 3-phosphoglyceraldehyde to form fructose-1,6-diphosphate that ultimately isomerizes to glucose. Each of these steps is catalysed by its specific enzyme. Assimilation of radioactive carbon dioxide for a very short period should thus give glucose labelled on carbon 3 and 4.

$$\begin{array}{c} {}^{*}COOH \\ | \\ HCOH \\ | \\ CH_2OP \end{array} \underset{NADP^{\oplus}}{\overset{ATP,\ NADPH}{\rightleftharpoons}} \begin{array}{c} {}^{*}CHO \\ | \\ HCOH \\ | \\ CH_2OP \end{array} \rightleftharpoons \begin{array}{c} {}^{*}CH_2OH \\ | \\ CO \\ | \\ CH_2OP \end{array} \underset{}{\overset{-H^{\oplus}}{\rightleftharpoons}} \begin{array}{c} {}^{\ominus *}CHOH \\ | \\ CO \\ | \\ CH_2OP \end{array}$$

$$\downarrow$$

$$\begin{array}{c} {}^{1}CH_2OP \\ | \\ {}^{2}CO \\ | \\ HO{}^{3}\overset{*}{C}H \\ | \\ H{}^{4}\overset{*}{C}OH \\ | \\ H{}^{5}COH \\ | \\ {}^{6}CH_2OP \end{array} \qquad (8)$$

Fructose-1,6-diphosphate

$$\begin{array}{c} CH_2OP \\ | \\ CO \\ | \\ HO\overset{*}{C}H \\ | \\ H\overset{*}{C}OH \\ | \\ HCOH \\ | \\ CH_2OP \end{array} \rightleftharpoons \begin{array}{c} CH_2OH \\ | \\ CO \\ | \\ HO\overset{*}{C}H \\ | \\ H\overset{*}{C}OH \\ | \\ HCOH \\ | \\ CH_2OP \end{array} \rightleftharpoons \begin{array}{c} CHO \\ | \\ HCOH \\ | \\ HO\overset{*}{C}H \\ | \\ H\overset{*}{C}OH \\ | \\ HCOH \\ | \\ CH_2OP \end{array} \rightleftharpoons \begin{array}{c} CHO \\ | \\ HCOH \\ | \\ HO\overset{*}{C}H \\ | \\ H\overset{*}{C}OH \\ | \\ HCOH \\ | \\ CH_2OH \end{array} \qquad (9)$$

Fructose-6-phosphate | Glucose-6-phosphate | Glucose

Equations (7)–(9) account for the production of one glucose molecule from carbon dioxide and ribulose-1,5-diphosphate but one ribulose-1,5-diphosphate molecule is regenerated for each molecule of carbon dioxide reduced. In order to make the photosynthesis of carbohydrates self-consistent Calvin proposed the cycle (10)–(11). The Calvin cycle means that of the six fructose-6-phosphate molecules formed from six ribulose-1,5-diphosphate molecules and six molecules of carbon dioxide, one fructose-6-phosphate molecule is diverted into one glucose molecule, whereas the other five fructose-6-phosphate molecules are recycled into six new ribulose-1,5-diphosphate molecules. By the action of the coenzyme thiaminpyrophosphate (RH) fructose-6-phosphate is cleaved in a retrobenzoin condensation fashion and recondensed with 3-phosphoglyceraldehyde to ribulose-5-phosphate via xylulose-5-phosphate as an intermediate, a reaction catalysed by the enzyme transketolase (10).

CH_2OH–CO–HOCH–HCOH–HCOH–CH_2OP (Fructose-6-phosphate) $\overset{R^\ominus}{\rightleftharpoons}$ [CH_2OH–C(R)(OH)–(H–O)CH–HCOH–HCOH–CH_2OP] $\rightleftharpoons$ R–$\overset{\ominus}{C}$(OH)–CH_2OH + CHO–HCOH–HCOH–CH_2OP (Erythrose-4-phosphate)

R–$\overset{\ominus}{C}$(OH)–CH_2OH + CHO–HCOH–CH_2OP $\rightleftharpoons$ [CH_2OH–C(R)(O H)–HOCH–HCOH–CH_2OP] $\overset{-R^\ominus}{\rightleftharpoons}$ (10)

CH_2OH–CO–HOCH–HCOH–CH_2OP (Xylulose-5-phosphate) $\rightleftharpoons$ CH_2OH–CO–HCOH–HCOH–CH_2OP (Ribulose-5-phosphate)

The other cleavage product, erythrose-4-phosphate, condenses with dihydroxyacetone phosphate to sedoheptulose-1,7-diphosphate which undergoes cleavage by transketolase. Ribose-5-phosphate is formed and the C_2-fragment gives xylulose-5-phosphate with 3-phosphoglyceraldehyde (11). Both pentoses rearrange finally to ribulose-1,5-diphosphate which is phosphorylated by ATP to ribulose-1,5-diphosphate.

$$
\begin{array}{c}
CH_2OP\\ CO\\ CH_2OH\\ +\\ CHO\\ HCOH\\ HCOH\\ CH_2OP
\end{array}
\rightleftharpoons
\begin{array}{c}
CH_2OP\\ CO\\ HOCH\\ HCOH\\ HCOH\\ HCOH\\ CH_2OP\\ \text{Sedoheptulose-}\\ \text{1,7-diphosphate}
\end{array}
\underset{}{\overset{-P,\ R^{\ominus}}{\rightleftharpoons}}
\left[\begin{array}{c}
CH_2OH\\ R{-}C{-}OH\\ H\,O{-}CH\\ HCOH\\ HCOH\\ HCOH\\ CH_2OP
\end{array}\right]
\rightleftharpoons
$$

$$
\begin{array}{c}
CH_2OH\\ R{-}\underset{\ominus}{C}{-}OH\\ +\\ CHO\\ HCOH\\ HCOH\\ HCOH\\ CH_2OP\\ \text{Ribose-5-}\\ \text{phosphate}
\end{array}
\rightleftharpoons
\begin{array}{c}
CH_2OH\\ CO\\ HCOH\\ HCOH\\ CH_2OP\\ \text{Ribulose-5-}\\ \text{phosphate}\\ \updownarrow\\ \text{Ribulose-1,5-}\\ \text{diphosphate}
\end{array}
\rightleftharpoons
\begin{array}{c}
CH_2OH\\ CO\\ HOCH\\ HCOH\\ CH_2OP\\ \text{Xylulose-5-}\\ \text{phosphate}
\end{array}
\overset{-R^{\ominus}}{\rightleftharpoons}
\qquad (11)
$$

$$
\left[\begin{array}{c}
CH_2OH\\ R{-}C{-}OH\\ HOCH\\ HCOH\\ CH_2OP
\end{array}\right]
\rightleftharpoons
\begin{array}{c}
CH_2OH\\ R{-}\underset{\ominus}{C}{-}OH\\ CHO\\ HCOH\\ CH_2OP
\end{array}
$$

The mechanism of transketolization is formulated in (12) and (13). The deprotonated form of thiamine pyrophosphate attacks the carbonyl group and the anion formed in the cleavage is stabilized. In this respect the coenzyme resembles the action of the cyanide group in the benzoin condensation or the thiazolium ion catalyst in the Stetter reaction. Thiamine pyrophosphate plays the same role in the decarboxylation of pyruvic acid (14).

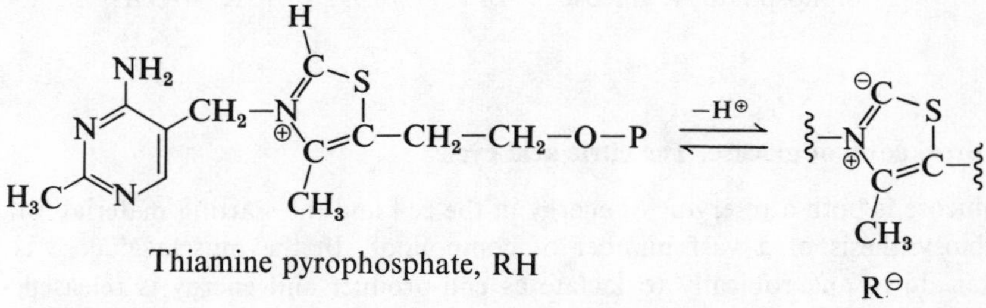

$\xrightarrow{-H^{\oplus}}$ (12)

Thiamine pyrophosphate, RH

$R^{\ominus}$

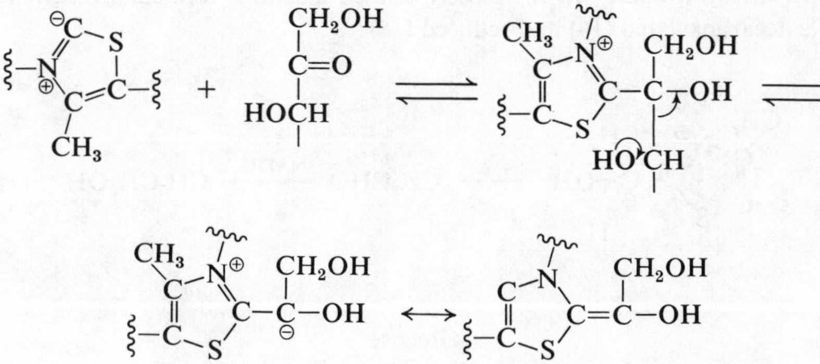

(13)

$-CO_2$

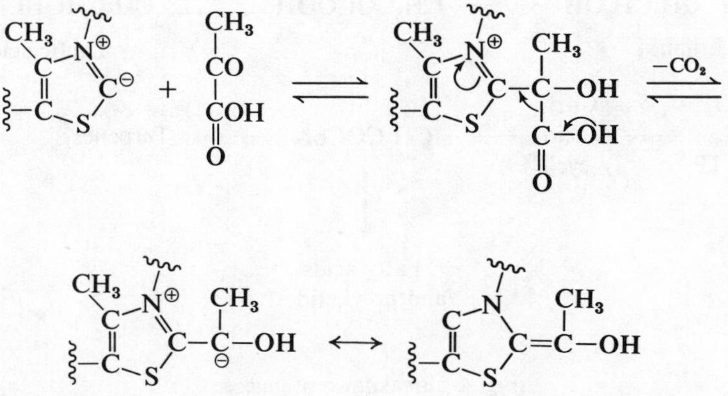

(14)

The reactions of the Calvin cycle can be summed up according to (15). Ribulose-1,5-diphosphate appears on both sides indicating that it is a true component of the photosynthesis and that another molecule is regenerated in the process.

$$\text{6 Ribulose-1,5-diphosphate} + 6\,CO_2 + 18\,\text{ATP} + 12\,\text{NADPH} + 12\,H^{\oplus} + 12\,H_2O \rightarrow \text{6 ribulose-1,5-diphosphate} + \text{glucose} + 18\,\text{P} + 18\,\text{ADP} + 12\,\text{NADP}^{\oplus} \quad (15)$$

2.4 Breakdown of glucose. The citric acid cycle

Glucose is both a reservoir of energy in the cell and the starting material for the biosynthesis of a vast number of compounds. In the muscle glucose is broken down anaerobically to lactate as end-product and energy is released. This process is called glycolysis, and we have seen in the previous section that the process is reversible. Lactate is brought by the blood to the liver where glucose is resynthesized. In the closely related alcoholic fermentation pyruvic acid is decarboxylated (14) and reduced (16).

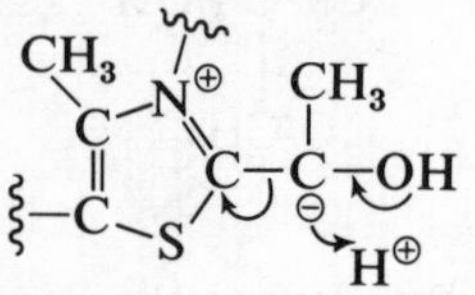

$$\text{(thiazolium adduct)} \longrightarrow CH_3CHO \xrightarrow{\text{NADH}} CH_3CH_2OH \quad (16)$$

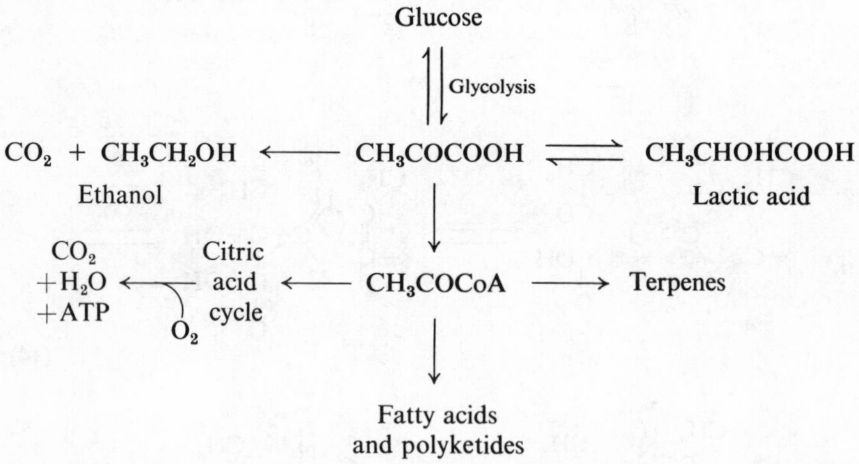

Fig. 4 Breakdown of glucose

By oxidative decarboxylation of pyruvic acid, acetyl coenzyme A is formed. This is the key compound in three vital processes (Fig. 4):

1. the citric acid cycle which coupled with oxidative phosphorylation, in essence is respiration;
2. fatty acid synthesis; and
3. terpene synthesis.

In this section we will examine in some detail the formation of acetyl coenzyme A and its combustion to carbon dioxide (17) in the citric acid cycle, a reaction that releases much more energy than glycolysis (18). In both processes about 50 per cent of free energy is conserved in ATP.

$$\text{Glucose} + 6\ O_2 + 36\ P + 36\ ADP \rightleftharpoons 6\ CO_2 + 36\ ATP + 42\ H_2O \qquad \Delta G^{O\prime} = -423\ \text{kcal/mol} \tag{17}$$

$$\text{Glucose} + 2\ P + 2\ ADP \rightleftharpoons 2\ \text{lactate} + 2\ ATP + 2\ H_2O \qquad \Delta G^{O\prime} = -32.4\ \text{kcal/mol} \tag{18}$$

The first step, decarboxylation, of the transformation of pyruvic acid to acetyl CoA is mediated by the coenzyme thiamine pyrophosphate according to (14). The enamine is oxidized by the coenzyme lipoic acid of the enzyme complex and the acetyl group is then transferred to coenzyme A which leaves the complex (19). The dihydrolipoic acid is reoxidized by FAD.

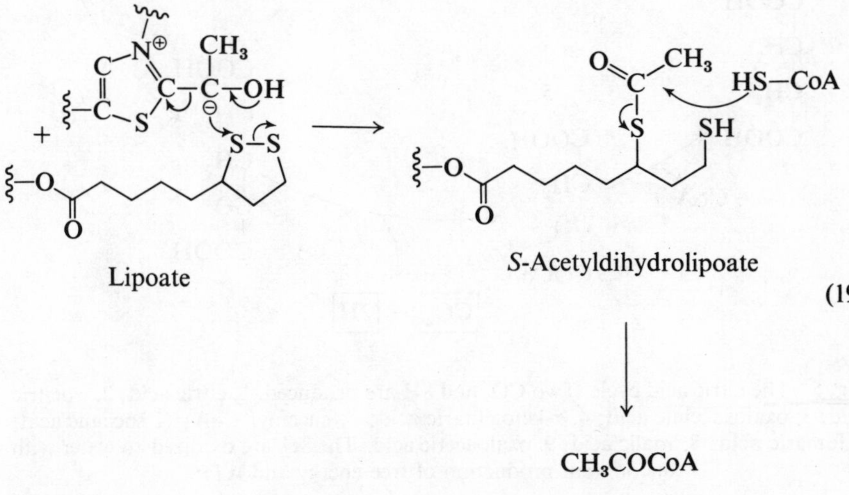

CH_3 OH + Lipoate → *S*-Acetyldihydrolipoate (+ HS—CoA, SH) → CH_3COCoA (19)

Lipoate

S-Acetyldihydrolipoate

CH_3COCoA

Acetyl CoA enters the citric acid cycle and condenses with oxaloacetic acid, 9, to form citric acid, 1. Elimination and addition of water give isocitric acid and oxidation of the hydroxyl group followed by decarboxylation give α-ketoglutaric acid, 4.

In much the same way as oxidative decarboxylation of pyruvic acid (14, 19),

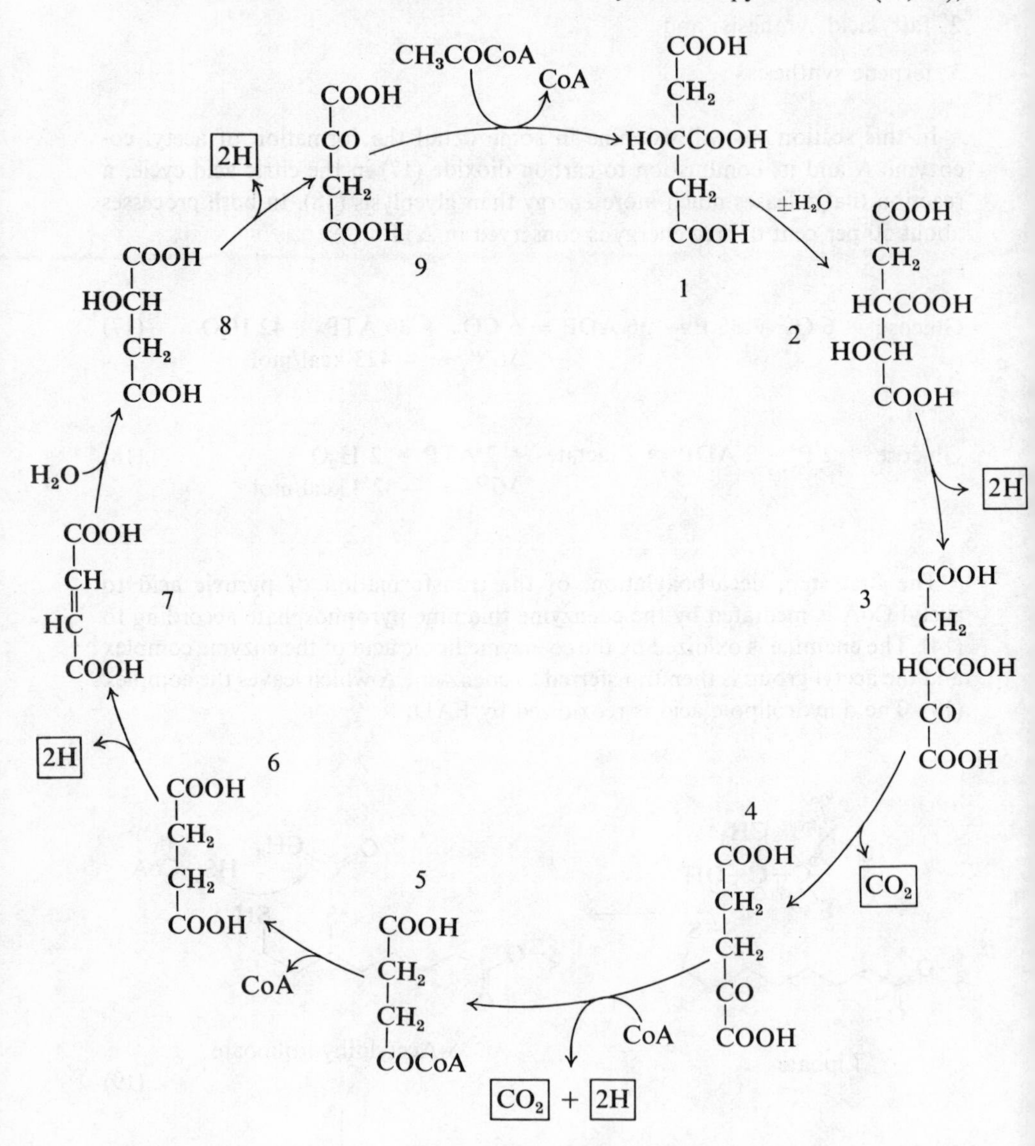

Fig. 5 The citric acid cycle. Two CO_2 and 8H are produced. 1, citric acid; 2, isocitric acid; 3, oxalosuccinic acid; 4, α-ketoglutaric acid; 5, succinyl CoA; 6, succinic acid; 7, fumaric acid; 8, malic acid; 9, oxaloacetic acid. The 8H are oxidized to water with concomitant production of free energy and ATP

α-ketoglutaric acid is decarboxylated to succinyl CoA, 5, which is dehydrogenated to fumaric acid, 7. Addition of water gives malic acid which on oxidation regenerates oxaloacetic acid and the circuit is completed (Fig. 5). For every turn of the cycle one molecule of acetyl CoA is consumed, and two moles of carbon dioxide are evolved together with eight hydrogens or eight electrons and eight protons. The hydrogens are oxidized to water with the production of large amounts of energy (17) part of which is conserved as ATP, a process called oxidative phosphorylation. The citric acid cycle was formulated by Krebs as the result of an ingenious piece of experimentation and reasoning, hence it is sometimes called the Krebs cycle.

2.5 Monosaccharides

Glucose is the precursor of other hexoses, disaccharides, polysaccharides and glycosides. Glucose-6-phosphate, fructose-6-phosphate and mannose-6-phosphate are directly interconvertible but transformation of glucose to other hexoses requires the assistance of a nucleoside triphosphate, usually uridine triphosphate, UTP, which is attacked at the α-P by the phosphate group of glucose-1-phosphate. Diphosphate is split off and uridine diphospho glucose is formed (20). UDP-glucose can be transformed to most other sugars by epimerizations, oxidations and reductions, or directly reacted with other alcohols or sugars to glycosides.

CH_2OH O HO OH OP OH + $PPPOCH_2$ O HN O O N HO OH →

Glucose-1-phosphate

Uridine triphosphate, UTP

(20)

CH_2OH O HO OH O–$PPOCH_2$ O HN O O N HO OH + PP

Uridine diphospho glucose, UDP-glucose

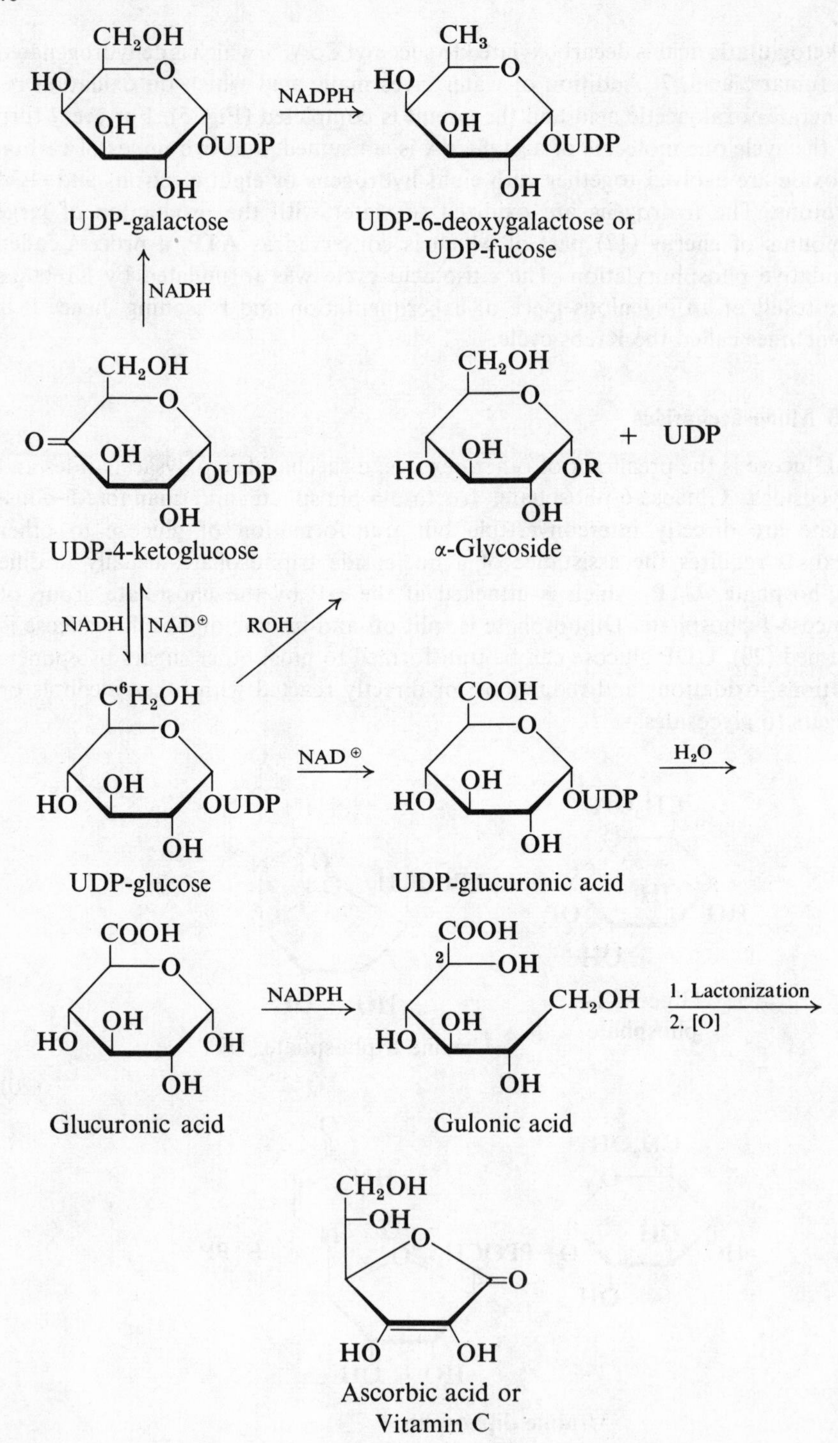

Fig. 6 Biosynthesis of galactose and ascorbic acid

These reactions have interesting mechanistic implications. Studies of the glucose–galactose interconversion demonstrate that no exchange of C^4–H or C^4–O takes place with the solvent, i.e. the ordinary S_N2 substitution with inversion of configuration at C^4 or a dehydration–hydration sequence are excluded. A redox process[3] involving $NAD^{\oplus}$ as oxidant is attractive for two reasons. It is in accordance with the lack of C^4–H exchange with the solvent and furthermore feeding 4-keto-*D*-glucose to the enzyme system containing 4–^{3}H–NADH results in the formation of 4–^{3}H-D-galactose but, strangely enough, 4–^{3}H-D-glucose was not detected. The redox process requires that the intermediate 4-keto sugar must rotate 180° before being reduced.[4] Fig. 6 shows the biosynthesis of ascorbic acid and fucose, a common constituent of cell walls in bacteria and algae, and the biosynthesis of galactose, an important sugar occurring in milk and a common constituent of many polysaccharides, e.g. agar and gum arabic. Hydrolysis of milk sugar, lactose, gives one mole of glucose and one mole or galactose.

Oxidation of C^6 of glucose to a carboxyl group gives glucuronic acid, the precursor of ascorbic acid, or vitamin C. Reduction of the aldehyde group gives gulonic acid which lactonizes to the five-membered gulonolactone. Oxidation of C^2 and enolization gives ascorbic acid. Ascorbic acid is a vitamin, an essential nutrient for human beings, who require *ca.* 70 mg a day, but are unable to synthesize it themselves. Deficiency leads to the symptoms of scurvy. The physiological processes in which ascorbic acid partakes are still unknown. It is excreted in human urine as the 2-*O*-sulphate. Ascorbic acid is produced in large quantities industrially from glucose.

Fig. 7 shows some rare sugar constituents of glycosidic antibiotics containing the amino function. It is introduced by oxidation of a hydroxyl group with $NADP^{\oplus}$ to a carbonyl group followed by transamination (section 6.4) and reduction.

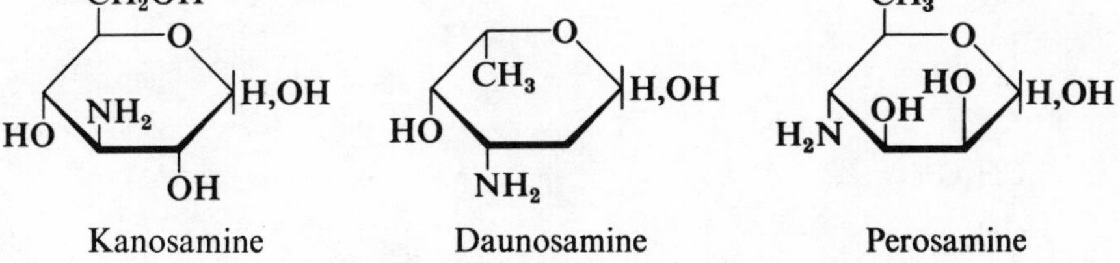

Kanosamine | Daunosamine | Perosamine

Fig. 7 Amino sugars from glycosidic antibiotics

The first branched sugars, apiose[5] (Fig. 11) and hamamelose[6] were isolated several years ago from parsley *Petroselium crispum*, and *Hamamelis virginiana*, respectively, and they were for a long period regarded as curiosities until in recent times branched sugars were detected in microorganisms as components of antibiotics,[7–9] i.e. mycarose, cladinose, streptose, and aldgarose.

Mycarose is biosynthesized from intact glucose in cell free extracts from *Streptomyces rimosus*, methionine acting as methyl donor.[9] The early biosynthetic studies were carried out in the 'prespectroscopic' period by using Kuhn–Roth oxidation, borohydride reduction, periodate cleavage, and Hunsdiecker decarboxylation (Fig. 8). The fact that TDP-6-deoxy-D-*xylo*hexos-4-ulose (TDP = thymidine diphosphate, section 8.2) acts as an intermediate precursor for rhamnose[10] and mycarose gives the key to the mechanisms of deoxygenation and *C*-methylation (Fig. 9). Oxidation of C^4–OH of TDP-glucose by NAD$^\oplus$ to a carbonyl group activates C^3–H and C^5–H. A carbanionic species is generated and C^6–OH is lost to yield an α,β-unsaturated ketone which is then reduced by NADH. The label of C^4–^2H of the substrate is recovered at C^6. The steric course of the hydride transfer is completely stereospecific and intramolecular and involves displacement of the hydroxyl group in a *trans* mode as shown by ^2H and ^3H labelling.[11] C^3–H and C^5–H undergo exchange with the solvent. Consequently C^2 is reduced before C^3 is methylated.

Formation of C^3-deoxy sugars can be formulated similarly by generation of a carbanion at C^2. The enzymatic formation of 2-keto-3-deoxy-6-P-gluconate from 6-P-gluconic acid has been carefully studied.[12] A carbanion is generated at C^2, facilitating the elimination of the C^3–OH. The intermediate enol rearranges subsequently to the 2-keto derivative (21).

In consistency with this mechanism one atom of ^3H was incorporated at C^3

$$\begin{array}{l} {}^1CHO \\ | \\ {}^2CH_2 \\ | \\ HO^3CCH_3 \\ | \\ HO^4CH \\ | \\ HO^5CH \\ | \\ {}^6CH_3 \end{array} \xrightarrow[\text{Roth}]{\text{Kuhn-}} \underset{(C^{3,\,3a},\ C^{5,\,6})}{2\ CH_3COOH} \xrightarrow{\text{Hunsdiecker}} \underset{(C^3,\ C^5)}{CO_2}$$

$$\downarrow NaBH_4$$

$$\begin{array}{l} CH_2OH \\ | \\ CH_2 \\ | \\ HOCCH_3 \\ | \\ HOCH \\ | \\ HOCH \\ | \\ CH_3 \end{array} \xrightarrow[\text{oxidation}]{\text{Periodate-}} \underset{(C^{5,\,6})}{CH_3CHO} + \underset{(C^4)}{HCOOH} \xrightarrow{Hg^{2\oplus}} \underset{(C^4)}{CO_2}$$

Fig. 8 Degradation of mycarose

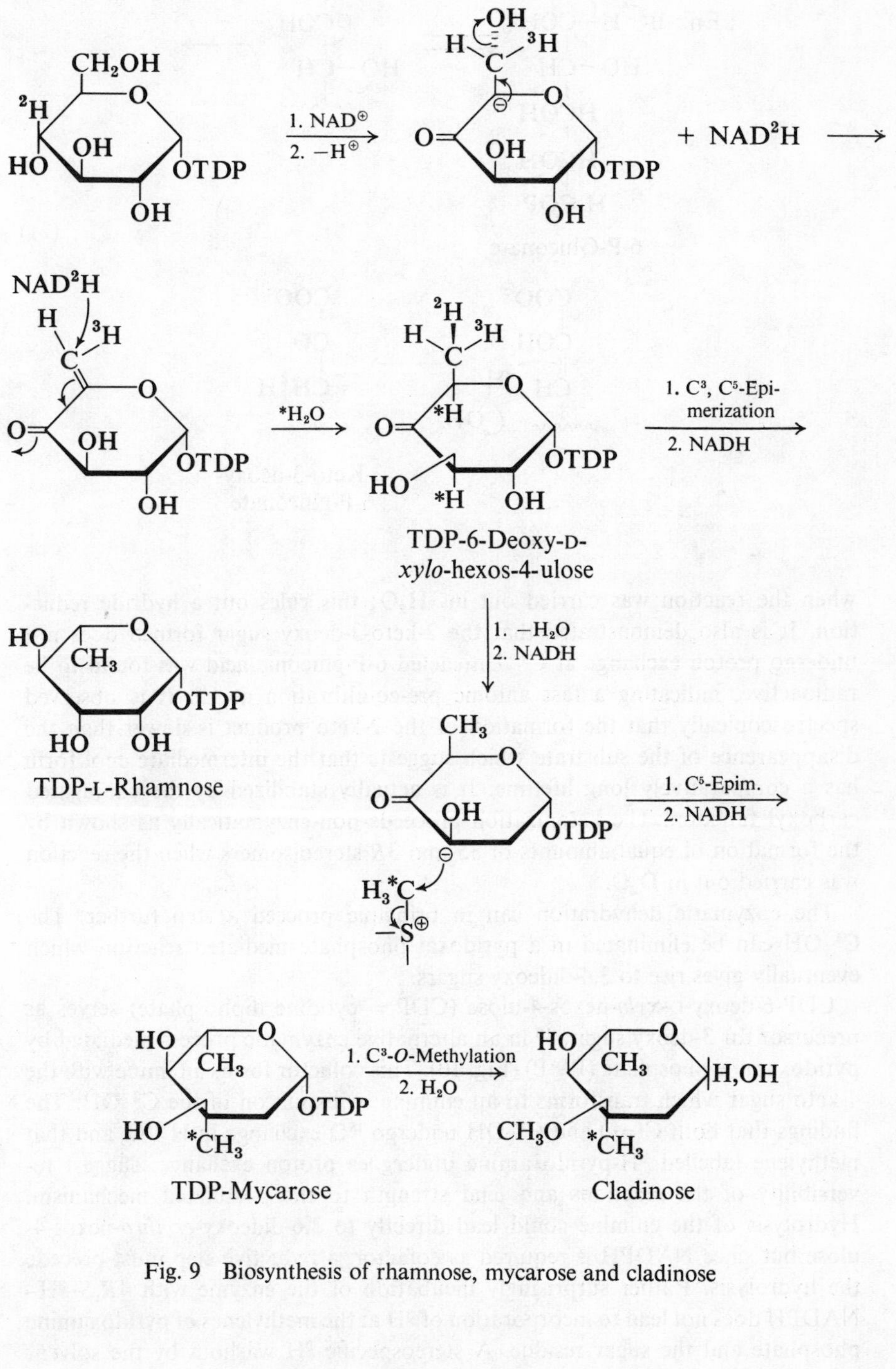
CH₂OH
²H
HO
OH
OTDP
OH
1. NAD⊕
2. −H⊕
OH
H
³H
O
+ NAD²H
NAD²H
*H₂O
TDP-6-Deoxy-D-
xylo-hexos-4-ulose
1. C³, C⁵-Epi-
merization
2. NADH
1. −H₂O
2. NADH
TDP-L-Rhamnose
CH₃
1. C⁵-Epim.
2. NADH
1. C³-O-Methylation
2. H₂O
H,OH
CH₃O
TDP-Mycarose
Cladinose

Fig. 9 Biosynthesis of rhamnose, mycarose and cladinose

$$
\begin{array}{ccc}
\begin{array}{c}
COO^{\ominus} \\
\text{Enz-B:}\ H{-}COH \\
HO{-}CH \\
HCOH \\
HCOH \\
H_2COP \\
\text{6-P-Gluconate}
\end{array}
& \rightleftharpoons &
\begin{array}{c}
COO^{\ominus} \\
{}^{\ominus}COH \\
HO{-}CH \\
\sim\sim
\end{array}
\longrightarrow
\end{array}
\qquad (21)
$$

$$
\begin{array}{ccc}
\begin{array}{c}
COO^{\ominus} \\
COH \\
\| \\
CH \quad {}^{3}H\ \ OH \\
\sim\sim
\end{array}
& \longrightarrow &
\begin{array}{c}
COO^{\ominus} \\
CO \\
CH^{3}H \\
\sim\sim \\
\text{2-Keto-3-deoxy-} \\
\text{6-P-gluconate}
\end{array}
\end{array}
$$

when the reaction was carried out in ^3H_2O; this rules out a hydride reduction. It is also demonstrated that the 2-keto-3-deoxy sugar formed does not undergo proton exchange at C^3. Unreacted 6-P-gluconic acid was found to be radioactive, indicating a fast anionic pre-equilibration at C^2. It is observed spectroscopically that the formation of the 2-keto product is slower than the disappearence of the substrate which suggests that the intermediate enol form has a comparatively long lifetime. It is actually stabilized by the conjugated carboxyl function. The ketonization proceeds non-enzymatically as shown by the formation of equal amounts of 3*S* and 3*R* stereoisomers when the reaction was carried out in D_2O.

The enzymatic dehydration can in principle proceed a step further. The C^4–OH can be eliminated in a pyridoxal phosphate mediated reaction which eventually gives rise to 3,4-dideoxy sugars.

CDP-6-deoxy-D-*xylo*-hexos-4-ulose (CDP = cytidine diphosphate) serves as precursor for 3-deoxy sugars [13] in an alternative enzymatic process mediated by pyridoxamine phosphate (PMP) (Fig. 10). This cofactor forms an imine with the 4-keto sugar which transforms to an enimine by expulsion of the C^3–OH. The findings that both $C^4{=}O$ and C^3–OH undergo ^{18}O exchange in $H_2{}^{18}O$, and that methylene labelled ^3H-pyridoxamine undergoes proton exchange, suggest reversibility of the reactions and lend strength to the postulated mechanism. Hydrolysis of the enimine could lead directly to 3,6-dideoxy-*erythro*-hexos-4-ulose but since NADPH is required as cofactor, a reductive step must precede the hydrolysis. Rather surprisingly incubation of the enzyme with 4*R*,*S*–^3H–NADPH does not lead to incorporation of ^3H at the methylenes of pyridoxamine phosphate and the sugar residue. A stereospecific ^3H washout by the solvent

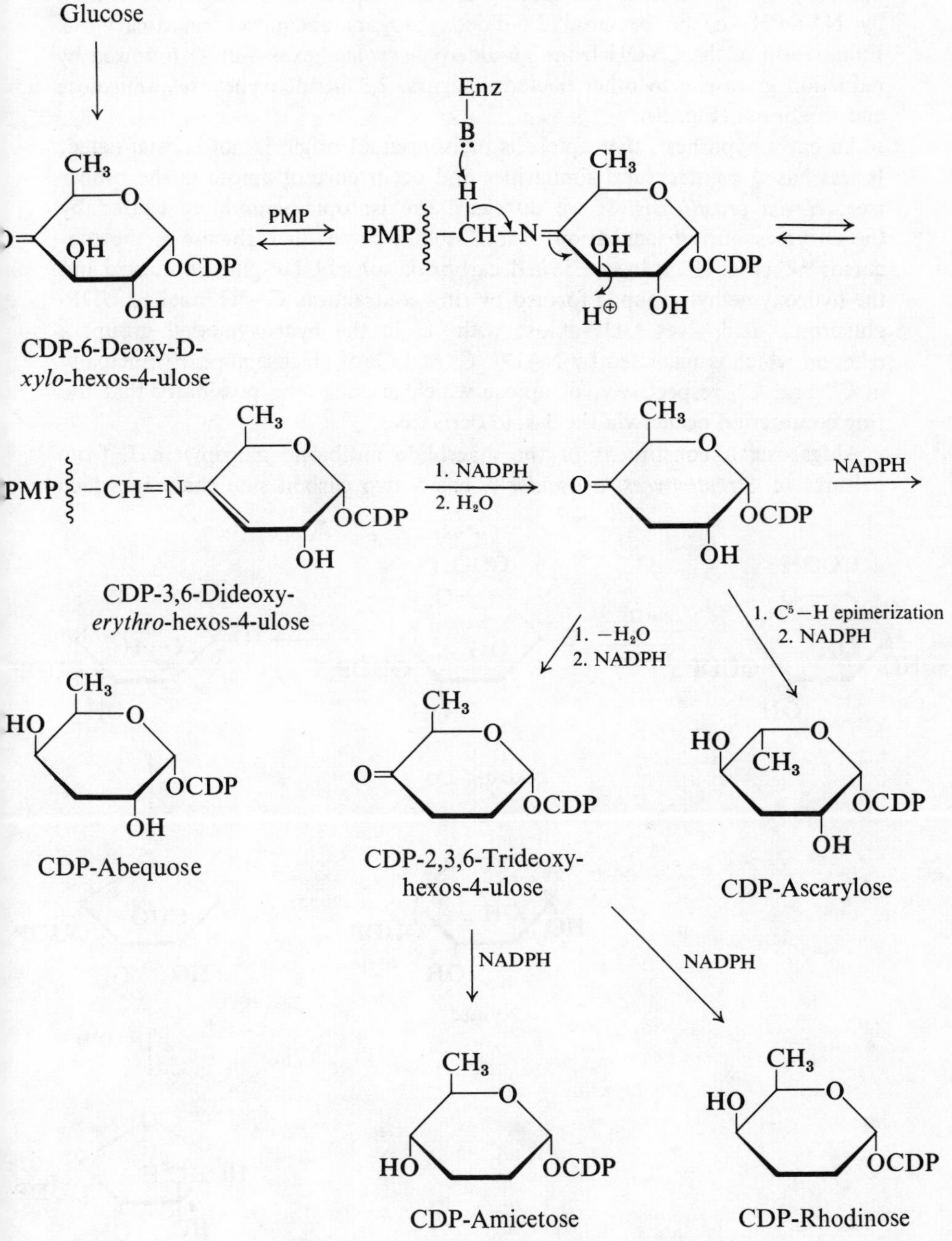

Glucose

CDP-6-Deoxy-D-*xylo*-hexos-4-ulose

PMP

CDP-3,6-Dideoxy-*erythro*-hexos-4-ulose

1. NADPH
2. H_2O

NADPH

1. $-H_2O$
2. NADPH

1. C^5-H epimerization
2. NADPH

CDP-Abequose

CDP-2,3,6-Trideoxy-hexos-4-ulose

CDP-Ascarylose

NADPH

NADPH

CDP-Amicetose

CDP-Rhodinose

Fig. 10 Biosynthesis of some deoxy sugars

has been offered as an explanation. The 4-keto function is eventually reduced by NADPH to the bacterial 3,6-dideoxy sugars abequose and ascarylose. Elimination of the C^2–OH from 3,6-dideoxy-*erythro*-hexos-4-ulose followed by reduction gives rise to other bacterial sugars, 2,3,6-trideoxyhexoses amicetose and rhodinose (Fig. 10).

An early hypothesis that apiose is of isoprenoid origin is not substantiated. It was based on structural similarities and occurrence of apiose in the rubber tree, *Hevea brasiliensis.* If we disregard the isotopic scrambling caused by the enzymes during incubation, tracer studies reveal that glucose is the precursor [14,15] (Fig. 11). It loses C^6 via decarboxylation of UDP-glucuronic acid and the hydroxymethyl group is formed by ring contraction. C^4–^3H labelled UDP-glucuronic acid gives UDP-apiose with ^3H in the hydroxymethyl group, a reaction which is mediated by $NAD^{\oplus}$. C^3 and C^4 of glucose appear principally in $C^{3'}$ and C^3, respectively, of apiose which excludes the possibility that the ring contraction occurs via the 3-keto derivative.

Aldgarose, a constituent of the macrolide antibiotic aldgamycin E from cultures of *Streptomyces lavendulae*[16], has a two carbon side chain in which

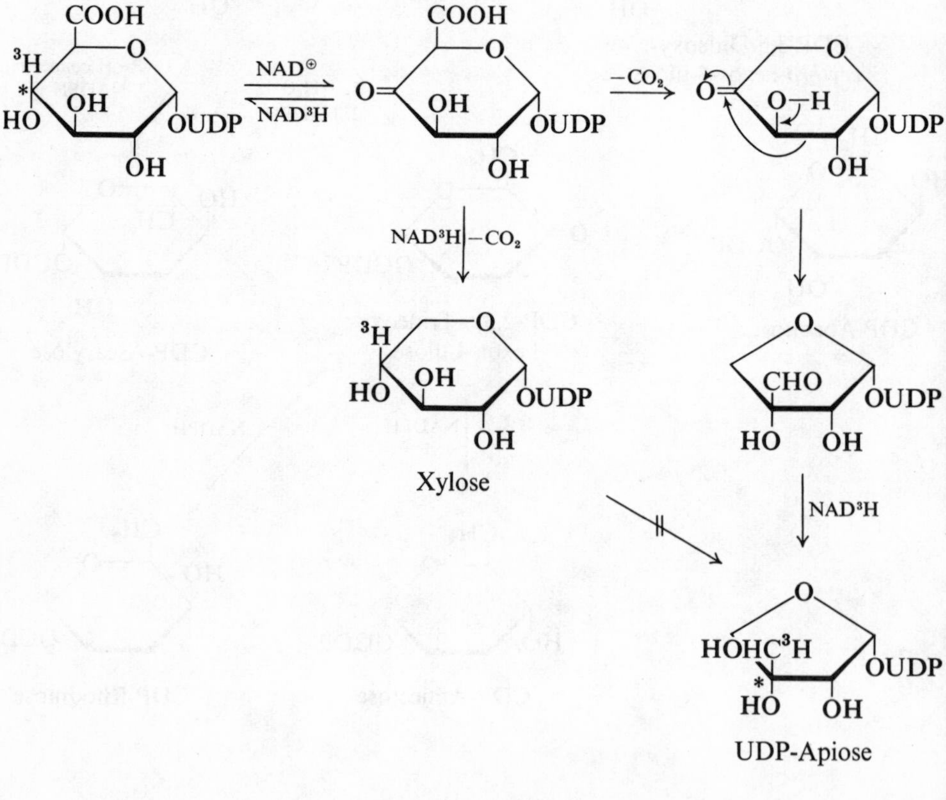

Fig. 11 Biosynthesis of UDP-apiose and xylose

labelled pyruvate, but not acetate, is incorporated; methionine and ethionine do not function as precursors. Glucose is exclusively incorporated in the hexose portion. The unique cyclic carbonate group derives from bicarbonate.[8] It is suggested that the two carbon fragment is introduced via a thiamine phosphate mediated decarboxylation of pyruvate. The enamine generated (14) condenses with a hexos-3-ulose (Fig. 12). This pathway is still not fully substantiated, since known thiamine deactivators have no influence on the incorporation of pyruvate in this enzymatic system. A pyridoxamine phosphate mediated decarboxylation of pyruvate is an alternative route to the two carbon fragment and alanine could conceivably be a precursor.

We have earlier met some pentoses as intermediates in the photosynthetic process. They appear also as products from the degradation of glucose via the phosphogluconate pathway. In this glucose-6-phosphate is dehydrogenated to 6-phosphogluconate and further oxidized and decarboxylated to ribulose-5-phosphate. β-Keto acids are known to decarboxylate with greatest ease and enzymes dexterously use this route to shorten the chain by one carbon atom. Isomerization gives the whole family of aldo- and ketopentoses (Fig. 13). Fig. 11 shows an alternative way to xylose from glucuronic acid.

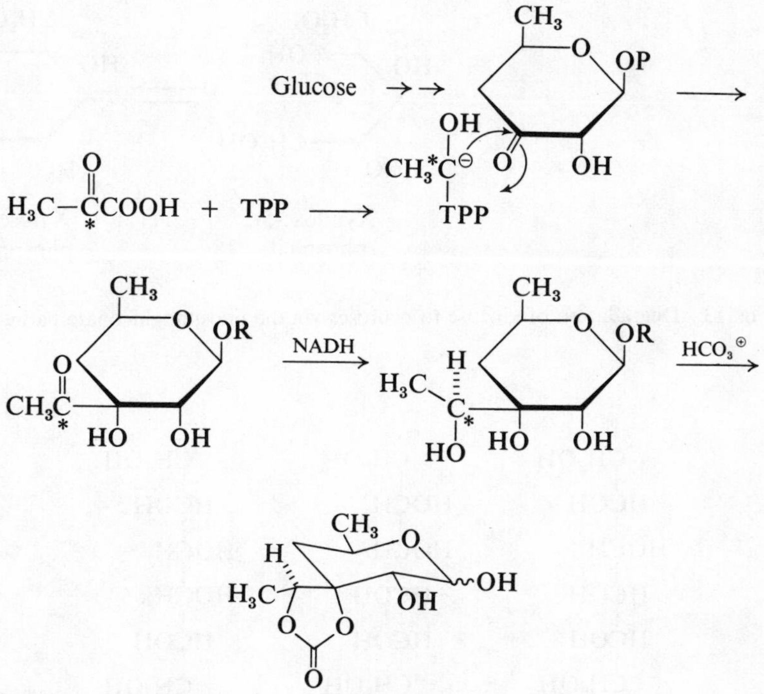

Fig. 12 Proposed biosynthesis of aldgarose

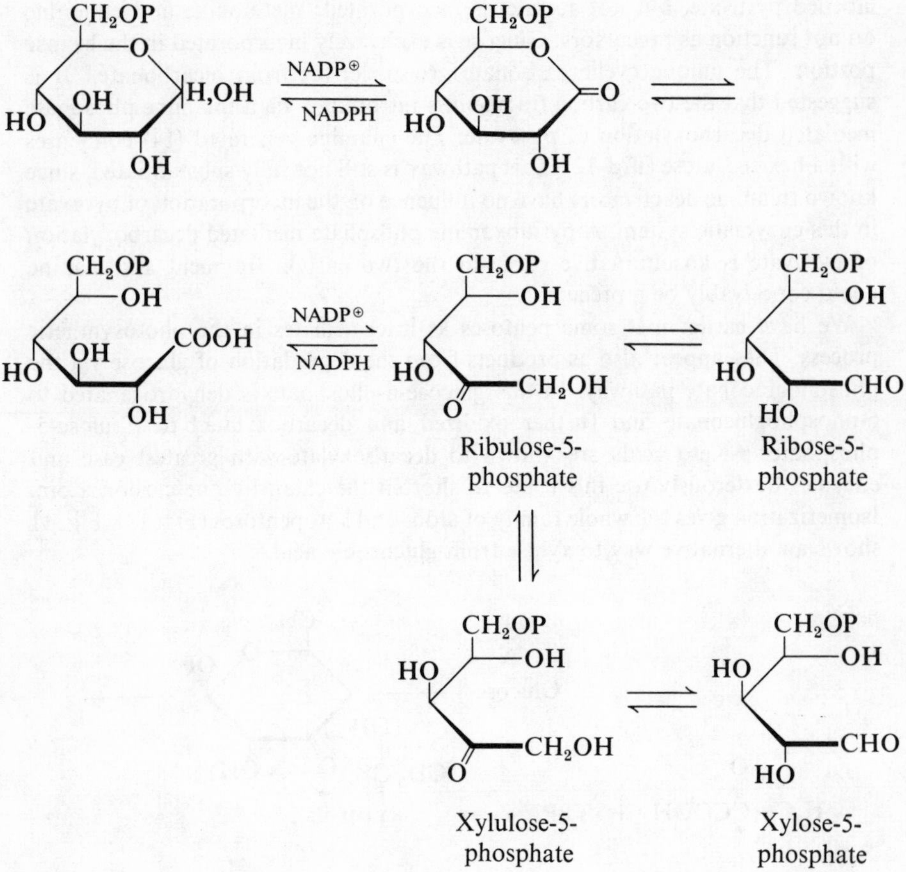

CH₂OP O H,OH OH HO OH — NADP⊕ / NADPH ⇌ CH₂OP O =O OH HO OH ⇌

CH₂OP —OH COOH OH HO OH ⇌ (NADP⊕ / NADPH) CH₂OP —OH HO —CH₂OH O ⇌ CH₂OP —OH HO —CHO HO

Ribulose-5-phosphate

Ribose-5-phosphate

CH₂OP —OH HO —CH₂OH O ⇌ CH₂OP —OH HO —CHO HO

Xylulose-5-phosphate

Xylose-5-phosphate

Fig. 13 Degradation of glucose to pentoses via the phosphogluconate pathway

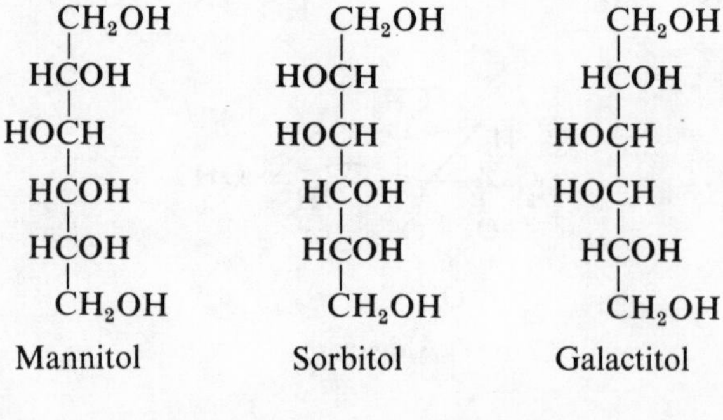

Mannitol	Sorbitol	Galactitol
CH_2OH	CH_2OH	CH_2OH
HCOH	HOCH	HCOH
HOCH	HOCH	HOCH
HCOH	HCOH	HOCH
HCOH	HCOH	HCOH
CH_2OH	CH_2OH	CH_2OH

Fig. 14 Structure of common sugar alcohols

Reduction of the carbonyl function of sugars in the open chain form leads to sugar alcohols. The most common sugar alcohols are sorbitol, mannitol, and galactitol obtained from glucose, mannose, and galactose, respectively (Fig. 14). Sorbitol was first isolated from the berries of mountain ash, *Sorbus aucuparia*. It occurs also in the red alga, *Bostrychia scorpioides* (14 per cent). Mannitol is widely distributed in nature, e.g. in many brown algae (*Fucus*, *Laminaria*, *Halidrys*) and in manna ash, *Fraxinus ornus*. Galactitol occurs in many plants and exudates. It has a plane of symmetry and is therefore optically inactive.

Related to the sugar alcohols are the cyclitols which are derived from glucose-6-phosphate, as shown by ^{14}C labelling.[17] A likely route is depicted in (22). C^5 is oxidized and C^6 condenses with C^1; epimerization leads to the other cyclitols. A synthesis of *myo*-inositol inspired by the proposed biosynthetic cyclization step has been carried out in the laboratory.[18]

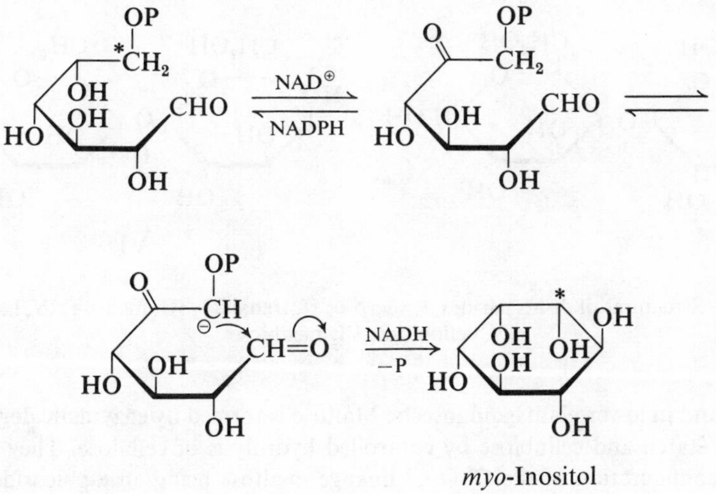

(22)

myo-Inositol

2.6 Disaccharides and glycosides

Disaccharides are formed from two monosaccharides joined by an acetal or ketal link. They belong to a group of glycosides where the alcohol is another sugar. They are classified according to their reducing power. Non-reducing are those with both carbonyl functions blocked as acetals, e.g. sucrose and trehalose. Examples of reducing sugars are lactose (4-*O*-β-D-galactopyranosyl-D-glucopyranose or in shorthand β-D-Gal*p*-1-4-D-G*p*), maltose, cellobiose, melibiose, etc. (Fig. 15). Sucrose is manufactured in large amounts from beet, *Beta vulgaris* and cane *Saccharum officinarum*. The world production in 1970 was *ca.* 70 million tons. It gives one mole of glucose and one mole of fructose on hydrolysis. Trehalose, like maltose and cellobiose, gives two moles of glucose on hydrolysis.

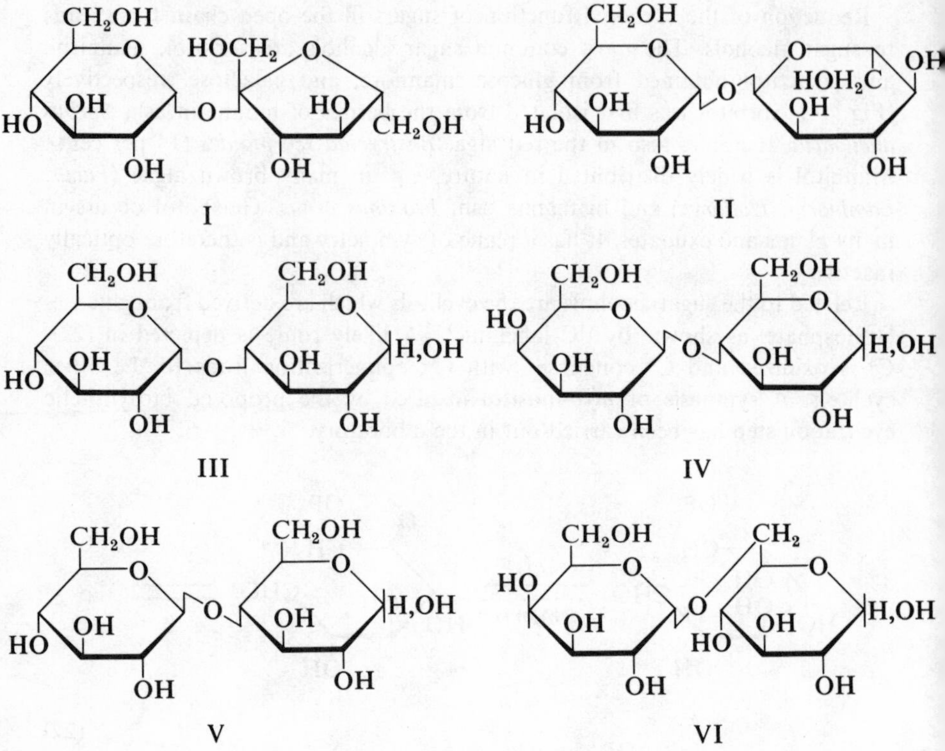

Fig. 15 Structures of disaccharides. I, sucrose; II, trehalose; III, maltose; IV, lactose; V, cellobiose; VI, melibiose

It is found in lower plants and insects. Maltose is formed by enzymatic degradation of starch and cellobiose by controlled hydrolysis of cellulose. They differ in the configuration of the C^1-O-C^4 linkage, maltose being an α-glucoside and cellobiose a β-glucoside. Lactose and melibiose give one mole of galactose and one mole of glucose on hydrolysis. They are both galactose glucosides; lactose forms a β-glucosidic linkage to the C^4 of glucose and melibiose an α-glucosidic linkage to the C^6 of glucose.

Sucrose is biosynthesized from glucose activated as UDP-glucose and fructose-6-phosphate (23). UDP functions as an effective leaving group. The configuration at C^1 of UDP-glucose and at C^1 in the glucose moiety of sucrose formed is α, i.e. the displacement proceeds with retention of configuration. A one-step S_N2 backside substitution at C^1 by fructose is therefore excluded. Two explanations have been advanced for the outcome of the reaction. UDP-glucose is absorbed on the enzyme surface and thus shielded from a backside attack. A carbonium ion at C^1, stabilized by the ring oxygen, is formed when UDP dissociates and fructose-6-phosphate enters from the same side. Alternatively the enzyme participates actively with e.g. an amino function by forming a covalent bond with C^1 with inversion of configuration and the disaccharide is formed in a second S_N2 displacement by attack of fructose (23). The exact nature of this glycosidic formation is still unknown.

CH_2OH, O, :B-Enz, OH, HO, O–UDP, OH → CH_2OH, O, $\overset{\oplus}{B}$-Enz, OH, HO, OH; OH, CH_2, O, HO, H–O, CH_2OP, HO → Sucrose (23)

On the other hand, the lactose synthesis proceeds by backside attack of glucose at C^1 of α-UDP-galactose thus forming the β-1,4-linkage (24).

CH_2OH, HO, OH, O–UDP, OH; H–O, CH_2OH, O, OH, H,OH, OH → Lactose (24)

The structures of the disaccharides were established by complete methylation followed by hydrolysis and identification of the methylated products. The structure of lactose was deduced from the following findings:

1. hydrolysis with dilute mineral acid gave the monosaccharides glucose and galactose;
2. lactose reduced Fehling's solution and it could be oxidized by bromine water to an acid which gave galactose and gluconic acid, i.e. lactose is a galactoside;
3. complete methylation of lactose gave octamethyllactose which on hydrolysis gave 2,3,4,6-tetra-*O*-methylgalactose and 2,3,6-tri-*O*-methylglucose (25); the acid gave 2,3,4,6-tetra-*O*-methylgalactose and 2,3,5,6-tetra-*O*-methylgluconic acid;
4. lactose was cleaved by β-glycosidase which is specific for β-glycosidic bonds.

These results show that IV (Fig. 15) represents the structure of lactose.

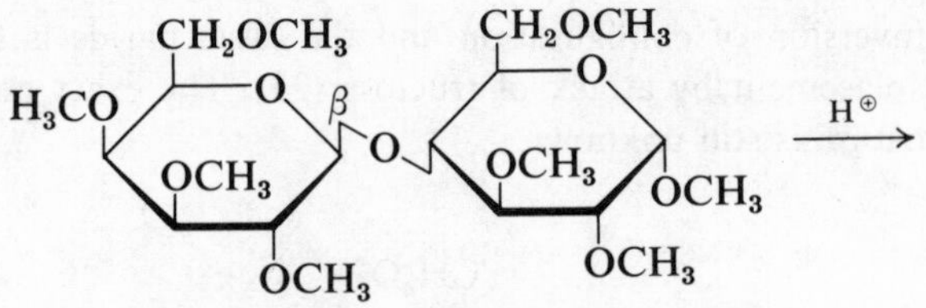

Octa-*O*-methyl-α-lactose

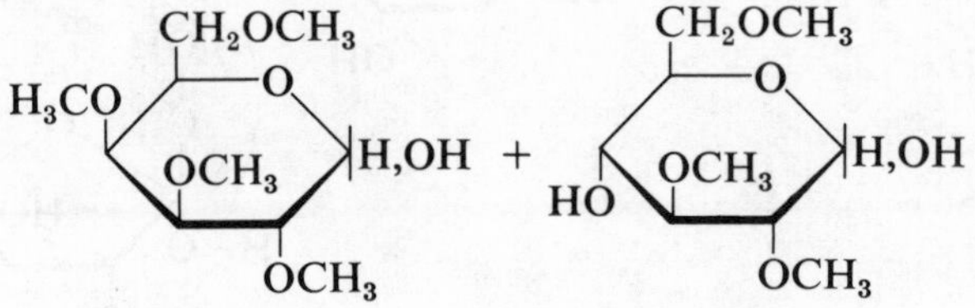

2,3,4,6-tetra-*O*-methylgalactose + 2,3,6-tri-*O*-methylglucose (25)

Glycosides of phenols or other alcohols are widely distributed in nature. Flavonoids and anthocyanins, pigments of flowers and berries, have their hydroxyls linked to sugars. The sharp flavour of mustard and horseradish is caused by an interesting thioglycoside, sinigrin, III (Fig. 16), which rearranges to an isothiocyanate on hydrolysis (26):

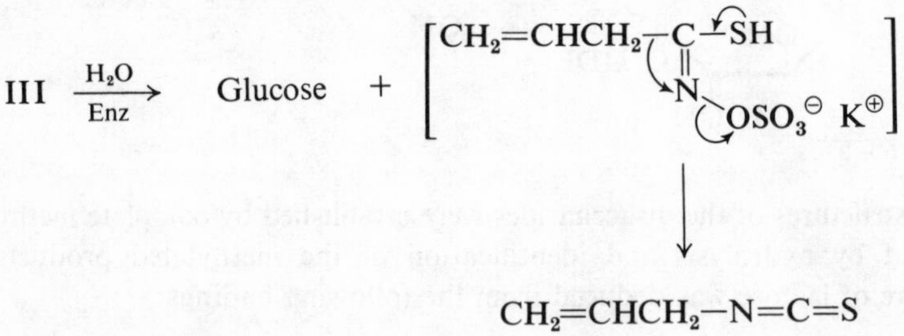

$$\text{III} \xrightarrow[\text{Enz}]{H_2O} \text{Glucose} + \left[CH_2{=}CHCH_2{-}C(SH){=}N{-}OSO_3^{\ominus}\ K^{\oplus}\right] \longrightarrow CH_2{=}CHCH_2{-}N{=}C{=}S \qquad (26)$$

Amygdalin belongs to the cyanogenetic group of glycosides. It occurs in bitter almonds, *Prunus amygdalus*, and is hydrolysed by the enzyme emulsin to benzaldehyde, hydrocyanic acid, and two D-glucose molecules. Further structural work revealed that it is the gentiobioside of benzaldehyde cyanohydrin, IV (Fig. 16).

The cardiac glycosides belong to the steroid glycosides present in the *Strophanthus* genus, the foxglove, *Digitalis purpurea*, and in the lily of the valley *Convallaria majalis*. Preparations from *Strophanthus* are used by African tribes as arrow poisons for hunting, and *Digitalis* found use in medieval ordeals as a test of innocence, often with a fatal outcome. The active agents have a powerful

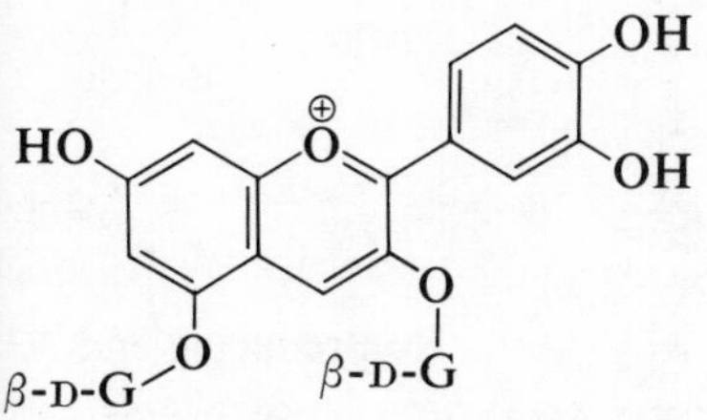

Fig. 16 Structures of some glycosides occurring in nature: I, pyrylium salt from red roses; II, quercitrin, an L-rhamnoside; III, sinigrin; IV, amygdalin; V, digitoxin

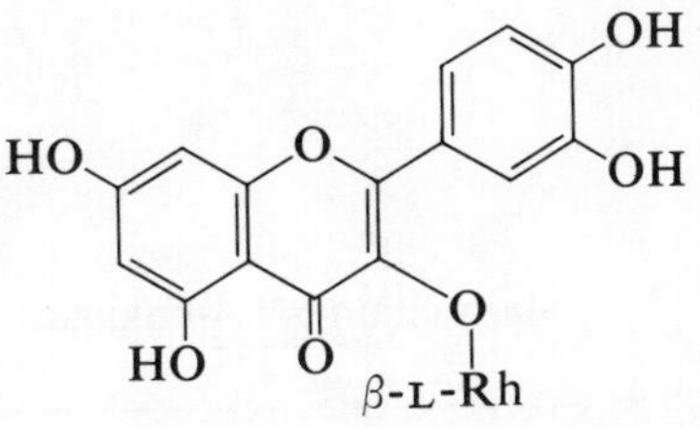

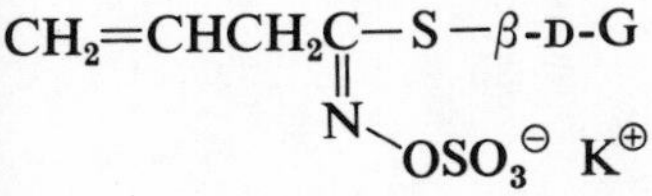

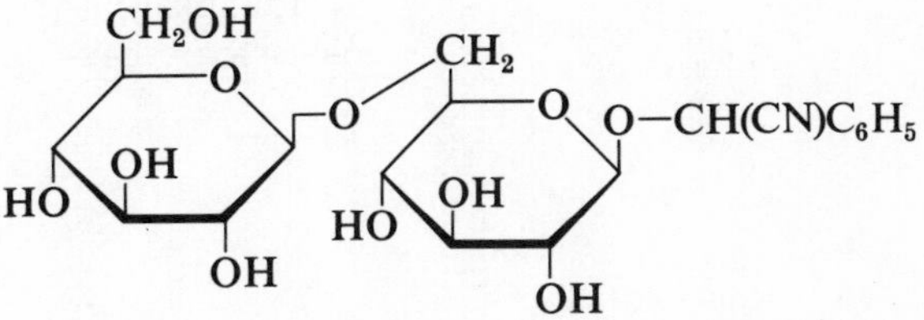

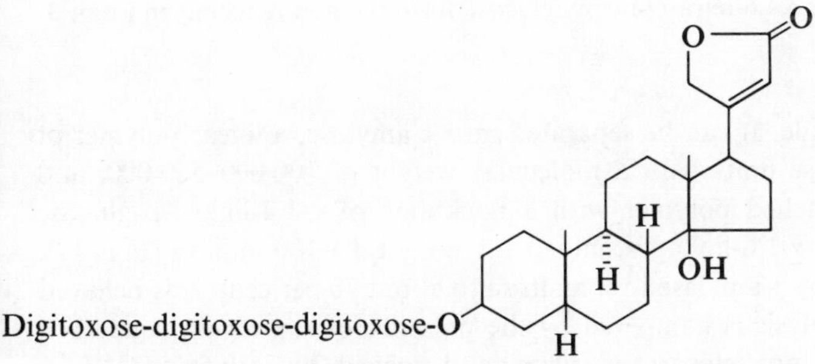

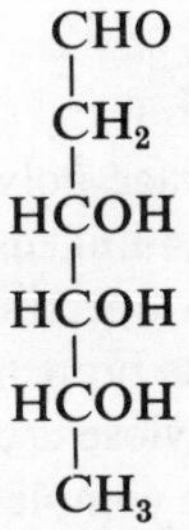

action on the heart muscle; *ca.* 0.1 mg of strophanthin injected into the blood stream stops the heart of a mouse after a couple of minutes. In very small dosages the compounds have a beneficial effect in the treatment of heart ailments.

The *N*-glycosides are of the utmost importance as they are structural units in coenzymes, nucleic acids, nucleotides, etc. This group of compounds is treated in Chapter 8.

2.7 Polysaccharides

Polysaccharides function either as storage (starch, glycogen, dextran, inulin) or as structural polysaccharides, i.e. cell wall constituents (cellulose, hemicellulose, chitin, pectin, alginic acids, hyaluronic acid). Starch is the most abundant

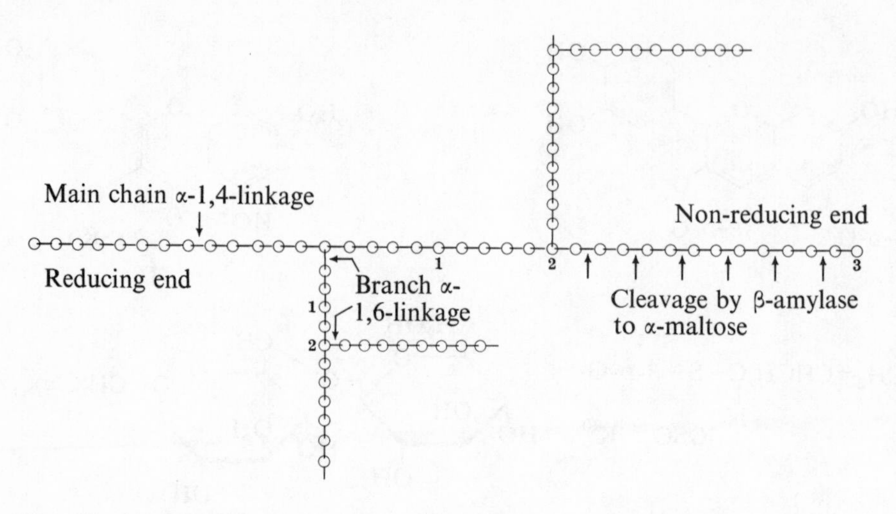

Fig. 17 Amylopectin model. Complete methylation and hydrolysis give 2,3,6-tri-*O*-methylglucose from 1 as main product, and small amounts of 2,3-di-*O*-methylglucose from 2 and 2,3,4,6-tetra-*O*-methylglucose from the non-reducing end unit 3

storage polysaccharide. It can be separated into α-amylose, a linear polymer of α-1,4-linked-D-glucose units with a molecular weight of 100 000–500 000, and amylopectin, a branched polymer with a backbone of α-1,4-linked-D-glucose units branched with α-1,6-linkages, molecular weight 10–100 million (Fig. 17). Amylose is cleaved by β-amylase to α-maltose to about 70 per cent. It is believed that complete hydrolysis is hampered by the presence of some β-linkages. The term β-amylase does not refer to the sugar bond cleaved, but—unfortunately—to a group of enzymes. Degradation of amylopectin by β-amylose stops at the branching points and gives a product of rather high molecular weight, known as limit dextrin. Complete degradation can be brought about by a special 1,6-α-glucosidase. The amylose chains are wound like a helix and can enclose large amounts of iodine forming an intense blue addition compound.

Glycogen is the reserve polysaccharide of animal cells and is especially abundant in the liver. It has the same general structure as amylopectin but is more branched. Glycogen or starch is synthesized in the cell according to (27). The reaction needs a primer,

$$\text{UDP-D-Glucose} + (\text{Glucose})_n \xrightarrow{\text{Enz}} (\text{Glucose})_{n+1} + \text{UDP} \qquad (27)$$

a polysaccharide having at least four glucose units, which is attacked at the non-reducing end.

Dextrans are highly branched polysaccharides of D-glucose produced by certain bacteria, e.g. *Leuconostoc mesenteroides* and *Betacoccus arabinosaceus*. The linkages in dextrans of different strains may vary: 1–2, 1–3, 1–4, and 1–6.

Dextran is used as blood plasma substitute. Inulin is a starch-like polysaccharide occurring particularly in the *Compositae* family. It is built up from β-1,2-linked D-fructose residues, and has a molecular weight of *ca.* 6000.

Cellulose is the most abundant cell wall constituent in the plant kingdom. It is made up linearly by β-1,4-linked D-glucose units and has a molecular weight of *ca.* one million. Complete methylation and hydrolysis give only 2,3,6-tri-*O*-methylglucose and minute amounts of 2,3,4,6-tetra-*O*-methylglucose from the end group, proving the pyran structure and no branching. α- or β-amylase does not attack cellulose and most mammals are unable to digest it, but some ruminants which have bacteria-produced cellulase in their intestinal tract, can do so. Cellulose is synthesized in plants from UDP- or GDP-glucose and a glucose polymer in the presence of cellulose synthase, *cf.* (27). Hemicelluloses are heteropolysaccharides containing pentoses (xylose, arabinose) and occur together with cellulose in wood and straw. Alginic acid is a gelatinous material present in the cell wall of most brown algae. It is extracted commercially from giant kelp *Macrocystis pyrifera*, *Laminaria digitata*, and *Ascophyllum nodosum* and used as an emulsifier in foodstuffs, pharmaceuticals, and cosmetics. Hydrolysis gives *ca.* two moles of D-mannuronic acid and one mole of L-guluronic acid which in the polymers are β-linked in 1,4-position. The shell of crustaceans (lobster) and the exoskeleton of insects contain a β-1,4-homopolymer, chitin, structurally similar to cellulose with *N*-acetyl-2-amino-2-deoxyglucose as the building block. Hyaluronic acid is a mucopolysaccharide occurring in the vitreous humour of the eye and in the synovial fluid in joints. Hydrolysis gives equal amounts of D-glucuronic acid and *N*-acetyl-D-glucosamine linked in β-1,3-position. This disaccharide is then linked in the β-1,4-position (Fig. 18).

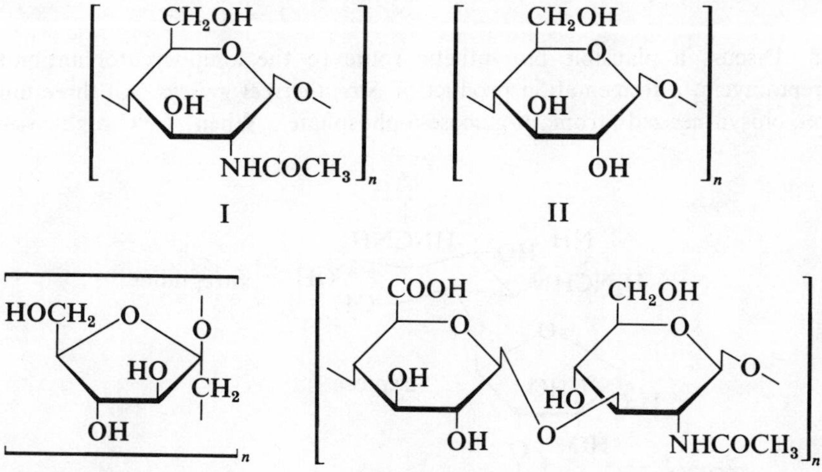

Fig. 18 Building units of polysaccharides: I, chitin; II, cellulose; III, inulin; IV, hyaluronic acid

2.8 Problems

2.1 A green plant is illuminated for a very short period in an atmosphere containing $^{14}CO_2$. In which positions will ribose-5-phosphate and sedoheptulose-7-phosphate be labelled?

2.2 A polysaccharide is exhaustively methylated and hydrolysed. Equal parts of 2,3,4-tri-*O*-methylgalactose and 2,3,6-tri-*O*-methylglucose are obtained. The polysaccharide has a reducing end group, and it is enzymatically degraded by an α-glycosidase. Suggest a structure. The molecular weight is 4.2×10^6. How many hexose residues does the polymer contain?

2.3 It is suggested that the epimerization of glucose to galactose proceeds via 3-keto-UDP-1-glucose. Discuss this reaction sequence assuming that the C^4–H is retained.

2.4 Propose a biosynthetic pathway for garosamine from TDP-glucose. The methyl groups are supposed to originate from methionine. (Grisebach, H. *Adv. Carbohydr. Chem. Biochem.* **35** (1978) 81).

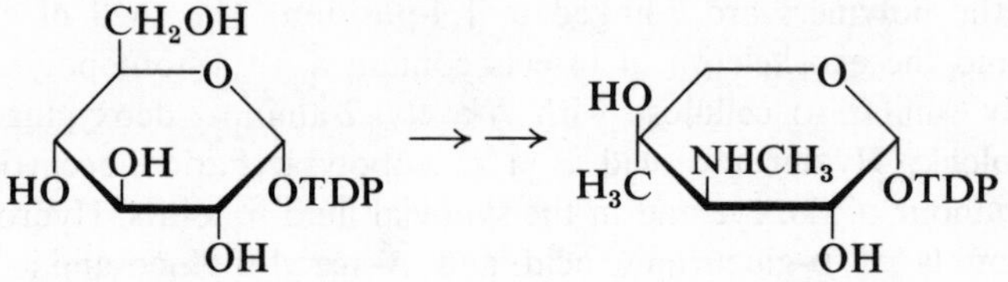

TDP-Garosamine

2.5 Discuss a plausible biosynthetic route to the aminocyclitol antibiotic streptomycin, a fermentation product of *Streptomyces griseus*. All three units are biosynthesized from D-glucose-6-phosphate. When 6-^{14}C-D-glucose-6-

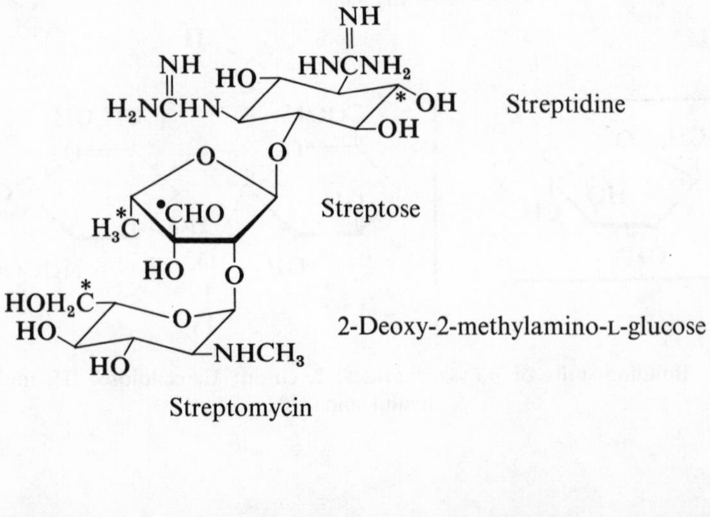

Streptidine

Streptose

2-Deoxy-2-methylamino-L-glucose

Streptomycin

phosphate was fed to the microorganism, the label was recovered at the starred positions. Streptose was labelled in the formyl and methyl groups when the microorganism was incubated with 3,6-di-^{14}C-D-glucose. The ^{13}C NMR spectrum of streptomycin incubated with 6-^{13}C-glucose showed three enhanced peaks at 13.4, 61.2, and 72.4 PPM. (Grisebach, H., *Adv. Carbohydr. Chem. Biochem.* **35** (1978) 81).

Bibliography

1. Durette, P. L. and Horton, D. *Carbohydr. Res.* **18** (1971) 57.
2. Perlin, A. S. in *Int. Rev. Sci. Org. Chem. Ser. II*, Hey, D. H. (Ed.). **7** (1976) 1. Butterworth, London.
3. Glaser, L. in *The Enzymes*, 3rd Edn. Boyer, P. (Ed.) **6** (1972) 355. Academic Press, New York.
4. Walsh, C. *Enzymatic Reaction Mechanisms*, Freeman, San Fransisco, 1979, p. 347.
5. Vongerichten, E. *Liebigs Ann.* **318** (1901) 121.
6. Fischer, E. and Freudenberg, K. *Chem. Ber.* **52** (1919) 177.
7. Umezawa, S. *Int. Rev. Sci. Org. Chem. Ser. II*, **7** (1976) 149. Butterworth, London.
8. Grisebach, H. *Adv. Carbohydr. Chem. Biochem.* **35** (1978) 81.
9. Grisebach, H. *Biosynthetic Patterns in Microorganisms and Higher Plants*, J. Wiley, New York, 1976, p. 66.
10. Glaser, L. and Zarkowsky, H. in *The Enzymes*, 3rd Edn., Boyer, P. (Ed.) **5** (1971) 465. Academic Press, New York.
11. Snipes, C. E., Brillinger, G. U., Sellers, L., Mascaro, L. and Floss, H. G. *J. Biol. Chem.* **252** (1977) 8113.
12. Meloch, H. P. and Wood, W. A. *J. Biol. Chem.* **239** (1964) 3505, 3517.
13. Rubenstein, P. and Strominger, J. L. *J. Biol. Chem.* **249** (1974) 3776, 3782, 3789.
14. Watson, R. R. and Orenstein, N. S. *Adv. Carbohydr. Chem. Biochem.* **31** (1975) 135.
15. Grisebach, H. and Dobereiner, U. *Biochem. Biophys. Res. Commun.* **17** (1964) 737.
16. Aschenbach, H. and Karl, W. *Chem. Ber.* **108** (1975) 759, 780.
17. Eisenberg, F. and Bolden, A. H. *Biochem. Biophys. Res. Commun.* **12** (1963) 72.
18. Kiely, D. E. and Sherman, W. R. *J. Am. Chem. Soc.* **97** (1975) 6810.

Chapter 3

The shikimic acid pathway

3.1 Biosynthesis of shikimic acid

A very large number of compounds exhibit a characteristic C_6-aromatic–C_3-side chain structure, e.g. aromatic amino acids, cinnamic acids, coumarins, flavonoids, lignin constituents, etc. It soon became evident that they must have some common origin. The biosynthesis of these compounds was elucidated by mutant studies of *Escherichia coli* and tracer studies particularly by Davis and Sprinson. Shikimic acid (1) was isolated as early as 1885 by Eykman from the Japanese plant *Illicium anisatum* long before we were aware of its biosynthetic significance. The name is derived from the Japanese name of the plant. The compound was later found to be widespread. The role of shikimic acid was revealed from the observation that it could replace the essential amino acids phenylalanine, tyrosine, and tryptophan in auxotropic *E. coli* mutants, i.e. it must be an intermediate in the biosynthetic sequence.

Erythrose-4-phosphate is a compound of far reaching importance in biosynthesis. We have seen that it is an intermediate in the regenerative process of ribulose-1,5-diphosphate in photosynthesis (section 2.3) and it appears in the pentose cycle. An analysis of the distribution of a ^{14}C label in shikimic acid, biosynthesized from specifically labelled ^{14}C-glucose in *E. coli* led to the proposition[1] that erythrose-4-phosphate starts the biosynthetic sequence leading to shikimic acid by condensation with phosphoenol pyruvic acid (PEP) to 3-deoxy-D-*arabino*-heptulosonic acid-7-phosphate (DAHP), (1). Elimination of phosphoric acid gives the ketone, formally in its enol form, that cyclizes to 3-dehydroquinic acid. Further elimination of water and reduction then gives shikimic acid. This is a sound, straightforward mechanism that was suggested at an early stage without detailed experimental proof. More recent studies revealed that the reaction is not quite so simple and they illustrate clearly that all facets of a reaction step have to be considered before acceptance can be given to a mechanistic interpretation. First, (1) implies that it is the P–O bond rather than the C–O bond that is broken in the PEP condensation with erythrose-4-phosphate. ^{18}O labelling shows that the reverse is true and that all the label is recovered in the phosphate liberated. The phosphate tended to be released before erythrose-4-phosphate was absorbed by the enzyme. These findings indicate an exchange of phosphate by a nucleophile of the enzyme and this complex then reacts with the sugar (2).[2] This formulation proved to be untenable because it implied that exchange of protons in the methyl of pyruvate should occur with the solvent. Only a low incorporation was observed and by using specifically tagged Z and E 3-^{3}H PEP it was shown that the condensation with erythrose-4-phosphate is stereospecific (3). Z-PEP gives rise to 3*S*-DAHP and E-PEP to 3*R*-DAHP, i.e. the *si* face of PEP adds to the *re* face of erythrose-4-phosphate.[3] (3) seems to account for most of the facts but it does not fully allow for the observation of the early release of phosphate and the idea of formation of an intermediate covalent bond between PEP and the enzyme.

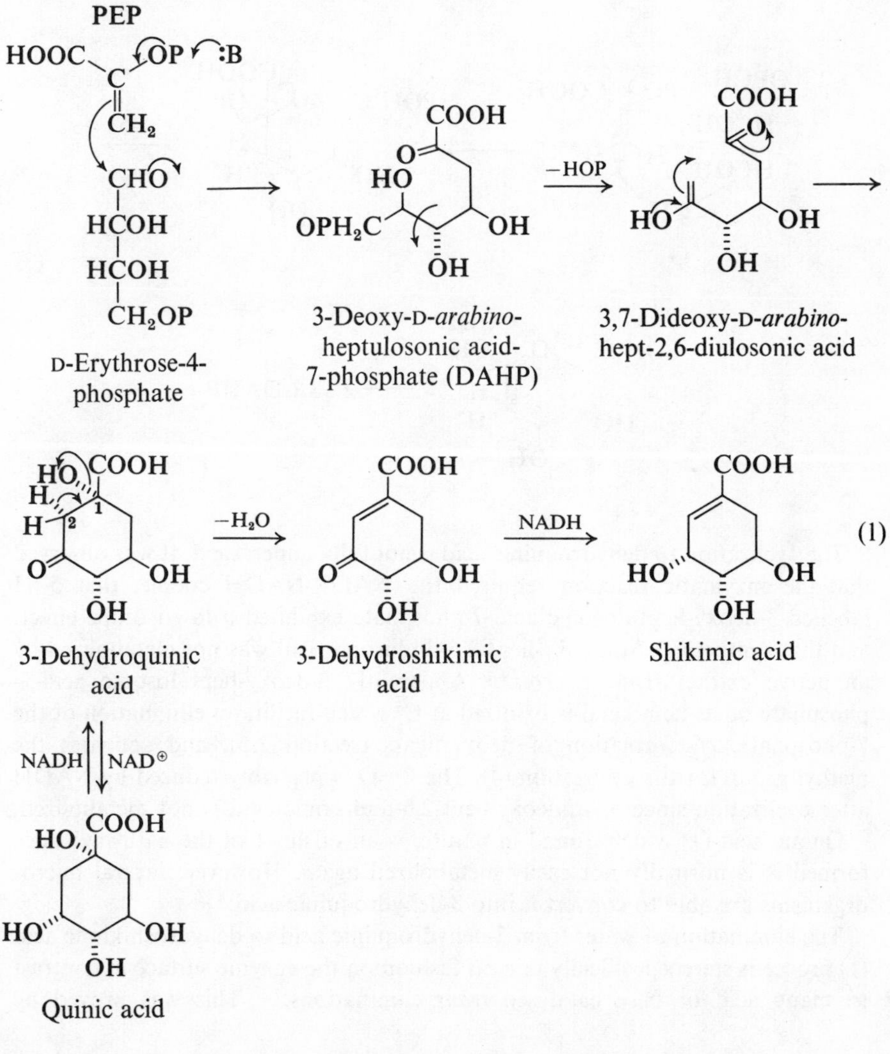

PEP

HOOC–C(OP)=CH₂ :B

CHO–HCOH–HCOH–CH₂OP

D-Erythrose-4-phosphate

→

3-Deoxy-D-*arabino*-heptulosonic acid-7-phosphate (DAHP)

−HOP →

3,7-Dideoxy-D-*arabino*-hept-2,6-diulosonic acid

→

3-Dehydroquinic acid

$-H_2O$ →

3-Dehydroshikimic acid

NADH →

Shikimic acid (1)

3-Dehydroquinic acid ⇅ NADH / NAD$^{\oplus}$

Quinic acid

(2)

(3)

$\longrightarrow$ 3*R*-DAHP

The cyclization to dehydroquinic acid is not fully understood. It was observed that the enzymatic reaction required the $NAD^{\oplus}$–NADH couple, that 5-^{3}H labelled 3-deoxy-heptulosonic acid-7-phosphate exhibited a low isotope effect, and that synthetic 3,7-dideoxy-hept-2,6-diulosonic acid was not metabolized by an active extract from *E. coli*.[4,5] Apparently 3-deoxy-heptulosonic acid-7-phosphate or its hemiketal is oxidized at C^5 which facilitates elimination of the 7-phosphate, *cf.* formation of deoxysugars (section 2.5), and activates the methyl group for the cyclization (4). The C^5=O is probably reduced by NADH after cyclization since 3,7-dideoxy-hept-2,6-diulosonic acid is not metabolized.

Quinic acid (1) widely found in nature, is an offshoot of the pathway; once formed it is normally not easily metabolized again. However, several microorganisms are able to convert it into 3-dehydroquinic acid.[6]

The elimination of water from 3-dehydroquinic acid to dehydroshikimic acid (1) proceeds stereospecifically in a *cis* fashion on the enzyme surface in contrast to many acid or base catalysed *trans* eliminations.[7,8] This was proved by

COOH, O=, HO, POH_2C, OH, OH ⇌ COOH, OH, O, POH_2C, 5, OH, OH $\xrightarrow[NADH]{NAD^{\oplus}}$ COOH, OH, O, H, $POCH_2$, O, OH $\xrightarrow{-HOP}$

COOH, OH, O, H_2C, O, OH ⇌ COOH, O, HO, O, OH ⟶ (4)

HO, COOH, O, O, OH $\xrightarrow{NADH}$ HO, COOH, O, OH, OH ⟶ Shikimic acid

stereospecific deuteration at C^2. The result can be rationalized by the supposition that the enzyme has a bifunctional action on the substrate. A basic centre assists the dissociation of the C^2-proton and an acid centre abstracts the hydroxyl group. There are indications that the dehydration occurs via Schiff's base formation between an amino group of the enzyme and the keto function of 3-dehydroquinic acid as demonstrated by inactivation of the enzyme by treatment with sodium borohydride.[9]

3.2 Aromatic amino acids

The further transformation of shikimic acid involves some remarkable reactions in biosynthesis. Shikimic acid is phosphorylated regioselectively at C^3 and then it reacts with PEP at C^5 (5). The phosphorylation serves the purpose of forming a more efficient leaving group for the following elimination. In contrast to the much debated reaction (3) the mechanism of the reversible (5) is supported by the observation of considerable hydrogen exchange in the methylene groups of 5-enolpyruvylshikimic-3-phosphate and recovered PEP and that ^{18}O in the C–O–P bond was recovered in the phosphate liberated.[10] The mechanism is not quite settled since one could argue in favour of an intermediary, covalent enzyme–PEP complex being the true transfer agent of enolpyruvate.[11]

Chorismic acid (5) is formed by a stereospecific phosphoric acid elimination. It was proved by specific labelling of C^6 that only H_R was eliminated and chorismic acid is thus formed in an overall 1,4-*trans* elimination.[12] This reaction

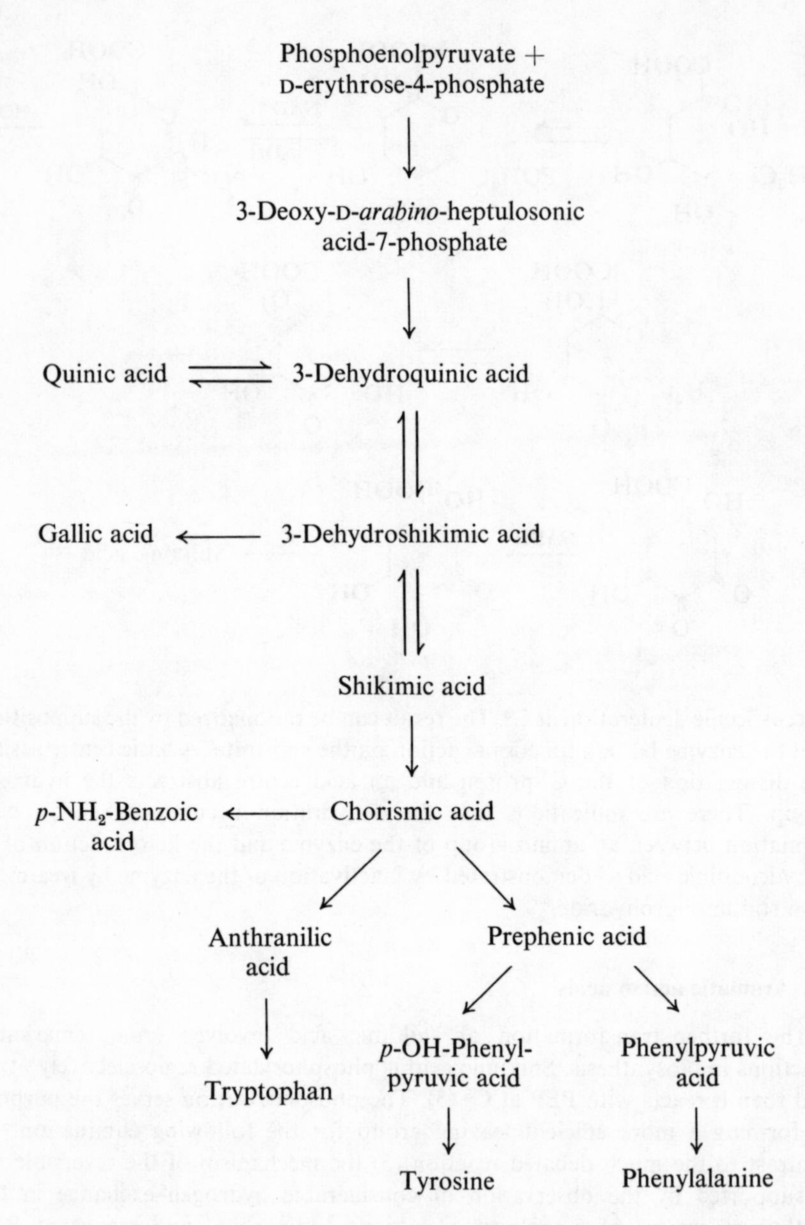

Fig. 1 Outline of shikimic acid pathway

is enzymatically controlled since most 1,4-conjugate eliminations in cyclohexene systems proceed predominantly in a *syn* fashion *in vitro*. Chorismic acid is an unstable intermediate, located at the branching point of the metabolic path (Fig. 1). One branch goes to anthranilic acid and indole derivatives, and the

(5)

Shikimic acid-3-phosphate

5-Enolpyruvylshikimic acid-3-phosphate

Chorismic acid

other via a Claisen rearrangement to prephenic acid, or properly *pre-phenylalanine*, and phenyl- and *p*-hydroxyphenylpyruvic acid (6). The transamination (section 6.4) can in some species occur already at the prephenic acid stage.[13] Biosynthesis of *p*-aminobenzoic acid (Fig. 3) is an offshoot from the stem. An outline of the shikimic acid pathway is given in Fig. 1.

Prephenic acid rearranges rapidly *in vitro* by the action of acids to phenylpyruvic acid but more slowly by alkali in an intramolecular redox reaction to β-*p*-hydroxyphenyl lactate[14] (7). Support for the unusual 9,4-hydride shift was produced by an investigation with a substrate labelled at the methine carbon.[15]

The transformation of chorismic acid to anthranilic acid is formulated in (8). Support for the Michael type concerted addition–elimination is found in the isolation of the related *trans*-2-amino-3-hydroxy-2,3-dihydrobenzoic acid[16] and *trans*-2,3-dihydroxy-2,3-dihydrobenzoic acid (Fig. 3) from natural sources. The latter compound is formed analogously by isomerization of chorismic acid into isochorismic acid followed by hydrolysis of the enolpyruvate group.[17] The further reactions of anthranilic acid are full of unexpected events before the indole nucleus is completed (Fig. 2). An enzyme complex promotes the reaction with 5-phosphoribosyl-1-pyrophosphate. The nucleotide, *N*-(5′-phosphoribosyl)-anthranilic acid, undergoes an Amadori rearrangement via the Schiff's base and cyclizes to indole-3-glycerylphosphate.[18] Alkylation with serine and expulsion of glyceraldehyde-3-phosphate complete the sequence.[19]

This last reaction requires pyridoxal which forms a Schiff's base with serine thus facilitating the elimination of water and the indole derivative adds to the aminoacrylate–pyridoxal complex formed. See section 6.4 for further mechanistic details of pyridoxal promoted reactions. Tryptophan is the precursor of indoleacetic acid, a plant growth hormone controlling cell elongation, and of the neurotransmitter 5-hydroxytryptamine (serotonin) (9). In a reaction, which essentially is a reversal of the serine alkylation, tryptophan is degraded to indole, pyruvic acid, and ammonia[20] (see Problem 3.1).

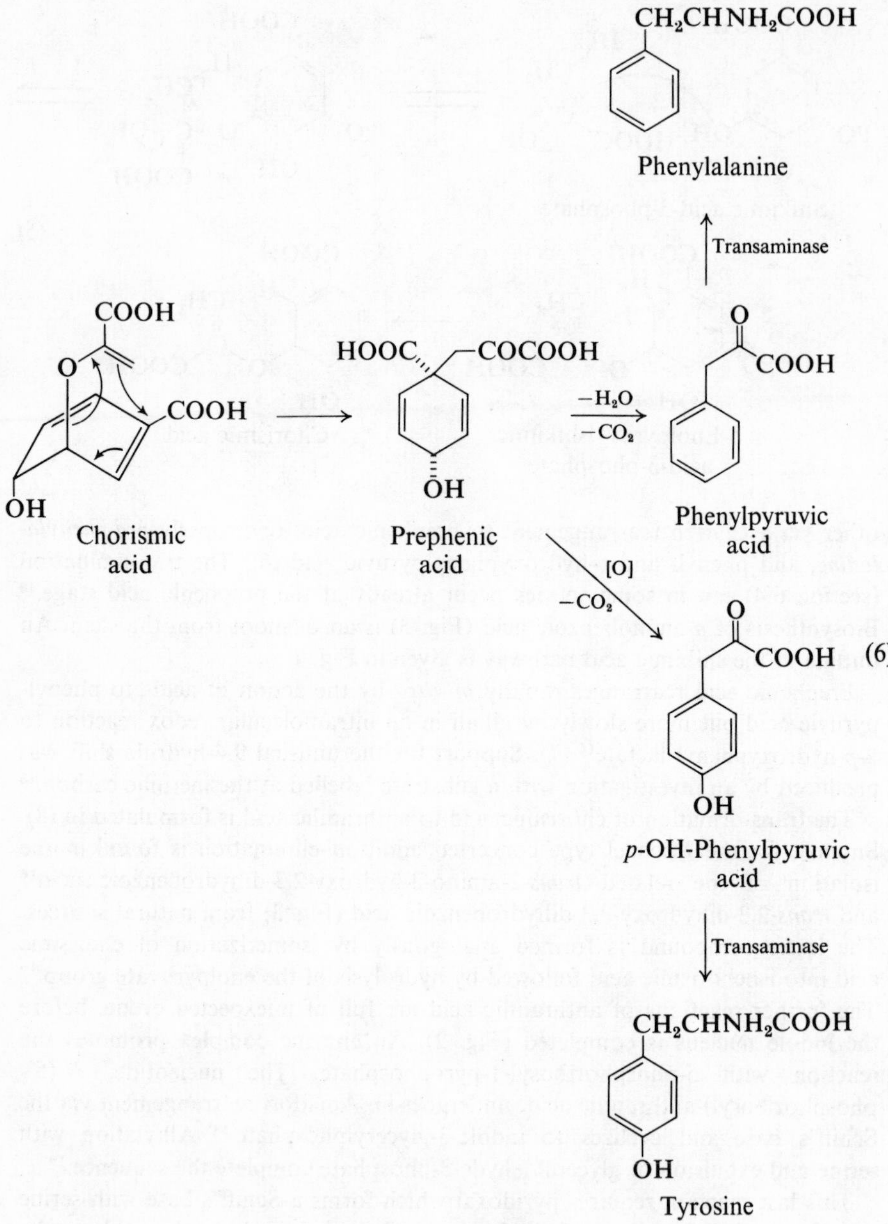

CH_2CHNH_2COOH

Phenylalanine

Transaminase

COOH

HOOC —COCOOH

COOH

$-H_2O$

$-CO_2$

O

COOH

OH

OH

Chorismic acid

Prephenic acid

Phenylpyruvic acid

[O]

$-CO_2$

O

COOH (6)

OH

p-OH-Phenylpyruvic acid

Transaminase

CH_2CHNH_2COOH

OH

Tyrosine

It is suggested that *p*-aminobenzoic acid, a component of the coenzyme folic, acid, is formed from isochorismic acid by addition of ammonia and elimination of water and pyruvate (Fig. 3, reaction *b*). This reaction bears close similarities to the formation of anthranilic acid. By feeding experiments with *Reseda lutea* it has

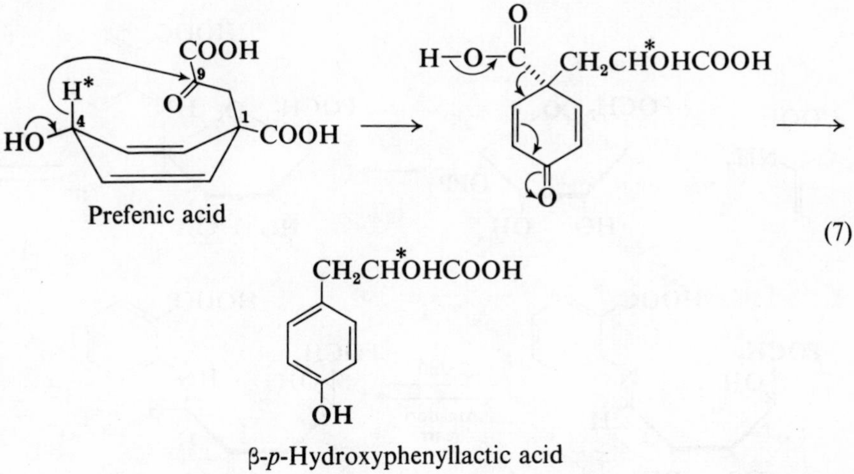

H* HO 4 O 9 COOH 1 COOH

Prefenic acid

→ H—O—C(=O) $CH_2\overset{*}{C}HOHCOOH$ O →

(7)

$CH_2\overset{*}{C}HOHCOOH$

OH

β-*p*-Hydroxyphenyllactic acid

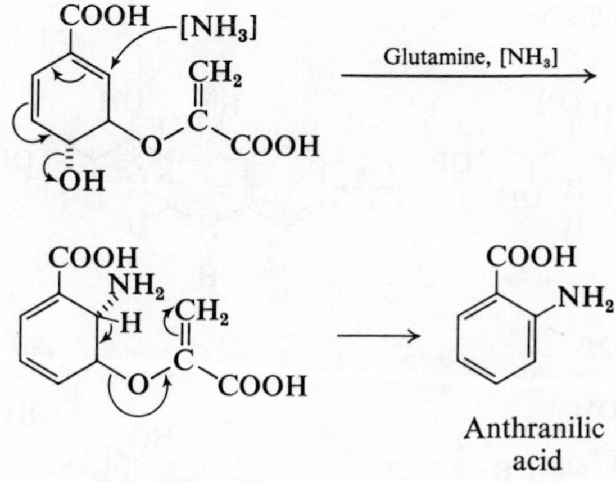

COOH $[NH_3]$ CH_2 C O COOH OH

Glutamine, $[NH_3]$ →

(8)

COOH NH_2 H CH_2 C O COOH → COOH NH_2

Anthranilic acid

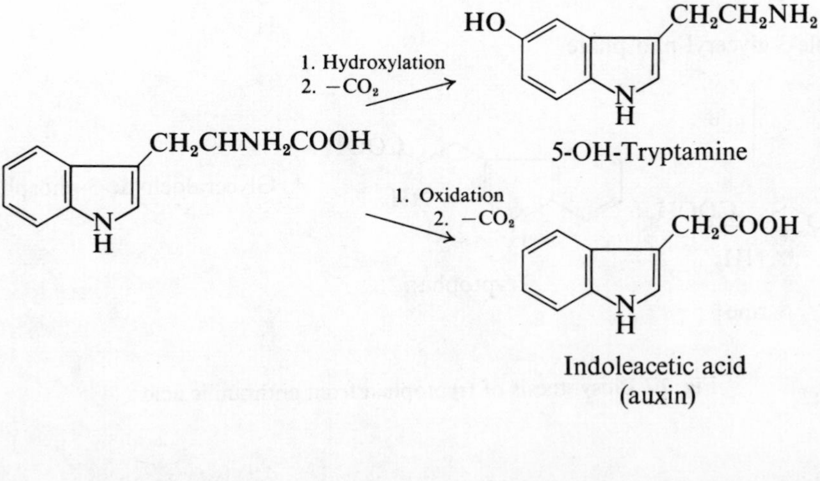

HO $CH_2CH_2NH_2$ N H

1. Hydroxylation
2. $-CO_2$

5-OH-Tryptamine

(9)

CH_2CHNH_2COOH N H

1. Oxidation
2. $-CO_2$

CH_2COOH N H

Indoleacetic acid (auxin)

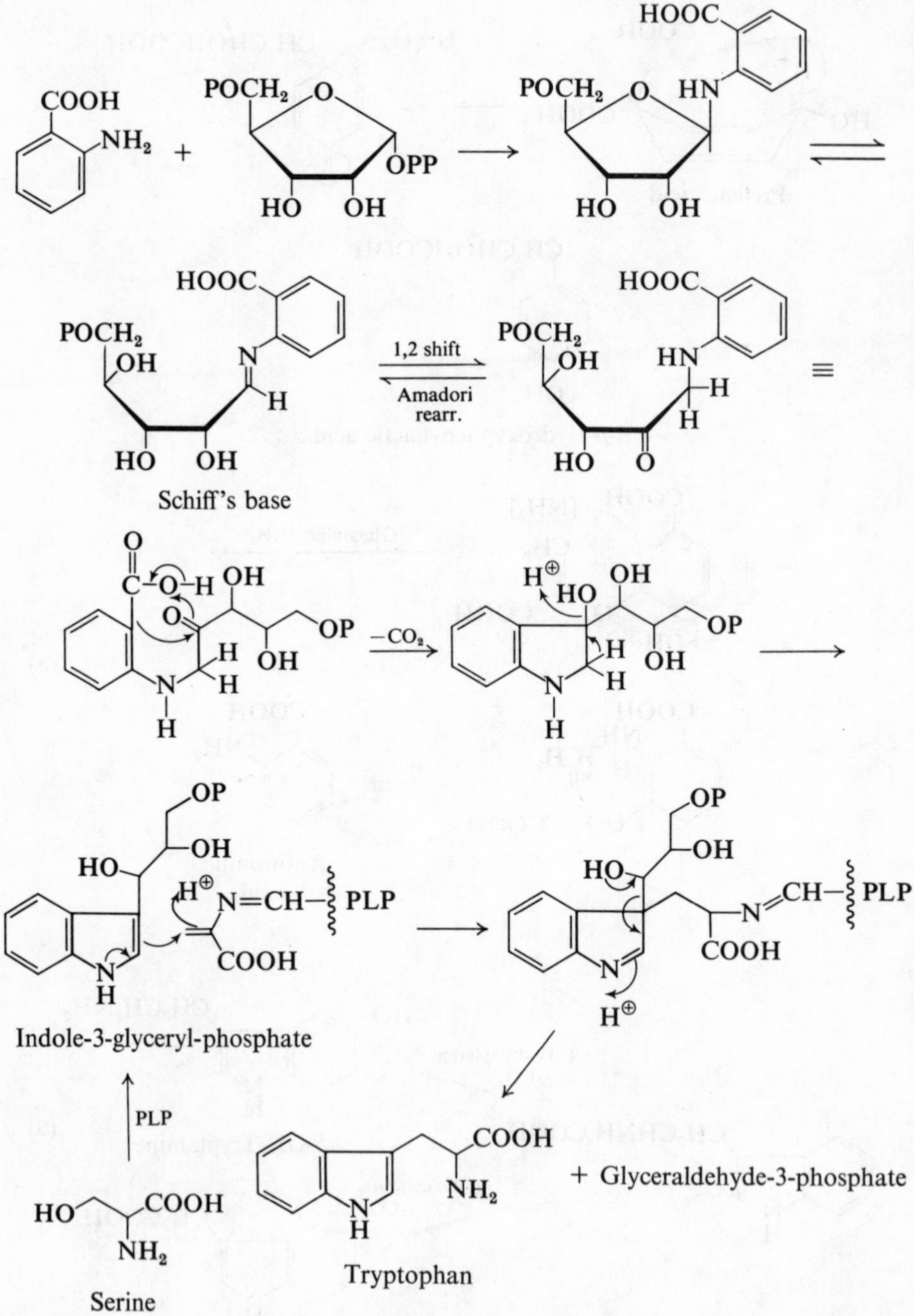

Fig. 2 Biosynthesis of tryptophan from anthranilic acid

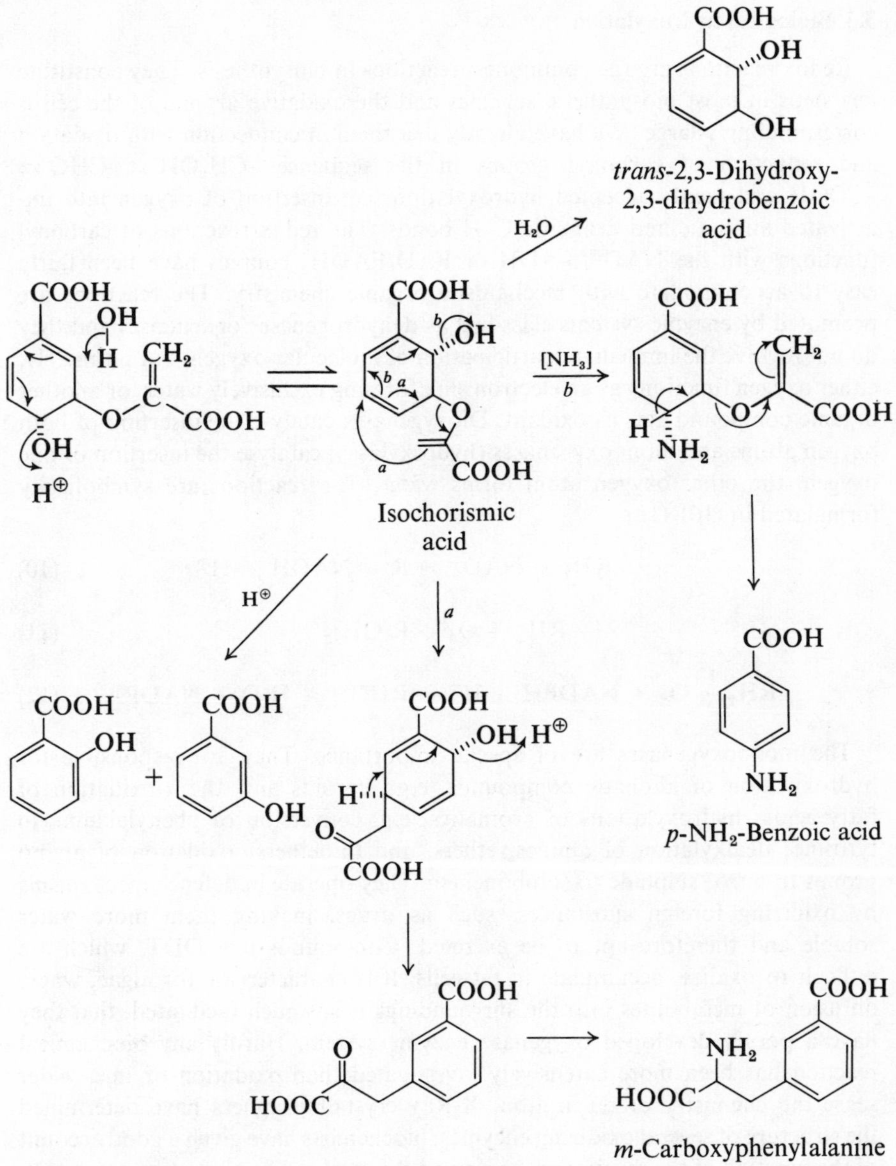

COOH
OH
OH
trans-2,3-Dihydroxy-2,3-dihydrobenzoic acid
H_2O
COOH
OH
H
CH_2
C
O
COOH
OH
$H^{\oplus}$
COOH
b
OH
b a
O
a
COOH
Isochorismic acid
$[NH_3]$
b
COOH
CH_2
H
NH_2
O
COOH
COOH
NH_2
p-NH_2-Benzoic acid
$H^{\oplus}$
a
COOH
OH
+
COOH
OH
COOH
OH, $H^{\oplus}$
H
O
COOH
COOH
O
HOOC
COOH
NH_2
HOOC
m-Carboxyphenylalanine

Fig. 3 Biosynthesis of *p*-aminobenzoic acid, *m*-carboxyphenylalanine and hydroxybenzoic acids

also been shown that labelled shikimate is built into *m*-carboxyphenylalanine.[21] This can be explained by a Claisen rearrangement of isochorismic acid (Fig. 3, reaction *a*), a reaction analogous to the chorismic–prephenic acid rearrangement.

3.3 Biological hydroxylation

Redox reactions are the commonest reactions in biosynthesis. They constitute key steps in most biosynthetic schemes and the oxidative arsenal of the cell is correspondingly large. We have already met them in connection with oxidation and reduction of carbonyl groups in the sequence $-CH_2OH \rightleftarrows -CHO \rightleftarrows$ $-COOH$ and tacitly accepted hydroxylation, i.e. insertion of oxygen into unactivated aliphatic and aromatic C–H bonds. The redox reactions of carbonyl functions with the $NAD^{\oplus}/NADH$ or $FAD/FADH_2$ couples have been fairly easy to accommodate with mechanistic organic chemistry. The reactions are promoted by enzyme systems classified as dehydrogenases or oxidases and they do not involve the immediate participation of molecular oxygen, but ultimately, either oxygen functions as an electron sink forming exclusively water, or another organic compound acts as oxidant. Dioxygenases catalyse the insertion of both oxygen atoms and monooxygenases (hydroxylases) catalyse the insertion of one oxygen, the other oxygen atom forms water. The reactions are symbolically formulated in (10)–(12).

$$RH_2 + NAD^{\oplus} \rightleftharpoons R + NADH + H^{\oplus} \tag{10}$$

$$RH_2 + O_2 \rightarrow R(OH)_2 \tag{11}$$

$$RH_2 + O_2 + NADPH + H^{\oplus} \rightarrow RHOH + H_2O + NADP^{\oplus} \tag{12}$$

The monooxygenases are of special importance. They are responsible for hydroxylation of aliphatic compounds, e.g. steroids and the α-oxidation of fatty acids; hydroxylations of aromatics, e.g. conversion of phenylalanine to tyrosine; dealkylation of amines, ethers, and thioethers; oxidation of amino groups to nitro; sulphide to sulphone, etc. They operate in defence mechanisms by oxidizing foreign substances, such as drugs, making them more water soluble and therefore apt to be excreted. Compounds like DDT, which are difficult to oxidize, accumulate in fat cells. It is characteristic for algae, where diffusion of metabolites into the surroundings is so much facilitated, that they have a poorly developed oxygenase enzyme system. Hardly any biochemical reaction has been more extensively investigated than oxidation or in a wider sense the chemistry of respiration. X-Ray crystallographers have determined the structure of several oxidizing enzymes, biochemists have given a good account of the function of the respiratory chain and the mechanism of electron transport, and mechanistic studies have revealed many secrets of substrate behaviour and product formation but, paradoxically we still cannot satisfactorily formulate the act of bond breaking and bond making of oxygen insertions, and the precise nature of the attacking oxygen species is still not clear.

We have to start out from the chemistry of oxygen and organic peroxides, and eventually contemplate on the unknown steps by analogy. Ground state oxygen, $^3O_2$, is a triplet with an uncoupled pair of electrons in different orbitals and with their spins parallel; 22.5 kcal above ground state lies singlet oxygen,

$^1O_2$, with paired electrons.[22] It is a shortlived and more reactive species with a lifetime of *ca.* 2 μs in water.[23] It has been inferred that oxygen, activated to the singlet state, could be responsible for certain oxidations in the living cell. It causes demethylation of amines, oxidizes sulphides to sulphoxides, forms allylic peroxides by an ene reaction with olefins, and adds in a Diels–Alder reaction to 1,3 dienes.[24] This last reaction explains in a simple way the formation of the peroxidic anthelmintic ascaridol[25] in *Chenopodium* spp. (13), a reaction

α-Terpinene → Ascaridol (13)

atypical for $^3O_2$. But the general reaction pattern of $^1O_2$ is different from that of $^3O_2$ and it is more difficult to accommodate with the oxidation products formed in nature. $^3O_2$ is therefore the most plausible oxidative agent. It is stepwise reduced by four electrons to water and all the reactive intermediates are known. The reduction starts by formation of the superoxide radical ion which gives well characterized salts with alkali metals. The ion is easily generated polarographically at a low reduction potential, $E_o\ (O_2/O_2^{\ominus}) = -0.33$ V,[26] and it can be kept for several days in aprotic solvents as a tetraalkylammonium salt (14). It decomposes rapidly in contact with water in a redox reaction. As a base it adds a proton and forms the reactive hydrogen peroxide radical (15), $pK_s = 4.9$,[27] i.e. at physiological pH only a few per cent are present in the acidic form. By protonation, which proceeds extremely rapidly, $k_{-15} = 4.7 \times 10^{10}$ l mol^{-1},[28] the chemical character of the superoxide ion is drastically changed. From being a good nucleophile and a mild reducing agent it is turned into a strong oxidant and this change in oxidation potential is of importance for our understanding of its reactions.

$$^3O_2 \underset{-e^{\ominus}}{\overset{e^{\ominus}}{\rightleftharpoons}} O_2^{\ominus\cdot} \qquad (14)$$

$$HO_2^{\cdot} \rightleftharpoons H^{\oplus} + O_2^{\ominus} \qquad (15)$$

$$O_2^{\ominus\cdot} + O_2^{\ominus\cdot} \overset{\text{slow}}{\rightleftharpoons} O_2^{2\ominus} + O_2 \qquad (16)$$

$$HO_2^{\cdot} + O_2^{\ominus\cdot} \rightarrow HO_2^{\ominus} + O_2 \qquad (17)$$

$$O_2^{\ominus\cdot} + Cu^{2\oplus} \rightleftharpoons Cu^{\oplus} + O_2 \qquad (18)$$

$$Cu^{\oplus} + HO_2^{\cdot} \rightleftharpoons Cu^{2\oplus} + HO_2^{\ominus} \qquad (19)$$

The superoxide ion undergoes a very slow disproportionation to peroxide and oxygen (16), predominantly caused by impurities and moisture. This electron transfer is speeded up enormously by protonation, $k_{17} = 8.5 \times 10^7$ l mole^{-1} s^{-1}.[27] Actually, (17) represents a reaction that leads to singlet oxygen. The electron transfer is facilitated by transition metal catalysis, e.g. by $Cu^{\oplus}$ in dismutase (18,19) and these reactions are nearly diffusion controlled, $k = 2.3 \times 10^9$ l mole^{-1}s^{-1}.[29,30,31] The superoxide ion is often described in the literature as an oxidant. This is a misunderstanding; it is basically a reductant but acts as oxidant in consequence of protonation in protic solvents. It is unreactive against cyt $Fe^{2\oplus}$ but reduces cyt $Fe^{3\oplus}$. The peroxide radical oxidizes cyt $Fe^{2\oplus}$ but is unreactive against cyt $Fe^{3\oplus}$.[32]

The superoxide ion is an efficient nucleophile (20) and reacts rapidly with alkyl bromides in dimethylsulphoxide with formation of peroxides and alcohols. The hydrogen peroxide radical carries all the characteristics of a shortlived reactive radical. It is a hydrogen abstractor (21). The O–H bond energy for homopolar fission of alkyl peroxides or hydrogen peroxide (22,23) is 89–90 kcal mole^{-1} which means that most H abstractions are approximately thermally neutral.[33] The C–H bond energies of the methylene groups in propane and 1,4-pentadiene are 95 and 80 kcal mole^{-1}, respectively. The hindrance for reaction (21) is the activation energy which has been measured in a few cases to *ca.* 10 kcal; k_{21} is consequently comparatively small, *ca.* 0.1–10 l mole^{-1} s^{-1}.[34] This is compensated

$$RBr + O_2^{\ominus\cdot} \rightarrow ROO^{\cdot} + Br^{\ominus} \tag{20}$$

$$HOO^{\cdot} + RH \rightarrow HOOH + R^{\cdot} \tag{21}$$

$$t\text{-BuOOH} \rightarrow t\text{-BuOO}^{\cdot} + H^{\cdot} \quad \Delta H = 90 \text{ kcal} \tag{22}$$

$$HOOH \rightarrow HOO^{\cdot} + H^{\cdot} \quad \Delta H = 89 \text{ kcal} \tag{23}$$

for by close orientation of reagent and substrate on the enzyme. Thus, the peroxide radical is the most likely candidate for the post as 'activated oxygen' in monooxygenase catalysis and (21) is suggested as the first step in aliphatic hydroxylation. The suggestion of the involvment of $O_2^{\ominus\cdot}$ in biological oxidation was put forward in the late fifties.[35] Haber and Weiss[36] pointed out several years earlier the intermediacy of hydroxyl and hydrogen peroxide radicals in the Fe^{2+} catalysed decomposition of hydrogen peroxide, the Fenton reaction (24–27). It is closely related to the reactions taking place at the cytochromes or other enzymes catalysing oxygenation.

$$Fe^{2+} + H_2O_2 \rightarrow Fe^{3\oplus} + HO^{\cdot} + HO^{\ominus} \tag{24}$$

$$HO^{\cdot} + H_2O_2 \rightarrow HOO^{\cdot} + H_2O \tag{25}$$

$$HOO^{\cdot} \rightleftharpoons H^{\oplus} + O_2^{\ominus\cdot} \tag{26}$$

$$Fe^{3\oplus} + O_2^{\ominus\cdot} \rightleftharpoons Fe^{2\oplus} + O_2 \tag{27}$$

The peroxide bond is weak with a homopolar dissociation energy of 51.0 kcal mole^{-1}. In Fenton's reaction it is cleaved by addition of an electron and the very reactive hydroxyl radical is formed. It is a strong oxidant (28) with weak acidic properties, pK_s = 11.8,[37] (29) and it abstracts hydrogen atoms from any aliphatic molecule in a nearly diffusion controlled reaction,[38] $k_{30} = 10^9$ l mole^{-1} s^{-1}, as a result of the high energy of the O–H bond. $\Delta H_{31} = 119$ kcal, i.e. (30) is always exothermic.

$$HO^{\cdot} \xrightarrow{e^{\ominus}} HO^{\ominus} \tag{28}$$

$$HO^{\cdot} \rightleftharpoons O^{\ominus\cdot} + H^{\oplus} \tag{29}$$

$$HO^{\cdot} + RH \rightarrow H_2O + R^{\cdot} \tag{30}$$

$$H_2O \rightarrow HO^{\cdot} + H^{\cdot} \tag{31}$$

$$H_2O_2 \xrightarrow{2\,e^{\ominus}} 2\,OH^{\ominus} \tag{32}$$

In view of the high chemical reactivity and the high oxidation potential of the hydroxyl radical, it is improbable that it is formed in the cell. This can be circumvented by cleavage of the peroxide bond in a two-electron transfer process to two hydroxyl ions (32). The hydroxyl radical is thus finally reduced to water either by electron transfer (28) or by hydrogen atom abstraction (25,30). Of importance for the following discussion is also the diffusion controlled reaction of $^3O_2$ with carbon centred radicals. This is one of the propagation steps in autooxidation (33,34).[39]

$$R^{\cdot} + O_2 \xrightarrow{\text{fast}} ROO^{\cdot} \tag{33}$$

$$ROO^{\cdot} + RH \rightarrow ROOH + R^{\cdot} \tag{34}$$

With the relevant chemistry of oxygen in mind, we now turn to enzymatic hydroxylation as it is enacted at the cytochromes. The prosthetic group is an $Fe^{2\oplus}$-porphyrin (P450, P for pigment and 450 for the wavelength of absorption band) that absorbs O_2 and gives it a partial negative charge thereby turning it into a nucleophile. Reaction with an electrophile $E^{\oplus}$, i.e. $H^{\oplus}$, or a strategically located effector group, a carboxylate[40], produces the peroxide radical (35,36) which is capable of abstracting a hydrogen atom from the substrate absorbed on the enzyme (38). The radical formed reacts immediately with dissolved oxygen forming a new peroxide radical (39). This is reduced by cyt $Fe^{2\oplus}$ to a peroxide by a one-electron transfer. In the absence of a substrate $EOO^{\cdot}$ is just reduced to the peroxide (37). Finally the peroxide is reduced by a two-electron

$$\text{cyt Fe}^{3\oplus} + e^{\ominus} \rightleftharpoons \text{cyt Fe}^{2\oplus} \tag{35}$$

$$\text{cyt Fe}^{2\oplus} + O_2 \rightleftharpoons \text{cyt Fe}^{2\oplus}O_2 \leftrightarrow \text{cyt Fe}^{3\oplus}O_2^{\ominus}$$

$$\overset{E^{\oplus}}{\rightleftharpoons} \text{cyt Fe}^{3\oplus} + O_2^{\ominus\cdot} \rightarrow \text{cyt Fe}^{3\oplus} + \text{EOO}^{\cdot} \tag{36}$$

$$\text{cyt Fe}^{3\oplus} + e^{\ominus} + \text{EOO}^{\cdot} \rightarrow \text{cyt Fe}^{3\oplus} + \text{EOO}^{\ominus} \tag{37}$$

$$\text{EOO}^{\cdot} + \text{RH} \rightarrow \text{EOOH} + \text{R}^{\cdot} \tag{38}$$

$$\text{R}^{\cdot} + O_2 \rightarrow \text{ROO}^{\cdot} \tag{39}$$

$$\text{cyt Fe}^{3\oplus} + \text{ROO}^{\ominus} + \text{H}^{\oplus} \rightarrow \text{RO}^{\ominus} + \text{OH}^{\ominus} + \text{cyt}^{\oplus}\text{Fe}^{4\oplus} \tag{40}$$

transfer process to the alcohol (40). Consequently, one oxygen atom is inserted into the hydrocarbon and the other is reduced to water. There is increasing evidence that the first intermediate in aliphatic hydroxylation is a peroxide. Since these compounds are unstable, they often decompose in the work-up procedure of natural products, but are now to an increasing extent isolated from natural sources. The oxygenation pattern of prostaglandins and the isolation of peroxidic intermediates illustrate the mechanism[41] (Fig. 4). In a lipoxygenase reaction C^{13}–H_S is removed from arachidonic acid; the radical formed reacts with oxygen and cyclizes. Another molecule of oxygen is added and a one-electron reduction gives the peroxide. Cleavage of the O–O bond is eventually accomplished by a peroxidase or by a bifunctional oxygenase. The resting enzyme has its haem iron in a high spin $Fe^{3\oplus}$ state (40). It ought to be noted that in the formation of the endoperoxide both oxygen atoms are incorporated, formally corresponding to the activity of a dioxygenase. The intermediacy of radicals is confirmed by adding 2-methyl-2-nitrosopropane as scavenger to a mixture of linoleic acid, oxygen and lipoxygenase.[42] 2-Methyl-2-nitrosopropane reacts fast with carbon centred radicals and competes with oxygen for the radical giving a stable nitroxide which can be recorded by ESR (41). The spectrum

13 9 COOH H_R H_S Linoleic acid $\xrightarrow{\text{EOO}^{\cdot}}$ H_R $\longleftrightarrow$ (41)

H_R $\longleftrightarrow$ H_R $\xrightarrow{t\text{-BuNO}}$ t-Bu $^{\cdot}$O–N H H_R

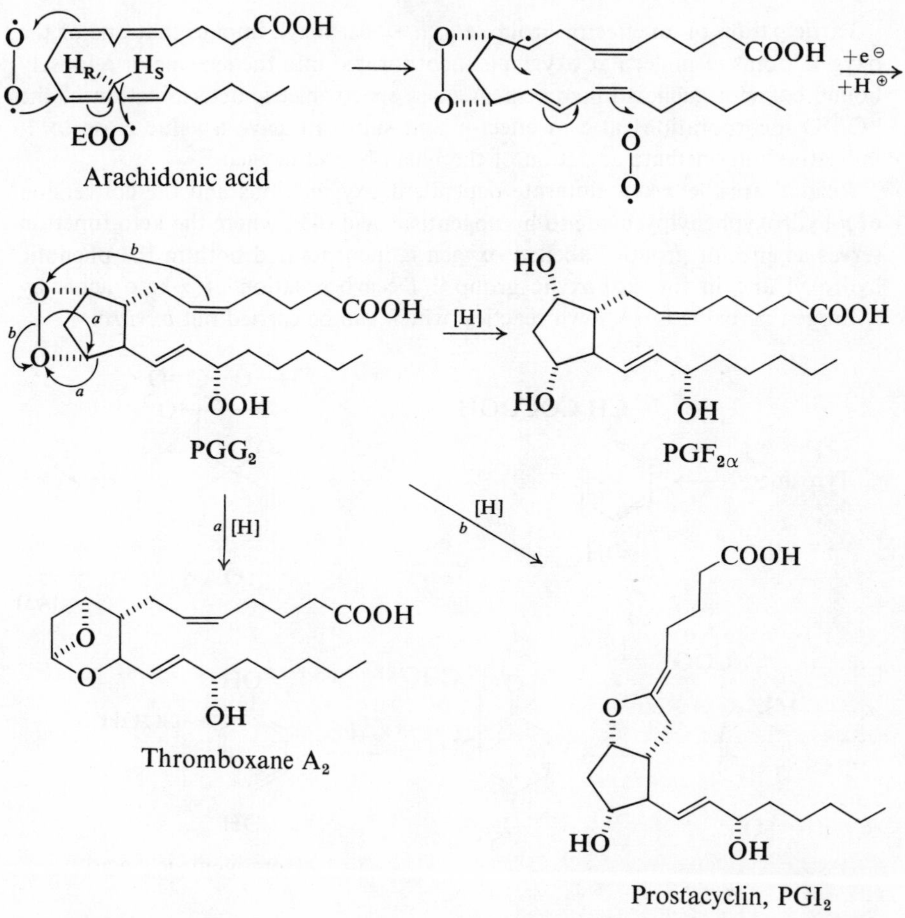

Fig. 4 Biosynthesis of prostaglandins and thromboxanes from arachidonic acid

gives a characteristic triplet from nitrogen and a smaller doublet from the proton at C^9 or C^{13} which vanishes on introducing deuterium at these positions. Deuteration of C^{11} did not change the spectrum, indicating that the more stable conjugated radical was trapped in accordance with the biosynthetic scheme for prostaglandins. The proposed pathway is mimicked in the laboratory in the serial cyclization of peroxy olefins (42).[43] *O*- or *N*-demethylation is explained by H abstraction from the methyl, actually the preferred point of attack in radical reactions, and the intermediate α-hydroxyamine or α-hydroxyether is easily hydrolysed to the free amine or alcohol.

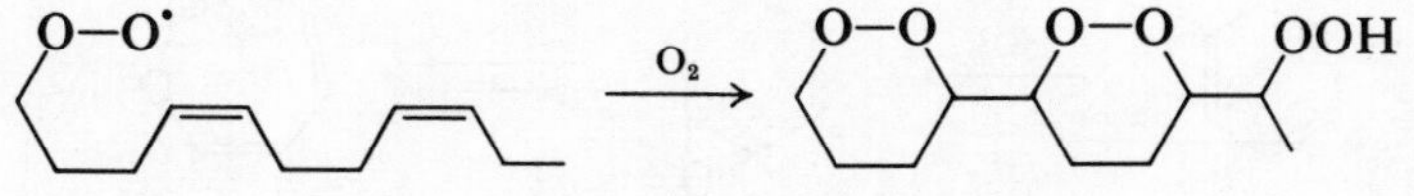

(42)

Participation of an effector group, e.g. E = carboxyl, implies that one of the oxygen atoms of molecular oxygen is incorporated into the enzyme or a loosely bound cofactor such as lipoic acid. A mass spectrometric determination of the $^{18}O/^{16}O$ incorporation ratio in effector and substrate gave a value close to 1, indicating intermediate acylation of the haem bound oxygen.[40]

Related are the α-ketoglutarate dependent oxygenations and the conversion of *p*-hydroxyphenylpyruvate to homogentisic acid (43), where the keto function serves as effector group. Labelled oxygen is incorporated both in the phenolic hydroxyl and in the carboxylic group.[44] Decarboxylation of α-keto acids by hydrogen peroxide is a known reaction which can be carried out *in vitro*.

Tyrosine → (p-hydroxyphenyl, $CH_2COCOOH$ / OH) —$^*O_2^{\ominus\cdot}$, $-e^{\ominus}$→ (H–O–C=O, O*, $O^{\ominus}$, O*, O) →

(43)

($COO^{\ominus}$, $^{\ominus}O$, O) → (O, $COO^{\ominus}$, H, $O^{\ominus}$) —$H^{\oplus}$→ (*OH, C^*OOH, OH)

Homogentisic acid

The formation of naturally occurring epoxides is explained by the reaction of olefins with peracids, anchored on the enzymes (44), by analogy with the *in vitro* epoxidation performed by peracids.

$$\text{Enz}-\overset{\text{O}}{\overset{\|}{\text{C}}}-\text{OOH} + \text{(olefin)} \longrightarrow \text{Enz}-\overset{\text{O}}{\overset{\|}{\text{C}}}-\text{OH} + \text{(epoxide)} \quad (44)$$

There is strong evidence for the intermediacy of epoxides in aromatic hydroxylation, e.g. in the biosynthesis of tyrosine from phenylalanine or *p*-coumaric acid and ferulic acid from cinnamic acid. In the benzene series these epoxides are unstable, but in higher condensed systems, such as naphthalene or pyrene,

(benzene) —[O]→ (benzene oxide, O) ⇌ (oxepin, O) (45)

epoxides have been isolated.[45] Epoxides of polycyclic aromatics display carcinogenic properties. That means, in fact, that the detoxicating mechanism leads to still more harmful derivatives. It is suggested that opening of the epoxide by nucleophilic amino groups in nucleic acids initiates chemical carcinogenesis. Benzene epoxide is shown to be in equilibrium with the oxepin isomer and the equilibrium is markedly shifted towards the seven-membered ring system (45). Related naturally occurring metabolites, which contain either the hexadiene structure or the oxepin structure, have been isolated. The substituents of the cyclohexadiene structure are always *trans* located (Fig. 5), which is typical for opening of an oxirane ring by a nucleophile.

The biosynthesis of gliotoxin[46] is illustrative for the process of aromatic epoxidation. The precursor is believed to be the cyclodipeptide formed from phenylalanine and serine. The arene oxide is opened in a *trans* fashion and a disulphide bridge is introduced. It has been established that all nine carbons of phenylalanine are incorporated in gliotoxin and it is shown that ^{14}C-labelled cyclo-(L-alanyl-L-phenylalanyl) is used by *Trichoderma viridis* to produce the 3-deoxy analogue of gliotoxin. Furthermore, a structurally related oxepin derivative, aranotin, has been isolated.

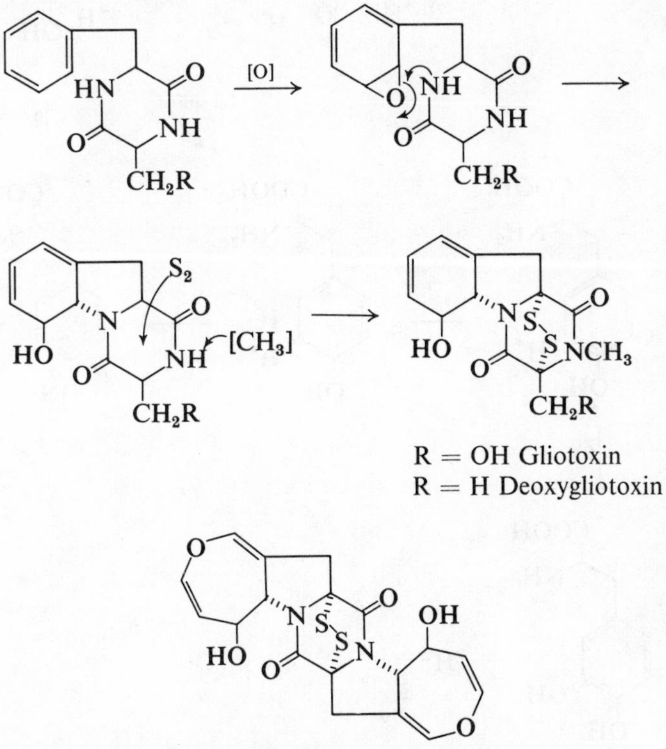

R = OH Gliotoxin
R = H Deoxygliotoxin

Aranotin

Fig. 5 Proposed biosynthesis of gliotoxin

An important characteristic of aromatic hydroxylation is the NIH shift, named after a research group at the National Institute of Health, USA. Phenylalanine labelled at C^4 with tritium or deuterium retained *ca.* 90 per cent of the label at C^3.[47] The NIH shift is not restricted to *para*-hydroxylation, but functions for *ortho*-hydroxylation as well. However, hydroxylation *ortho* or *para* to an existing hydroxyl group leads to loss of label.[48] The mechanism is depicted in Fig. 6. Two paths lead to loss of label. In path a the label is directly eliminated by aromatization and in path b in the enolization step. The high retention indicates high migratory aptitude of the proton and an unusually high isotope effect. It has been argued that the enolization therefore is enzymatically controlled. Support for the intermediacy of epoxides in aromatic hydroxylation has been

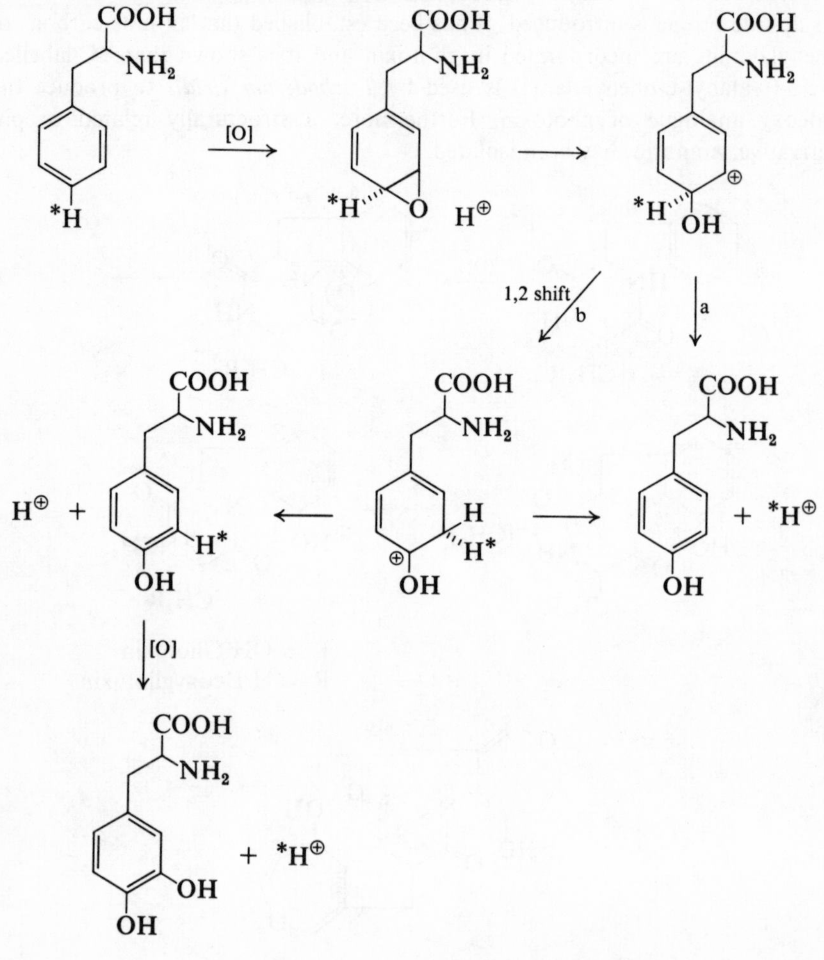

COOH NH₂ *H —[O]→ COOH NH₂ *H O H⊕ → COOH NH₂ *H OH ⊕

1,2 shift b — a

COOH NH₂ H⊕ + H* OH ← COOH NH₂ H H* ⊕ OH → COOH NH₂ OH + *H⊕

[O]

COOH NH₂ OH OH + *H⊕

Fig. 6 Mechanism of aromatic hydroxylation

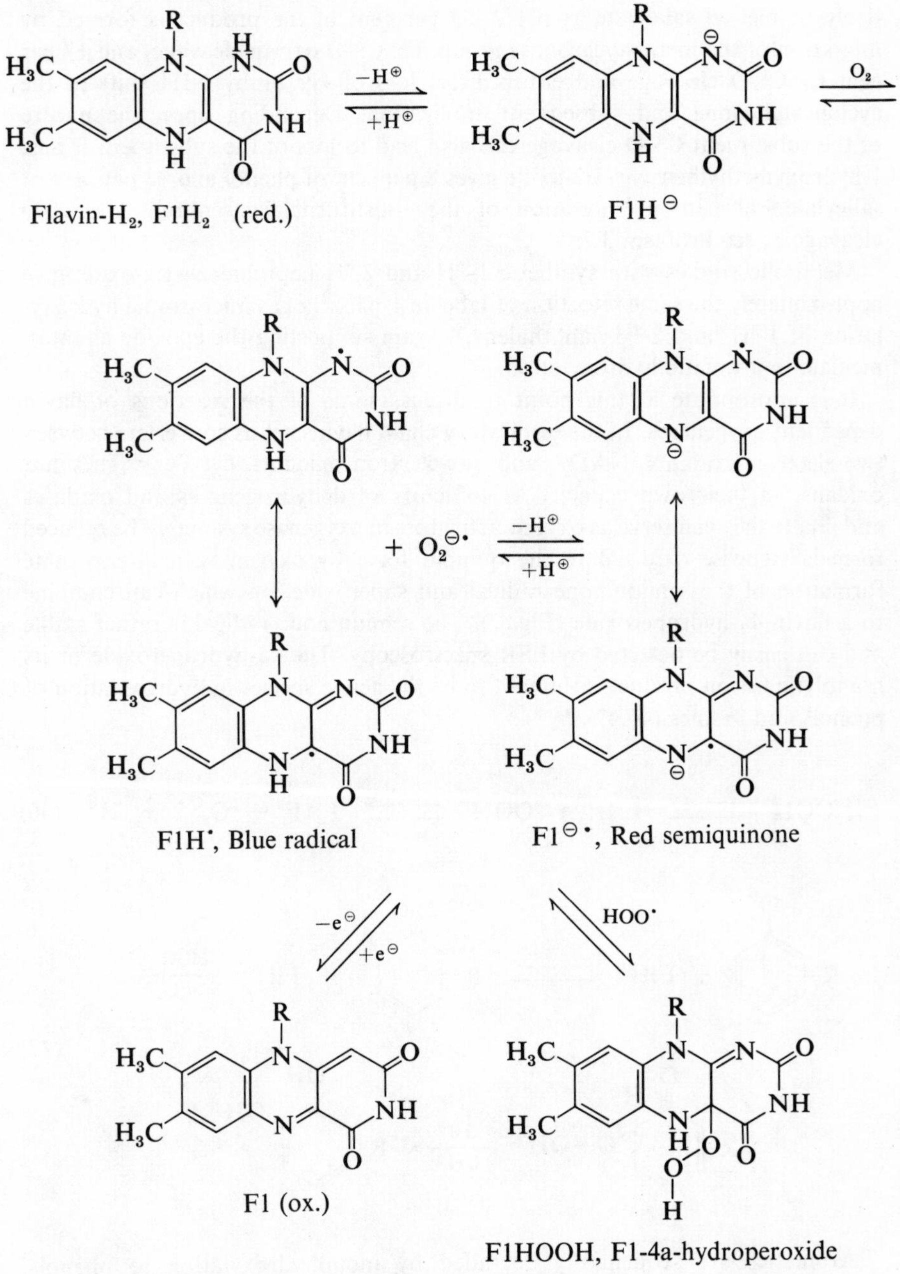

Flavin-H_2, FlH_2 (red.) $\underset{+H^{\oplus}}{\overset{-H^{\oplus}}{\rightleftharpoons}}$ Fl$H^{\ominus}$ $\overset{O_2}{\rightleftharpoons}$

Fl$H^{\bullet}$, Blue radical + $O_2^{\ominus\bullet}$ $\underset{+H^{\oplus}}{\overset{-H^{\oplus}}{\rightleftharpoons}}$ $Fl^{\ominus\bullet}$, Red semiquinone

$-e^{\ominus}$ / $+e^{\ominus}$ $\qquad$ $HOO^{\bullet}$

Fl (ox.) $\qquad$ FlHOOH, Fl-4a-hydroperoxide

Fig. 7 Stepwise one-electron transfers; oxidation of flavin-H_2 by oxygen

gained from studies of the mechanism of aromatization of some substituted benzene-1,2-oxides.[49] 1-Methoxycarbonylbenzene-1,2-oxide rearranges quantitatively to methyl salicylate at pH 7. 83 per cent of the product is formed by migration of the methoxycarbonyl group, i.e. C^2–O oxiran cleavage, and 17 per cent by C^1–O cleavage and either direct loss of $H^{\oplus}$ or by NIH shift to the cyclohexadienone and subsequent enolization. Depending upon the nature of the substituent C^2–O cleavage can also lead to loss of the substituent. Thus, 1-hydroxymethylbenzene-1,2-oxide gives 8 per cent of phenol and 92 per cent of salicylalcohol without migration of the substituent. i.e. entirely via C^1–O cleavage 1; see Problem 3.7.

Metabolic studies with synthetic 1–^{2}H and 2-^{2}H naphthalene-1,2-oxide give approximately the same retention of label in 1-naphthol as microsomal hydroxylation of 1-^{2}H- and 2-^{2}H-naphthalene,[50] again supporting the epoxide as intermediate in aromatic hydroxylation.

It is appropriate at this point to discuss some of the reactions of flavin dependent oxygenases. In the respiratory chain flavins act as converters between two-electron oxidants, $NAD^{\oplus}$, and one-electron oxidants, cyt $Fe^{3\oplus}$. They are oxidants in their own capacity as cofactors of dehydrogenases and oxidases and finally they can serve as oxygen activators in oxygenase systems. The reduced form is stepwise oxidized to the quinoid form by oxygen with intermediate formation of the semiquinone radical and superoxide ion which can combine to a flavin-4a-hydroperoxide (Fig. 7). The semiquinone radical is rather stable and can easily be detected by ESR spectroscopy. The 4a-hydroperoxide or its homolytic fission product is thought to be the active species in hydroxylation of phenols and indoles (46,47).[51]

$$\text{FlHOOH} \rightleftharpoons \text{FlH}^{\bullet} + {}^{\bullet}\text{OOH} \rightleftharpoons \text{FlH}^{\bullet} + \text{O}_2^{\ominus\bullet} + \text{H}^{\oplus} \qquad (46)$$

$$\text{R-C}_6\text{H}_4\text{-O}^{\ominus} + \text{FlH}^{\bullet} \rightleftharpoons \text{R-cyclohexadienone(H)}^{\bullet} + \text{FlH}^{\ominus} \xrightarrow[\text{or O}_2]{\text{HOO}^{\bullet}} \text{R-cyclohexadienone(H)(O-OH)} \xrightarrow[2\,\text{H}^{\oplus}]{2\,\text{e}^{\ominus}} \text{R-C}_6\text{H}_3(\text{OH})_2 \qquad (47)$$

Aromatics are sequentially degraded by monohydroxylation to phenols, *ortho*-dihydroxylation to catechols and finally ring opening by iron-dependent dioxygenases via a dienone hydroperoxide (48). The ring is cleaved either between the hydroxyl groups or next to one of them. The mechanism is not known with certainty. We know that both oxygen atoms of molecular oxygen are incorporated and spectral data indicates coordination of the phenolic group with the ferric centre of the cofactor. The ferric ion abstracts an electron and the catechol radical formed combines with oxygen to a hydroperoxide or a highly strained dioxetane which rearranges in an obscure way.[52] The intermediacy of the cyclodienone peroxide is supported by isolation of such derivatives and the finding that they indeed are cleaved to dicarbonyl compounds. The heteroaromatic 3-hydroperoxy-2,3-dimethylindolenine rearranges to 2-acetamidoacetophenone on treatment with base[51] (49), a reaction that is analogous to the one occurring at one stage of the biosynthesis of kynurenine (50).

(48)

3-Hydroperoxy-2,3-dimethylindolenine → 2-Acetamidoacetophenone (49)

Tryptophan → → Kynurenine (50)

The literature on biological oxidation mechanisms is enormous, full of paper chemistry and more or less well founded hypotheses. This presentation is centred on the known chemistry of the $O_2^{\ominus\cdot}/HOO^{\cdot}$ couple. The active species in iron catalysed hydroxylation is still heavily debated. An attractive candidate for a number of hydroxylations is the iron coordinated oxygen atom, the oxene,[52a] formed by cleavage of the peroxide (51). This oxygen is expected to react as the isoelectronic carbene and nitrene species, i.e. by insertion into C–H bonds and addition to double bonds. In horseradish peroxidase, the so-called green compound I seems to be consistent with a $cyt^{\oplus}Fe^{4\oplus} = O \rightleftarrows cyt\ Fe^{5+} = O$ formulation. This is one oxidation equivalent above the species formulated in (51). In model experiments a green intermediate was obtained by oxidation of a porphyrin-Fe^{3+} complex with peracid. The intermediate is reported to effect epoxidation of olefins and the oxygen is exchangeable with added $H_2{}^{18}O$. This excludes a metal coordinated peroxy structure and supports the oxene structure as active species in certain oxygenations. Still, another structure[52b] has been proposed for HRP compound I based on X-ray studies of similar carbene complexes. It is suggested that the electron deficient oxygen in $cyt^{\oplus}Fe^{4\oplus} = O$ is no longer stabilized by iron but requires support from a lone pair of the pyrrole nitrogen and the oxygen is therefore considered to be inserted into an iron-N bond. The last word is apparently not written on this subject yet.

$$\text{cyt Fe}^{2\oplus} + O_2 \longrightarrow \text{cyt Fe}^{2\oplus}O_2 \longleftrightarrow \text{cyt Fe}^{3\oplus}{-}O{-}O^{\ominus}\ H^{\oplus} \longrightarrow$$

$$\text{cyt Fe}^{4\oplus}{=}O \longleftrightarrow \text{cyt}^{\oplus}\text{Fe}^{3\oplus}{=}O \longleftrightarrow \text{cyt}^{\oplus}\text{Fe}^{4\oplus}O^{\ominus} \qquad (51)$$

Ferryloxene

3.4 Cinnamic and benzoic acids

The ubiquitously distributed cinnamic acids arise from phenylalanine by enzymatic elimination of ammonia followed by aromatic hydroxylation and methylation. It was earlier believed that the biosynthesis took the route via phenylpyruvic acid which was reduced and dehydrated, reactions that *per se* have ample precedents, but when it was discovered that enzymes were able to eliminate ammonia directly from the amino acid, this route was considered to represent the major pathway. This reaction is categorized as a concerted α,β-elimination assisted by a basic centre on the enzyme abstracting the β-proton (52). The elimination takes place in the stereoelectronically most favourable

B: ; H_S, H, $COO^{\ominus}$, H_5C_6, H_R, $NH_3^{\oplus}$ $\longrightarrow$ H_5C_6, H, H_R, COOH (52)

trans-planar conformation by exclusive loss of the pro-3*S* hydrogen leading directly to *trans*-cinnamic acid.[53] The reaction is analogous to the classical Hofmann elimination. A closer investigation of the phenylalanine ammonia lyase, PAL, revealed that the reaction is reversible which means that the re-addition of ammonia must take place opposite to the polarity of the double bond. Chemical inhibition of the enzyme is achieved by carbonyl reagents such as sodium cyanide and sodium borohydride. Reduction of the enzyme with tritiated sodium borohydride gives on subsequent hydrolysis alanine with tritium predominantly confined to the methyl group.[54] These observations suggest that the active site of the enzyme is a dehydroalanine residue probably generated from serine (Fig. 8). The amino group of serine forms a Schiff's base with another subunit of the enzyme activating the α-proton and resulting in elimination of water. The β-carbon of the dehydroalanine residue reacts in a Michael-type reaction with nucleophiles, such as hydride ions from sodium borohydride or cyanide ions blocking its activity. The amino group of the substrate binds in the

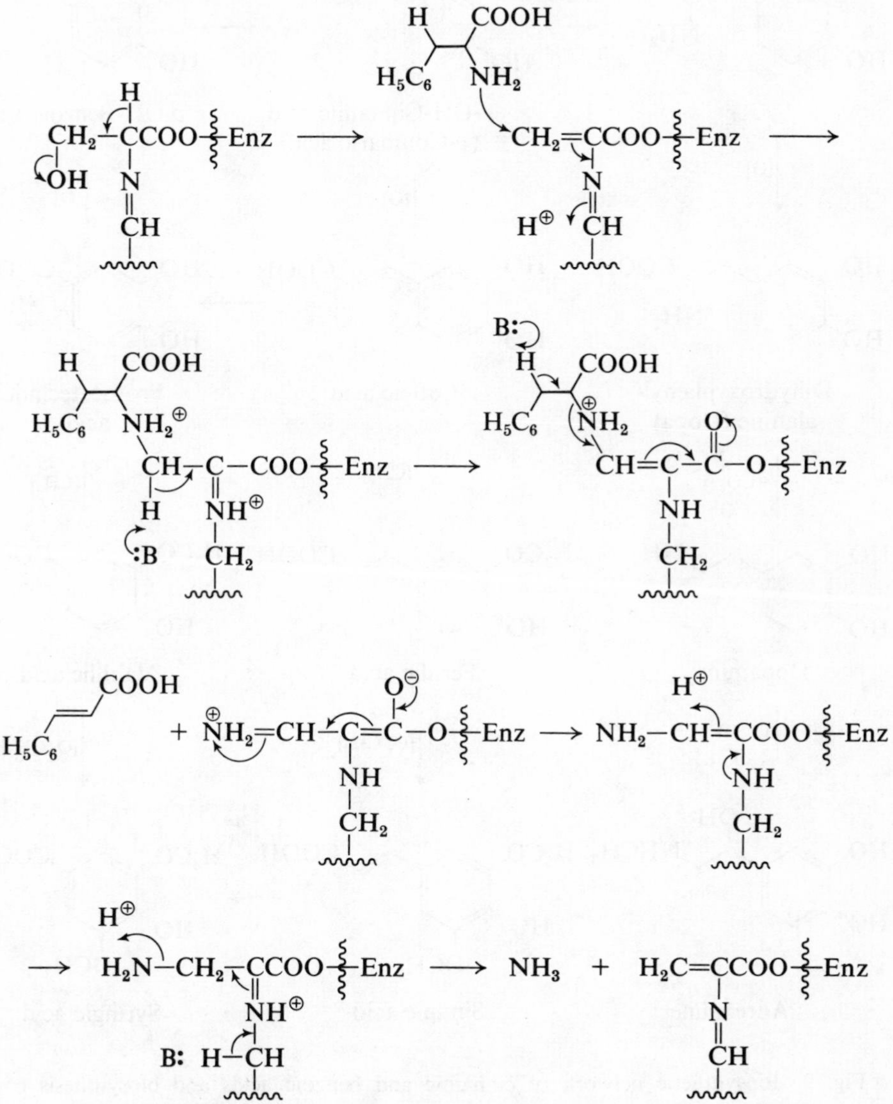

Fig. 8 Hypothetical mechanism for PAL activity

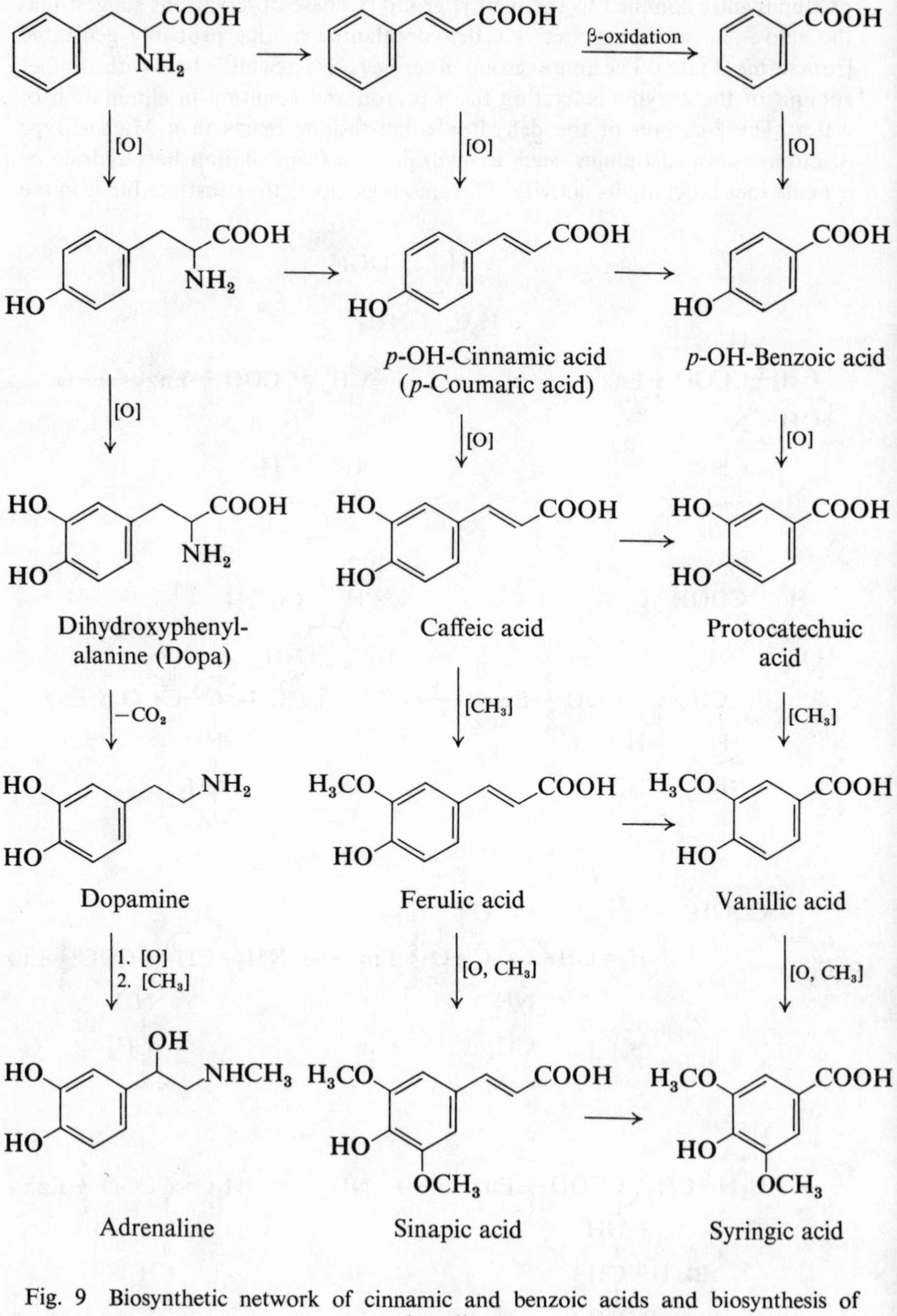

COOH NH$_2$ → COOH β-oxidation → COOH

[O] [O] [O]

HO COOH NH$_2$ → HO COOH → HO COOH

p-OH-Cinnamic acid
(*p*-Coumaric acid)

p-OH-Benzoic acid

[O] [O] [O]

HO HO COOH NH$_2$ | HO HO COOH → HO HO COOH

Dihydroxyphenyl-
alanine (Dopa)

Caffeic acid

Protocatechuic
acid

$-CO_2$ [CH$_3$] [CH$_3$]

HO HO NH$_2$ | H$_3$CO HO COOH → H$_3$CO HO COOH

Dopamine

Ferulic acid

Vanillic acid

1. [O]
2. [CH$_3$]

[O, CH$_3$] [O, CH$_3$]

HO HO OH NHCH$_3$ | H$_3$CO HO OCH$_3$ COOH → H$_3$CO HO OCH$_3$ COOH

Adrenaline

Sinapic acid

Syringic acid

Fig. 9 Biosynthetic network of cinnamic and benzoic acids and biosynthesis of adrenaline

same way to the dehydroalanine residue and this could improve its ability as leaving group as demonstrated by shuffling around charges and protons in the prosthetic group (Fig. 8). Expulsion of ammonia in the final step regenerates the dehydroalanine residue making it ready for action in the next cycle.

Shortening of the side chain of cinnamic acids by β-oxidation is one of the main routes leading to benzoic acids (section 4.6).[55] The exact timing of the aromatic hydroxylation varies from plant to plant but in general, the hydroxylation is more effective at the C_6C_3 level than at the C_6C_1 level in higher plants. We can arrive, in principle, at vanillic acid, either along the route coumaric acid → caffeic acid → ferulic acid, or via coumaric acid → *p*-OH-benzoic acid → protocatechuic acid (Fig. 9).[56] There is also a shortcut to the hydroxylated benzoic acids. Dehydration and dehydrogenation of 3-dehydroshikimic acid give protocatechuic acid and gallic acid directly.[57] It was found that labelled glucose was a better precursor of gallic acid than labelled phenylalanine in *Geranium pyrenaicum*, an observation that speaks in favour of the direct route.[58] On the other hand phenylalanine was metabolized more effectively than glucose to gallic acid in *Rhus typhina* suggesting that phenylalanine → cinnamic acid →

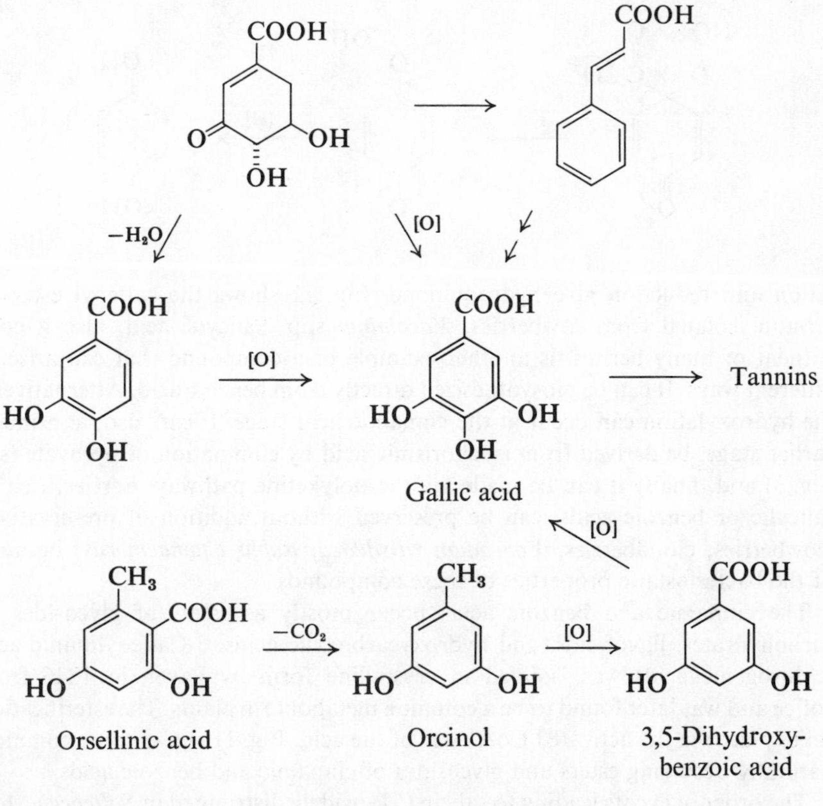

Fig. 10 Biosynthesis of gallic acid, the major constituent of gallotannins

p-OH-cinnamic acid → caffeic acid → 3,4,5-trihydroxycinnamic acid → gallic acid is the preferred pathway here.[59] *Epicoccum nigrum* on the other hand, produces gallic acid from orsellinic acid via the polyketide pathway by decarboxylation and oxidation (Fig. 10).

The side chain can eventually be completely eliminated as in the biosynthesis of the hydroquinone glycoside arbutin, a metabolite of *Bergenia crassifolia*. p-Hydroxybenzoic acid was shown to be an efficient precursor for arbutin. The hydroxylation of benzoic acid follows the general scheme developed in the preceding section. p-Hydroxybenzoic acid is oxidized by $Fe^{3\oplus}$ to the phenol radical which then couples with oxygen or a peroxide radical (53). Decarboxy-

COO⊖ Fe³⊕ → COO⊖ ⟷ COO⊖ —O₂ or HOO·→

O⊖ O. O

(53)

HO O OH OH
O C—O⊖ O [H]
→ →
O O⊖ OH

lation and reduction give hydroquinone. Fig. 11 shows the caffeoyl ester of arbutin isolated from cowberries, *Vaccinium* spp. Salicylic acid, also a constituent of many berries, is another example of a compound that can arise in different ways. It can be biosynthesized directly from benzoic acid. Alternatively, the hydroxylation can occur at the cinnamic acid stage. It can also, at a much earlier stage, be derived from isochorismic acid by elimination of pyruvate (see Fig. 3) and, finally it can be made via the polyketide pathway. Berries, rich in salicylic or benzoic acids, can be preserved without addition of preservatives (cowberries, cloudberries, *Vaccinium vitisildaea*, *Rubus chamaemorus*) because of the bacteriostatic properties of these compounds.

The cinnamic and benzoic acids occur mostly as esters of glycosides of carbohydrates, flavonoids, and hydroxycarboxylic acids. 3-Caffeoylquinic acid (chlorogenic acid) was isolated in crystalline form by Payen in 1846 from coffee and was later found to be a common metabolite in plants. The esterification takes place via the activated CoA ester of the acid. Fig. 11 shows some common naturally occurring esters and glycosides of cinnamic and benzoic acids.

The order of events leading to salicin (54), widely distributed in *Salicaceae*, has been looked at in some detail. ^{14}C-Labelled phenylalanine, cinnamic acid and

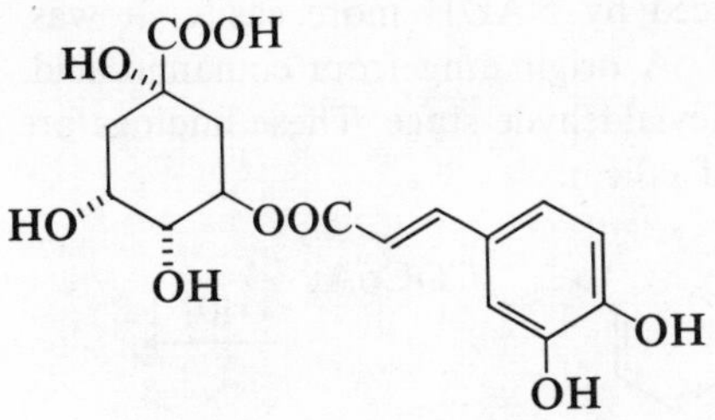

3-Caffeoylquinic acid

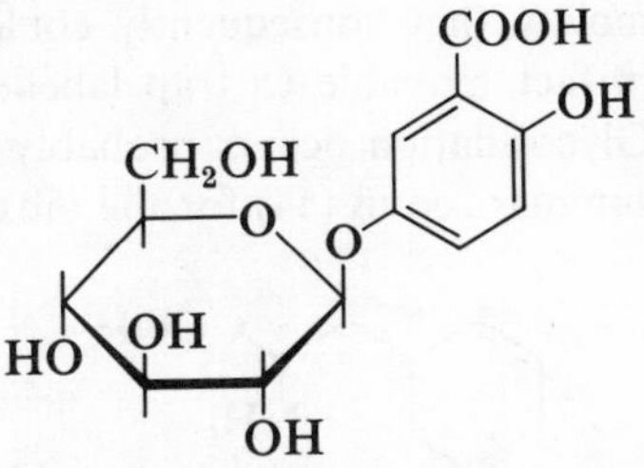

Gentisic acid 5-*O*-β-D-glycoside

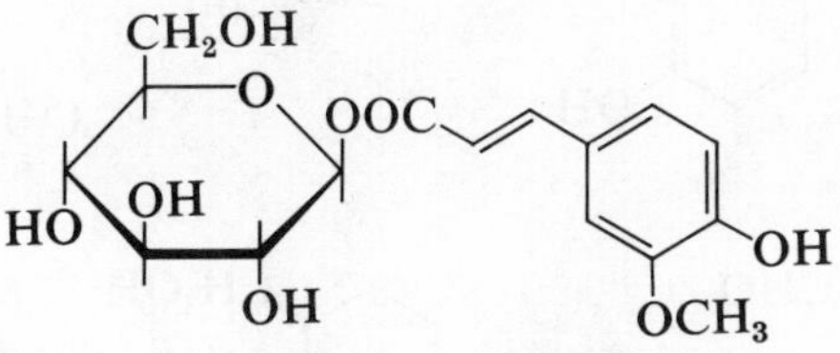

1-Feruloyl-β-D-glucose

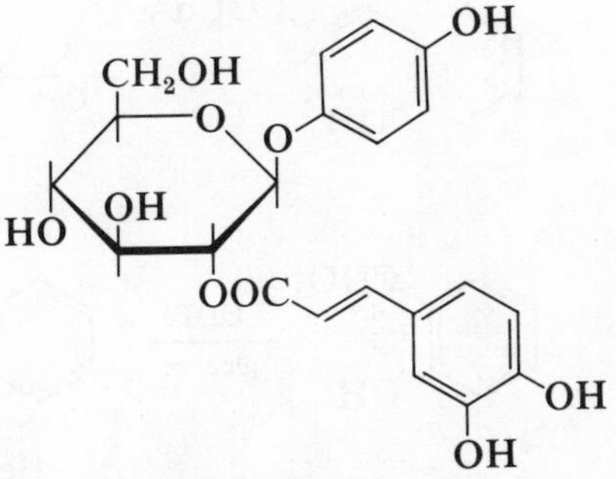

2-Caffeoylarbutin

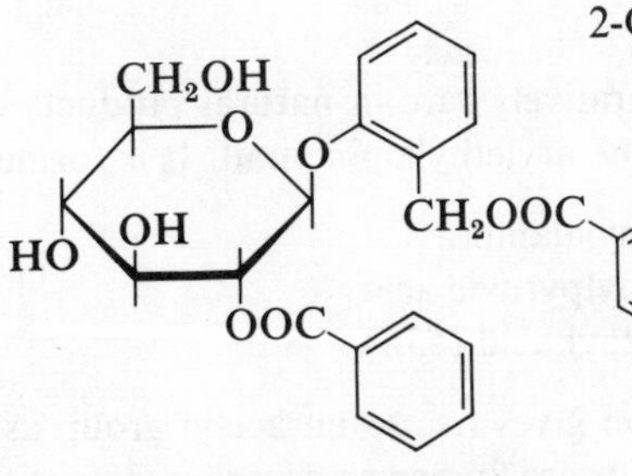

ω-Salicylsalicin-2-benzoate

Fig. 11 Naturally occurring esters and glycosides of cinnamic and benzoic acids

o-coumaric acid were incorporated into salicyl alcohol in *Salix purpurea* with increasing order of efficiency, whereas salicylic acid was poorly incorporated.[60] In another series of investigations with *Vanilla planifolia* the synthesis of vanillin was studied. Phenylalanine, cinnamic acid, and ferulic acid were incorporated with increasing ease, whereas labelled vanillic acid was poorly converted which means that in higher plants free benzoic acids are poorly reduced. It was also found that glycoside formation occurs at a higher oxidation level than salicylic alcohol since administration of this compound to the plant led to glycosidation of the $-CH_2OH$ group.[61] β-Oxidation of cinnamic acids according to the classical fatty acid degradation scheme should give rise to acetyl-CoA and benzoyl-CoA, representing a more energy-rich carboxyl moiety that consequently could be reduced by NADH more easily. It was, in fact, possible to trap labelled acetyl-CoA originating from cinnamic acid. Glycosidation occurs probably at the salicylaldehyde stage. These findings are summarized in (54) for the biosynthesis of salicin.

COOH
NH_2
$-NH_3$
COCoA
[O]

COCoA
OH
[O]
CoA
COCoA
OH
[H]

(54)

CHO
OH
UDP
glucose
CHO
O—G
CH_2OH
OG

Helicin

Salicin

The C_6C_2 unit is comparatively rare in natural products except in alkaloids which frequently contain the arylethylamine unit. It is formed by:

1. decarboxylation of phenylalanine;
2. decarboxylation of phenylpyruvic acid;
3. decarboxylation of benzoyl acid (55).

The last mentioned reaction gives rise to an acetyl group as in pungenin from *Picea pungens*.[62] β-Keto acids are known to decarboxylate with exceptional ease. The one carbon degradation sequence (55b) parallels the enzymatic degradation of pyruvic acid (section 2.3, equation(14)).

Vanillic acid is occasionally detected in mammalian urine. This compound is biosynthesized from adrenaline according to (56). The *meta*-hydroxyl is methylated, oxidation gives mandelic aldehyde, and further oxidation and decarboxylation eventually lead to vanillic acid. In plants vanillic acid is biosynthesized from ferulic acid by β-oxidation. The biosynthesis of adrenaline is shown in Fig. 9.

Tannins are considered to be excretion products of many plants, but are probably involved in the defence mechanism against parasites and grazing animals. Bark and heartwood of trees (e.g. *Quercus*) are rich in these compounds and so are gallnuts which develop as the result of attack by certain insects. The structures of tannins are very complex. The water soluble, hydrolysable tannins are polygalloyl or ellagoyl glycosides. There are often as many as nine gallic acids per glucose unit, i.e. at least one of the gallic acid residues has its hydroxyls

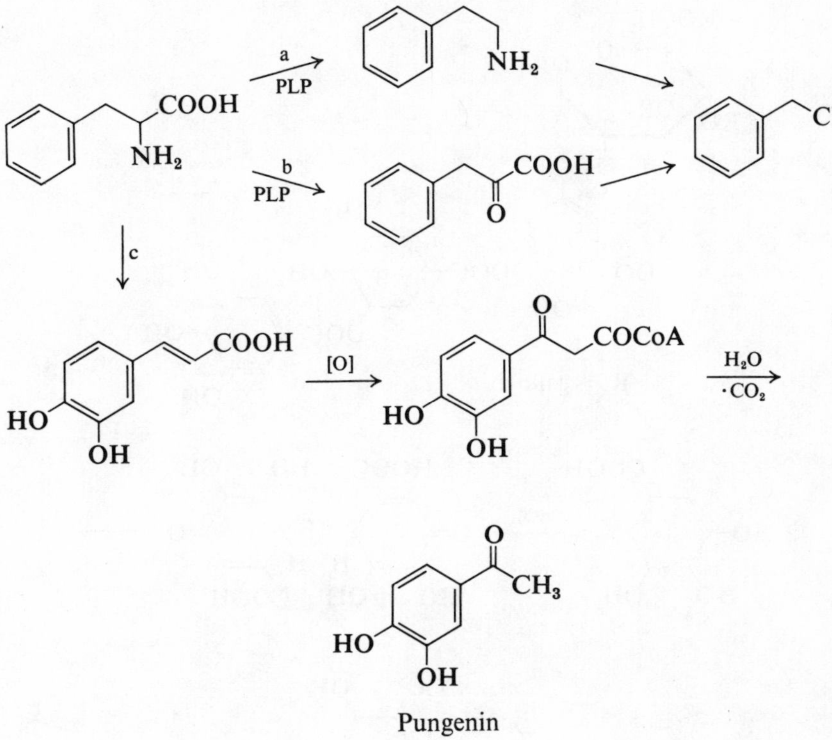

COOH NH$_2$ —a, PLP→ NH$_2$ → CHO

—b, PLP→ COOH, O →

↓ c

HO, OH, COOH —[O]→ O, COCoA, HO, OH —H_2O, $\cdot CO_2$→ (55)

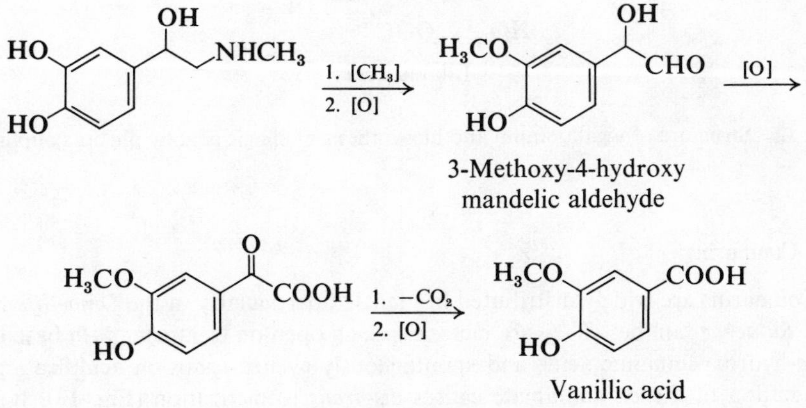

O, CH_3, HO, OH

Pungenin

HO, HO, OH, $NHCH_3$ —1. [CH_3], 2. [O]→ H_3CO, HO, OH, CHO —[O]→

3-Methoxy-4-hydroxy
mandelic aldehyde

(56)

H_3CO, HO, O, COOH —1. $-CO_2$, 2. [O]→ H_3CO, HO, COOH

Vanillic acid

esterified by other gallic acid molecules. This link is called a depside bond. Ellagic acid is a dimer of gallic acid formed by phenol oxidation (Fig. 12). Gallotannins give glucose and gallic acid on hydrolysis with acids or enzymes. Scheele isolated gallic acid in 1787 by this latter method from gallnut. Large amounts of oak and quebracho wood were formerly used in the vegetable tanning procedure but it is now to a great extent replaced by mineral tanning.

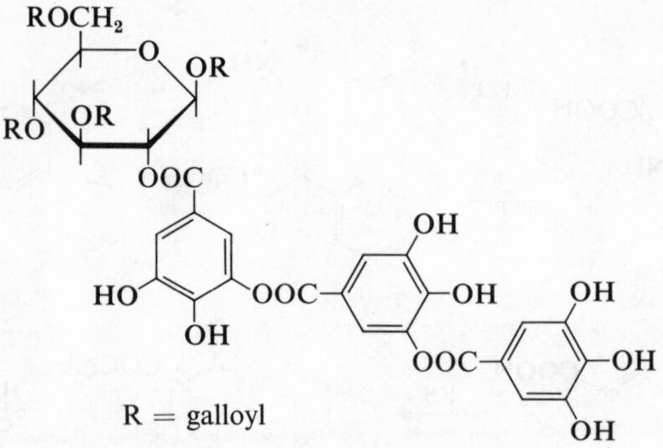

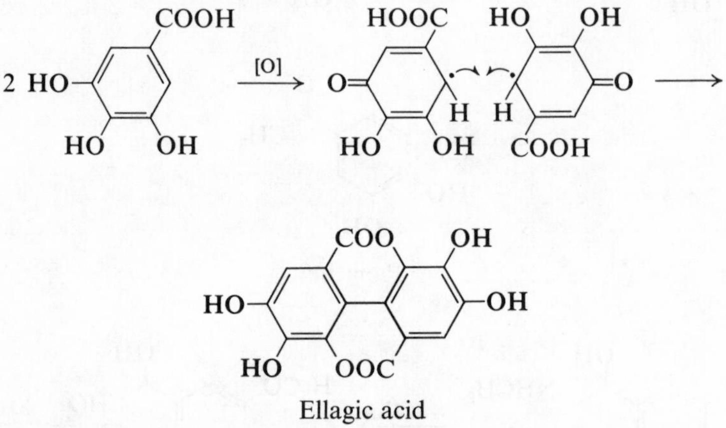

ROCH$_2$ OR RO OR OOC HO OH OOC OH OH OH OOC OH OH OH

R = galloyl

2 HO COOH HO OH [O] HOOC HO OH O O H H HO OH COOH

COO OH HO OH HO OOC

Ellagic acid

Fig. 12 Structure of a gallotannin and biosynthesis of ellagic acid by phenol coupling

3.5 Coumarins

Coumarins are widely distributed in plants,[63] particularly in the *Umbelliferae* and *Rutaceae* families. They are lactones which open on treatment with base to *cis-o*-hydroxycinnamic acids and spontaneously cyclize again on acidification. Irradiation of the *cis*-cinnamate causes *cis–trans* isomerization (Fig. 13). It is

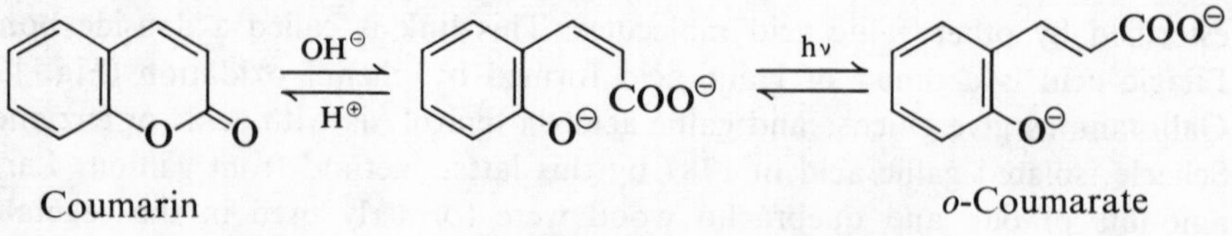

Coumarin $OH^{\ominus}$ $H^{\oplus}$ $COO^{\ominus}$ $O^{\ominus}$ hν $COO^{\ominus}$ $O^{\ominus}$ *o*-Coumarate

Fig. 13 *cis–trans* Isomerization of coumarin

RO 7, O, O
R = H Umbelliferone
R = CH_3 Herniarin

RO, HO, O, O
R = H Asculetin
R = CH_3 Scopoletin

Fig. 14 Structures of some coumarin derivatives

evident that coumarins derive from shikimic acid via cinnamic acids and organisms possessing an enzyme system capable of *o*-hydroxylating cinnamate are thus able to biosynthesize coumarins. Most coumarins are hydroxylated at C^7. Fig. 14 shows some representative derivatives. It is found by tracer experiments

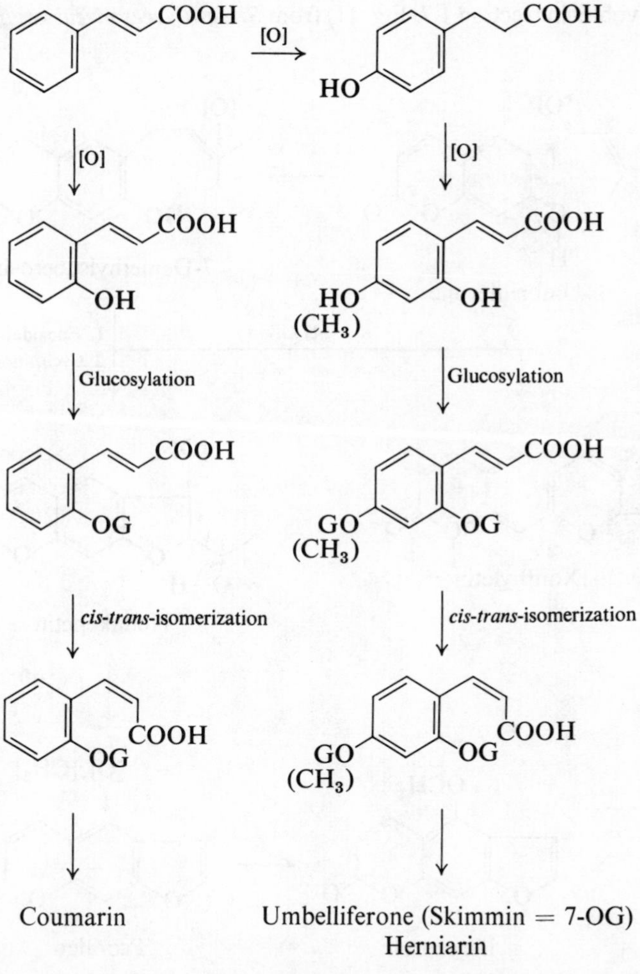

COOH
[O]
COOH
HO
[O]
[O]
COOH
OH
COOH
HO
(CH_3)
OH
Glucosylation
Glucosylation
COOH
OG
COOH
GO
(CH_3)
OG
cis-trans-isomerization
cis-trans-isomerization
COOH
OG
COOH
GO
(CH_3)
OG

Coumarin

Umbelliferone (Skimmin = 7-OG)
Herniarin

Fig. 15 Biosynthesis of coumarin, umbelliferone, and herniarin. G = glucose

in lavender, *Lavandula officinalis*, that glucose, phenylalanine, and cinnamic acid are efficiently incorporated into both coumarin itself and herniarin. *o*-Coumaric acid and *o*-coumarylglycoside are selectively converted into coumarin, whereas *p*-coumaric acid and 2-glucosyloxy-4-methoxycinnamic acid are selectively converted into herniarin. That places cinnamic acid at a branching point of the biosynthetic sequence. The first step to herniarin is a *p*-hydroxylation of cinnamic acid and the first step to coumarin is a *o*-hydroxylation. It was also found that in the intact cell herniarin is predominantly present as a glycoside, presumably the *cis*-2-β-glycoside. The stage at which methylation occurs has not been determined with certainty. The results are summarized in Fig. 15.[64] Studies of the biosynthesis of umbelliferone gave analogous results.

The mechanism of the *o*-hydroxylation was the subject of some dispute when it was demonstrated that the ring oxygen of the coumarin residue in the antibiotic novobiocin (section 1.7, Fig. 11) from *Streptomyces niveus* originated from

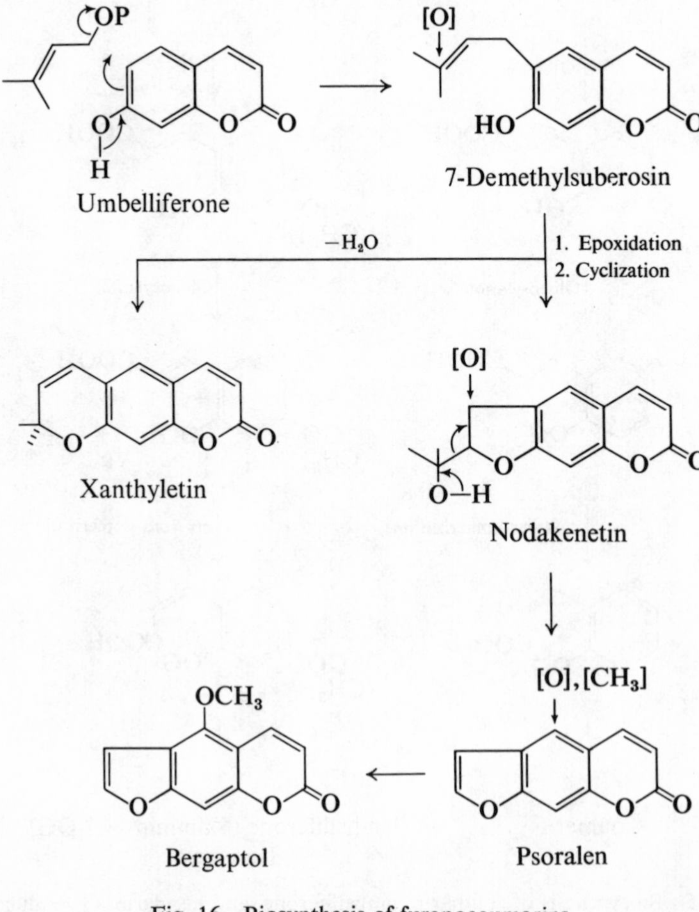

Fig. 16 Biosynthesis of furanocoumarins

the carboxyl group[65] as schematically formulated in (57) and by the use of ^{14}C it was also shown that the coumarin residue originated from tyrosine, i.e. the 7-OH is introduced at a much earlier stage. However, when the spirolactone was fed to *Lavandula officinalis*, poor incorporation was observed.[66] The fact that the coumarins predominantly occur as glycosides in the intact plant cell favours direct *ortho*-hydroxylation as opposed to oxidative cyclization in the micro-organism. Different pathways are apparently followed in moulds and in higher plants.

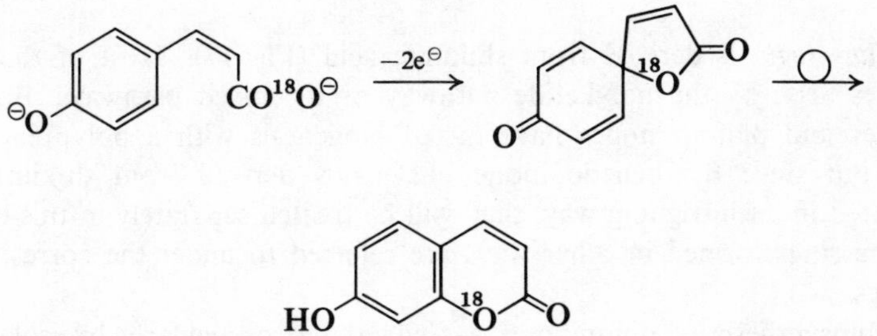

(57)

A series of furanocoumarins are known. The two carbon chain is of isoprenoid origin and formed by cleavage of a three carbon fragment. Feeding studies have established the pathway shown in Fig. 16.

3.6 Quinones

The quinones derive their name from the simplest member, *p*-benzoquinone, obtained 1838 by Woskresensky as an oxidation product of quinic acid. They structurally encompass pigments, antibiotics, vitamin K, coenzymes (ubiquinones and plastoquinones). The last mentioned quinones function in metabolism as one-electron transfer agents by virtue of their property to form reversibly rather stable semiquinone radicals on reduction (58).

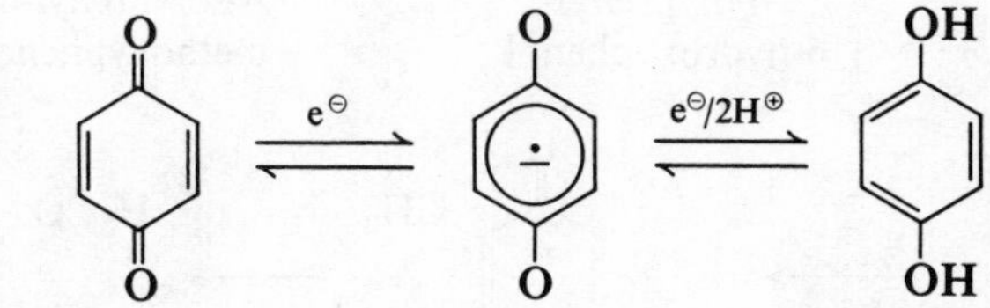

(58)

Benzoquinone Semiquinone Hydroquinone

Quinone biosynthesis shows a very diversified picture. It often differs in moulds and higher plants and the gross structure of a compound often gives poor guidance as to the origin of the compound. Plumbagin and 7-methyljuglone in *Plumbago europaea* come from the polyketide pathway, but juglone

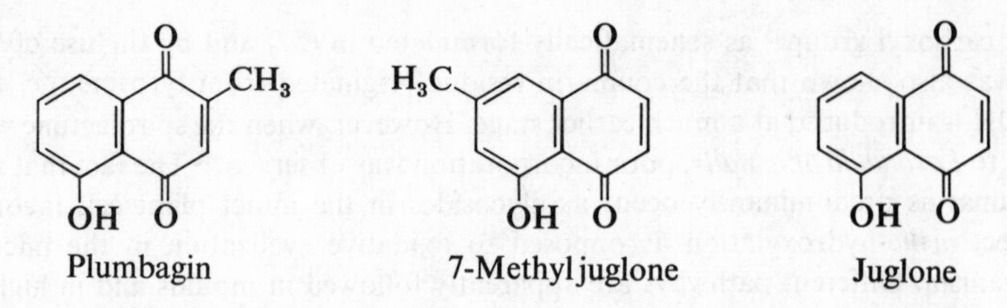

Fig. 17 Compounds of similar structure but of different biogenesis

in *Juglans regia* is derived from shikimic acid (Fig. 17). Most of the higher quinones arise by the polyketide pathway or by mixed pathways. Both ubiquinones and platoquinones have mixed biogenesis with a polyprenoid side chain, but since the benzoquinone nucleus is derived from shikimic acid, elaborated in an intriguing way, they will be treated separately in this chapter. Quinone rings formed in other ways are referred to under the corresponding headings.

The biosynthesis of ubiquinone was elucidated particularly by isolation of

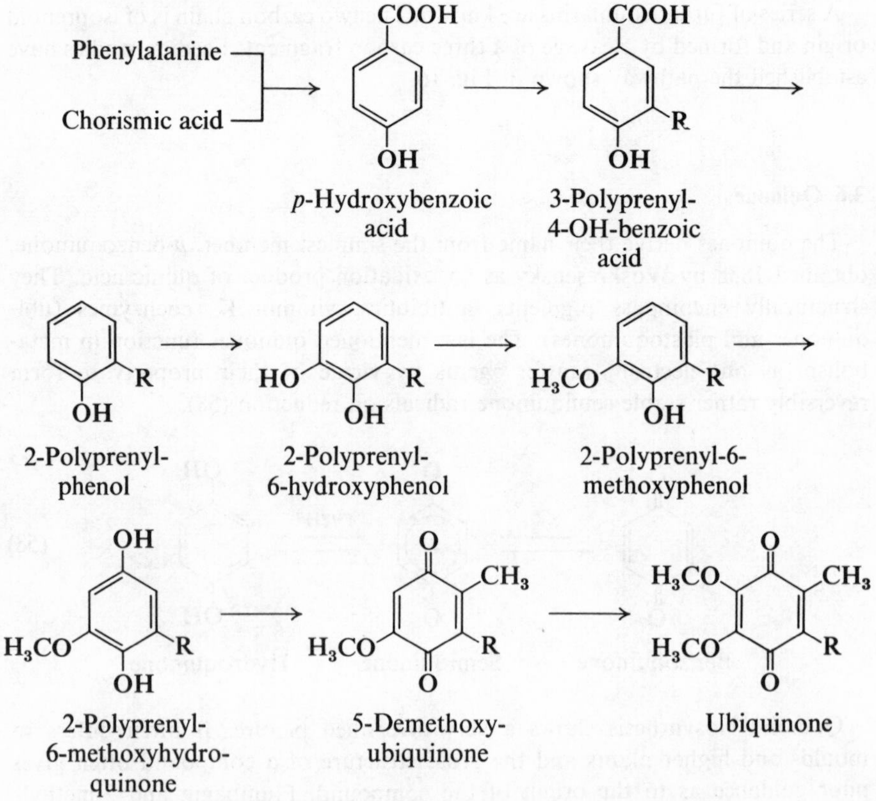

Phenylalanine
Chorismic acid
p-Hydroxybenzoic acid
3-Polyprenyl-4-OH-benzoic acid
2-Polyprenylphenol
2-Polyprenyl-6-hydroxyphenol
2-Polyprenyl-6-methoxyphenol
2-Polyprenyl-6-methoxyhydroquinone
5-Demethoxyubiquinone
Ubiquinone

Fig. 18 Biosynthesis of ubiquinones. *R* = polyprenyl chain

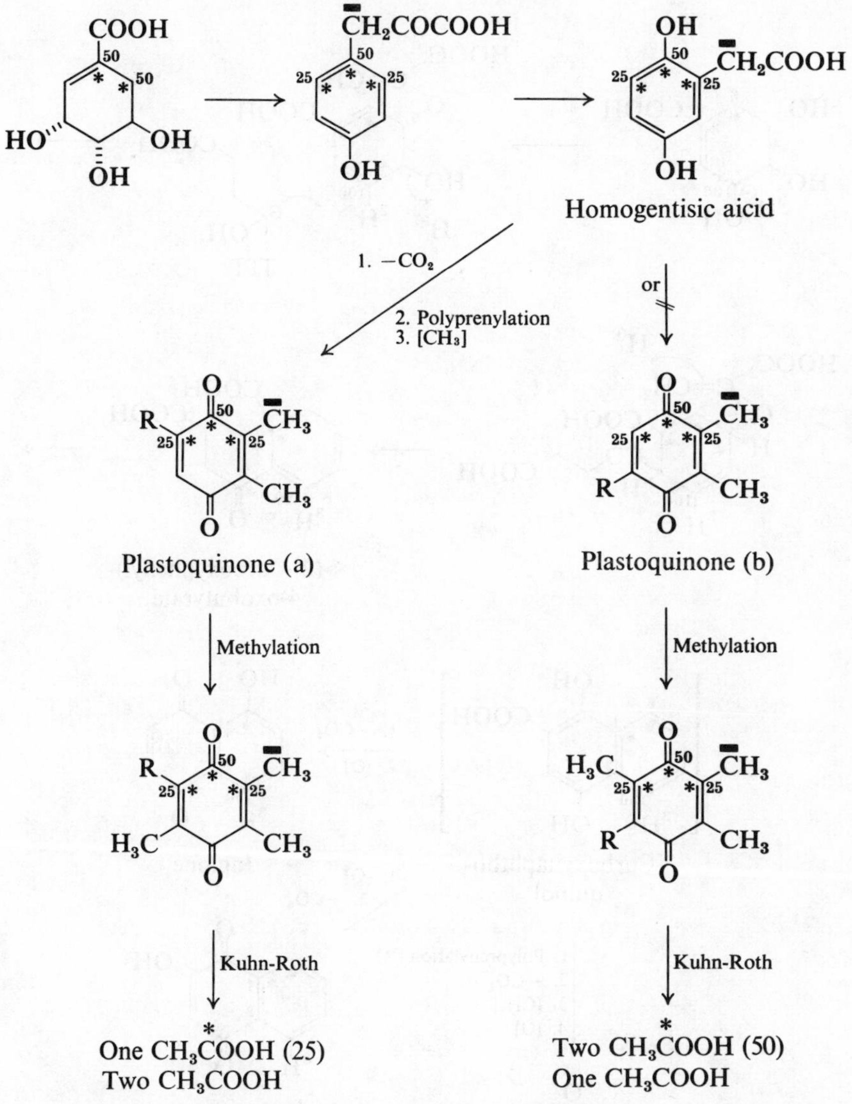

Homogentisic aicid

1. $-CO_2$

2. Polyprenylation

3. $[CH_3]$

or

Plastoquinone (a)

Plastoquinone (b)

Methylation

Methylation

Kuhn-Roth

Kuhn-Roth

One $CH_3\overset{*}{C}OOH$ (25)
Two CH_3COOH

Two $CH_3\overset{*}{C}OOH$ (50)
One CH_3COOH

Fig. 19 Biosynthesis and degradation of plastoquinones. R = polyprenyl

metabolites in the photosynthetic bacterium *Rhodospirillum rubrum*[67] and by the use of mutants of *Escherichia coli.*[68] The intermediates accumulated by the different mutants were isolated and identified by mass spectrometry and NMR spectrometry. *p*-Hydroxybenzoic acid formed by elimination of pyruvic acid from chorismic acid in bacteria or by degradation of phenylalanine in plants and mammals, is alkylated by polyprenyl phosphate (Fig. 18). The phenolic acid is decarboxylated to polyprenylphenol, hydroxylated and *O*-methylated in the 6-position. At this stage *p*-hydroxylation occurs; *cf.* the biosynthesis of

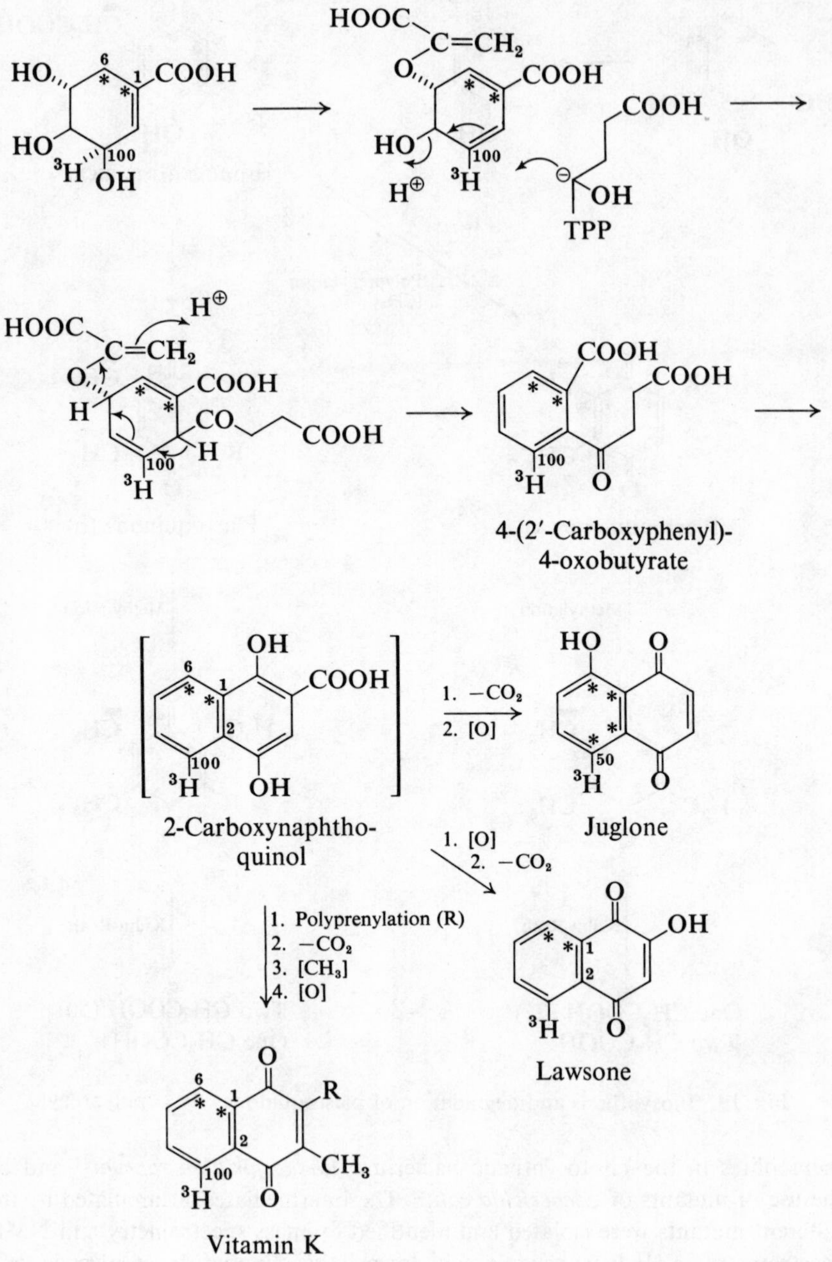

4-(2′-Carboxyphenyl)-4-oxobutyrate

2-Carboxynaphthoquinol

1. $-CO_2$
2. [O]

Juglone

1. [O]
2. $-CO_2$

Lawsone

1. Polyprenylation (R)
2. $-CO_2$
3. $[CH_3]$
4. [O]

Vitamin K

Fig. 20 A possible biosynthesis of vitamin K, lawsone and juglone

arbutin in *Salix*, where oxidation to hydroquinone either is coincident with decarboxylation or immediately follows decarboxylation of *p*-hydroxybenzoic acid. The hydroxylation involves molecular oxygen as shown by use of $^{18}O_2$.[69] The *C*- and *O*-methyls are derived from methionine.[70] The structurally related plastoquinones have a somewhat different history. They derive from tyrosine or *p*-hydroxypyruvate which oxidatively rearrange to homogentisate followed by polyprenylation, methylation, and decarboxylation (Fig. 19). The mechanism of this rather unusual rearrangement is discussed in section 3.3. The order of events is not known with certainty, but toluquinol and gentisyl alcohol are not intermediates. The exact orientation of the two methyl groups with respect to the polyprenyl chain was solved by the use of 1,6-^{14}C shikimic acid.[71] The plastoquinone was completely methylated and degraded according to Kuhn–Roth. For substitution pattern (a) 25 per cent of the activity should be recovered as acetic acid and for substitution pattern (b) 50 per cent. Finally, if both routes (a) and (b) are followed, we expect 37.5 per cent recovery. The measured activity was 23 per cent indicating prenylation in the *meta* position to the α-carbon of homogentisate. It was shown by use of C^2H_3 labelling that one of the methyls originated from methionine.

In vitamins K (menaquinones) all seven carbon atoms of shikimic acid are retained (Fig. 20). The remaining three carbon atoms originate from the three

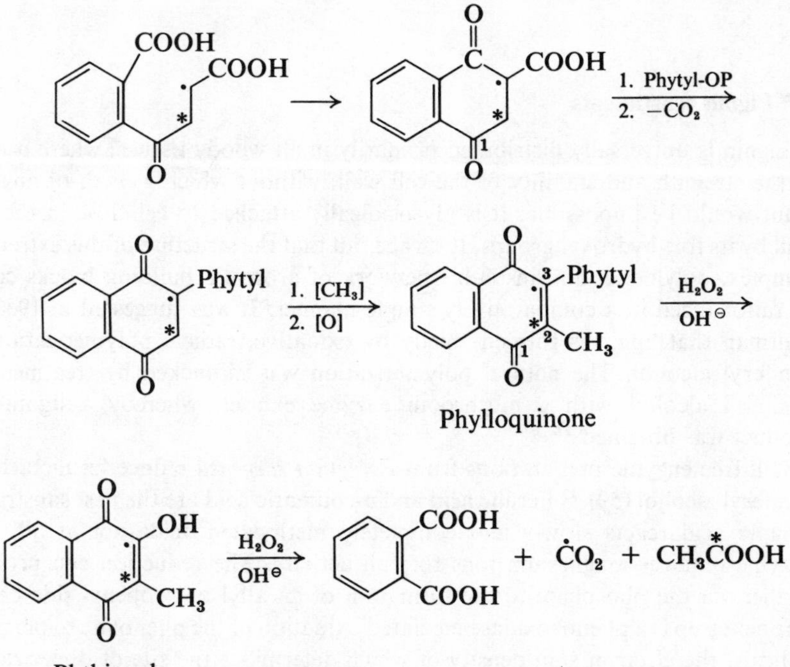

Fig. 21 Biosynthesis and degradation of phylloquinone

central carbons in glutamic acid or 2-ketoglutaric acid.[72] It was also observed that 4-(2′-carboxyphenyl)-4-oxobutyrate was efficiently incorporated, but the likely intermediate 2-carboxynaphthoquinol was poorly incorporated. Experiments with 1,6-^{14}C or 3-^{3}H labelled shikimic acid have shown that the ring junction occurs at $C^{1,2}$ in vitamins K, juglone, and lawsone in both plants and bacteria.[73] The observed activities obtained by degradation show that the biosynthesis of juglone involves a symmetrical intermediate, probably naphthoquinone, whereas no such intermediate appears on the route to lawsone and vitamins K. Hence the hydroxylation and prenylation takes place at the position of the carboxyl group. However, the related 2-carbomethoxy-3-prenyl-1,4-naphthoquinone has been isolated from *Galium mollugo* showing that *ortho*-prenylation takes place here[74] (*cf.* biosynthesis of alizarin, section 4.10). Much paper chemistry has accumulated to explain the mechanism of formation of 4-(2′-carboxyphenyl)-4-oxobutyrate. Both pyridoxal phosphate and thiamine pyrophosphate mediated condensations have been invoked. A plausible mechanism is depicted in Fig. 20.

In a related investigation[75] with maize seedlings, *Zea mays*, the labels of 2-^{14}C-4-(2′-carboxyphenyl)-4-oxobutyrate and 3-^{14}C-4-(2′-carboxyphenyl)-4-oxobutyrate were incorporated at C^3 and C^2, respectively, of phylloquinone showing asymmetric alkylations (Fig. 21). The phytyl group is introduced at the position of the carboxyl group.

3.7 Lignin constituents

Lignin is universally distributed primarily in all woody tissues, where it adds to the strength and stability of the cell wall, without which growth of any tall plant would be impossible. It is glycosidically attached to cellulose in the cell wall by its free hydroxyl groups. It turned out that the structure of this extremely complex, polymeric, and insoluble network of aromatic building blocks could be rationalized in a comparatively simple manner. It was suggested in 1933 by Erdtman that lignin is formed chiefly by oxidative, radical polymerization of coniferyl alcohol. The natural polymerization was mimicked by treatment of coniferyl alcohol with a mushroom enzyme extract, whereby a lignin-like product was obtained.[76]

Cell-free enzyme preparations from *Forsythia suspensa* reduce ferulic acid to coniferyl alcohol (59).[77] Ferulic acid and *p*-coumaric acid are the best substrates. Sinapic acid reacts slowly and completely methylated acids not at all. The enzyme requires magnesium ions for full activity. The reduction can proceed further via the phosphate to the formation of an allyl or propenyl side chain. The next step is a phenol oxidase mediated oxidation of the phenolate to phenoxy radicals, the electron spin density of which determines the site of dimerization or polymerization. Using the resonance formalism, one can predict the active sites (Fig. 22). This is an example of a general type of reaction called phenol

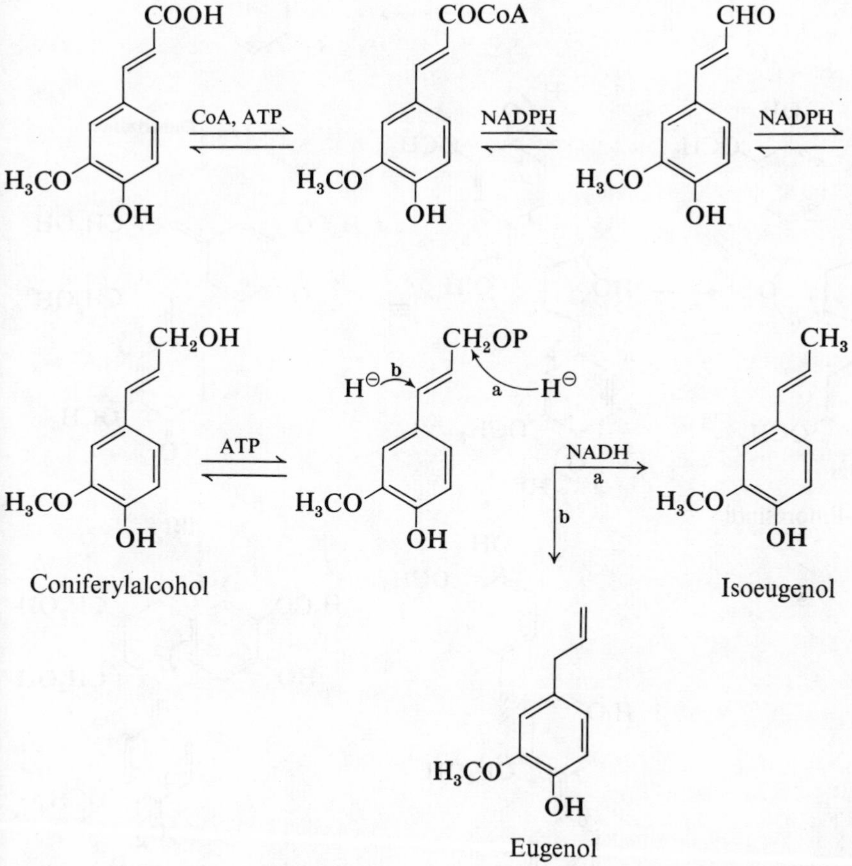

Coniferylalcohol

Isoeugenol

(59)

Eugenol

oxidation which we shall encounter on several occasions. The radicals can pair at several positions and if we first consider the presence of several precursors and second the different possibilities for radical coupling of the building blocks in the course of polymerization, we arrive at an unlimited number of slightly different lignins. A great number of dimers have been isolated and identified

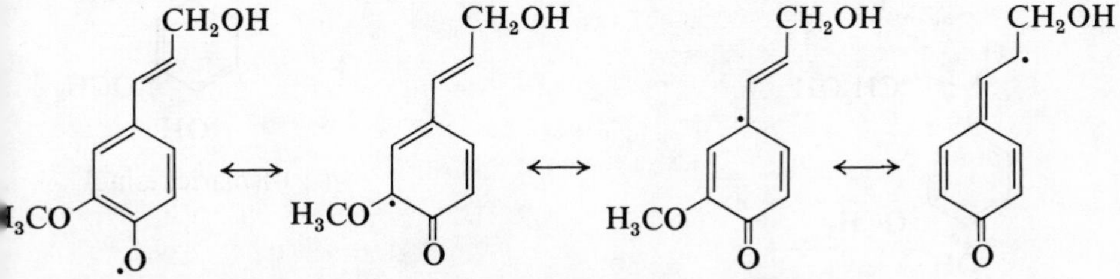

Fig. 22 Resonance structures of the coniferyl radical

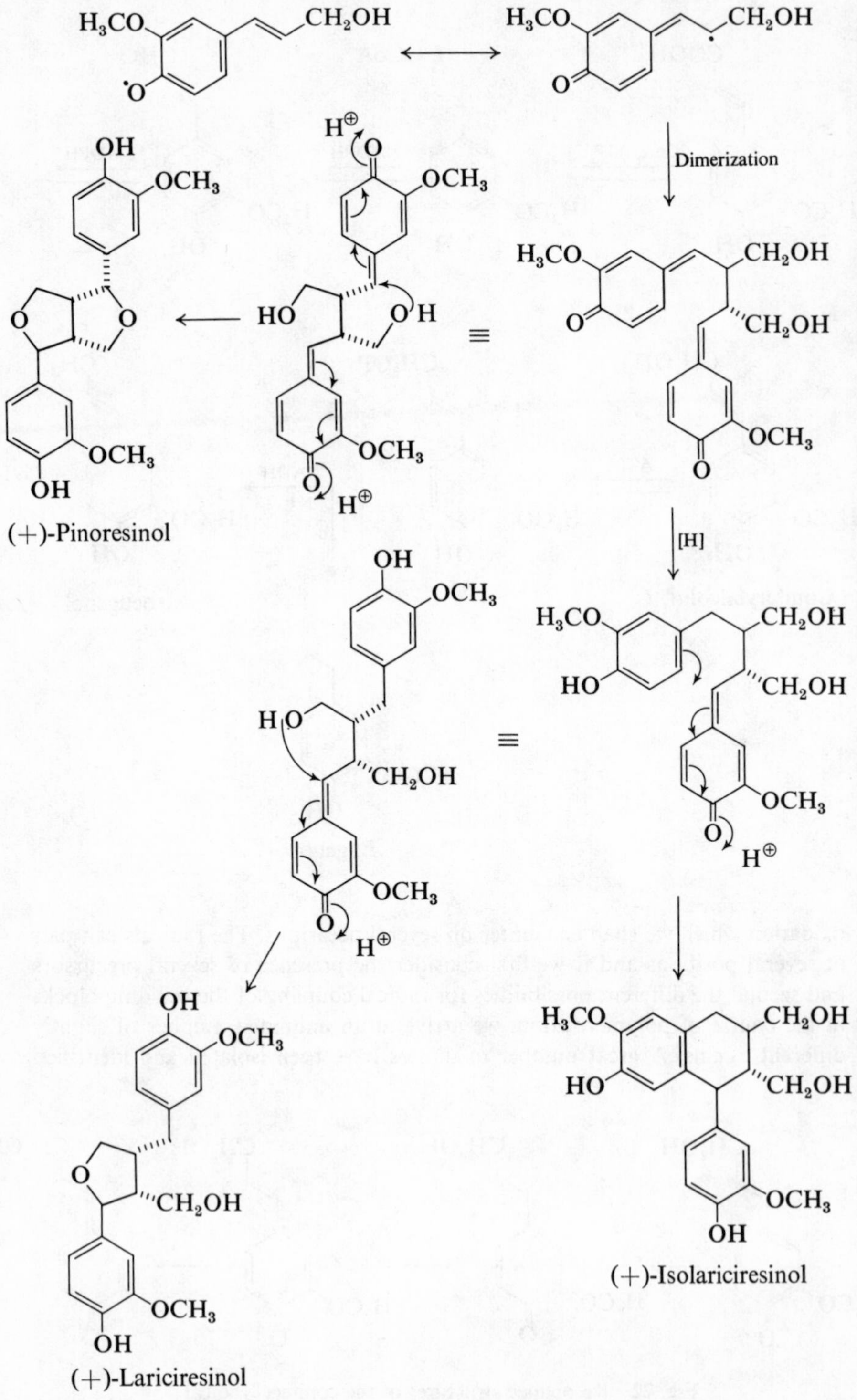

H_3CO CH_2OH $\dot{O}$ ⟷ H_3CO $\dot{}CH_2OH$ O

Dimerization

$H^{\oplus}$ O OCH_3 HO O H ≡ H_3CO CH_2OH O CH_2OH OCH_3 O $H^{\oplus}$ OCH_3

OH OCH_3 O O OCH_3 OH

(+)-Pinoresinol

[H]

OH OCH_3 H O CH_2OH OCH_3 O $H^{\oplus}$ ≡ H_3CO CH_2OH HO CH_2OH OCH_3 O $H^{\oplus}$

OH OCH_3 O CH_2OH OCH_3 OH

(+)-Lariciresinol

H_3CO CH_2OH HO CH_2OH OCH_3 OH

(+)-Isolariciresinol

Fig. 23 Biosynthesis of some lignans from coniferyl alcohol

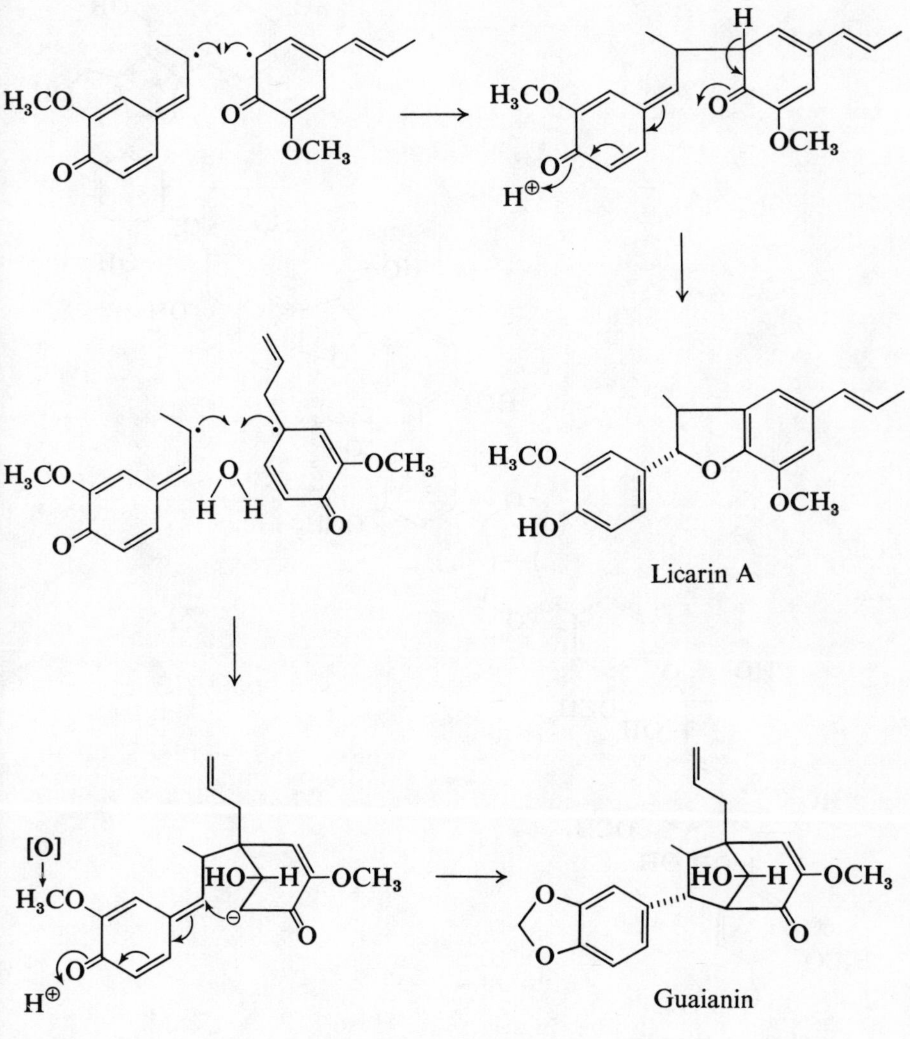

Licarin A

Guaianin

Fig. 24 Synthesis of lignans from eugenol and isoeugenol

from nature. Radical combination is probably the preferred reaction and not addition of one radical to the double bond of another cinnamyl alcohol. In Figs 23 and 24 the general principle of dimer formation is illustrated, and in Fig. 25 a lignin model is depicted.[78] By the term lignans we understand dimers formed by reactions involving the side chains.

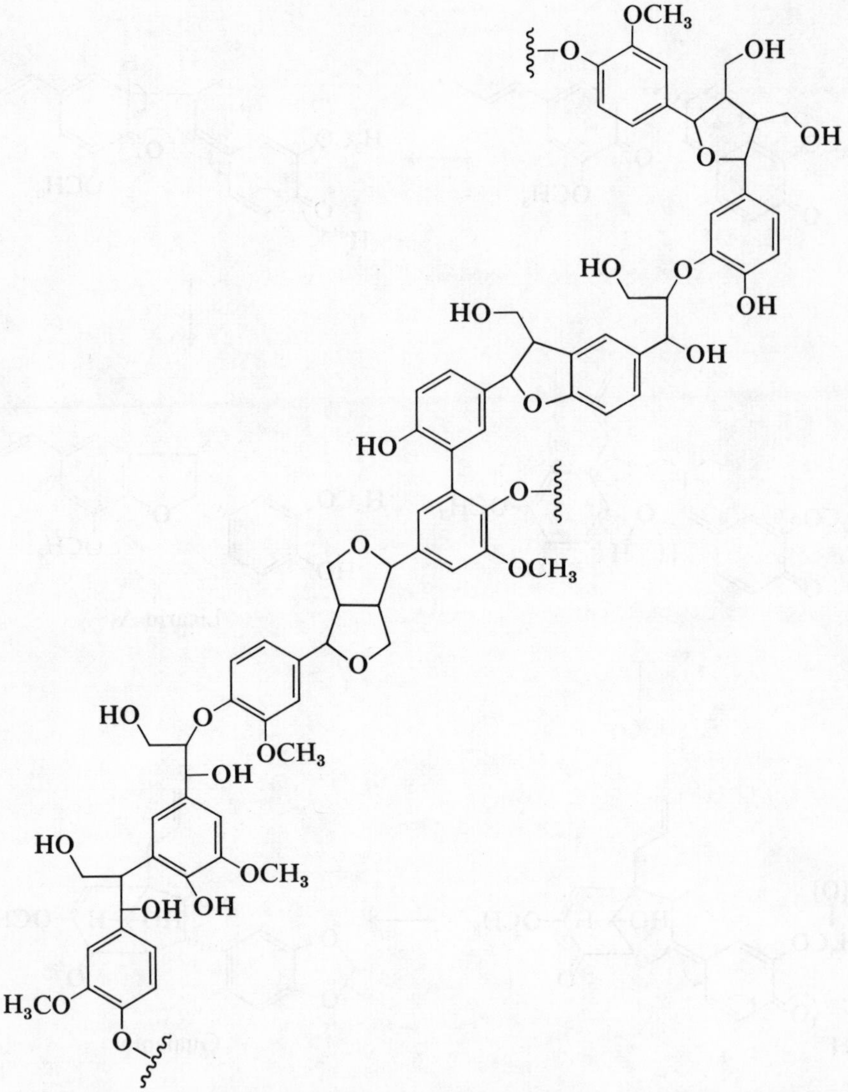

Fig. 25 Hypothetical lignin model

3.8 Total synthesis

Synthesis is no longer practised for the purpose of proving or confirming structures. This is a cumbersome procedure in our days of advanced spectroscopy. Nevertheless syntheses of all kinds of naturally occurring compounds are constantly executed more for the purpose of exercising the noble art of total synthesis. Woodward has with brilliance been able to solve the immense problems surrounding the synthesis of such complicated molecules as strychnine,

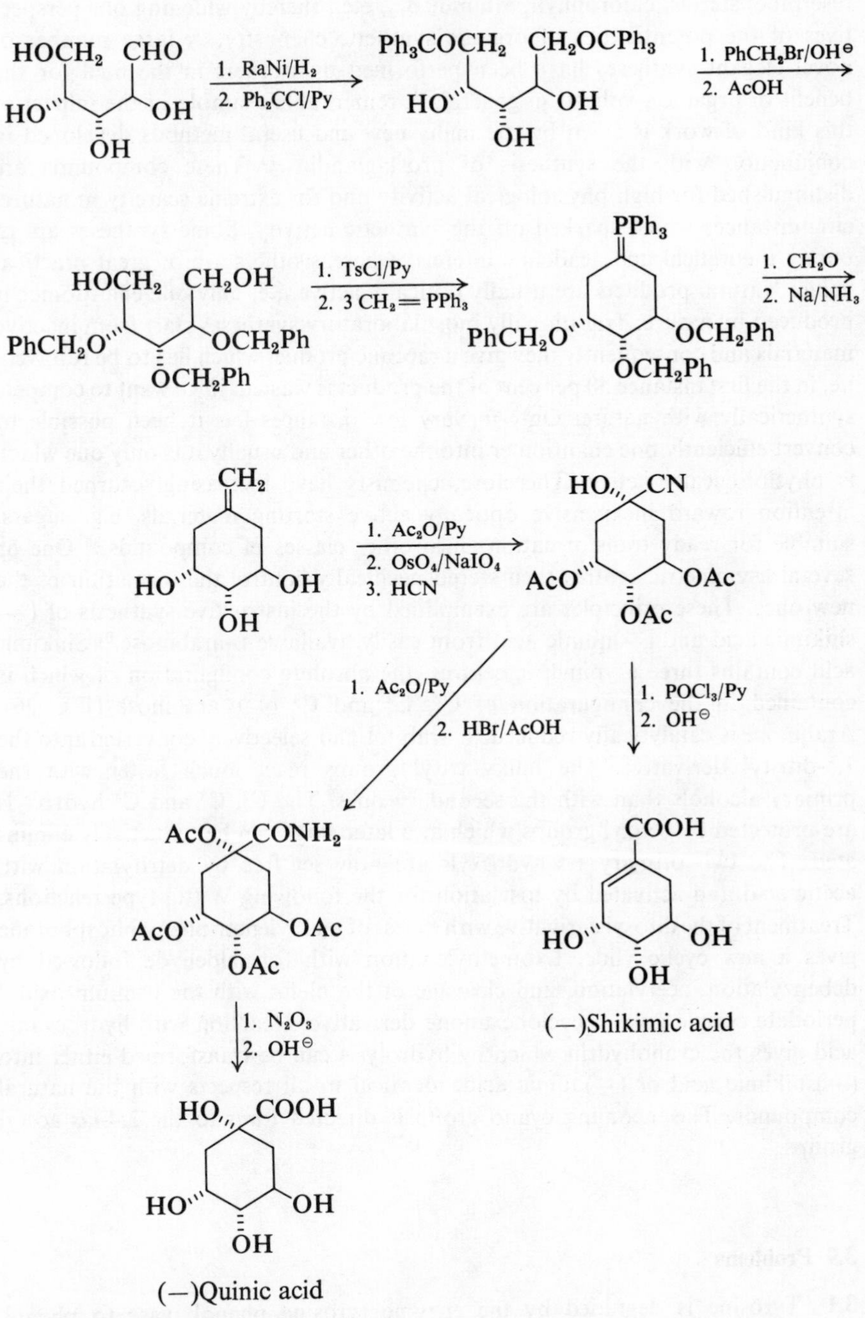

$HOCH_2$ CHO, HO, OH, OH → 1. RaNi/H_2 2. Ph_3CCl/Py → Ph_3COCH_2 CH_2OCPh_3, HO, OH, OH → 1. $PhCH_2Br/OH^{\ominus}$ 2. AcOH

$HOCH_2$ CH_2OH, $PhCH_2O$, OCH_2Ph, OCH_2Ph → 1. TsCl/Py 2. 3 $CH_2{=}PPh_3$ → PPh_3, $PhCH_2O$, OCH_2Ph, OCH_2Ph → 1. CH_2O 2. Na/NH_3

CH_2, HO, OH, OH → 1. Ac_2O/Py 2. $OsO_4/NaIO_4$ 3. HCN → HO, CN, AcO, OAc, OAc

→ 1. Ac_2O/Py 2. HBr/AcOH → AcO, $CONH_2$, AcO, OAc, OAc → 1. N_2O_3 2. $OH^{\ominus}$ → HO, COOH, HO, OH, OH

(−)Quinic acid

→ 1. $POCl_3$/Py 2. $OH^{\ominus}$ → COOH, HO, OH, OH

(−)Shikimic acid

Fig. 26 Total synthesis of (−)quinic acid and (−)shikimic acid

reserpine, sterols, chlorophyll, vitamin B_{12}, etc., thereby widening our perspectives of the potentialities of organic synthetic chemistry. A large number of novel elegant syntheses have been performed by workers in the field for the benefit of organic synthesis in general. A remarkable example of the impact of this kind of work is given by the many new and useful methods developed in conjunction with the synthesis of prostaglandins.[79] These compounds are distinguished for high physiological activity and for extreme scarcity in nature, circumstances which sparked off the synthetic activity. Some syntheses are of purely theoretical and academic interest, whereas others are of great practical value. Natural products are usually optically active, i.e. only one enantiomer is produced by nature. Traditionally most laboratory syntheses start from inactive materials and consequently they give a racemic product which has to be resolved, i.e. in the first instance 50 per cent of the product is wasted, if we want to compete synthetically with nature. Only in very few instances has it been possible to convert efficiently one enantiomer into the other and usually it is only one which is physiologically active. Therefore, chemists have increasingly turned their attention toward inexpensive optically active starting materials, e.g. sugars, suitable for ready transformations into other classes of compounds.[80] One or several asymmetric centres then stereospecifically control the formation of the new ones. These principles are exemplified by the instructive synthesis of (—) shikimic acid and (—)quinic acid from easily available D-arabinose.[81] Shikimic acid contains three asymmetric centres, the absolute configuration of which is contained in the configuration of C^2, C^3 and C^4 of D-arabinose (Fig. 26). Arabinose is catalytically reduced to arabitol and selectively converted into the 1,5-ditrityl derivative. The bulky trityl groups react much faster with the primary alcohols than with the secondary ones. The C^2, C^3 and C^4 hydroxyls are protected by benzyl groups which at a later stage can be reductively eliminated. The two primary 1,5-hydroxyls are now set free by detritylation with acetic acid and activated by tosylation for the following Wittig-type reactions. Treatment of the ditosyl derivative with excess of methylenetriphenylphosphorane gives a new cyclic ylide. Exomethylenation with formaldehyde followed by debenzylation, acetylation, and cleavage of the olefin with the osmium oxide/periodate couple gives a cyclohexanone derivative. Reaction with hydrocyanic acid gives the cyanohydrin which by hydrolysis can be transformed either into (—)shikimic acid or (—)quinic acid, identical in all respects with the natural compounds. The incoming cyano group is directed *trans* to the 3,4-*cis* acetyl groups.

3.9 Problems

3.1 Tyrosine is degraded by the enzyme tyrosine phenol lyase to phenol, pyruvic acid, and ammonia. Formulate a mechanism for this fragmentation under the supposition that pyridoxal serves as cofactor. The reactions are actually shown to be reversible. (Yamada, H., Kumagai, H., Kushima, N.,

Torrii, H., Enei, H. and Okumura, S. *Biochem. Biophys. Res. Commun.* **46** (1972) 370).

3.2 Formulate some intermediate steps for the biosynthesis of betanidin, the red pigment in beetroot, from two molecules of tyrosine. Dopa is readily incorporated and by use of ^{15}N-tyrosine it is demonstrated that the intact amino acid is incorporated in both fragments. The labels of 1-^{14}C-3,5-$^{3}H_2$-^{15}N-tyrosine show up at the marked positions. Monodecarboxylation of betanidin leads to loss of labelled C^{19} and migration of $\Delta^{17,18}$ to $\Delta^{14,15}$. (Fischer, N. and Dreiding, A. S. *Helv. Chim. Acta* **55** (1972) 649; Liebisch, H. W., Matscheiner, B. and Schütte, H. R. *Z. Pflanzenphysiol.* **61** (1969) 269).

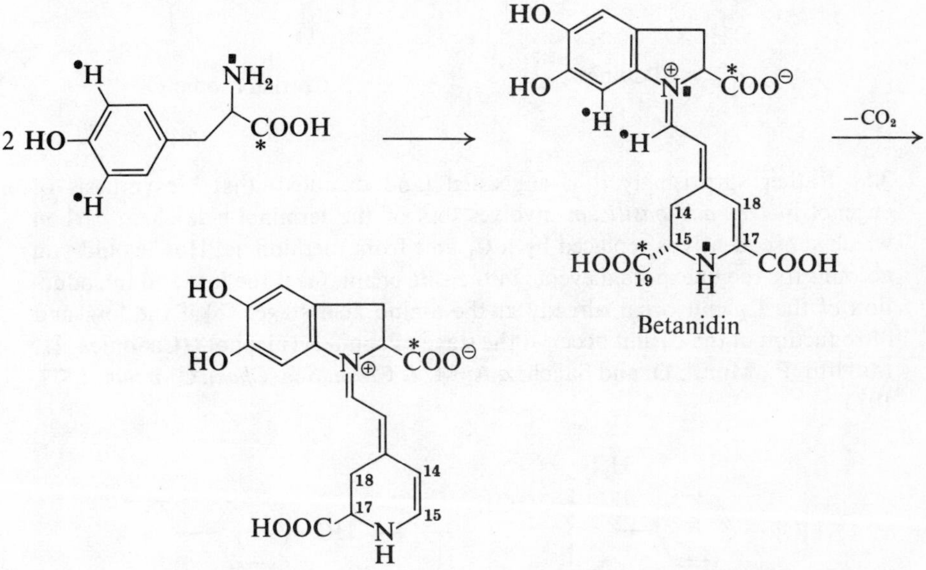

Betanidin

3.3 Several lignans possess the general structure I with different configurations. Another lignan, futoenone, was found to have the spiroenone structure. Rationalize the biosynthetic formation of these compounds. (Gottlieb, O. R. *Prog. Chem. Org. Nat. Prods.* **35** (1978) 1).

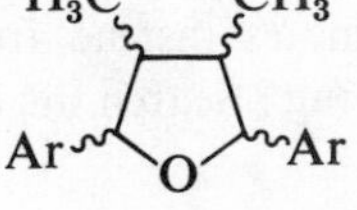

I

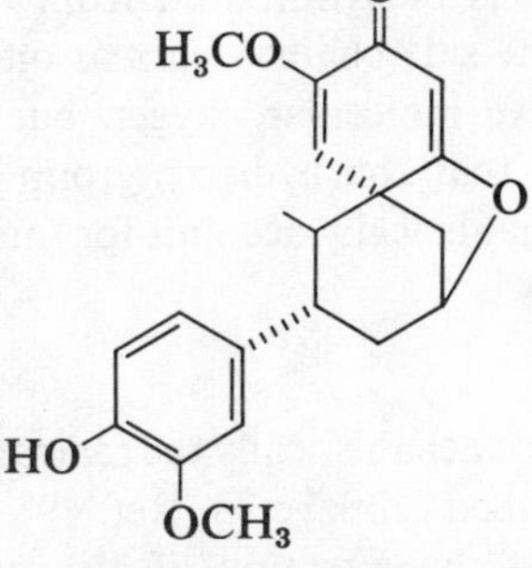

Futoenone

3.4 Suggest biosynthetic pathways for the plant naphthoquinones alkannin and cordiachrome C. Geranyl phosphate is a precursor. (Schmid, H. V. and Zenk, M. H. *Tetrahedron Lett.* **1971**, 4151. Moir, M. and Thomson, R. H. *J. Chem. Soc. Perkin I*, **1973**, 1352, 1556; Inouye, H., Ueda, S., Inoue, K. and Matsumura, H. *Phytochemistry* **18** (1979) 1301).

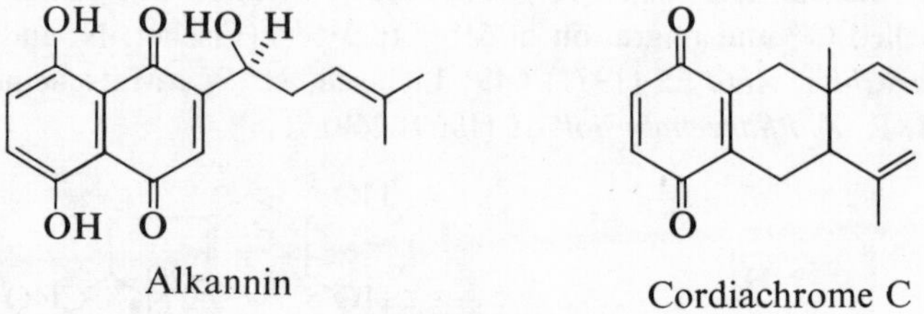

OH O HO H

OH O

Alkannin

O

O

Cordiachrome C

3.5 Rather surprisingly it is suggested (and disputed) that biosynthesis of eugenol in *Ocimum basilicum* involves loss of the terminal side chain carbon which subsequently is replaced by a C_1 unit from methionine. How would you account for the unexpected events in the side chain, (a) if the loss and introduction of the C_1 unit occur already at the amino acid stage; (b) if the loss and introduction of the C_1 unit occur at the stage of coniferyl alcohol? (Canonica, L., Manitto, P., Monti, D. and Sanchez, A. M. *J. Chem. Soc. Chem. Commun.* **1979**, 1073).

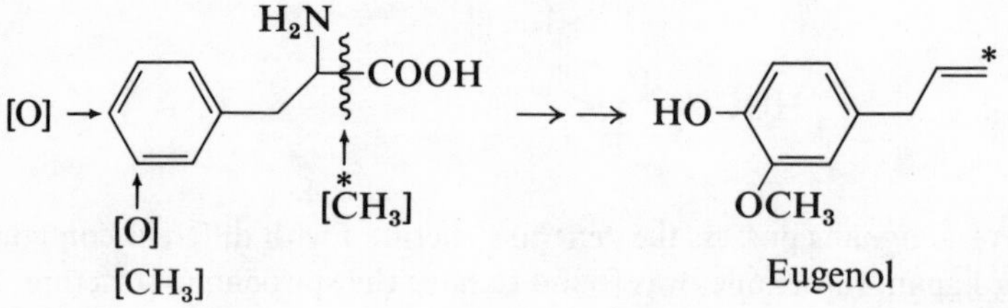

H_2N

COOH

[O] →

→ → HO

*

↑

*

[CH_3]

↑

[O]

[CH_3]

OCH_3

Eugenol

3.6 Adrenaline is biosynthesized from tyrosine (Fig. 9, section 3.4). Oxygen insertion into the side chain occurs at one stage. Many of these insertions are known to involve molecular oxygen but let us suppose that an investigation gives the result that the hydroxy group originates instead from water. How would you mechanistically account for this finding? Search the literature for the correct answer.

3.7 Formulate mechanistically the rearrangements of the 1-substituted benzene-1,2-oxides described on page 78 (Ref. 49). Suggest a suitable experiment for the determination of the direction of cleavage of the oxiran ring in 1-methoxy-carbonylbenzene-1,2-oxide.

Bibliography

1. Sprinson, D. B. *Adv. Carbohydr. Chem.* **15** (1961) 235.
2. DeLeo, A. B. and Sprinson, D. B. *Biochem. Biophys. Res. Commun.* **32** (1968) 373.
3. Floss, H. G., Onderka, D. K. and Carroll, M. *J. Biol. Chem.* **247** (1972) 736.
4. Sprinson, D. B., Rothschild, J. and Sprecher, M. *J. Biol. Chem.* **238** (1963) 3170.
5. Adlersberg, H. and Sprinson, D. B. *Biochemistry* **3** (1964) 1855.
6. Cain, R. B., Bilton, B. J. and Darrah, J. A. *Biochem. J.* **108** (1968) 797.
7. Hanson, K. B. and Rose, I. A. *Proc. Nat. Acad. Sci.* **50** (1963) 981.
8. Turner, M. J., Smith, B. W. and Haslam, E. *J. Chem. Soc. Perkin I* **1975,** 52.
9. Butler, J. R., Alworth, W. L. and Nugent, M. J. *J. Am. Chem. Soc.* **96** (1974) 1617.
10. Bondinell, W. E., Vnek, J., Knowles, P. F., Sprecher, M. and Sprinson, D. B. *J. Biol. Chem.* **246** (1971) 6191.
11. Zemell, R. J. and Anwar, R. A. *J. Biol. Chem.* **250** (1975) 4959.
12. Hill, R. K. and Newcome, G. R. *J. Am. Chem. Soc.* **91** (1969) 5893.
13. Patel, N., Pierson, D. L. and Jensen, R. A. *J. Biol. Chem.* **252** (1977) 5839.
14. Plieninger, H. *Angew. Chem. Int. Ed.* **1** (1962) 367.
15. Danishefsky, S. and Hirama, M. *Tetrahedron Lett.* **1977,** 4565.
16. McCormick, J. R. D., Reichenthal, J., Kirsch, U. and Sjolander, N. O. *J. Am. Chem. Soc.* **84** (1962) 3711.
17. Young, I. G., Jackman, L. M. and Gibson, F. *Biochim. Biophys. Acta* **177** (1969) 381.
18. Lingens, F. *Angew. Chem. Int. Ed.* **7** (1968) 350.
19. Tatum, E. L. and Bonner, D. *Proc. Nat. Acad. Sci.* **30** (1944) 30.
20. Metzler, D. A., Ikawa, M. and Snell, E. E. *J. Am. Chem. Soc.* **76** (1954) 648.
21. Larsen, P. O., Onderka, D. K. and Floss, H. G. *Biochem. Biophys. Acta* **381** (1975) 397.
22. Herzberg, G. *Nature* **133** (1934) 759.
23. Merkel, P. B. and Kearns, D. R. *J. Am. Chem. Soc.* **94** (1972) 7244.
24. Wasserman, H. H. and Murray, R. W. *Singlet Oxygen.* Academic Press, New York, 1979.
25. Schenck, G. O. and Ziegler, K. *Naturwissenschaften* **32** (1945) 157.
26. Wood, P. M. *FEBS Letters* **22** (1974) 44.
27. Behar, D., Czapski, G., Rabani, J., Dorfman, L. M. and Schwarc, H. A. *J. Phys. Chem.* **74** (1970) 3209.
28. Divisek, J. and Kastening, B. *J. Electroanal. Chem.* **65** (1975) 603.
29. Rigo, A., Tomat, R. and Rotilio, G. *J. Electroanal. Chem.* **57** (1974) 291.
30. Ross, F. and Ross, A. B. *Selected Specific Rates of Reactions of Transients from Water in Aqueous Solution*, Nat. Bur. Stands., Washington, June 1977.
31. Afanas'ev, J. B. *Russian Chem. Revs.* **48** (1979) 527. Eng. Transl.
32. Butler, J., Jayson, C. G. and Swallow, A. J. *Biochem. Biophys. Acta* **408** (1975) 215.
33. Golden, D. M. and Benson, S. W. *Chem. Revs.* **69** (1969) 125.
34. Howard, J. A. in *Free Radicals II*, Kochi, J. (Ed.), John Wiley, New York, 1973, p. 3.
35. Fridovich, I. and Handler, P. *J. Biol. Chem.* **233** (1958) 1578.
36. Haber, F. and Weiss, J. *Proc. Roy. Soc.* **A 412** (1934) 332.
37. Weeks, J. L. and Rabani, J. *J. Phys. Chem.* **70** (1966) 2100.
38. Dorfman, L. M. and Adams, G. E. *Reactivity of the Hydroxyl Radical in Aqueous Solution*, Nat. Bur. Stands., Washington, June 1973.
39. Pryor, W. A., *Free Radicals*, McGraw-Hill, New York, 1966, p. 287.
40. Sligar, S. G., Kennedy, K. A. and Pearson, D. C. *Proc. Natl. Acad. Sci.* **77** (1980) 1240.
41. Hamberg, M., Svensson, J., Wakabayashi, T. and Samuelson, B. *Proc. Natl. Acad. Sci.* **71** (1974) 345.

42. deGroot, J. J. M. C., Garssen, G. J., Vliegenhart, J. F. G. and Boldingh, J. *Biochem. Biophys. Acta* **326** (1973) 279.
43. Porter, N. A., Roe, A. N. and McPhail, A. T. *J. Am. Chem. Soc.* **102** (1980) 7576.
44. Lindblad, B., Lindstedt, G. and Lindstedt, S. *J. Am. Chem. Soc.* **92** (1970) 7446.
45. Jerina, D. M., Daly, J. W., Witkop, B., Zalzman-Nirenberg, P. and Udenfriend, S. *J. Am. Chem. Soc.* **90** (1968) 6525.
46. Bu'Lock, J. D. and Ryles, A. P. *Chem. Commun.* **1970,** 1404.
47. Guroff, G., Reifsnyder, A. and Daly, J. W. *Biochem. Biophys. Res. Commun.* **24** (1966) 720.
48. Nagatsu, T., Lewitt, M. and Udenfriend, S. *J. Biol. Chem.* **239** (1964) 2910.
49. Chao, H. S. J. and Berchtold, C. H. *J. Am. Chem. Soc.* **103** (1981) 898.
50. Boyd, D. R., Daly, J. W. and Jerina, D. M. *Biochemistry* **11** (1972) 1961.
51. Muto, S. and Bruice, T. C. *J. Am. Chem. Soc.* **102** (1980) 7559.
52. (a) Hamilton, G. in *Molecular Mechanisms of Oxygen Activation*, Hayashi, O. (Ed.), Academic Press, New York, 1974, p. 405; (b) Groves, J. T., Haushalter, R. C., Nakamura, M., Nemo, T. E. and Evans, B. J. *J. Am. Chem. Soc.* **103** (1981) 2884; Chevrier, B., Weiss, R., Lange, M., Chottard, J.-C. and Mansuy, D. *J. Am. Chem. Soc.* **103** (1981) 2899.
53. Hanson, K. R., Wightman, R. H., Staunton, J. and Battersby, A. R. *Chem. Commun.* **1971**, 185.
54. Hanson, K. R. and Havir, E. A. *Arch. Biochem. Biophys.* **141** (1970) 1.
55. Geissman, T. A. and Hinreiner, E. *Bot. Rev.* **18** (1952) 165.
56. Zenk, M. H. and Müller, G. *Z. Naturforsch.* **19B** (1964) 398.
57. Gross, S. R. *J. Biol. Chem.* **233** (1958) 1146.
58. Conn, E. E. and Swain, T. *Chem. Ind.* **1961,** 592.
59. Zenk, M. H. *Z. Naturforsch.* **19B** (1964) 83.
60. Zenk, M. H. *Phytochemistry* **6** (1967) 245.
61. Pridham, J. B. and Saltmarsh, M. J. *Biochem. J.* **87** (1963) 218.
62. Neish, A. C. *Can. J. Bot.* **37** (1959) 1085.
63. Murray, R. D. H., *Prog. Chem. Org. Nat. Prod.* **35** (1978) 199.
64. Brown, S. A. in *Biosynthesis of Aromatic Compounds*, G. Billek (Ed.), Pergamon Press, Oxford, 1969, p. 15.
65. Bunton, C. A., Kenner, G. W., Robinson, M. J. T. and Webster, B. R. *Tetrahedron* **19** (1963) 1001.
66. Austin, D. J. and Meyers, M. B. *Phytochemistry* **4** (1965) 245.
67. Friis, P., Daves Jr., G. D. and Folkers, K. *J. Am. Chem. Soc.* **88** (1966) 4754.
68. Gibson, F. *Biochem. Soc. Trans.* **1** (1973) 317.
69. Uchida, K. and Aida, K., *Biochem. Biophys. Res. Commun.* **46** (1972) 130.
70. Jackman, L. M., O'Brien, I. G., Cox, G. B. and Gibson, F. *Biochem. Biophys. Acta* **141** (1967) 1.
71. Whistance, G. R. and Threlfall, D. R. *Phytochemistry* **10** (1971) 1533.
72. Campbell, I. M. *Tetrahedron Lett.* **1969,** 4777.
73. Leduc, M. M., Dansette, P. M. and Azerad, R. G. *European J. Biochem.* **15** (1970) 428.
74. Heide, L. and Leistner, E. *J. Chem. Soc. Chem. Commun.* **1981,** 334.
75. Hutson, K. G. and Threlfall, D. R. *Phytochemistry* **19** (1980) 535.
76. Freudenberg, K. and Richtzenhain, H. *Chem. Ber.* **76** (1943) 997.
77. Mansell, R. L., Gross, G. G., Stöckigt, J., Franke, H. and Zenk, M. H. *Phytochemistry* **13** (1974) 2427.
78. Harkin, J. M. *Fortschritte Chem. Forsch.* **6** (1966) 100.
79. Caton, M. P. L. *Tetrahedron Reports* 68. *Tetrahedron* **35** (1979) 2705.
80. Fraser-Reid, B. and Anderson, R. C. *Prog. Chem. Org. Nat. Prod.* **39** (1980) 1.
81. Bestman, H. J. and Heid, H. A. *Angew. Chem. Int. Ed.* **10** (1971) 336.

Chapter 4

The polyketide pathway

4.1 Introduction

Acetic acid or its biosynthetic equivalent, acetyl CoA, occupies a central position in the synthesis of natural compounds (see Fig. 7, Chapter 1). Linear Claisen condensation leads to β-keto esters (1a), which either by reduction and repeated condensation give fatty acids or by further direct condensation give polyketides. They, in turn, can cyclize to a vast number of aromatics. Acetyl CoA is also the starting material for terpenoids. In a branched condensation the keto function of acetoacetyl CoA reacts with another acetyl CoA molecule to form β-hydroxy-β-methylglutaryl CoA which is transformed into the 'active' isoprene unit (1b) and ultimately to the terpenoids. This pathway is separately treated in Chapter 5. From the intermediate formation of the β-keto ester, both in the biosynthesis and in the β-oxidative degradation of fatty acids, it may be concluded one process is just the reversal of the other. However, later research showed that they are significantly different. They proceed in different compartments of the cell with different sets of enzyme systems: biosynthesis in the cytosol, β-oxidation in the mitochondria. The presence of citrate was found to stimulate the synthesis, but was not required for the breakdown and unexpectedly carbon

$$CH_3COCH_2COCoA + CH_3COCoA \xrightarrow[\text{branched}]{b} (HO)(H_3C)C(CH_2COOH)(CH_2COCoA) \rightarrow \text{Isoprene unit}$$

(a: attack at the thioester carbon; b: attack at the keto carbon)

$$CH_3COCH_2COCoA + CH_3COCoA \xrightarrow{\text{a linear}} CH_3COCH_2COCH_2COCoA \tag{1}$$

$CH_3COCH_2COCH_2COCoA$ → Cyclization → Aromatics

$CH_3COCH_2COCH_2COCoA$ → Reduction → Fatty acids

dioxide was essential for the synthesis. By isotopic labelling it could be shown that it was not incorporated. Biosynthesis employs NADPH for reduction, whereas oxidation employs $NAD^{\oplus}$ and in biosynthesis (3*R*)-3-hydroxy acids occur as intermediates, whereas (3*S*)-3-hydroxy acids are formed in the β-oxidation sequences. The simple condensation step (1a, which does occur, but is not the common pathway) has thus to be modified. In the subsequent sections a detailed account is given of the metabolism of fatty acids, olefins, acetylenes, some branched derivatives and aromatics.

4.2 Fatty acids, fats

Acetyl CoA is formed in the cell either by degradation of fatty acids, decarboxylation of pyruvic acid obtained via glycolysis or degradation of certain amino acids. The thio group has a twofold effect. It activates the carbonyl function as an electrophile and the methyl group as a nucleophile, i.e. acetyl CoA is considerably more active than an ordinary ester in a Claisen condensation, but still not sufficiently reactive to be utilized by the fatty acid synthetase complex for the purpose. A further activation of the α-carbon is necessary and this is brought about by carboxylation of acetyl CoA to malonyl CoA. The reaction is mediated by ATP and biotin as a carbon dioxide carrier (2,3). The cofactor biotin is anchored by its carboxyl group to the ε-amino group of a lysine residue of the enzyme. Bicarbonate forms in a reversible reaction with ATP a reactive mixed anhydride of carbonic and phosphoric acids which carboxylates biotin to N^1-carboxybiotin. In agreement with (2) the label of $HC^{18}O_3^{\ominus}$ ends up in the inorganic phosphate formed.[1] N^1-carboxybiotin is an unstable intermediate, the existence of which was established by transformation into its methyl ester by diazomethane, identical to authentic N^1-methoxycarbonylbiotin.[2,3] Stereochemical investigations carried out on 2-^{3}H labelled propionyl CoA show that the pro-*R* hydrogen is lost and is replaced by carbon dioxide with retention of configuration.[4] No hydrogen exchange occurs since rate of ^{3}H release is equal to rate of carboxylation. It was also demonstrated by use of chiral acetyl CoA that the carboxylation occurs with retention of configuration but as a result of the small isotope effect of the reaction combined with some enzyme catalysed non-specific hydrogen exchange the discrimination between (*R*) and (*S*) acetate turned out to be low.[5] A concerted cyclic mechanism has been suggested (3). Intuitively one could now be led to believe that the ready formation of a carbanion from malonate is the immediate cause of the condensation (4). However, no isotopic scrambling with the solvent was observed in the methylene group of dideuteriomalonyl CoA, nor was any isotope effect on the rate of condensation observed and this eliminates the suggested mechanism.[6] Incidentally, the lack of hydrogen exchange in this case contrasts with the findings involving the chiral acetyl CoA discussed above and illustrates the problems and difficulties involved in this kind of enzymatic research. One can conclude that the condensation occurs with concomitant exothermic decarboxylation generating the anion, so giving an extra push to the reaction (7). The condensation takes place on a

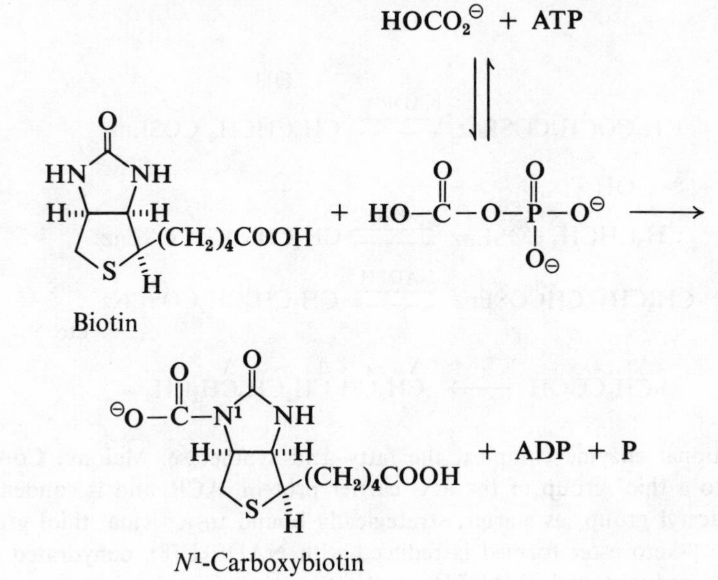

$$HOCO_2^{\ominus} + ATP$$

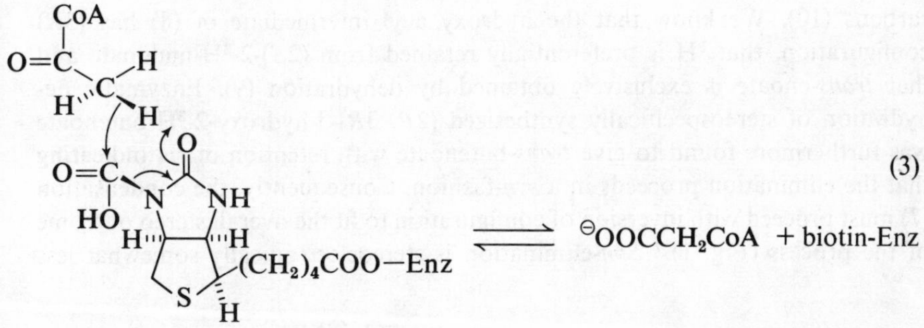

$$\text{Biotin} + HO{-}C(=O){-}O{-}P(=O)(O^{\ominus}){-}O^{\ominus} \longrightarrow \qquad (2)$$

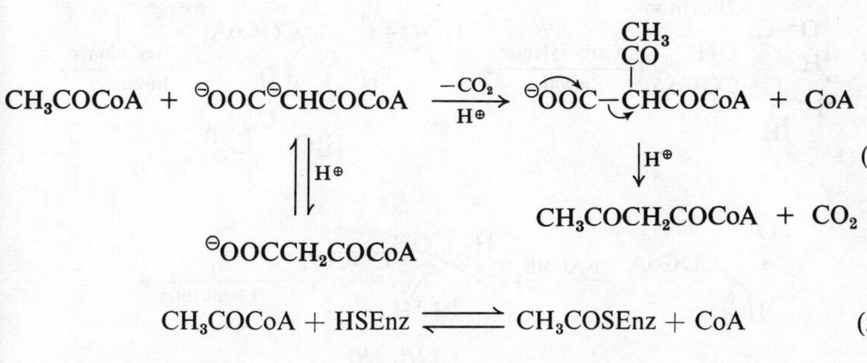

$$N^1\text{-Carboxybiotin} + ADP + P$$

Biotin

N^1-Carboxybiotin

$$\text{CoA–CO–CH}_3 + {}^{\ominus}O{-}CO{-}N^1\text{-biotin–}(CH_2)_4COO{-}Enz \rightleftharpoons {}^{\ominus}OOCCH_2CoA + \text{biotin-Enz} \qquad (3)$$

$$CH_3COCoA + {}^{\ominus}OOC^{\ominus}CHCOCoA \xrightarrow[H^{\oplus}]{-CO_2} {}^{\ominus}OOC{-}CH(COCH_3)COCoA + CoA \qquad (4)$$

$${}^{\ominus}OOC^{\ominus}CHCOCoA \overset{H^{\oplus}}{\rightleftharpoons} {}^{\ominus}OOCCH_2COCoA$$

$${}^{\ominus}OOC{-}CH(COCH_3)COCoA \xrightarrow{H^{\oplus}} CH_3COCH_2COCoA + CO_2$$

$$CH_3COCoA + HSEnz \rightleftharpoons CH_3COSEnz + CoA \qquad (5)$$

$${}^{\ominus}OOCCH_2COCoA + HSEnz \rightleftharpoons {}^{\ominus}OOCCH_2\ COSEnz + CoA \qquad (6)$$

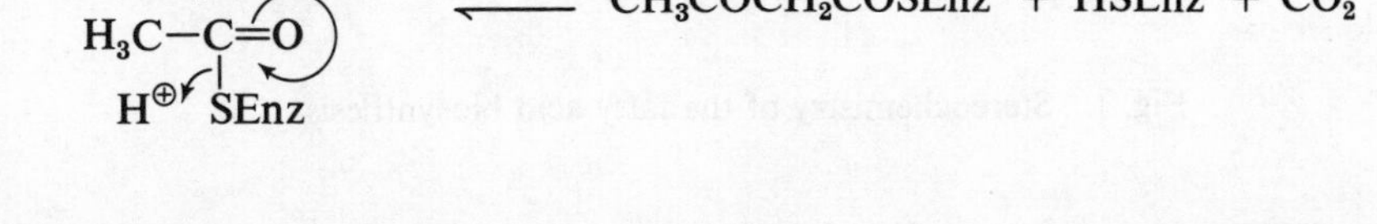

$${}^{\ominus}OOC{-}CH_2COSEnz + H_3C{-}C(=O){-}SEnz\ (H^{\oplus}) \rightleftharpoons CH_3COCH_2COSEnz + HSEnz + CO_2 \qquad (7)$$

$$CH_3COCH_2COSEnz \xrightleftharpoons{NADPH} CH_3\overset{OH}{\overset{|}{C}}HCH_2\ COSEnz \qquad (8)$$

$$CH_3\overset{OH}{\overset{|}{C}}HCH_2\ COSEnz \xrightleftharpoons{-H_2O} CH_3CH{=}CHCOSEnz \qquad (9)$$

$$CH_3CH{=}CHCOSEnz \xrightleftharpoons{NADPH} CH_3CH_2CH_2COSENz \qquad (10)$$

etc.

$$n\overset{\Delta}{C}H_3\overset{*}{C}OOH \longrightarrow \overset{\Delta}{C}H_3\overset{*}{C}H_2\overset{\Delta}{C}H_2\overset{*}{C}H_2\overset{\Delta}{C}H_2\overset{*}{C}H_2- \qquad (11)$$

multifunctional enzyme complex, the fatty acid synthetase. Malonyl CoA is attached to a thiol group of the acyl carrier protein, ACP, and is condensed with the acetyl group, as starter, strategically bound to a vicinal thiol group (5–7). The β-keto ester formed is reduced with NADPH (8), dehydrated (9), and finally reduced with NADPH to the homologue containing two more carbons (10). We know that the hydroxy acyl intermediate of (8) has (3*R*) configuration, that ^3H is preferentially retained from (2*S*)-2-^3H-malonate and that *trans*-enoate is exclusively obtained by dehydration (9). Enzymatic dehydration of stereospecifically synthesized (2*R*, 3*R*)-3-hydroxy-2-^3H-butenoate was furthermore found to give *trans*-butenoate with retention of ^3H indicating that the elimination proceeds in a *syn*-fashion. Consequently, the condensation (7) must proceed with inversion of configuration to fit the overall stereo outcome of the process (Fig. 1).[7] *Syn*-elimination is stereoelectronically somewhat less

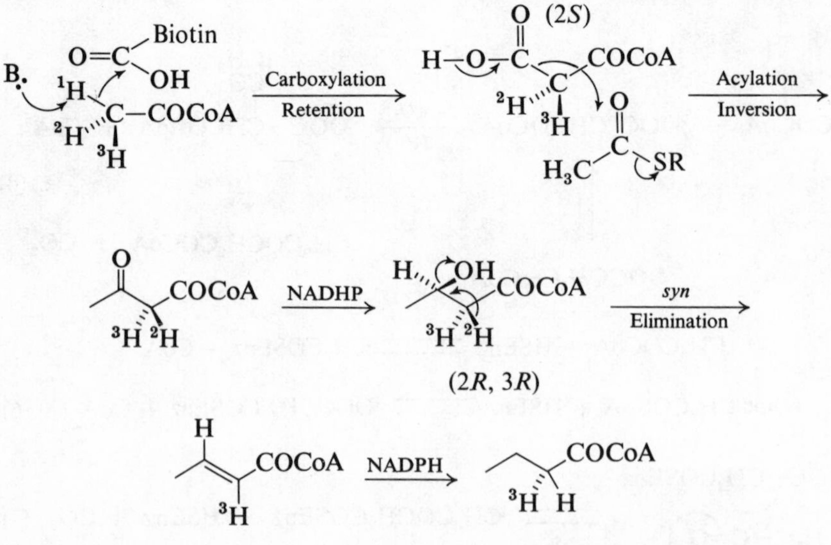

Fig. 1 Stereochemistry of the fatty acid biosynthesis

favourable than *trans*-elimination. Nevertheless, it occurs occasionally in conjunction with enzyme mediated reactions, *cf.* the *syn*-elimination of water from 5-dehydroquinic acid to 5-dehydroshikimic acid (section 3.1).

The butyryl CoA can now act as starter and be processed in reactions (5)–(10). This scheme has been verified by feeding the organism with labelled acetic acid (11). For the synthesis of palmitic acid, the most frequently occurring fatty acid, one acetyl CoA as a starter, seven malonyl CoA units, and fourteen NADPH, are needed. It is rather remarkable that in most organisms the chain elongation practically stops at C_{16}. The enzyme accepts readily a C_{14} CoA ester as starter but reluctantly a C_{16} CoA. The successive addition of two carbon units to acetyl CoA explains why even numbered acids are commonest. However, small amounts of odd numbered acids do occur. They normally originate from propionic acid as starter. We shall later meet several cases where other acids are used as starters. Table 1 shows the structure of some naturally occurring fatty acids. The first number refers to the number of carbon atoms in the acid, the second to the number of double bonds and the bracketed number to the position and structure of the double bond.

Table 1 Naturally occurring fatty acids

Structure	Name
$CH_3(CH_2)_8COOH$ 10 : 0	Capric acid
$CH_3(CH_2)_{10}COOH$ 12 : 0	Lauric acid
$CH_3(CH_2)_{12}COOH$ 14 : 0	Myristic acid
$CH_3(CH_2)_{14}COOH$ 16 : 0	Palmitic acid
$CH_3(CH_2)_{16}COOH$ 18 : 0	Stearic acid
$CH_3(CH_2)_7CH{=}CH(CH_2)_7COOH$ 18 : 1 (9c)	Oleic acid
$CH_3(CH_2)_7CH{=}CH(CH_2)_7COOH$ 18 : 1 (9tr)	Elaidic acid
$CH_3(CH_2)_4(CH{=}CHCH_2)_2(CH_2)_6COOH$ 18 : 2 (9c, 12c)	Linoleic acid
$CH_3CH_2(CH{=}CHCH_2)_3(CH_2)_6COOH$ 18 : 3 (9c, 12c, 15c)	Linolenic acid
$CH_3(CH_2)_4(CH{=}CHCH_2)_4(CH_2)_2COOH$ 20 : 4 (5c, 8c, 11c, 14c)	Arachidonic acid

Lipids is the collective name for compounds soluble in hydrocarbons. They include fats, waxes, phosphoglycerides but also hydrocarbons of quite different biogenesis such as steroids. The term is reserved by most biochemists for compounds yielding fatty acids and glycerol on hydrolysis, i.e. fats, and they function as an energy depot. They are biosynthesized in steps from fatty acid CoA and glycerol-3-phosphate (12). The most common acids are palmitic, stearic, and oleic acids. In some lipids one or several hydroxyl groups are *O*-alkylated by a long chain alcohol. The phospholipids (lecithins) are important constituents of cell membranes. They contain a betain head group, e.g. phosphatidyl choline, in which one of the OH groups of phosphatidic acid (12) is esterified by choline, $(CH_3)_3N^{\oplus}CH_2CH_2OH$ (Fig. 2).

Sphingolipids, e.g. galactocerebroside, are closely related compounds containing sphingosine, with the amino function acylated by a fatty acid. The hydroxyl

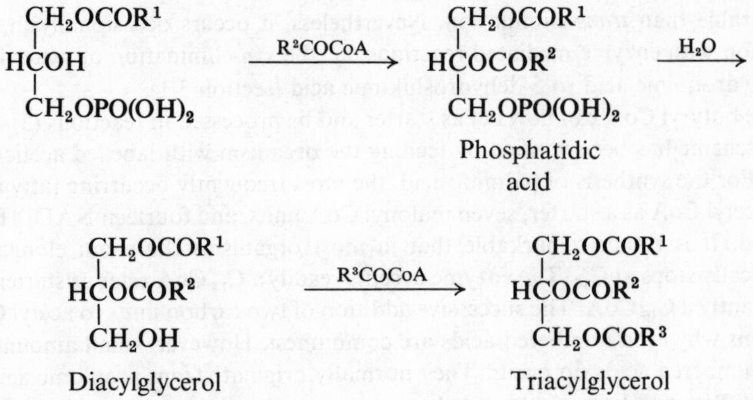

$$\begin{array}{l} CH_2OCOR^1 \\ | \\ HCOH \\ | \\ CH_2OPO(OH)_2 \end{array} \xrightarrow{R^2COCoA} \begin{array}{l} CH_2OCOR^1 \\ | \\ HCOCOR^2 \\ | \\ CH_2OPO(OH)_2 \end{array} \xrightarrow{H_2O}$$

Phosphatidic acid

$$\begin{array}{l} CH_2OCOR^1 \\ | \\ HCOCOR^2 \\ | \\ CH_2OH \end{array} \xrightarrow{R^3COCoA} \begin{array}{l} CH_2OCOR^1 \\ | \\ HCOCOR^2 \\ | \\ CH_2OCOR^3 \end{array} \qquad (12)$$

Diacylglycerol — Triacylglycerol

group is functionalized by phosphoric acid or a sugar. They are found in cerebral membranes and nerve endings where they function as impulse transmitters.

Waxes are defined as esters of fatty acids with long chain alcohols. They function as water repellant coatings on skin, feathers, fruits, etc.

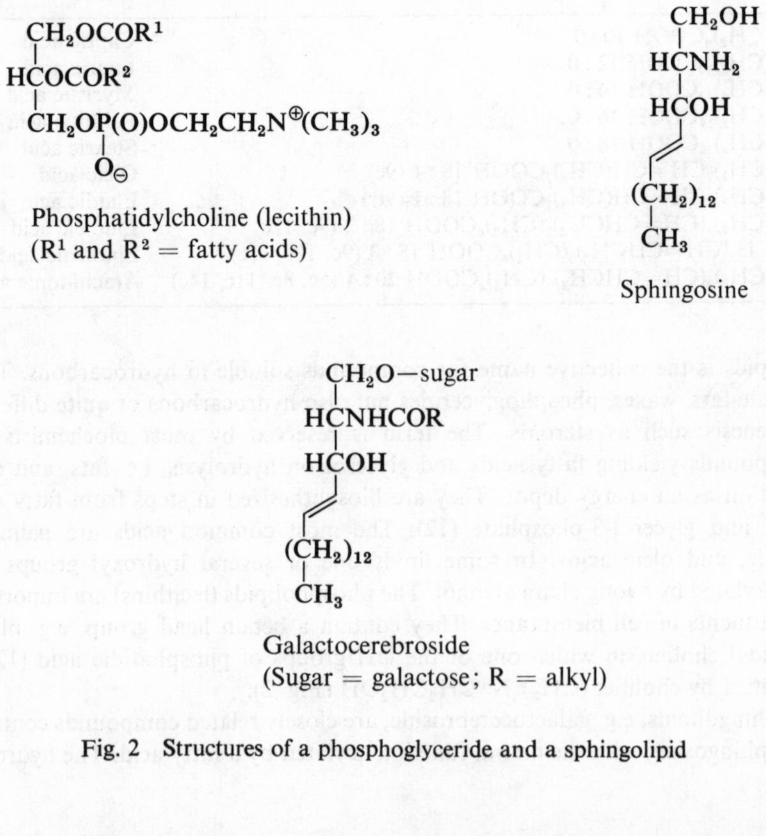

$$\begin{array}{l} CH_2OCOR^1 \\ | \\ HCOCOR^2 \\ | \\ CH_2OP(O)OCH_2CH_2N^{\oplus}(CH_3)_3 \\ \quad | \\ \quad O_{\ominus} \end{array}$$

Phosphatidylcholine (lecithin)
(R^1 and R^2 = fatty acids)

$$\begin{array}{l} CH_2OH \\ | \\ HCNH_2 \\ | \\ HCOH \\ | \\ CH{=}CH \\ | \\ (CH_2)_{12} \\ | \\ CH_3 \end{array}$$

Sphingosine

$$\begin{array}{l} CH_2O{-}sugar \\ | \\ HCNHCOR \\ | \\ HCOH \\ | \\ CH{=}CH \\ | \\ (CH_2)_{12} \\ | \\ CH_3 \end{array}$$

Galactocerebroside
(Sugar = galactose; R = alkyl)

Fig. 2 Structures of a phosphoglyceride and a sphingolipid

4.3 Branched fatty acids

Branched fatty acids are formed, either by priming the reaction with a branched starter, e.g. isobutyryl CoA, or α-methylbutyryl CoA (13), or by condensation with an alkylated malonyl CoA which arise by carboxylation of homologues of acetic acids (14,15).

$$\underset{\displaystyle CH_3}{CH_3CH_2\overset{|}{C}HCOSEnz} + n\,CH_3COCoA \quad \underset{\displaystyle CH_3}{CH_3CH_2\overset{|}{C}H(CH_2)_{2n}COOH} \tag{13}$$

$$CH_3CH_2COCoA + {}^{\ominus}OOC\text{—biotin—Enz} \longrightarrow CH_3CH\begin{matrix} COO^{\ominus} \\ COCoA \end{matrix} \tag{14}$$

Methylmalonyl CoA

$$H\text{—}O\text{—}\underset{\displaystyle O}{\overset{\|}{C}}\text{—}\underset{\displaystyle CH_3}{\overset{|}{C}}HCOSEnz \;\;(+\; H^{\oplus},\; CH_3C(=O)SEnz) \longrightarrow CH_3CO\underset{\displaystyle CH_3}{\overset{|}{C}}H\text{—}COSEnz \tag{15}$$

Thus, an unusual fatty acid is synthesized in the uropygial gland of the goose (Fig. 3). It is derived from one acetyl CoA and four methylmalonyl residues.[8] When ^{14}C-labelled 2-cyclopentenylcarboxylic acid, derived from cyclopentenylglycine, was administered to plants belonging to the *Flacourtiaceae*, labelled cyclopentenyl fatty acids such as chaulmoogric acid, used in treatment of leprosy, were synthesized.[9] The polyketide chain of the macrolide antibiotics from *Streptomyces* is derived from propionate as in erythromycin,[10] III. The oxygenation of erythromycin deserves some comment. 1-^{18}O-Propionate is incorporated at the alternating odd $C^{1,3,5,9,11,13}$ atoms, excluding a pathway via dehydration–rehydration sequence, since this involves an exchange with hydroxyl groups from water. However, the tertiary hydroxyl groups at $C^{6,12}$ are derived from molecular oxygen (*cf.* Problem 4.6).

Tuberculostearic acid or 10-methylstearic acid IV and sterculic acid V are synthesized in a different manner in that the methyl group comes from methionine.[11] Labelling experiments [12] in cultures of *Lactobacillus plantarum* with 9,10-$^2H_2$-oleate and 8,8,11,11-$^2H_4$-oleate showed no scrambling or loss of label, ruling out rapid equilibration of edge or corner protonated cyclopropyl species and also any mechanism involving allylic activation of the double bond. Incubation with ^2H-C-methionine produced *ca.* 10 per cent of 19 $^2H_1$-dihydrosterculic acid which indicated that exchange in the ^2H_3C species occurred at some stage of the biosynthesis. It was established that the exchange did not take place in the methyl group of methionine prior to alkylation. The findings, which are incompatible with intermediate exomethylene formation, might be explained by

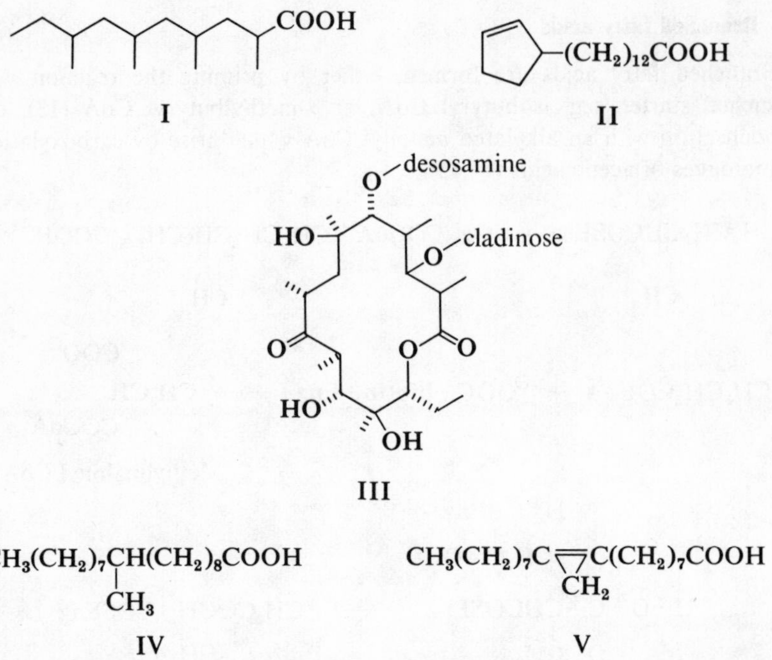
COOH
I
(CH2)12COOH
II
desosamine
HO
cladinose
O
O
O
HO
OH
III
CH3(CH2)7CH(CH2)8COOH
CH3
IV
CH3(CH2)7C=C(CH2)7COOH
CH2
V

Fig. 3 Branched fatty acids: I, metabolite in the uripygial gland of the goose; II, chaulmoogric acid; III, erythromycin A (seven propionate units); IV, tuberculostearic acid; V, sterculic acid

reversible cyclopropane group formation as shown in (16). The intermediate carbonium ion is either reduced to IV, path a, or rearranges to the cyclopropane. This intramolecular electrophilic aliphatic substitution has ample analogues in terpene chemistry. Dehydrogenation of the cyclopropane gives ultimately the cyclopropene V.

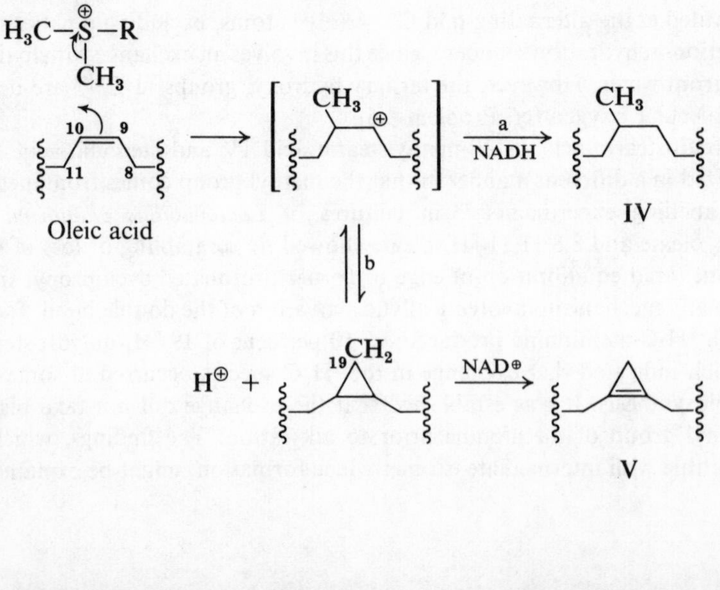
H3C—S—R
CH3
10
9
11
8
Oleic acid
CH3
CH3
a
NADH
IV
b
H⊕ +
19
CH2
NAD⊕
V

(16)

4.4 Olefinic acids

Unsaturated fatty acids are common and constitute a large part of the lipids. They can be formed by more than one route.[13] It is easy to believe that they arise directly in the course of biosynthesis by incomplete reduction (10), but this is a minor pathway, and it is in principle anaerobic. Scheme (17) has been proposed for the synthesis of oleic acid from decanoic acid in anaerobic bacteria.[14]

$$\underset{\text{Decanoic acid}}{CH_3(CH_2)_7CH_2COCoA} + CH_3COCoA \rightarrow$$

$$CH_3(CH_2)_7CH_2COCH_2COCoA \xrightarrow{[H]} CH_3(CH_2)_7CH_2\overset{\overset{OH}{|}}{C}HCH_2COCoA \qquad (17)$$

$$\xrightarrow{-H_2O} CH_3(CH_2)_7CH \overset{c}{=} CH{-}CH_2COCoA \xrightarrow{+3\ \text{Acetate}}$$

$$\underset{\text{Oleic acid}}{CH_3(CH_2)_7CH \overset{c}{=} CH(CH_2)_7COOH}$$

It ought to be noted that here the elimination of water gives specifically the unconjugated *cis* duble bond. Elongation of the unsaturated acid with three acetate units gives oleic acid. In most organisms oleic acid is formed by desaturation of stearic acid in the presence of molecular oxygen, FAD, NADPH, and a specific monooxygenase. The mechanistic details of this fundamental reaction is not known, just open to speculation. With reference to our discussion on biological hydroxylation (section 3.3) we can assume that a peroxy radical abstracts a proton from C^9 or C^{10} of stearic acid and forms a carbon centred radical which reacts with oxygen to a hydrogen peroxide. Reduction and stereocontrolled elimination of the elements of water could then give oleic acid. From a mechanistic point of view it is not an obligatory prerequisite that desaturation goes via elimination of water. The carbon radical initially formed could e.g. react directly with FAD to olefin and $FADH^{\bullet}$ (*cf.* section 4.6). The 9-pro-*R* and 10-pro-*R* hydrogens are specifically removed. Further desaturation of oleic acid leads to linoleic and linolenic acids.[15,16] It is remarkable that polyunsaturated acids rarely have their double bonds in conjugation and that they occur as the thermodynamically less stable *cis* isomer. Arachidonic acid, a precursor of the important prostaglandin hormones, first detected in seminal plasma, is biosynthesized from oleic acid according to Fig. 4.[17] The prostaglandins possess a manifold of biological effects. They control blood pressure and renal blood flow, contractions of smooth muscle, gastric acid secretion and platelet aggregation and have found clinical use in the regulation of pregnancy, in treatment of ulcers, heart failure, thrombosis, etc. Dietary investigations revealed that some fatty acids, e.g. linoleic acid, are essential for the normal development of mammals which are able to synthesize saturated and monosaturated acids but not particular polyunsaturated acids. They have to be supplied in the diet. Their mechanism of action is explained, at least in part, by the fact that they are precursors of the prostaglandins.

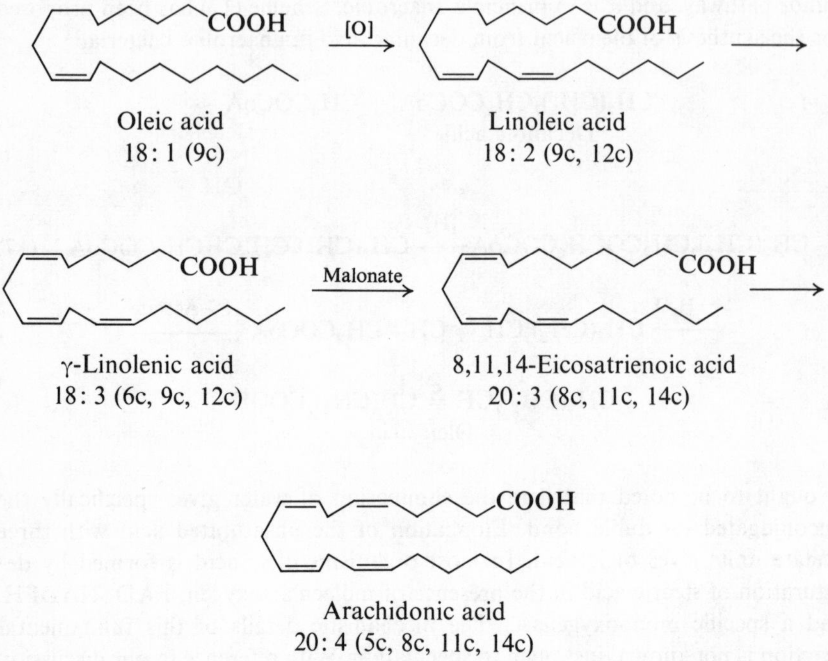

COOH [O] → COOH →

Oleic acid
18:1 (9c)

Linoleic acid
18:2 (9c, 12c)

COOH Malonate → COOH →

γ-Linolenic acid
18:3 (6c, 9c, 12c)

8,11,14-Eicosatrienoic acid
20:3 (8c, 11c, 14c)

COOH

Arachidonic acid
20:4 (5c, 8c, 11c, 14c)

Fig. 4 Biosynthesis of arachidonic acid

4.5 Acetylenic compounds

An array of acetylenes and allenes are produced naturally, particularly in *Compositae*, *Umbelliferae*, and in some fungi of the group *Basidiomycetes*, albeit in small quantities.[18] They are often very unstable, sensitive to heat, light, and oxygen. Their isolation and identification have been facilitated by their characteristic UV spectra. There is convincing evidence that acetylenes are of polyketide origin arising by further desaturation of fatty acids. The structural relationship between a number of acetylenes isolated from the same plant suggests a plausible biosynthetic pathway. For example several acetylenic acids, VI–XI, have been isolated from *Santalum acuminatum*[19,20] and it appears to be a sound working hypothesis to assume that they are formed from oleic acid by sequential dehydrogenation (Table 2). The order of events is not known but as the number of known acetylenic compounds increases the biosynthetic pathway becomes clearer.

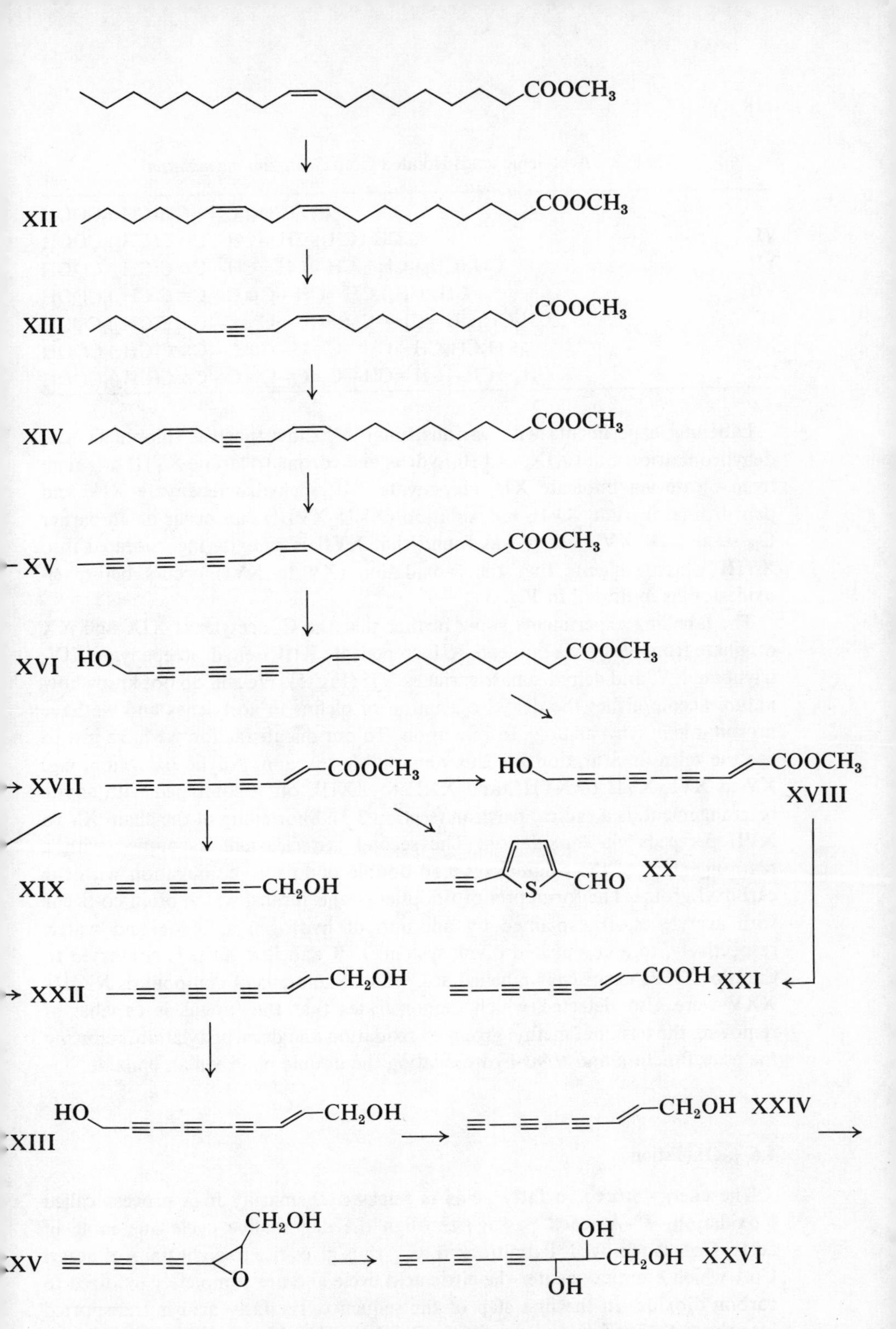

Fig. 5 Biosynthesis of acetylenic compounds

Table 2 Acetylenic acids isolated from *Santalum acuminatum*

	$CH_3(CH_2)_7CH{=}CH(CH_2)_7COOH$
VI	$CH_3(CH_2)_5CH{=}CH{-}C{\equiv}C(CH_2)_7COOH$
VII	$CH_3(CH_2)_3CH{=}CH{-}CH{=}CH{-}C{\equiv}C(CH_2)_7COOH$
VIII	$CH_3(CH_2)_3CH{=}CH{-}C{\equiv}C{-}C{\equiv}C(CH_2)_7COOH$
IX	$CH_3CH_2CH{=}CH{-}CH{=}CH{-}C{\equiv}C{-}C{\equiv}C(CH_2)_7COOH$
X	$CH_3CH_2CH{=}CH{-}C{\equiv}C{-}C{\equiv}C{-}C{\equiv}C(CH_2)_7COOH$
XI	$CH_2{=}CH{-}CH{=}CH{-}C{\equiv}C{-}C{\equiv}C{-}C{\equiv}C(CH_2)_7COOH$

Labelling experiments with various fungi[21–23] show that the thiophene XX, dehydromatricarianol XIX, and 10-hydroxydehydromatricariate XVIII originate from oleate via linoleate XII, crepenyate XIII, dehydrocrepenyate XIV and dehydromatricariate XVII. ω-Oxidation (XVII–XVIII) can occur at an earlier C_{18} stage (XV–XVI) but it was found that XVII is effiently incorporated into XVIII which suggests that the β-oxidation (XV to XVI) occurs before ω-oxidation as indicated in Fig. 5.

The labelling experiments show further that the C_8 acetylenes XIX and XX originate from oleate via linoleate XII, crepenyate XIII, dehydrocrepenyate XIV, triynoate XV, and dehydromatricariate XVII (Fig. 5). We still do not know how nature accomplishes the dehydrogenation of olefins to acetylenes and we have no suitable *in vitro* analogy to rely upon. To our dissatisfaction we have just to use the term desaturation for this remarkable reaction. Allylic oxidation, e.g. XV to XVI, XVII to XVIII, and XXII to XXIII, often combined with allylic rearrangement, is a radical reaction (section 3.3). Shortening of the chain XV to XVII preceeds via β-oxidation. The second last cleavage is coupled with a rearrangement of the original *cis*-9,10 double bond into conjugation with the carboxyl group. The formation of thiophenes and furans, which often co-occur with acetylenes, is explained by addition of hydrogen sulphide and water, respectively, to a conjugated diyne system. 1-^{14}C-labelled XVII is converted to the triol XXVI exclusively labelled at C^1. Small amounts of compounds XXIII–XXV were also detected which demonstrates that the fungus is capable of removing the terminal methyl group by oxidation and decarboxylation, reducing the ester function and *trans*-hydroxylating the double bond via an epoxide.

4.6 β-Oxidation

The energy stored in fatty acids is released chemically in a process called β-oxidation. [13b] At each passage through the degradative cycle one mole of acetyl CoA is removed. Palmitic acid, C_{16}, thus gives rise to eight units of acetyl CoA which eventually enter the citric acid cycle and are completely oxidized to carbon dioxide. In the first step of the sequence, the fatty acid is transported into the mitochondria and esterified with CoA (18) which activates the α-C protons, and prepares the molecule for the first FAD-mediated dehydrogenation.

$$RCOOH + ATP + CoA \rightarrow RCOCoA + AMP + PP \quad (18)$$

The mechanism of this reaction deserves comment since it represents one of the commonest redox reactions in biosynthesis. First, it is worth noting that NADPH is reported to serve as donor in the saturation step, whereas FAD serves as oxidant in the reverse α,β-desaturation of the fatty acid. Since both are coupled in the respiratory chain, it is hard to tell which is doing what. It is generally accepted that flavins act as electron transfer agents rather than hydride transmitters on the grounds that flavins can pass electrons down the respiratory chain, $FADH_2$ is sensitive to oxygen contrary to NADH, and finally the steps in the electron transfer mechanism are thermodynamically reasonable, e.g. for reduction of a carbonyl group (19).

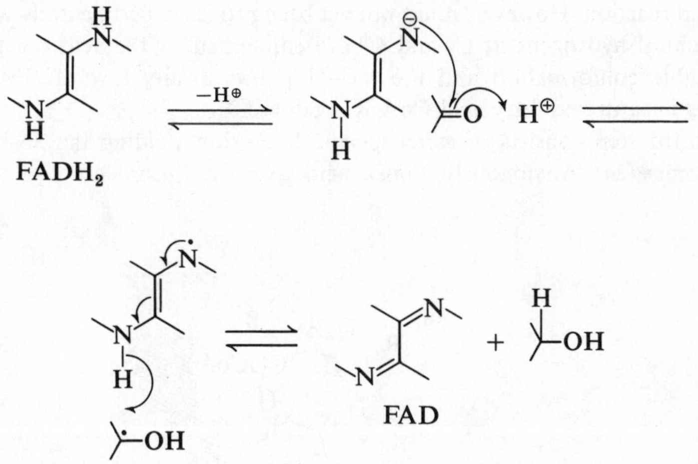

$FADH_2$ $\underset{}{\overset{-H^{\oplus}}{\rightleftharpoons}}$ … $H^{\oplus}$ $\rightleftharpoons$ … $\rightleftharpoons$ FAD + H–C–OH (19)

It is possible to formulate the NADH reduction in terms of single electron transfers, e.g. (20), but NADH is usually regarded as a hydride shift reagent (21). In that respect the reaction has several analogies in non-enzymatic organic oxidations, e.g. the Cannizzaro (22) and the Meerwein–Ponndorf–Verley–Oppenauer reactions (23). No exchange of protons with the solution is observed

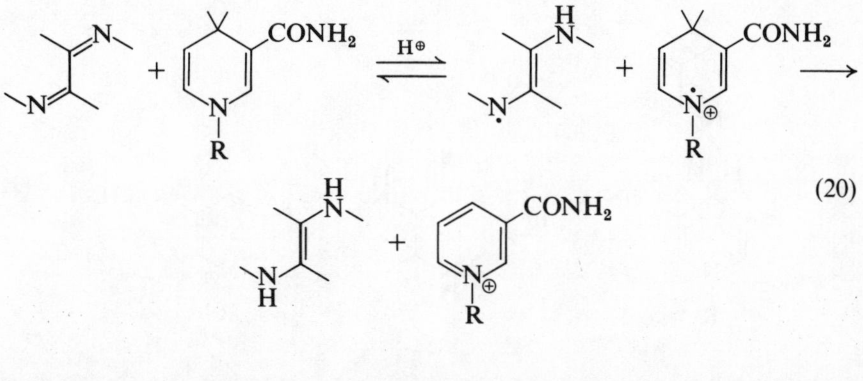

(20)

in the stereospecific NADH reduction or in (23) and (24) which is taken as an argument for a hydride shift. The hydride ion is a very strong base that reacts with water with development of hydrogen. Consequently, no free hydride ion is involved in the hydride shift. It ought to be pointed out that the direct transfer of hydrogen from NADH to the oxidant by no means excludes a radical pathway because hydrogen abstraction by a radical leads to the same result. The situation is different for the FAD/$FADH_2$ couple because H–N protons equilibrate extremely rapidly with protonated solvents.

We are now ready to discuss the mechanism of the second step, the desaturation of fatty acids. A base removes the activated α-C–H of a fatty acid CoA generating the anion which is oxidized by FAD, anchored at the enzyme, to the neutral α-Ċ-radical and flavin radical anion. H abstraction from the β-carbon completes the dehydrogenation. The dehydrogenation is depicted in Fig. 6 as a concerted reaction. However, it has not yet been proved experimentally which of the prochiral hydrogens at C^2 and C^3 are eliminated. If the acid reacts in its most stable conformation and the β-C–H points axially towards FAD, the α,β-*trans* unsaturated fatty acid CoA is produced.

The third step consists of stereospecific hydration yielding the 3*S*-hydroxy stereoisomer (24). Analogously, the *cis* acid gives the *R*-isomer.

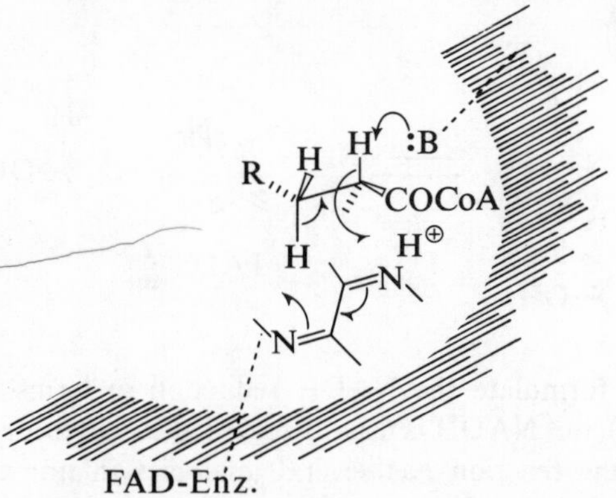

Fig. 6 Suggested mechanism for dehydrogenation of fatty acid CoA by FAD. Only the active quinonoid section of FAD is depicted

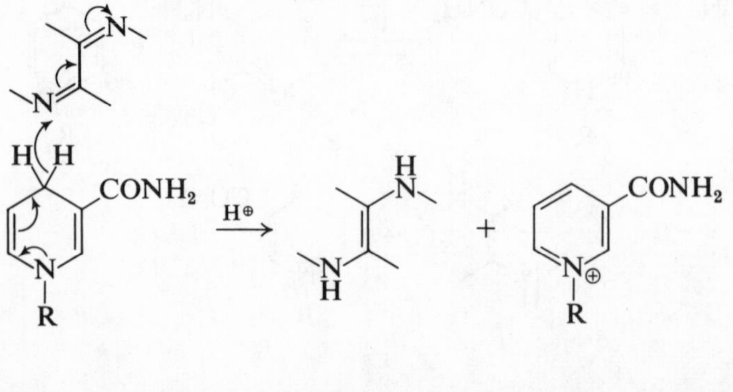

(21)

$$C_6H_5CHO \underset{}{\overset{OH^{\ominus}}{\rightleftharpoons}} C_6H_5C(O^{\ominus})(H)OH + C_6H_5CH{=}O \xrightarrow{OH^{\ominus}} C_6H_5COO^{\ominus} + C_6H_5CH_2OH \quad (22)$$

$$(RO)_2Al{-}O{-}CHR_2 + R^1COR^2 \rightleftharpoons (RO)_2Al{-}O{-}C(R^1)(H){-}R^2 + R_2CO \quad (23)$$

$$RCH{=}CHCOCoA \underset{}{\overset{H_2O}{\rightleftharpoons}} RCH(OH)CH_2COCoA \quad (24)$$

The 3*S*-hydroxy fatty acid is oxidized by NAD⊕ to the 3-keto fatty acid (25). In the final step the 3-keto fatty acid undergoes thiolysis by another CoA molecule to acetyl CoA and a new fatty acid CoA, shorter by two carbon atoms (26). The overall equation for one cycle of β-oxidation starting from palmitic acid is (27).

$$RCH(OH)CH_2COCoA \underset{NADH}{\overset{NAD^{\oplus}}{\rightleftharpoons}} RCOCH_2COCoA \quad (25)$$

$$RCOCH_2COCoA + HSCoA \rightleftharpoons RCOCoA + CH_3COCoA \quad (26)$$

$$C_{15}H_{31}COCoA + FAD + NAD^{\oplus} + CoA + H_2O \rightarrow C_{13}H_{27}COCoA + FADH_2 + NADH + CH_3COCoA + H^{\oplus} \quad (27)$$

NADH and $FADH_2$ are reoxidized via the respiratory chain by coenzyme Q and their hydrogens end up eventually as water by combination with oxygen.

Part of the energy developed is thereby transformed to chemical energy as ATP. If we also consider these transformations, complete degradation of palmitic acid CoA to acetyl CoA is given by (28).

$$C_{15}H_{31}COCoA + 7\ CoA + 7\ O_2 + 35\ P + 35\ ADP \rightarrow 8\ CH_3COCoA + 35\ ATP + 42\ H_2O \quad (28)$$

Odd-carbon fatty acids give propionyl CoA in the last β-oxidation step. It is eventually converted to succinyl CoA in a vitamin B_{12} mediated process.

4.7 Cyclization of polyketides to aromatics

Reduction after each condensation step leads to fatty acids as described above. If further condensation occurs before reduction takes place, intermediate β-polyketoesters of various chain length are formed. These compounds are very reactive and they undergo Claisen or aldol condensations to form aromatics because of their bifunctionality. The poly-β-ketoester (polyketide) is temporarily stabilized by chelation or hydrogen bonding on the enzyme surface until the assembly is accomplished. The cyclization is then guided by the special topology of the enzyme. The activated methylene groups give rise to carbanions or enolates by removal of protons and the polarized carbonyl group has carbonium ion character. The tetraketide can cyclize in several ways. Path a leads to the acetophenone derivative xanthoxylin, paths b and c to the pyrone derivatives, and path d to orsellinic acid, all known from natural sources (29).

$CH_3COCH_2COCH_2COCH_2COCoA$ —a→ Xanthoxylin (HO, OH, OH, $COCH_3$ substituted benzene)

d → Orsellinic acid (CH_3, COOH, HO, OH substituted benzene)

c → 2-methyl-4-pyrone with CH_2COOH (H_3C, O, CH_2COOH)

b → HO-substituted 2-pyrone with CH_2COCH_3 side chain

(29)

Model cyclizations of this kind have been carried out chemically in the laboratory thus mimicking the biosynthetic process.[24] Acylation of the acetoacetate dianion gives selectively the β-diketoester. By special carbonyl protection methods it is possible to repeat the procedure and synthesize poly-β-ketoesters. The unstable phenyl derivative undergoes base catalysed aldol condensation to the orsellinic acid analogue (30).

$$CH_3COCH_2COOC_2H_5 \xrightarrow{2\,\overset{\ominus}{N}R_2} \overset{\ominus}{C}H_2CO\overset{\ominus}{C}HCOOC_2H_5 \xrightarrow{CH_3COOC_2H_5}$$

$$CH_3COCH_2COCH_2COOC_2H_5 \xrightarrow[C_6H_5COCl]{base} \qquad (30)$$

$$C_6H_5COCH_2COCH_2COCH_2COOC_2H_5 \xrightarrow[C_2H_5OH]{KOH}$$ [benzene ring bearing OH, $COOC_2H_5$, C_6H_5 and HO]

A great variety of metabolites are formed depending upon several factors:

1. the starter or chain initiating unit;
2. the number of acetyl CoA units involved, or occassionally other esters such as propionyl or butyryl CoA;
3. the mode of cyclization;
4. the condensation of separately synthesized polyketides; and
5. the secondary processes, such as halogenation, alkylation, redox reactions, rearrangements, etc.

These factors determine the class of compounds biosynthesized. One distinguishes traditionally between simple benzenoids, tropolones, condensed aromatic systems, quinones, flavonoids, phenolic coupling products, and non-aromatic polyketides. A classification according to the numbers of units is also convenient; thus, one speaks of tetra-, penta-, hexaketides, etc. and the two systems overlap to a large extent.

4.8 Confirmation of the acetate hypothesis. The use of NMR in biosynthetic studies

The acetate hypothesis[25] received its first confirmation by an investigation of the labelling pattern of 6-methylsalicylic acid formed by *Penicillium griseofulvum*. 1-^{14}C-labelled acetic acid was added to the nutrient solution. By a special degradation technique each carbon of the molecule was separately isolated (as CO_2 etc.) and analysed for radioactivity (Fig. 7). The carbon dioxide obtained by decarboxylation and the acetic acid from Kuhn–Roth degradation

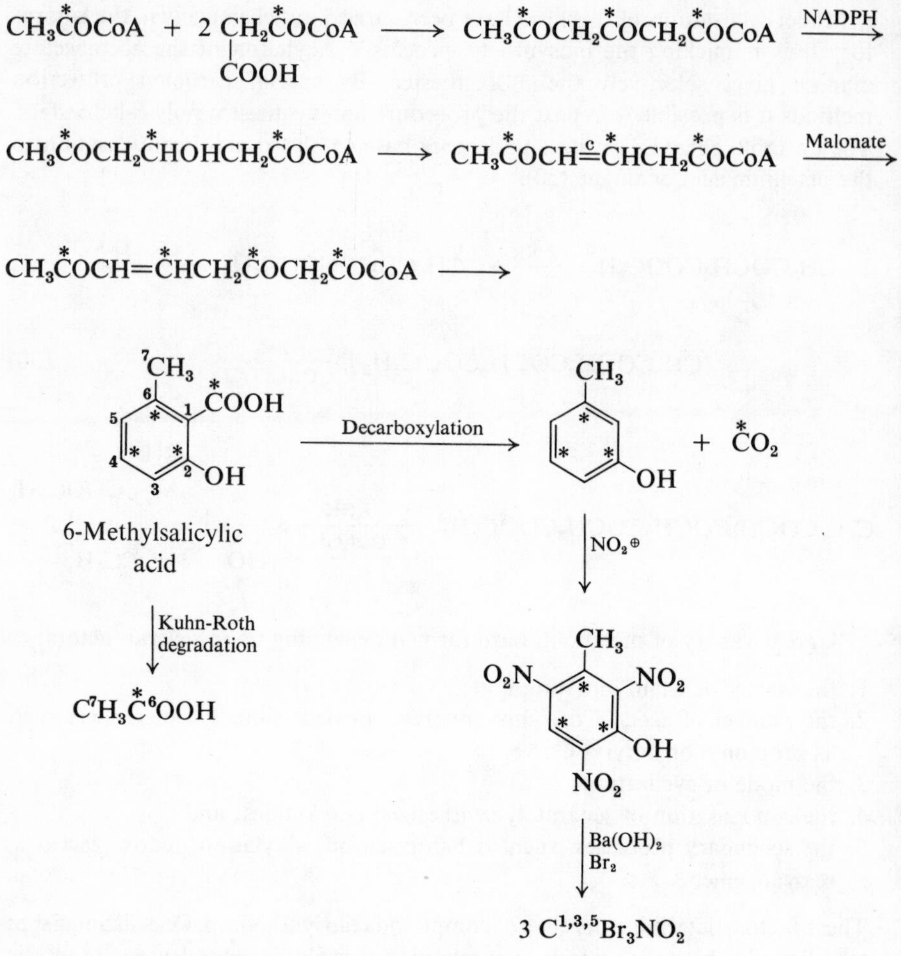

$$CH_3\overset{*}{C}OCoA + 2\ \overset{*}{C}H_2COCoA\ (COOH) \longrightarrow CH_3\overset{*}{C}OCH_2\overset{*}{C}OCH_2\overset{*}{C}OCoA \xrightarrow{NADPH}$$

$$CH_3\overset{*}{C}OCH_2\overset{*}{C}HOHCH_2\overset{*}{C}OCoA \longrightarrow CH_3\overset{*}{C}OCH\overset{c}{=}\overset{*}{C}HCH_2\overset{*}{C}OCoA \xrightarrow{Malonate}$$

$$CH_3\overset{*}{C}OCH=\overset{*}{C}HCH_2\overset{*}{C}OCH_2\overset{*}{C}OCoA \longrightarrow$$

6-Methylsalicylic acid $\xrightarrow{\text{Decarboxylation}}$ m-cresol $+ \overset{*}{C}O_2$

6-Methylsalicylic acid $\xrightarrow{\text{Kuhn-Roth degradation}}$ $C^7H_3\overset{*}{C}{}^6OOH$

m-cresol $\xrightarrow{NO_2^{\oplus}}$ trinitro-m-cresol $\xrightarrow{Ba(OH)_2,\ Br_2}$ $3\ C^{1,3,5}Br_3NO_2$

Fig. 7 Biosynthesis and chemical degradation of 6-methylsalicylic acid

were found to be active, whereas the bromopicrin was inactive. The activity of $C^{2,4}$ was calculated by difference and the relative intensity was found to be close to 1.0. In another experiment with ^{18}O-labelled acetic acid, the 6-methylsalicylic acid produced contained ^{18}O-labelled oxygens demonstrating that the original oxygens are retained throughout the biosynthesis in complete accordance with the hypothesis.[26] Some washout of ^{18}O was noted as a result of exchange with the medium.

We have seen earlier that the starter normally is acetyl CoA, but that the chain elongating units consist of malonyl CoA. This was demonstrated in an experiment with 2-^{14}C-malonate administered to *Penicillium urticae*.[27] The 6-methylsalicylic acid was isolated, degraded, and analysed. In this case activity was absent in acetic acid from the Kuhn–Roth degradation, i.e. in $C^{6,7}$ and all the activity accumulated in the bromopicrin, supporting the mechanism of polyketide formation.

One more detail in the synthesis of 6-methylsalicylic acid requires comment. The compound lacks a hydroxyl group at C^4. The reduction occurs before cyclization since, according to experience, aromatic deoxygenation is not carried out by organisms producing the polyketides. Metabolites are frequently modified by 'missing' hydroxyl functions. The exact timing of the reduction and dehydration has been the object to extensive research. 6-Methylsalicylic acid synthase is susceptible to inhibitors of *cis*-unsaturated fatty acid synthase. The inhibitor still allows NADPH reduction and synthesis of stearic acid to continue and these observations have been taken as evidence for a proposal that the reduction takes place at the 6-carbon stage rather than at the 8-carbon stage. Then follow *cis* elimination, condensation with a C_2 unit and cyclization (Fig. 7).[28]

Alternariol, produced by an enzyme extract from the mould *Alternaria tenuis*, is made up of seven acetate units and the labelling pattern conforms with theory (31).[29]

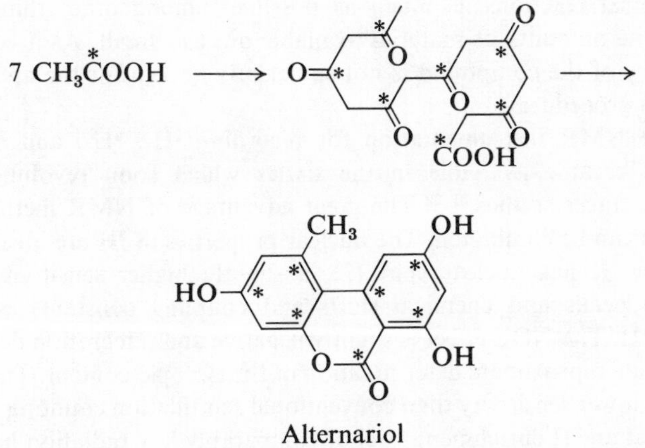

$$7\ CH_3\overset{*}{C}OOH \longrightarrow \longrightarrow$$

(31)

CH₃ OH HO OH O O

Alternariol

The relative radioactivity of the starred carbon atoms is close to 1.0, the starter being somewhat higher. The other positions were devoid of any activity. The uniform distribution of activity at the starred positions indicates that the whole polyketide chain is constructed in one complete sequence without leaving the protein surface. Islandicin showed the labelling pattern depicted in (32) when *Penicillium islandicum* was fed with 1-^{14}C-acetate.[30] Three secondary reactions have taken place here, decarboxylation, reduction, and hydroxylation.

In the conventional ^{14}C-labelling method it is necessary to have a degradative method allowing separation and isolation of all the carbons of the molecule. Increasing complexity of a metabolite rapidly raises serious, if not insurmountable problems. The degradative methods, the isolation, and the purification of defined fragments become more complicated and time-consuming and small amounts of radioactive impurities seriously affect the measurements. Modern

decarboxylation

$8\ CH_3\overset{*}{C}OOH \longrightarrow$ [polyketide chain: COOH, [H], [O], [O]] $\longrightarrow$ (32)

Islandicin

structural determinations rely heavily on spectroscopic methods and try to avoid chemical reactions as much as possible, among other things because frequently the amounts of material available are too small. As a consequence the chemistry of the compound is not sufficiently known for the application of degradation procedures.

Adequate NMR instrumentation for recording ^{2}H-, ^{3}H-, and ^{13}C-labelled compounds became available in the sixties which soon revolutionized the biosynthetic tracer studies.[31,32] The great advantage of NMR methods is that degradation can be eliminated. The nuclear properties of ^{3}H are ideal for NMR spectroscopy. It has nuclear spin 1/2, a slightly higher sensitivity than ^{1}H, gives sharp peaks and chemical shifts and coupling constants very similar to those of ^{1}H. The NOE effect is slightly negative and integration can therefore be used for an approximate determination of the isotopic content. The ^{3}H NMR method has lower sensitivity than conventional scintillation counting but spectra can be run at an ^{3}H enrichment level of comparably low radiation hazard.

Deuterium, nuclear spin 1, has the advantage of being an inexpensive, stable isotope having low natural abundance, 0.016 per cent, allowing higher dilution of the precursor in comparison with ^{13}C-labelled compounds. It exhibits short relaxation times and no NOE effect and can therefore be accurately integrated. However, the sensitivity is low (1/100 of ^{1}H), the chemical shift scale and coupling constants are only 1/6 the value of ^{1}H and the line width is broader than that of ^{1}H. The spectral resolution is consequently low (see Fig. 9).

The stable ^{13}C, natural abundance 1.1 per cent, has a nuclear spin 1/2 and a positive NOE effect. It has a large chemical shift range (200 PPM) in comparison with ^{1}H (10 PPM) and gives narrow line widths. The spectra are consequently well resolved. ^{13}C has unfortunately a low sensitivity (1/60 or ^{1}H), which can in part be compensated for by proton decoupling, causing collapse of the multiplets to single sharp peaks in combination with intensity enhancement due to the positive NOE effect. Progress in the refinement of instrumental techniques and application of Fourier transformation allow direct detection of the ^{13}C nuclei in natural abundance on samples as small as 1–10 mg and an accumulation time of a few hours. By proton noise decoupling all ^{13}C–^{1}H coupling information is lost. By off-resonance technique residual couplings are observed giving information on the number of hydrogens attached to the carbon atom. Thus, a quaternary carbon appears as a singlet, a methine carbon as a narrow doublet, a methylene gives a triplet and a methyl a quartet centred at the δ value of the singlet in the noise decoupled spectrum. ^{13}C NMR is used with success in tracing the fate of hydrogen in biosynthesis. In proton decoupled ^{13}C NMR spectra a cabon atom attached to one ^{2}H (nuclear spin 1) appears as a triplet of equal line intensity, a $C^{2}H_2$ group as a quintet (1: 2: 3: 2: 1:) and a $C^{2}H_3$ as a septet (1: 3: 6: 7: 6: 3: 1) with an upfield shift of *ca.* 0.4 PPM for each ^{2}H. The spectral analysis of overlapping signals can be simplified by running ^{13}C–^{2}H decoupled spectra. The weakness of the ^{13}C–^{2}H signal, which is due to loss of NOE enhancement and inefficient ^{13}C–^{2}H relaxation limits the use of this technique. On the other hand, reduction of the ^{13}C signal strength indicates that the carbon atom is ^{2}H-labelled.

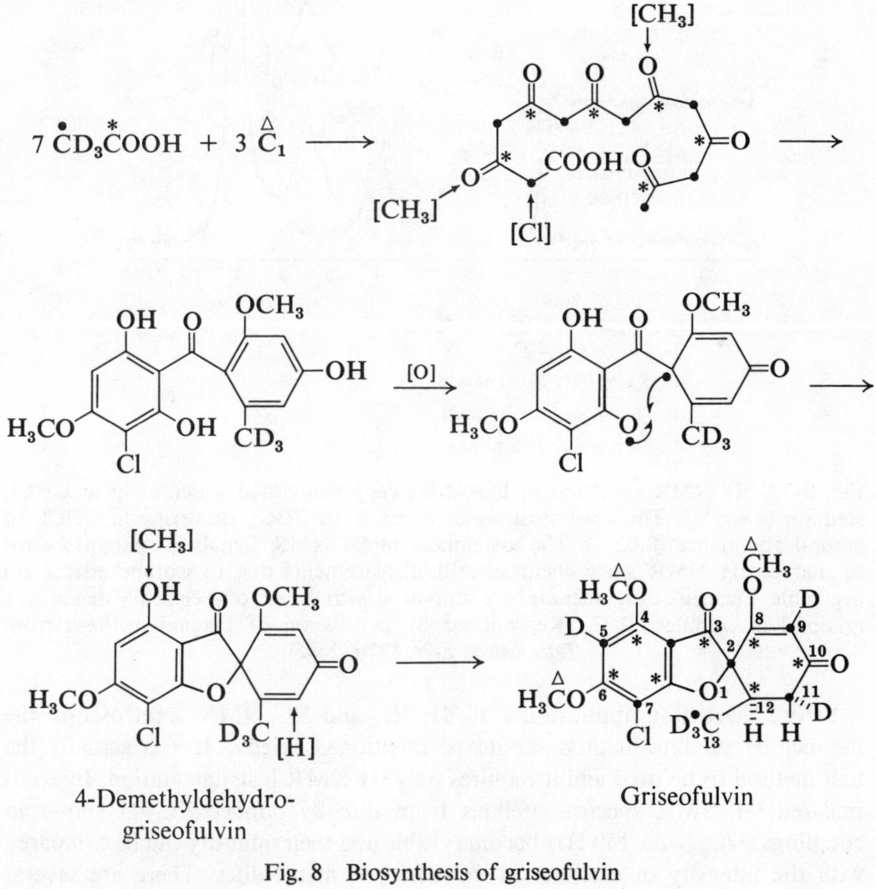

Fig. 8 Biosynthesis of griseofulvin

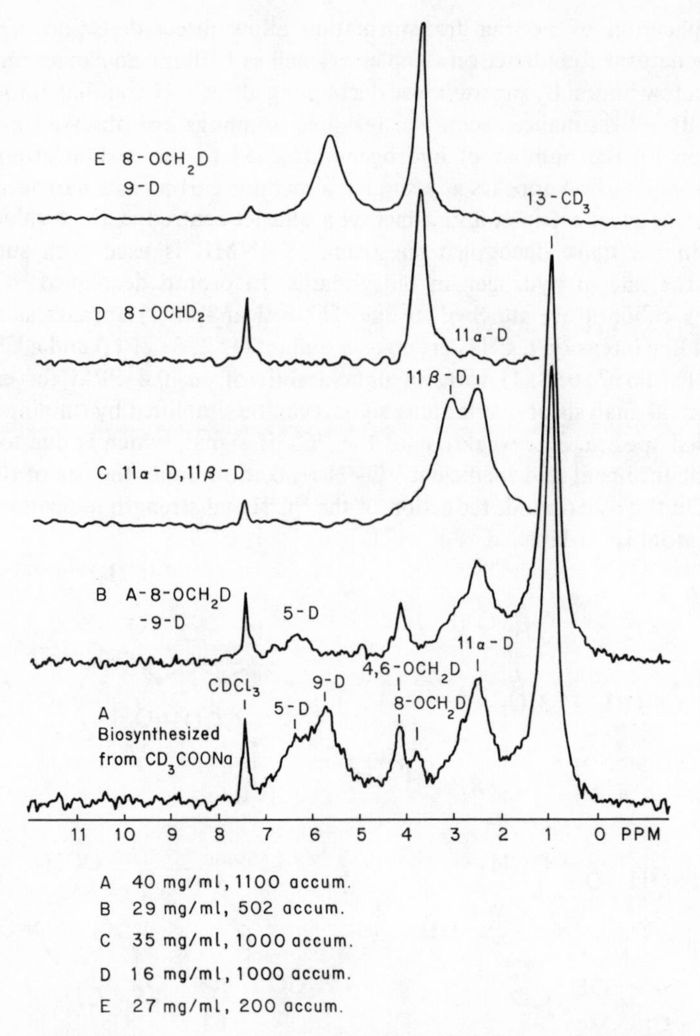

Fig. 9 A ^{2}H NMR spectrum of biosynthetically deuterated griseofulvin in $CHCl_3$ solution (4 w/v%). The lowermost signal arises from $CDCl_3$ occurring in $CHCl_3$ of natural abundance (0.02%). The assignment of ^{2}H NMR signals is straightforward to that of ^{1}H NMR since chemical-shift displacements due to isotope effects are negligible. The peak assignment is here supported with the aid of specifically deuterated griseofulvin samples, B–E. (Reproduced by permission of Pergamon Press from *Tetrahedron Lett*. **1976,** 2695)

Before discussing applications of ^{2}H, ^{3}H, and ^{13}C NMR spectroscopy the indirect ^{1}H satellite method should be mentioned briefly. It was actually the first method to be used and it requires only ^{1}H NMR instrumentation. In accumulated ^{1}H NMR spectra satellites from directly bonded ^{13}C–^{1}H spin–spin couplings ($^{1}J_{^{1}H^{13}C}$ *ca.* 150 Hz) become visible and their intensity can be compared with the intensity of satellites in ^{13}C enriched metabolites. There are several restrictions connected with this method. The individual ^1H NMR signal has to be well defined and the satellite peak must not overlap with other peaks of the spectrum. Carbons not directly bound to hydrogens escape detection. It had earlier been shown by the conventional ^{14}C technique that griseofulvin was formed by cyclization of a linear C_{14} polyketide (Fig. 8). The *O*-methyl groups come from the C_1 pool. Direct observation of the ^1H satellites in griseofulvin produced from 2-^{13}C-acetate (dotted carbons) showed that the intensity of C^5H, C^9H, and $C^{13}H$ satellites had increased in agreement with the result obtained by the chemical study. Satellites from the unlabelled methoxy groups were used as internal standards. The signals from $C^{11}H_2$ were too broad to be measured.[33]

The biosynthesis of griseofulvin has also been studied by ^2H NMR spectroscopy.[34] Griseofulvin produced from 2-$^2H_3$-acetate is expected to be deuterated at C^5, C^9, C^{11}, and C^{13}. The ^2H NMR spectrum (Fig. 9). shows that these positions are indeed labelled. Some washout was noted but label is also incorporated to some extent in the methoxy groups at C^4, C^6, and C^8. C^5 is exclusively labelled in the α-position demonstrating that reduction of the intermediate dehydrogriseofulvin takes place specifically in a *trans* fashion. The finding that C^{13} contained a C^2H_3 residue proves that this is the starter end of the C_{14} chain. The ^2H NMR spectrum is assigned by comparison with specifically labelled derivatives. As can be seen from the spectrum, the line width is much broader than that of a corresponding ^1H spectrum.

^{13}C NMR spectroscopy is the most important physical method for biosynthetic studies. Presuming that the shifts of all carbon atoms can be assigned, it is possible to identify the labelled positions by measuring the relative intensities of the decoupled ^{13}C peaks in ^{13}C enriched samples and at natural abundance. Shanorellin is a benzoquinone pigment synthesized by *Shanorella spirotricha.* By administration of 1-^{13}C-acetate, or ^{13}C-formate to the culture medium, ^{13}C enriched shanorellins are obtained (33). Table 3 shows the results of the ^{13}C NMR recordings in complete agreement with an earlier, conventional ^{14}C

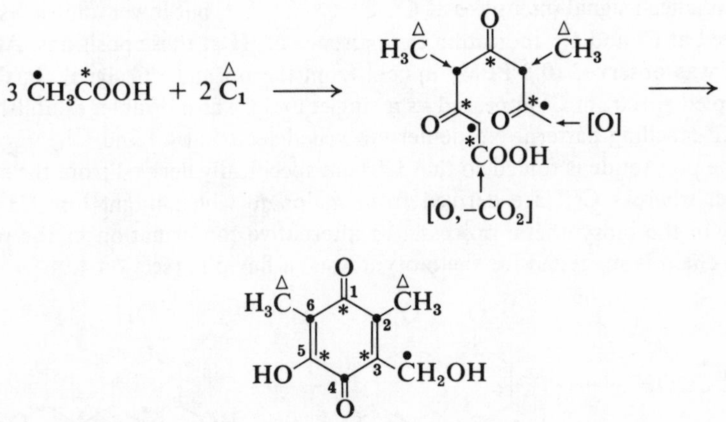

(33)

Shanorellin

Table 3 ^{13}C NMR data for labelled shanorellin, δ. Positions for enhanced intensity

Position	1-^{13}C-acetate	2-^{13}C-acetate	^{13}C-formate
C_1	187.7	—	—
C_2	—	146.8	—
C_3	137.8	—	—
C_4	—	183.3	—
C_5	152.0	—	—
C_6	—	117.4	—
CH_2OH	—	54.8	—
CH_3	—	—	12.0
CH_3	—	—	7.8

study.[35] Chromic acid oxidation of shanorellin labelled with 1-^{14}C-acetate gave inactive acetate ($CH_3C^{2,6}OOH$), whereas samples labelled with 2-^{14}C-acetate and $^{14}CH_3$ methionine gave acetic acid labelled in the carboxy and methyl groups, respectively. The proposal has been made that methylation and oxidation of shanorellin occur at an early stage prior to cyclization. Palmitoleic acid produced by *Saccharomyces cerevisiae* supplemented by 2-^{13}C-acetate gave a spectrum showing enhanced signal intensity for every second carbon atom in agreement with theory.[36] This experiment demonstrates the strength of the ^{13}C NMR method in that the very similar C^3–C^7 carbon atoms appear well resolved in the ^{13}C NMR spectrum (34).

8 $\overset{*}{C}H_3COOH$ ⟶ [palmitoleic acid chain with labelled carbons *] COOH (34)

Loss of ^{13}C signal strength due to deuteration was used in an investigation of the origin of skytalone produced by the fungus *Phialophora lagerbergii* (35).[37] The compound, synthesized in a culture medium containing 2-^{13}C-2-$^{2}H_3$-acetate gave enhanced signal intensities at C^2, C^4, C^5, C^7, C^{8a}, but lower intensities than expected at C^4 and C^5, indicating the presence of ^{2}H at these positions. At C^4 a triplet was observed (0.3 PPM) upfield from the normal ^{13}C signal. In the ^{2}H decoupled spectrum C^5 appeared as a singlet and C^4 as a doublet establishing a $^1H^2HC^4$-labelling pattern. No deuterium was detected at C^2 and C^7, suggesting that the polyketide is folded so that $C^{4,5}$ are specifically derived from the starter acetate, whereas $C^{2,7}$ are derived from malonate which might lose ^{2}H more readily in the biosynthetic process. An alternative conformation of the penta-ketide chain is suggested for the biosynthesis of flaviolin (section 4.9).

5 $^2H_3{}^{13}\overset{*}{C}OOH$ ⟶ [pentaketide: O, O, O, OC, O, ^2C, H_3 CoA] ⟶ [skytalone: OH, O, 8a, 7, 2, HO, OH, 4, 5, ^2H, ^1H ^2H] (35)

Chartreusin is a complex isocoumarin glycoside antibiotic produced by *Streptomyces chartreusis*. ^{13}C NMR analysis of chartreusin aglycone, biosynthesized from 1-^{13}C-, 2-^{13}C-, and 1,2-^{13}C-acetate, revealed details of its formation which would be difficult to extract from any other method.[38] From recordings of spectra from singly labelled precursors the positions of the labelled carbons could be assigned confirming that the metabolite is acetate-derived. The carbon skeleton cannot be formed by direct cyclization of one polyketide chain. Several modifications have taken place. Condensation of two or more separate polyketides or cyclization of one decaketide, e.g. according to (36) have been suggested. The labelling pattern as determined by the intensities of the ^{13}C NMR spectrum is in agreement with the suggested folding of the decatide. The use of doubly labelled acetate in conjunction with ^{13}C–^{13}C spin–

Decaketide $\overset{\square}{C}H_3\overset{\bullet}{C}OOH$

(36)

Undecaketide

Chartreusin aglycone

spin couplings, $^1J_{^{13}C^{13}C}$, also gave information as to which intact C_2 units are present. Selective ^{13}C decoupling and ^{13}C off-resonance techniques showed $^1J_{^{13}C^{13}C}$ couplings between $C^{3,4}$, $C^{4a,5}$, $C^{6,7}$, $C^{8,8a}$, $C^{2,1'}$, $C^{3',4'}$, C^{5,CH_3}, and $C^{6',7'}$. This is in agreement with the cyclization of an undecaketide to a benzpyrene type intermediate. Decarboxylation and bond cleavages with loss of three carbon atoms, rotation around $C^{2,1'}$ and lactonization gives the aglycone, route (36b). Alternatively (36a) gives an isolated $C^{4'}$ from the decarboxylation that should not give rise to any $C^{3',4'}$ coupling, and analysis of the remaining couplings anticipated gives a pattern different from that observed.

Mollisin (37)

The same method has been used to unveil details of the biosynthesis process leading to mollisin.[39] Condensation of two tetraketides, pathway (37a,b) has been suggested but ^{13}C NMR studies of mollisin obtained from fermentations of *Mollisia caesia* fed with doubly labelled ^{13}C acetate showed $^{1}J_{c2,12}$ 45.0 Hz, $^{1}J_{c13,14}$ 47.5 Hz, $J_{c6,7}$ 61.3 Hz, $J_{c3,4}$ 52.5 Hz, but no couplings with $C^{11,7}$. This eliminates pathways a and b and favours pathway c, but a single octaketide, which is cleaved, is also conceivable, pathway d. These studies provide compelling demonstrations of how the doubly labelling technique can give insights into the conformatory processes at the enzymatic level unavailable by other methods.

4.9 Derivation of structure

In a polyketide every second carbon is oxygenated which will give a basic pattern of *meta*-hydroxy substitution in the aromatic product. Hence, this is an indication that we are dealing with an acetate-derived metabolite. Shikimic acid derived metabolites are often characterized by having their hydroxy groups in *ortho* positions. Many secondary transformations can, of course, mask the biogenesis, and a metabolite can be produced by different routes in different organisms. 2,5-Dihydroxybenzoic acid is an example of the latter case. It is produced from phenylalanine in *Primula acaulis*, but from acetate in *Penicillium griseofulvum*. As already noted (section 3.5), structure proves to be an illusory guide for biosynthesis in the naphthoquinone and anthraquinone series. Having said this, we can turn to some classical examples, where biosynthetic principles have been used successfully for structural determination. By using the oxygenation pattern and the positions of alkyl and carbonyl groups as markers, it is clear that certain structures are more likely than others. For example two structures were considered for the naphthoquinone flaviolin, XXVII or XXVIII, from the available data. If biosynthetic considerations are taken into account, XXVII is the most probable structure and this was later confirmed (Fig. 10).[40]

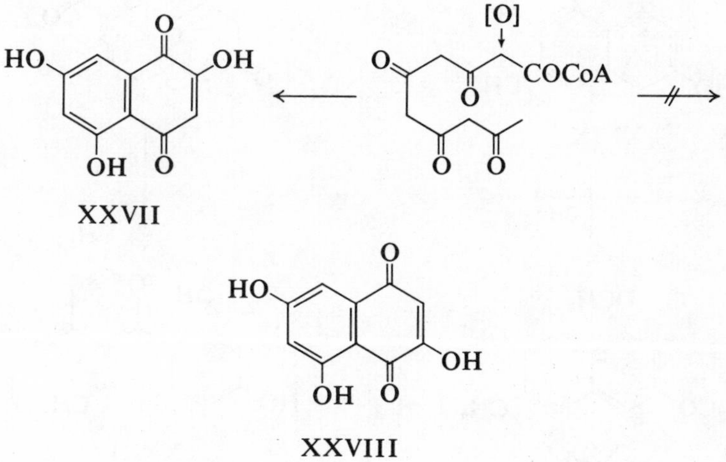

HO OH O OH O [O] COCoA O O O O

XXVII

HO O OH OH O

XXVIII

Fig. 10 Biosynthetic derivation of the most probable structure XXVII for flaviolin

The revision of the structure of eleutherinol illuminates the potency of the acetate rule (Fig. 11).[41] Structure XXIX was first proposed for eleutherinol, primarily on the basis that the naphthol XXX was claimed to be formed on alkaline hydrolysis. It is not possible to fold a polyketide chain in harmony with structure XXIX without making certain provisions. An extra carbon has to be introduced and the carboxyl group has to be reduced to methyl, route a, or the oxygenation pattern has to be drastically changed in conjunction with the introduction of one extra carbon, route b. Both possibilities are unlikely because the molecule has to undergo a series of rather unusual reactions. Suppose that instead we start out from a polyketide with two more carbons, route c; decarboxylation and cyclization then give XXXI which on degradation

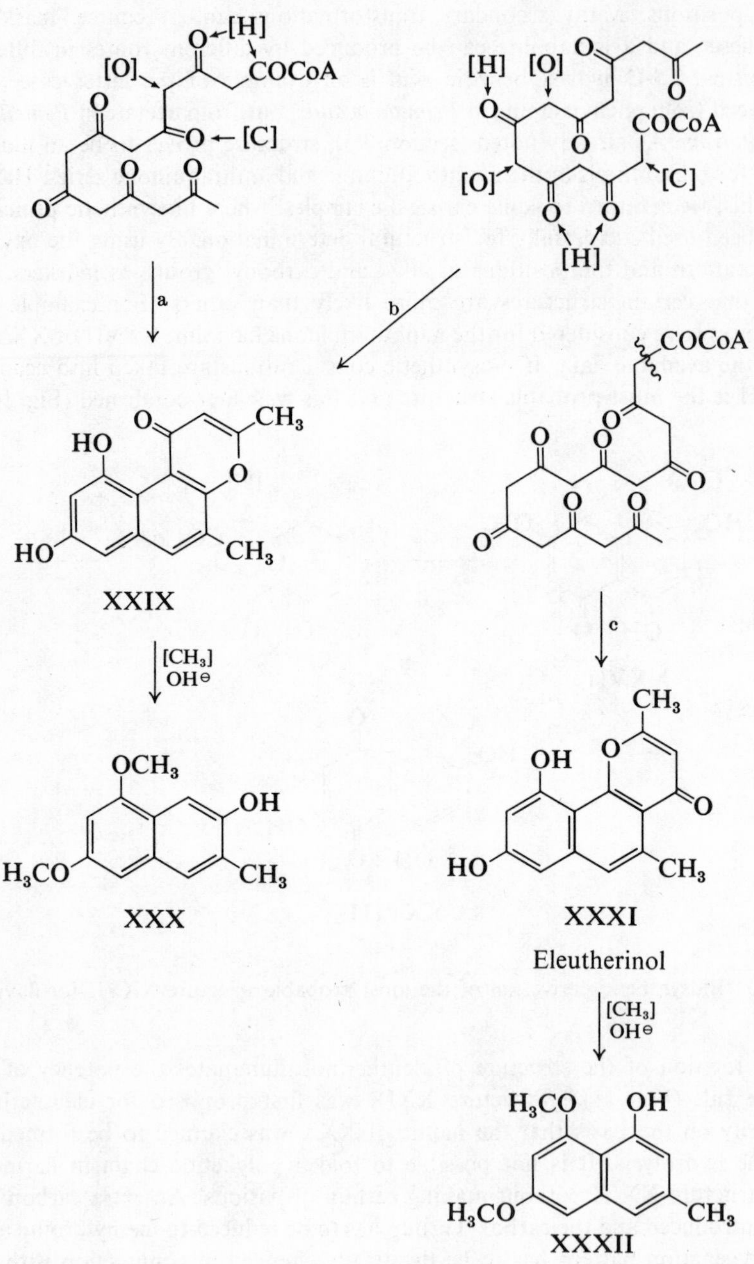

Fig. 11 Biosynthetic derivation of the structure of eleutherinol

gives a different naphthol XXXII. This folding gives directly the basic oxygenation pattern and the methyl in the naphthol correctly positioned. Generation of the other methyl by decarboxylation is a straightforward reaction. However, the structure of the naphthol has to be revised. Later synthetic work confirmed that XXXII was indeed the correct structure for the naphthol and that XXXI hence represents the correct structure for eleutherinol.

4.10 Anthraquinones, anthracyclinones, and tetracyclines

Anthraquinones are the largest group of quinones. They have been used as mordant dyes, e.g. alizarin from *Rubia tinctorum* and purgatives, e.g. emodin from *Rheum*, *Rumex*, or *Rhamnus* spp. They lost their importance, like so many other natural dyes, with the development of the synthetic dye industry. Anthraquinones are widely spread in lower and higher plants and occur also in the animal kingdom. They are present as glycosides in young plants.

It is difficult to systematize the biosynthesis of quinones because it shows up such a diversified picture.[42] Benzoquinones originate either from shikimic acid, polyketides, or mevalonate. Naphthoquinones may either by completely synthesized from acetate, e.g. in flaviolin XXVII or in the polyhydroxylated spinochromes discovered in the calcareous parts of sea urchins, or they originate from mixed biosynthesis, with e.g. one ring deriving from shikimic acid and the missing carbons of the other ring coming from acetate via glutamate (section 3.6), or one ring may come from a polyketide and the other from mevalonate. The polyketide pathway leading to anthraquinones and tetracyclines is well documented. Annelation of another mevalonate to the naphthoquinone skeleton constitutes another route.

As a general rule fungal anthraquinones and plant anthraquinones with

COCoA

Cycl.

HO, CH_3, OH, COOH, OH, O, O

1. $-H_2O$
2. [O]

1. $-CO_2$
2. $-H_2O$
3. [O]

(38)

HO, O, CH_3, COOH, OH, O, OH

?

HO, O, CH_3, OH, O, OH

Endocrocin

Emodin

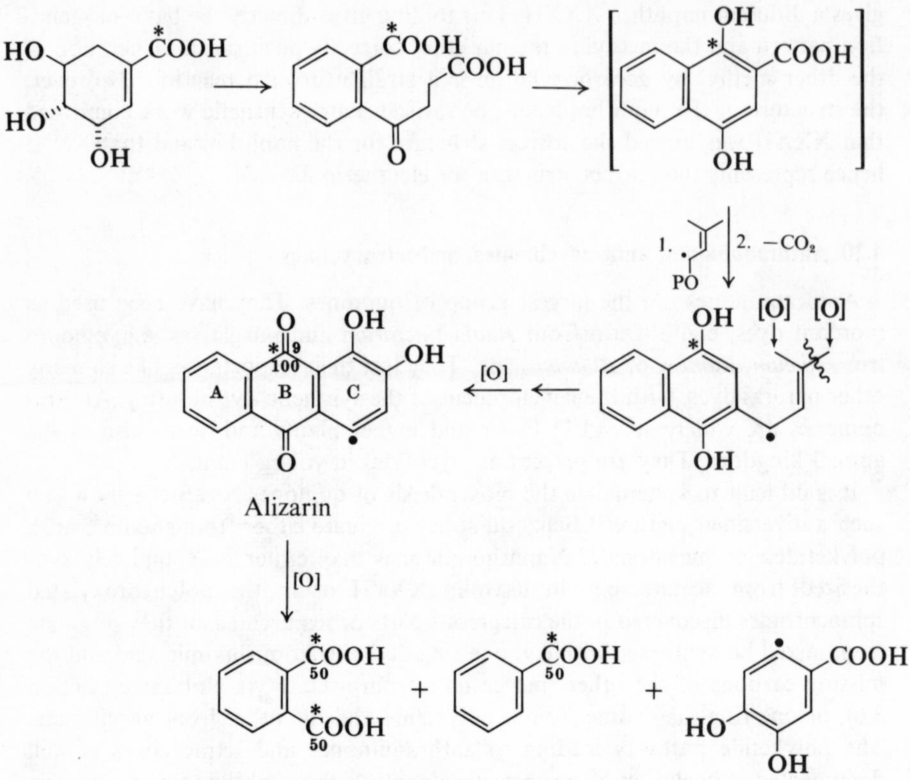

Fig. 12 Biosynthesis of alizarin

hydroxy groups are derived from polyketides, whereas plant anthraquinones devoid of hydroxy groups in one ring, e.g. alizarin, come from mixed pathways.

Emodin is constructed from one acetate and seven malonate units, but since it is questionable if endocrocin is an intermediate, the decarboxylation may occur at an earlier stage (38). There are produced numerous anthraquinones by the octaketide pathway conforming to the basic emodin structure. They arise via different folding, *O*-methylation, side chain oxidation, nuclear hydroxylation or elimination of hydroxyl groups, chlorination, dimerization via phenol oxidation, etc.

The historical dye alizarin has a more complicated biosynthesis. *Rubia tinctorum* was fed with 7-^{14}C-shikimic acid and alizarin was isolated and degraded to phthalic, veratric and benzoic acids.[43] The label was recovered only in the carboxy groups. Phthalic acid had the same specific activity as alizarin, benzoic acid 50 per cent, and veratric acid no activity, which shows that the carboxyl group of shikimic acid is located at C^9 and no symmetrical intermediate is involved during the construction of ring C. The remaining three carbon atoms of ring B come from glutamic acid. 5-^{14}C-mevalonic acid is incorporated in ring

C. Dimethylallyl pyrophosphate alkylates selectively in the *meta* position relative to the labelled carbonyl group, and one of the methyls is lost as carbon dioxide (Fig. 12).

Anthracyclinones and tetracyclines belong to a group of linear tetracyclic, highly active antibiotics, some of which have found use in the treatment of cancer, e.g. daunomycin. They are produced in cultures of *Streptomyces* spp. The order of the modifications of the polyketide chain has been determined by a series of mutant studies. Amidomalonyl CoA is the starter of the nonaketide chain. Methylation at C^6, reduction at C^8, and cyclization to the fully aromatic system occur next followed by hydroxylations at C^4, C^6 and C^{12a} and amination (Fig. 13). We have here examples of other starters, propionic acid for pyrromycinone and daunomycinone, and amidomalonic acid for tetracycline.

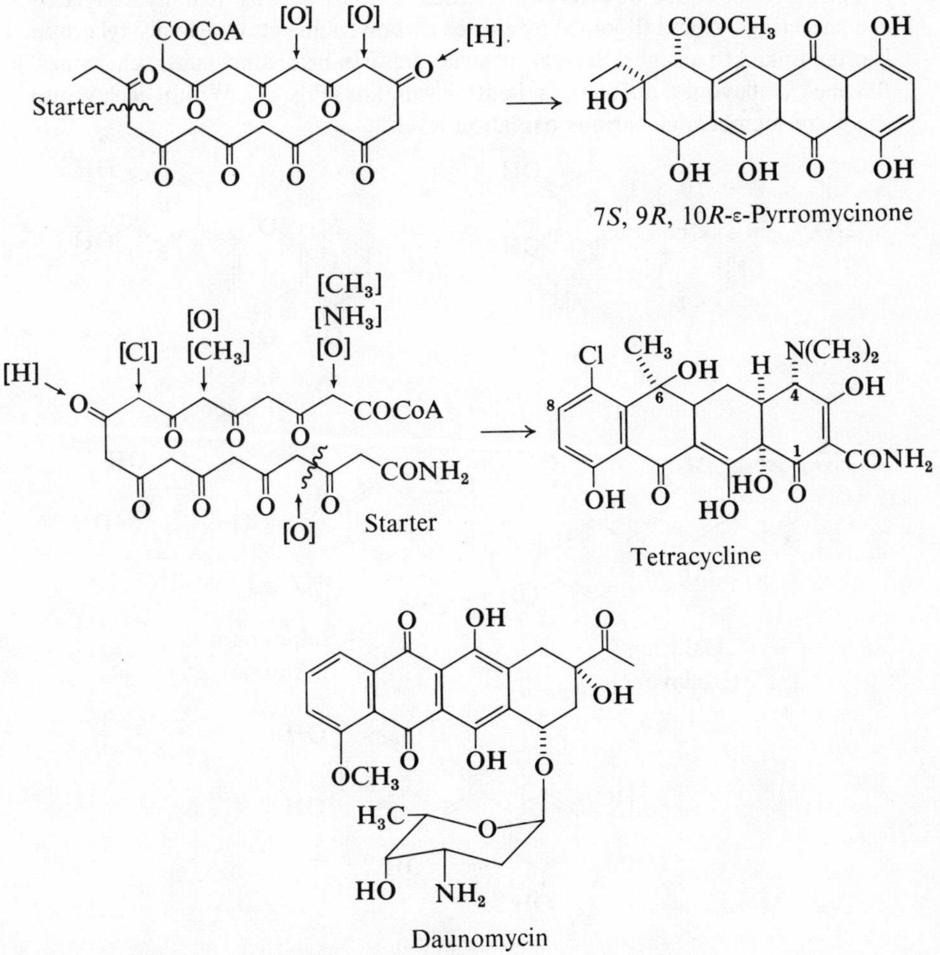

7*S*, 9*R*, 10*R*-ε-Pyrromycinone

Tetracycline

Daunomycin

Fig. 13 Biosynthesis of tetracyclines

4.11 Flavonoids

No colouring matters are so conspicious as the flavonoids in contributing to the beauty and splendour of flowers and fruits in nature. The flavones give yellow or orange colours, the anthocyanins red, violet or blue colours, i.e. all the colours of the rainbow but green. The occurrence of this numerous class of oxygen heterocycles is restricted to higher plants and ferns. Mosses contain a few flavonoid types but they are absent in algae, fungi and bacteria. The hydroxylation and methylation patterns appear to be genetically controlled, i.e. the distribution of flavonoids is a useful auxiliary tool for classification purposes.[44] Biologically the flavonoids play a major role in relation to insects pollinating or feeding on plants but some flavonoids have a bitter taste, repelling certain caterpillars from feeding on leaves.

The flavonoids are structurally characterized by having two hydroxylated aromatic rings, A and B, joined by a three carbon fragment. One hydroxyl group is often linked to a sugar. Several substructures can be distinguished: chalcones, flavones, isoflavones, aurones, and anthocyanidins (Fig. 14). Within each group there are members at various oxidation levels.

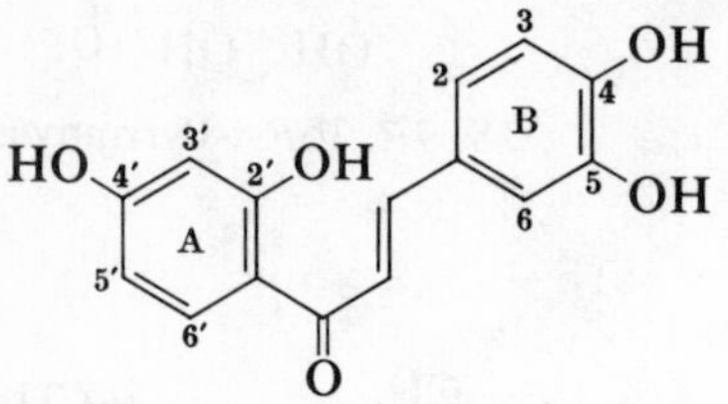

Butein
(Chalcone)

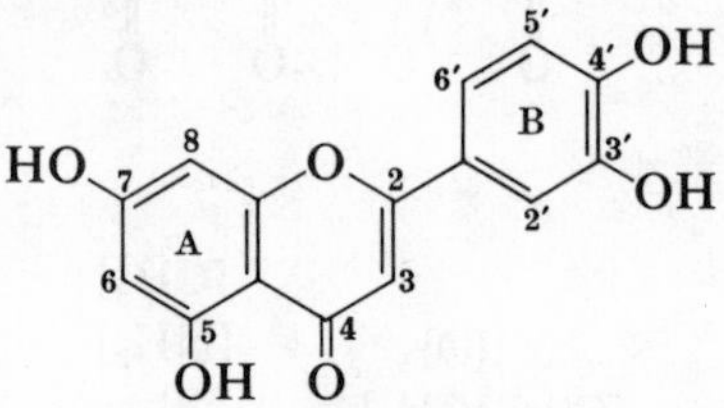

Luteolin
(Flavone)

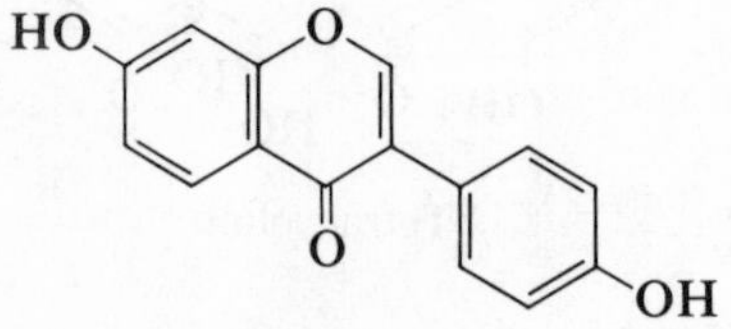

Daidzein
(Isoflavone)

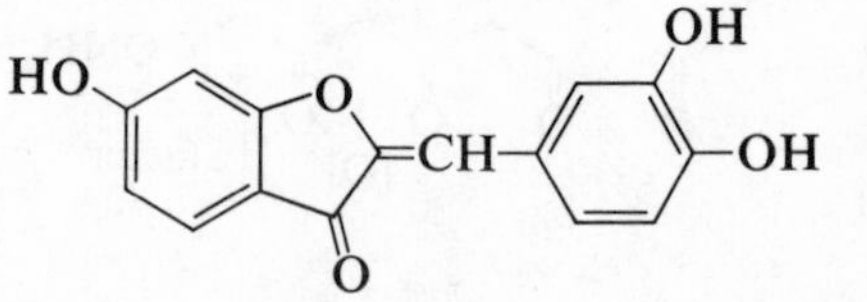

Sulphuretin
(Aurone)

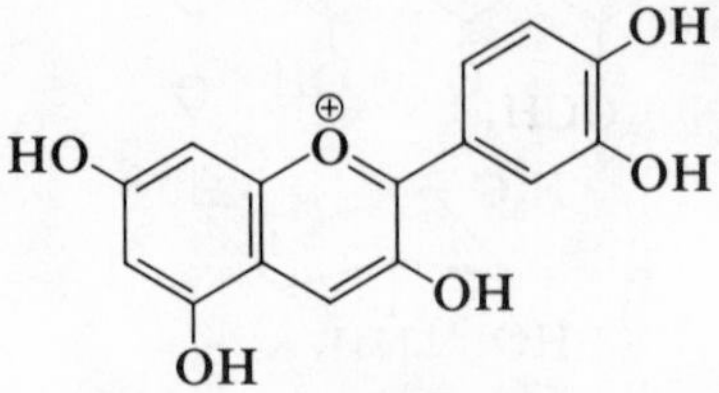

Cyanidin
(Anthocyanidin)

Fig. 14 Structures of flavonoid compounds

2′-Hydroxy-substituted chalcones cyclize easily to flavanones, the structure of which is stabilized by hydrogen bonding at C^4O and C^5O. The structural determination was first accomplished by alkaline degradation, which gave e.g. acetic acid, acetophlorophenone, *p*-hydroxybenzaldehyde and phloroglucinol according to a retrocondensation mechanism (39). Anthocyanidins were related to 3-hydroxyflavones by reduction of the carbonyl group followed by treatment with acid (40). The basic skeleton arises from three malonyl CoA units and a

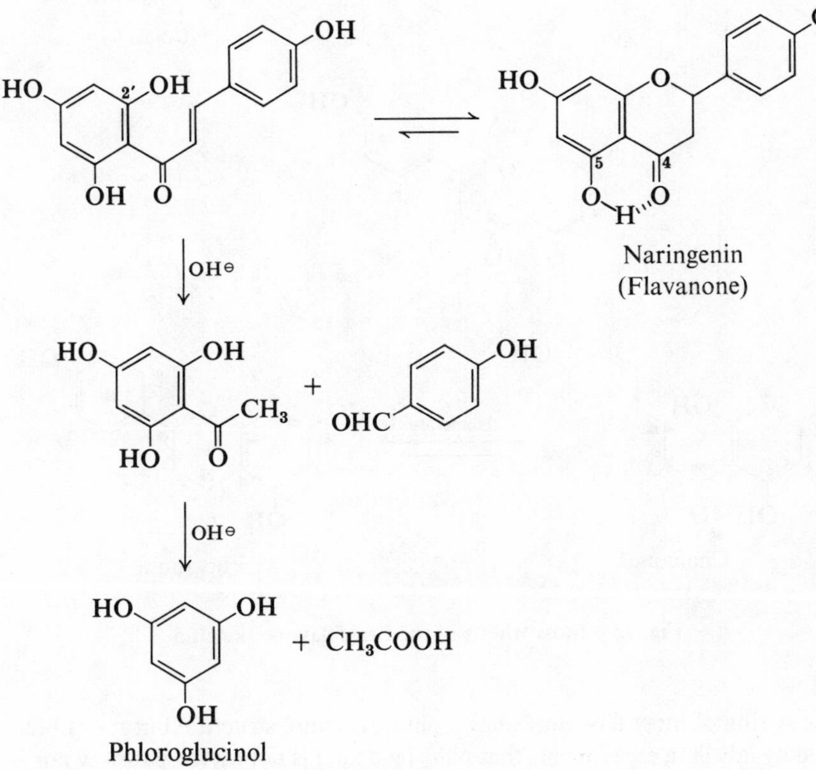

HO, 2′ OH, OH, OH O ⇌ HO, O, OH, 5, 4, O–H···O

Naringenin
(Flavanone)

$OH^{\ominus}$

HO, OH, CH_3, HO O + OH, OHC (39)

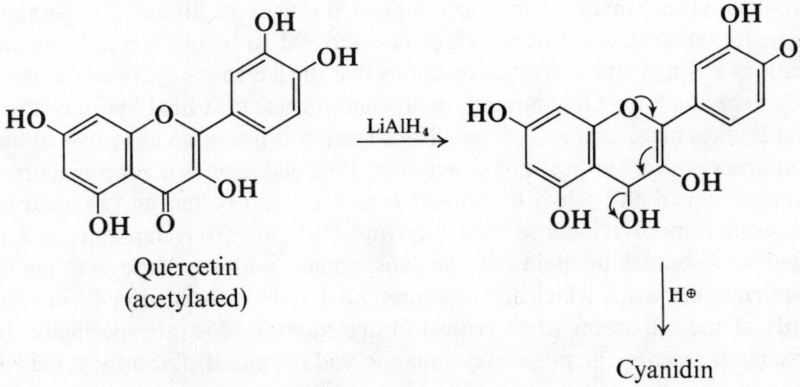

$OH^{\ominus}$

HO, OH, OH + CH_3COOH

Phloroglucinol

HO, O, OH, OH, OH, OH O $\xrightarrow{LiAlH_4}$ HO, O, OH, OH, OH, OH OH (40)

Quercetin
(acetylated)

$H^{\oplus}$

Cyanidin

cinnamoyl CoA as starter which is incorporated intact. Specifically labelled acetic acid and phenylalanine were incorporated according to Fig. 15.[45] The cyclization of the chalcones are enzymatically controlled since the flavanones are optically active.

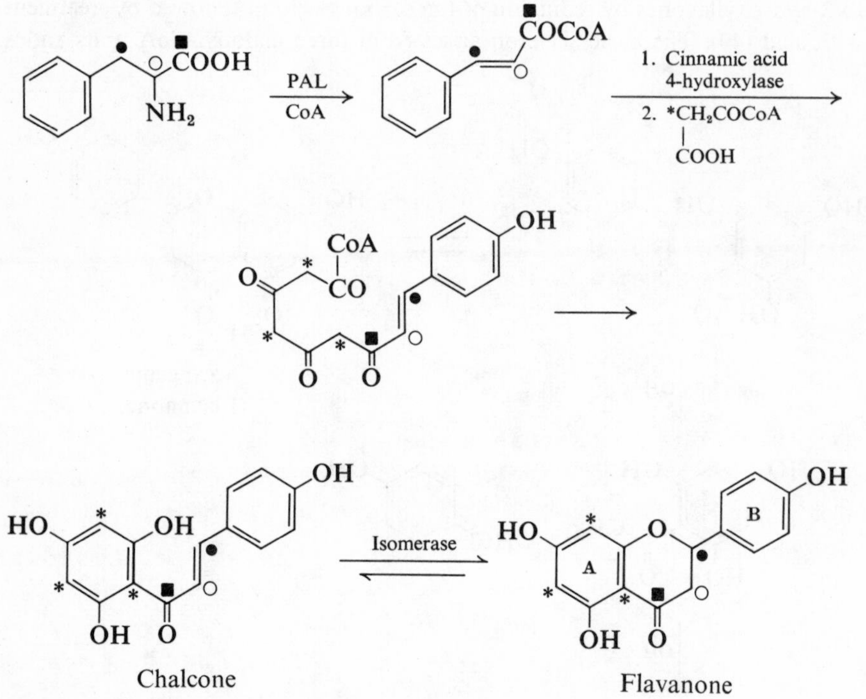

PAL
CoA
1. Cinnamic acid 4-hydroxylase
2. $^*CH_2COCoA$ | COOH
Isomerase
Chalcone
Flavanone

Fig. 15 Biosynthesis of the basic flavone skeleton

The A ring of most flavonoids has a phloroglucinol structure but it has been proved by labelling experiments that phloroglucinol is not on the pathway nor is phloroglucinyl cinnamate. It was once suggested that some kind of biosynthetic Fries rearrangement could occur which later proved to be incorrect. There are indications arising from enzymatic reactions that the flavanone synthase complex is rather specific for *p*-OH-cinnamic acid which means that final hydroxylation of ring B takes place at the C_{15} stage. The situation is not quite unequivocal but several observations support this conclusion. Phenylalanine is a good precursor but cinnamic acid and *p*-hydroxycinnamic acid are still better, and in contrast caffeic acid is poorly incorporated implying that *p*-hydroxycinnamic acid is situated at a branching point of the biosynthetic pathway.[46] Several plants contain cinnamic acids which are not represented in the co-occurring flavonoids. The pH of the cell seems to be critical in determining substrate specificity. In *Haplopappus gracilis* e.g., *p*-hydroxycinnamic acid is utilized efficiently at pH 8.0

as precursor for eriodictyol, whereas caffeic acid is utilized for the purpose only at pH 6.5–7.0.[47] It is an interesting idea to think of pH as a factor determining the metabolic ratio in different plant cells.

O-Methylation and glycosidation are modifications taking place at the final stage.

The anthocyanidins are biosynthesized from flavanones via dihydroflavonols according to (41).

Eriodictyol

[O]

Dihydroquercetin

[O]

Quercetin

$H^{\oplus}$

Cyanidin

(41)

The flavanones or chalcones are also precursors for the isoflavones.[45] The exact nature of this rearrangement in unknown, but a plausible mechanism is oxidation of the anion to a diradical, combination to the cyclopropanoid intermediate and proton elimination (42). It is unlikely that the oxidation proceeds to a carbonium ion α to the carbonyl group followed by a carbonium ion rearrangement as has been proposed. The rotenoids, which are used as fish poisons, are structurally related to the isoflavones. Tracer experiments in seedlings of *Amorpha fruticosa* showed that rotenone is biosynthesized from formononetin by extra hydroxylation and methylation in ring B, hydrogen abstraction

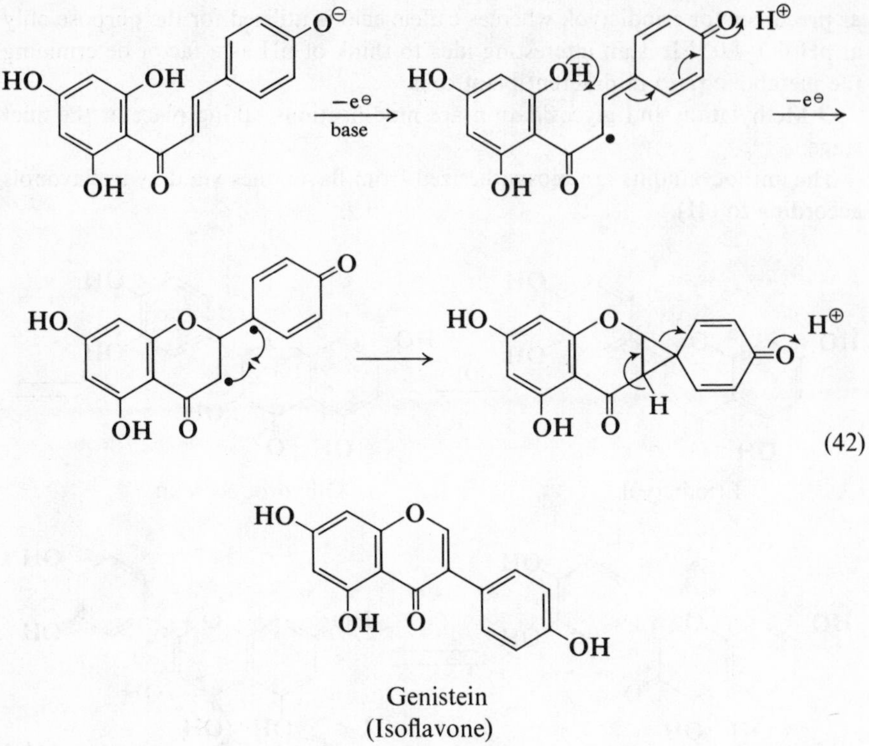

HO OH O⊖ OH O $\xrightarrow[\text{base}]{-e^{\ominus}}$ HO OH O H⊕ OH O $\xrightarrow{-e^{\ominus}}$

HO O O OH O ⟶ HO O H⊕ O H OH O

(42)

HO O OH O OH

Genistein
(Isoflavone)

from the methyl (section 3.3) followed by radical cyclization and isoprenylation to rotenoic acid. Radical Michael type addition is a favoured and common reaction. Epoxidation of the isoprenoid double bond of rotenoic acid, cyclization via opening of the epoxide, and finally dehydration of the tertiary alcohol, actually discovered as a natural constituent, gives rotenone (Fig. 16).[48]

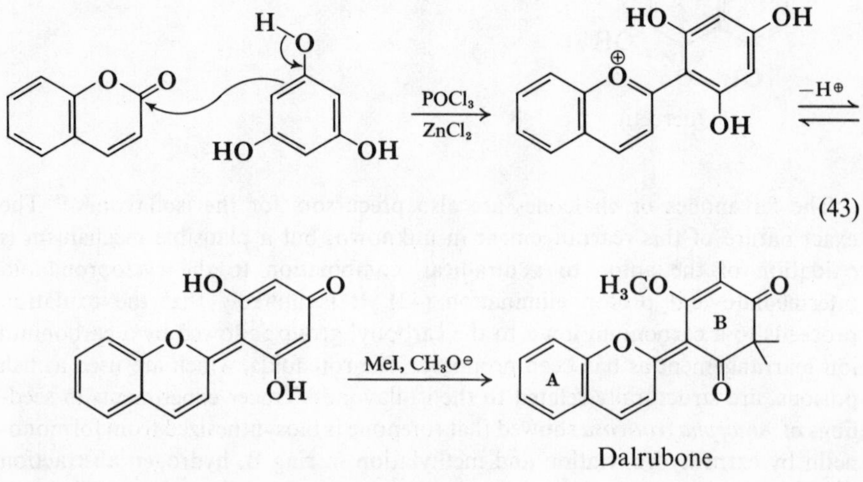

O O H O HO OH $\xrightarrow[ZnCl_2]{POCl_3}$ HO OH O⊕ OH $\xrightleftharpoons{-H^{\oplus}}$

(43)

HO O O OH $\xrightarrow{MeI,\ CH_3O^{\ominus}}$ H_3CO O A O B O

Dalrubone

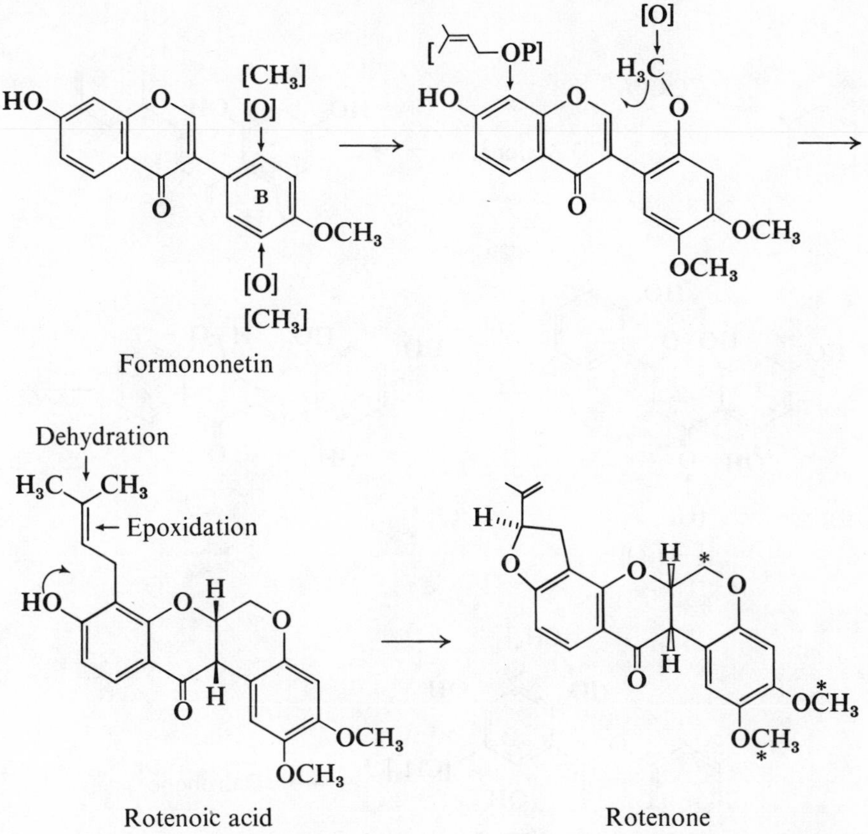

Formononetin

Rotenoic acid

Rotenone

Fig. 16 Biosynthesis of rotenone. The starred carbon atoms originate from methionine

Dalrubone, isolated from *Dalea emoryi*, has an unusual oxygenation pattern in that rings A and B appear to be reversed.[49] A biomimetic synthesis (43) from co-occurring coumarin and phloroglucinol suggests that this flavonoid may arise via a different pathway.[50]

However, the reversal can be explained by a 1,3-carbonyl transposition in the chalcone in the normal biosynthesis, effected by β-oxidation, reduction and elimination of water (Fig. 17). This pathway is supported by feeding experiments in *Glycyrrhiza echinata* which produces transposed chalcones.[51]

The flavonoids are degraded by several routes[52] (Fig. 18). The first step is an elimination of the sugar by a glycosidase. Ring A is usually broken down to carbon dioxide and B gives rise to benzoic acids.[53] The cleavage of the pyrone ring is effected by a monooxygenase at C^{4a} (route a). Oxidation at $C^{1'}$ gives rise to chromones[54] and there is indication that certain chromones are *post mortem* products[55] (route b).

These findings illuminate the long standing problem concerning the true

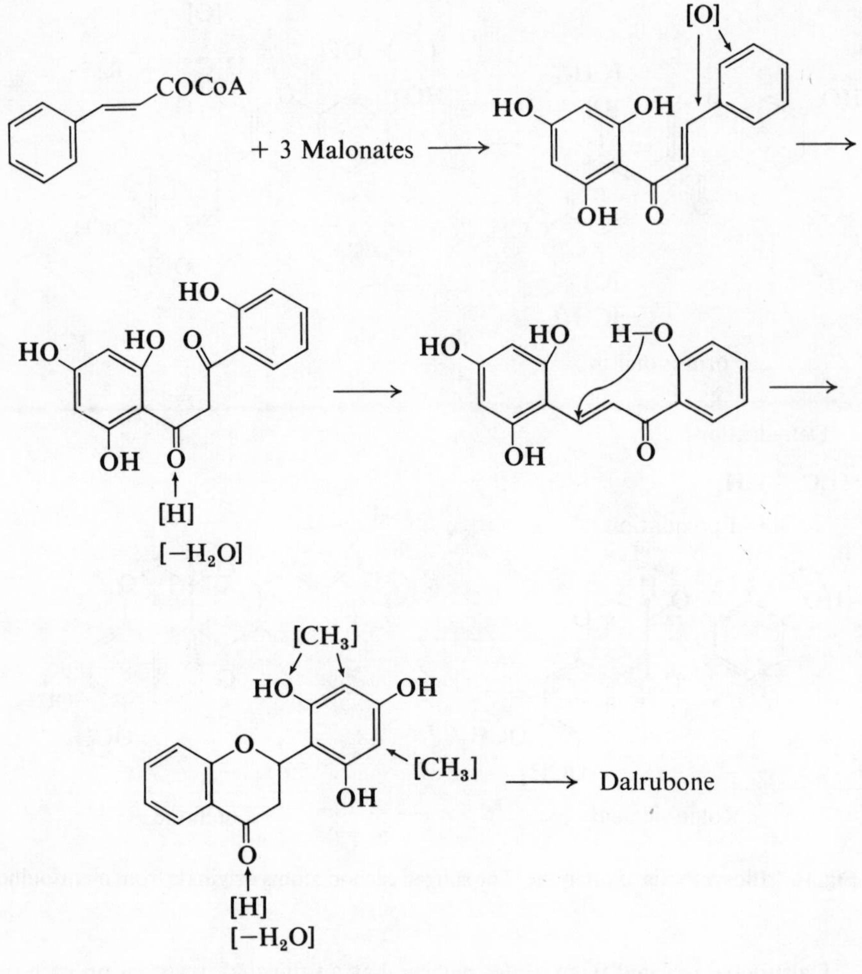

Fig. 17 Proposed biosynthesis for dalrubone

origins of a metabolite. Is it produced by a direct biosynthetic route or is it simply a secondary degradation product—an artefact? In this case it seems unnecessary to speculate on alternative routes to chromones missing a substituent at C^2.

The flavonoids are thus produced by folding the polyketide as shown in Fig. 15. A slightly different folding gives stilbenes. Feeding experiments with *Pinus resinosa* have shown that pinosylvin originates from the same C_6–C_3–C_6 unit by cyclization according to Fig. 19.[56] Most stilbenes have lost the carboxyl group, although it remains in hydrangenol which like the antimicrobial plant constituent resveratrol derives from *p*-hydroxycinnamic acid and three malonate units.[57,58]

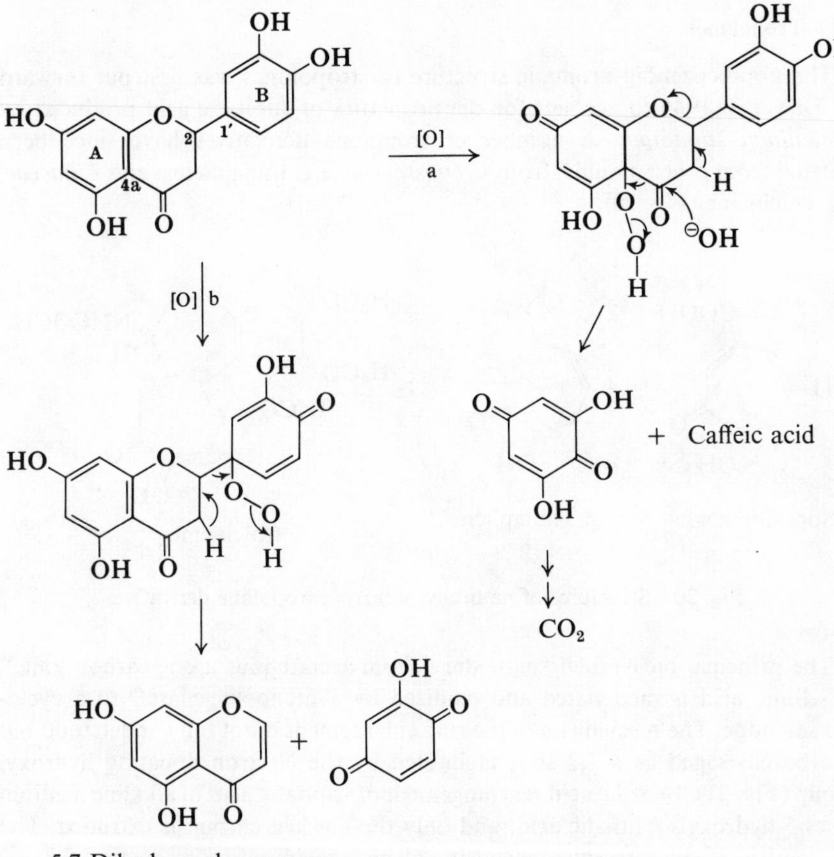

[O]
a

[O] b

+ Caffeic acid

CO_2

5,7-Dihydroxychromone

Fig. 18 Flavonoid metabolism

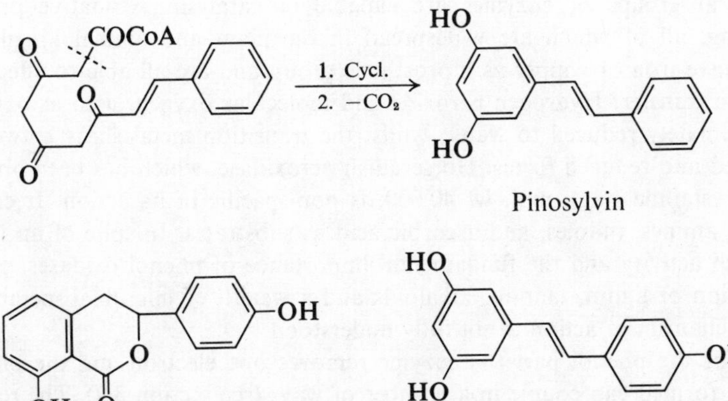

COCoA

1. Cycl.
2. $-CO_2$

Pinosylvin

Hydrangenol

Resveratrol

Fig. 19 Naturally occurring stilbene derivatives

4.12 Tropolones

The non-benzenoid aromatic structure for tropolones was first put forward by Dewar in 1945 to account for the properties of stipitatic acid produced by *Penicillium stipitatum*. A number of tropolone derivatives have since been isolated from other moulds, from *Cupressaceae*, e.g. thujaplicins, and *Liliaceae*, e.g. colchicine (Fig. 20).

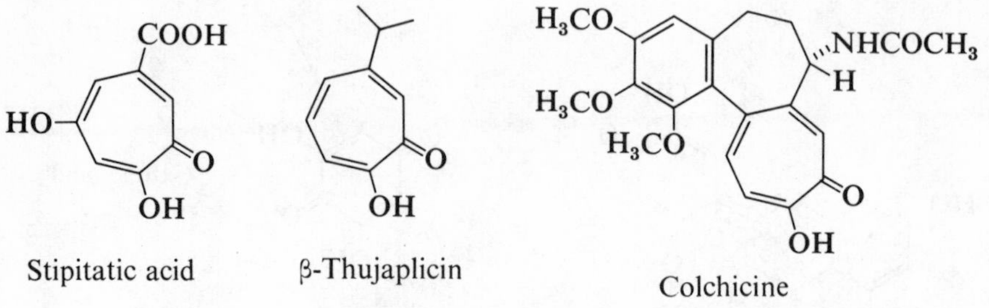

Stipitatic acid β-Thujaplicin Colchicine

Fig. 20 Structures of naturally occurring tropolone derivatives

The principal biosynthetic path starts from acetate plus a one carbon unit.[59] Orsellinic acid is methylated and oxidized by a monooxygenase[60] to a cyclohexadienone. The mechanism of the ring enlargement is not fully understood but can be envisaged as a 1,2 shift facilitated by the electron donating hydroxyl group (Fig. 21). Benzilic acid rearrangement of stipitatic acid in alkaline medium gives 5-hydroxyisophthalic acid, and only the marked carbon is extruded. The mechanism of the ring enlargement in colchicine is discussed in section 7.3.

4.13 Oxidative coupling of phenols

Several groups of enzymes are capable of catalysing oxidative phenolic coupling, all of which are widespread in the plant and animal kingdoms.[61] They have iron or copper as a prosthetic group and are all able to effect one-electron transfer. Hydrogen peroxide and molecular oxygen, used as oxidants, are ultimately reduced to water, whilst the transition metal shifts between its oxidized and reduced forms. Horseradish peroxidase, which has been obtained in a crystalline state, mol. wt 40 000, is non-specific in its action. It can use iodide, amines, indoles, and ascorbic acid as substrates. In spite of an intense research activity and the fundamental importance of phenol oxidases, e.g. for formation of lignin, tannins, alkaloids and a wealth of microbial metabolites, the mechanism of action is not fully understood.

As for the phenol part the enzyme removes one electron and the phenoxy radical formed can couple in a number of ways (see section 3.7). The reaction can be simulated *in vitro* by oxidation with ferricyanide as in (44) showing the oxidation of *p*-cresol to the so-called Pummerer's ketone. When this ketone was

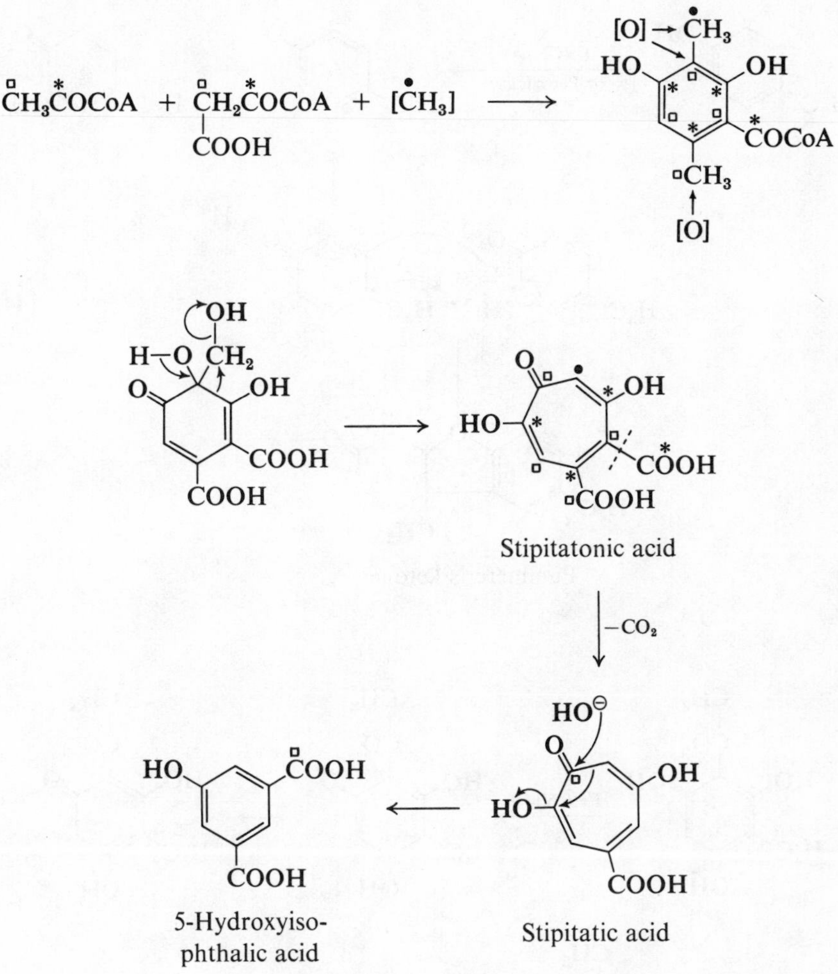

Fig. 21 Biosynthesis of tropolones

prepared from *p*-cresol or dehydrogriseofulvin was prepared from griseophenone (Fig. 8), by the action of cell free phenol oxidase preparations racemic products were obtained. The latter reaction is relevant to the biogenetic scheme proposed for griseofulvin. The intact organism *Penicillium griseofulvum* produces optically active griseofulvin. This suggests that the enzyme solely acts as producer of phenoxy radicals which then can couple uninfluenced by the topology of the enzyme. In the intact cell, on the other hand, the phenol oxidase must be integrated into a larger enzyme complex also catalysing the condensation and cyclization of the polyketide.[62]

It has been argued that the dimerization is not the result of a coupling of two radicals, but rather of an attack of one radical on another phenol. The new

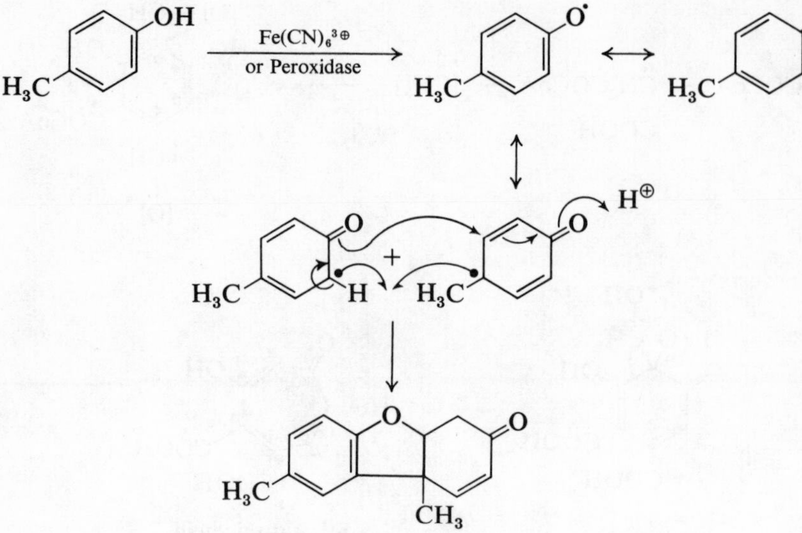

OH

$Fe(CN)_6^{3\oplus}$ or Peroxidase

H_3C — O• ⟷ O, •H

$H^\oplus$

O + O

H_3C •H H_3C • (44)

Pummerer's ketone

CH_3 CO HO OH H_3C OH

$Fe(CN)_6^{3\oplus}$

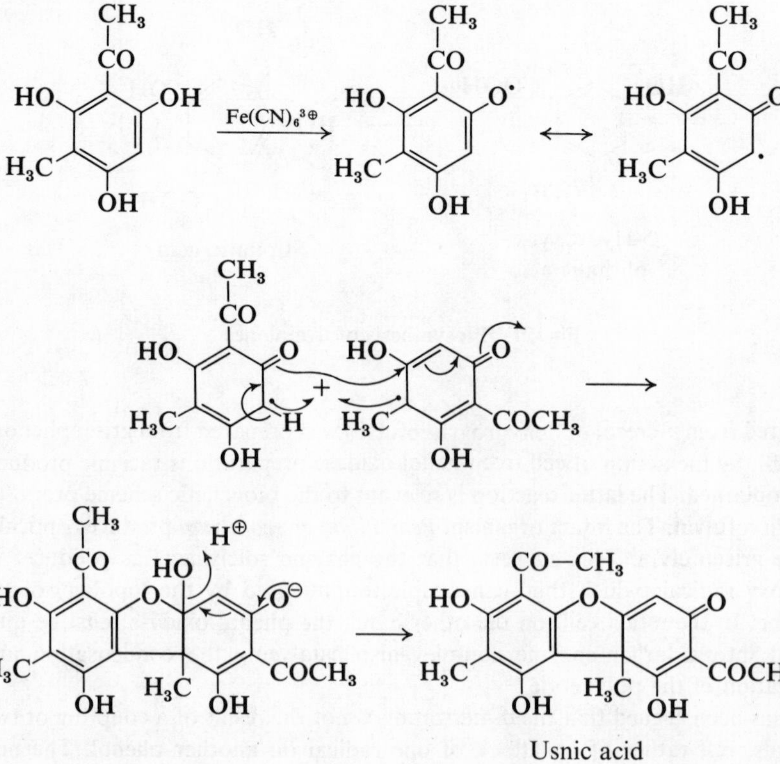

(45)

Usnic acid

radical is then oxidized. This seems not to be the case because oxidation of a phenol in the presence of a large excess of 1,2-dimethoxybenzene does not give rise to any crossed products.

The principle of oxidative phenol coupling is convincingly demonstrated in the synthesis of usnic acid from two moles of methylphloracetophenone (45).[63] Incorporation experiments with labelled precursors later confirmed that the biosynthetic machinery followed a similar path.[64] The introduction of *C*-methyl is an early step because phloroacetophenone is not utilized as substrate in the biosynthesis.

Picrolichenic acid from the lichen *Pertusaria amara* is built up in the same way by oxidation of two moles of 2,4-dihydroxy-6-pentyl-benzoic acid (46).[65] We have encountered this reaction earlier in some tannins e.g. in ellagic acid (section 3.4).

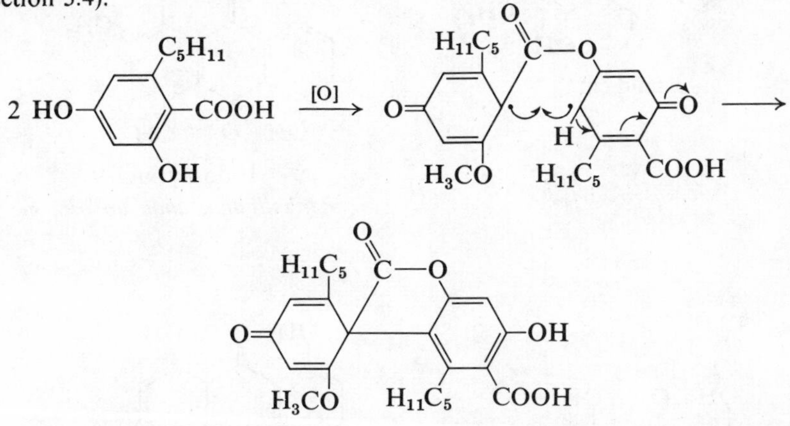

(46)

Picrolichenic acid

A number of dimeric naphthalenes and anthracenes occur in nature which evidently arise via phenol oxidation (47–50). Insects of the *Aphididae* family, many of which are serious pests on cultivated crops, produce a series of pigments[66] having the dihydroxyperylenequinone nucleus. Hypericin[67] is a photo-

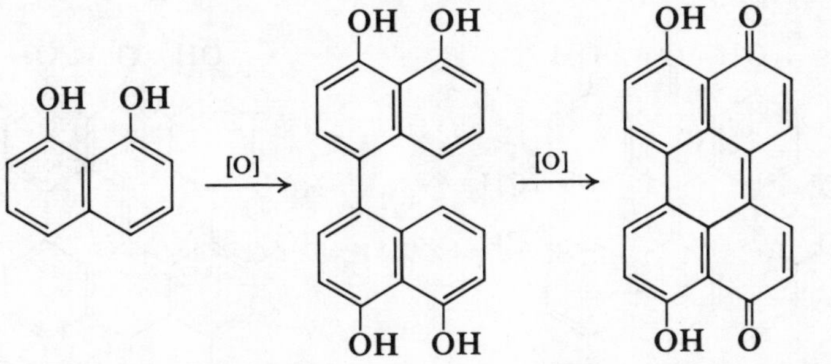

(47)

3,10-Dihydroxyperylene-4,9-quinone
Daldinia concentrica[68,69]

dynamic pigment produced by *Hypericum perforatum*. It originates from emodin anthrone which is dimerized and further oxidized to photohypericin and finally hypericin. This process can be mimicked by passing air into an alkaline solution of emodin anthrone.

Xanthones[72] are widely distributed among several families, e.g. *Gentianaceae*,

Islandicin $\xrightarrow{[O]}$ Iridoskyrin (48)

Iridoskyrin
Penicillium islandicum[70,71]

Emodin anthrone $\xrightarrow{[O]}$ $\xrightarrow{[O]}$ (49)

Protohypericin $\xrightarrow{[O]}$ Hypericin

Hypericin
Hypericum perforatum

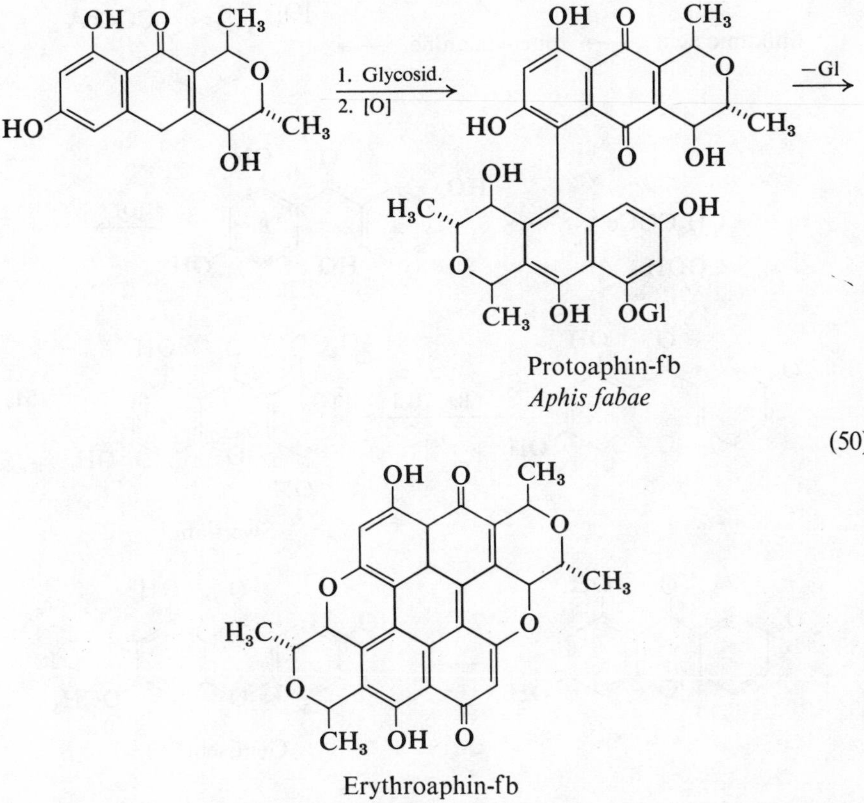

1. Glycosid.
2. [O]

−Gl

Protoaphin-fb
Aphis fabae

(50)

Erythroaphin-fb

Guttiferae, *Moraceae*, *Polygalaceae*, and are of chemotaxonomic interest. Ring A contains characteristically hydroxyl groups at C^5 and/or C^7 and originates from shikimic acid.[73] A hydroxyl group is probably first introduced into the *meta* position in the benzoic acid ring. Phenol oxidation and cyclization then automatically locates the OH group at C^5 or C^7 (51). It is also demonstrated that the polyhydroxybenzophenone is incorporated and co-occurrence is known in a few cases. The *meta*-oxygenation pattern of ring B indicates its acetate origin and it has been shown by radio labelling that acetate is incorporated.[74] We cannot dismiss another route to the 5- or 7-OH substituted xanthones from *p*-hydroxybenzoic acid via the spirointermediate which undergoes a 1,2 shift (52)[75] The spiro-structure is contained in several fungal metabolites, e.g. griseofulvin, where the 1,2 shift is inhibited by substitution. In some cases the whole framework is formed from one polyketide, e.g. in griseoxanthone C which is of fungal origin.[76] The oxygenation pattern is also different and no phenolic oxidation is called for in this case (53).

Thyroxine is an iodine-containing thyroid hormone which controls the oxygen consumption in the tissue. Too high a production of thyroxine causes an increase of the metabolic rate and is recognized as Basedow's disease.

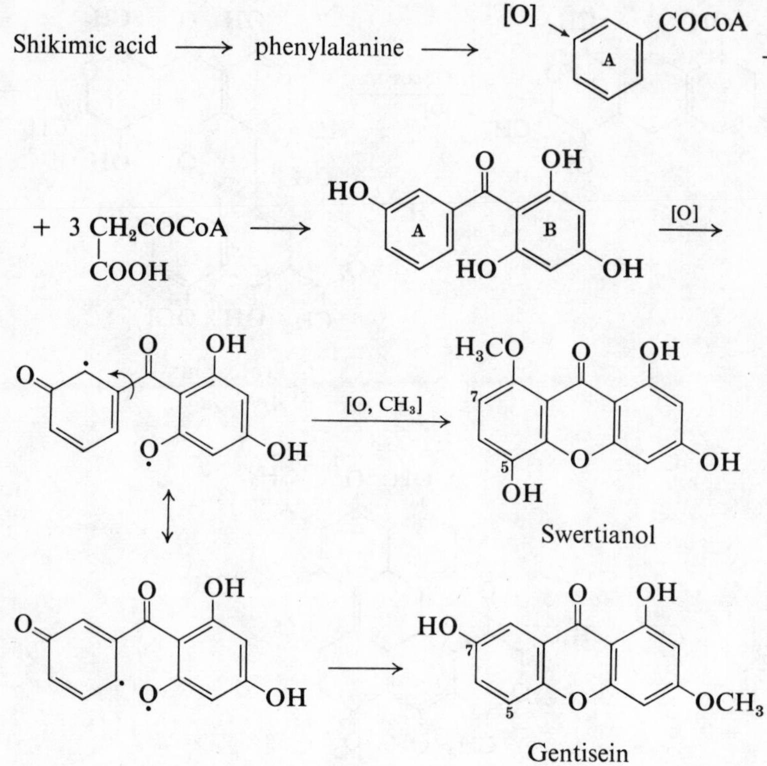

Shikimic acid ⟶ phenylalanine ⟶ [O] → (A) COCoA +

+ 3 CH_2COCoA / COOH ⟶ HO (A) C(=O) (B) OH, HO, OH —[O]⟶

(51)

—[O, CH_3]⟶ H_3CO, O, OH, 7, 5, OH, OH

Swertianol

⟶ HO, 7, 5, O, OH, OCH_3

Gentisein

(52)

HO, O, OH, HO, OH —[O]⟶ ⟶

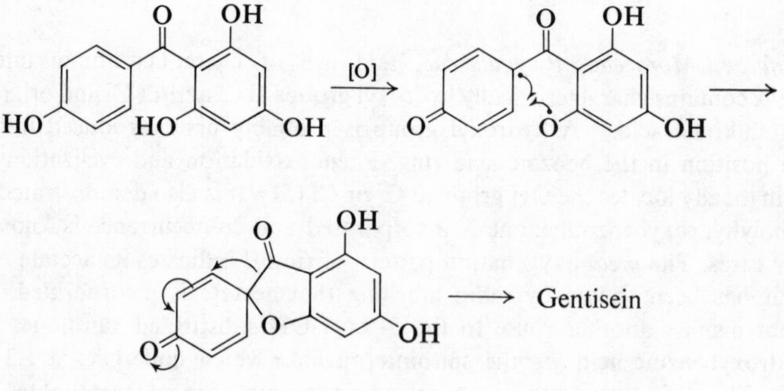

⟶ Gentisein

(53)

CH_3 CO, O, OC CoA, O O O O ⟶ CH_3, O, OH, HO, O, OCH_3

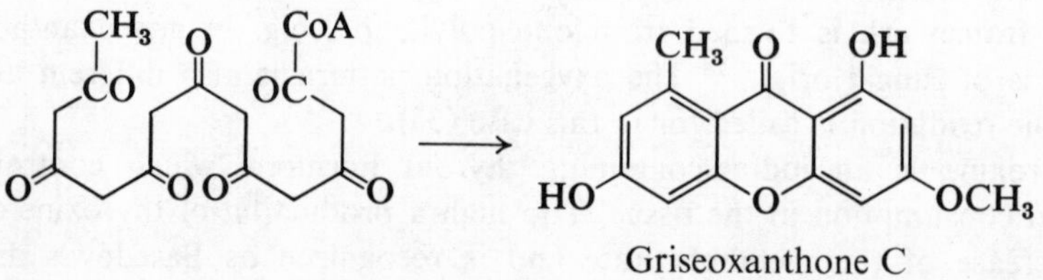

Griseoxanthone C

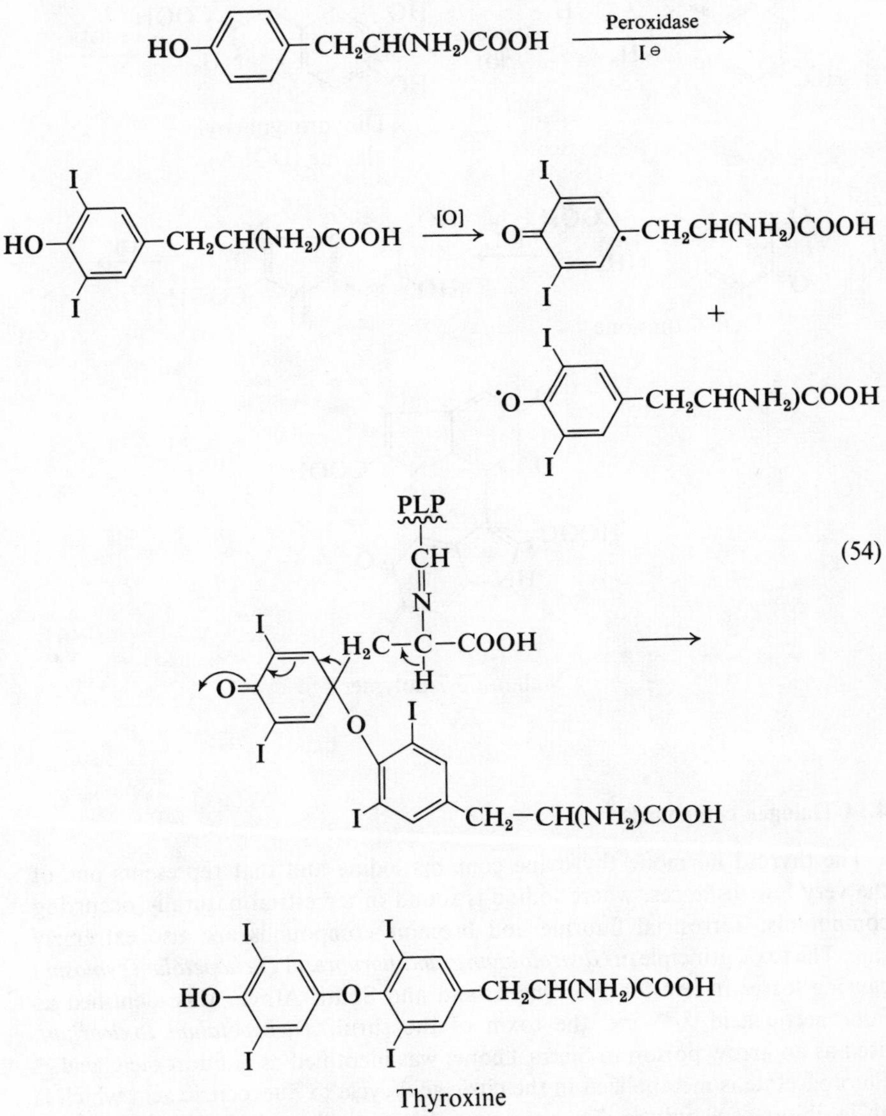

HO–C_6H_4–$CH_2CH(NH_2)COOH$ $\xrightarrow[I^{\ominus}]{\text{Peroxidase}}$

HO–$C_6H_2I_2$–$CH_2CH(NH_2)COOH$ $\xrightarrow{[O]}$ O=$C_6H_2I_2$•–$CH_2CH(NH_2)COOH$ + •O–$C_6H_2I_2$–$CH_2CH(NH_2)COOH$

PLP

CH

N

H_2C–C–COOH

H

O

CH_2–$CH(NH_2)COOH$ $\longrightarrow$

(54)

HO–$C_6H_2I_2$–O–$C_6H_2I_2$–$CH_2CH(NH_2)COOH$

Thyroxine

Thyroxine deficiency causes myxoedema. It is formed by electrophilic iodination of tyrosine which then couples with another iodinated tyrosine molecule with loss of a side chain, presumably in a pyridoxal phosphate mediated reaction (54).

Peroxidases are responsible for the darkening of cut fruits or vegetables, whereby catechols are rapidly oxidized forming darkly coloured polymers. Darkening of the skin by the action of sunshine is caused by formation of melanin, a polymer built up by enzymatic oxidation of tyrosine, predominantly through 3,7-linkage of the intermediate indole derivative (55).

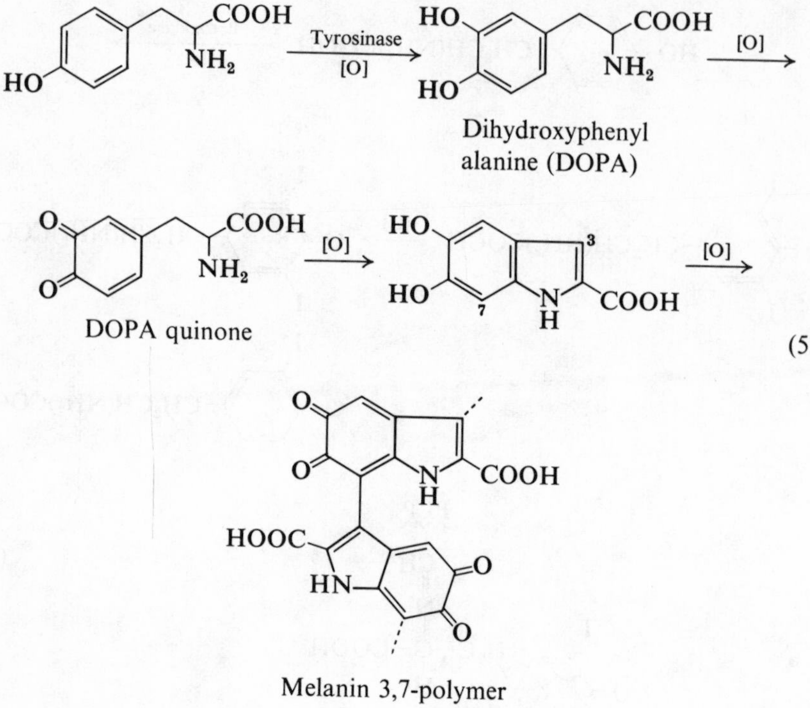
COOH
NH2
HO
Tyrosinase
[O]
HO
HO
COOH
NH2
[O]
Dihydroxyphenyl
alanine (DOPA)
O
O
COOH
NH2
DOPA quinone
[O]
HO
HO
3
7
N
H
COOH
[O]
O
O
N
H
COOH
HOOC
HN
O
O
Melanin 3,7-polymer

(55)

4.14 Halogen compounds

The thyroid hormone thyroxine contains iodine and that represents one of the very few instances, where iodine is found in terrestrial naturally occurring compounds. Terrestrial fluorine and bromine compounds are also extremely rare. The toxic principle in *Gastrolobium grandiflorum* and *Dichapetalum cymosum* causing losses in livestock in Queensland and South Africa, was identified as fluoroacetic acid,[77,78] and the toxin of the shrub *Dichapetalum toxicarium*, used as an arrow poison in Sierra Leone, was identified as ω-fluorooleic acid.[79] Fluoroacetate is metabolized in the citric acid cycle to fluorocitric acid which is an inhibitor of aconitase. Thus the organism commits suicide by carrying out lethal syntheses. Various chloro compounds have been isolated, especially from microorganisms. We have already met some representatives of the group, chlorotetracycline and griseofulvin. Fig. 22 gives further examples of naturally occurring terrestrial halogen compounds.[80] By replacing chloride with bromide in the culture medium several organisms start producing the brominated analogues, e.g. griseofulvin. In a strict sense these metabolites may not be considered natural but it is not unreasonable to argue that they do occur in nature, yet in quantities which have escaped detection.

The recent active research on marine natural products has uncovered a great number of chlorine, bromine, and iodine containing compounds,[81–83] many of which have unusual structures, and it is quite clear that the metabolism of marine organisms has been adapted to the unique saline aqueous environment.

The genus *Laurencia* of Rhodophyta proved to be a rich source of structurally interesting compounds, mostly of terpenoid nature but several C_{15} acetylenic cyclic ethers containing halogen of polyketide origin have been isolated as well. From the intestinal tract of certain gastropods, sea hares, several algal metabolites have been isolated which gave information about the dietary habits of these animals. The red alga, *Asparagopsis taxiformis*, contains an array of

FCH_2COOH

Fluoroacetic acid
Gastrolobium grandiflorum

$FCH_2(CH_2)_7CH{=}CH(CH_2)_7COOH$

ω-Fluorooleic acid
Dichapetalum toxicarium

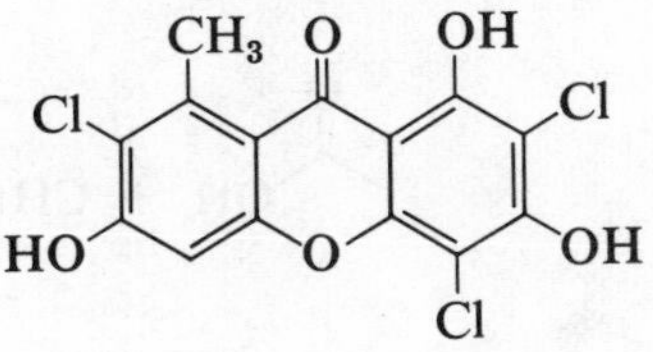

Xanthone from
Lecanora spp.

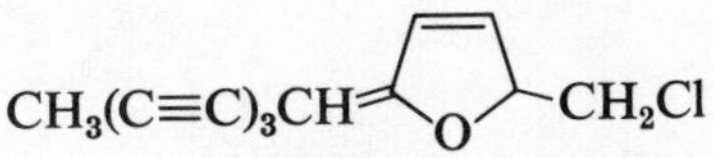

$CH_3(C{\equiv}C)_3CH{=}$ ... CH_2Cl

Polyyne from
Gnaphalium sp.

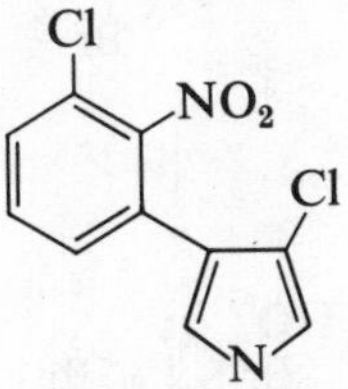

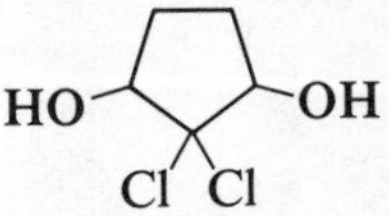

Caldariomycin
Caldariomyces fumago

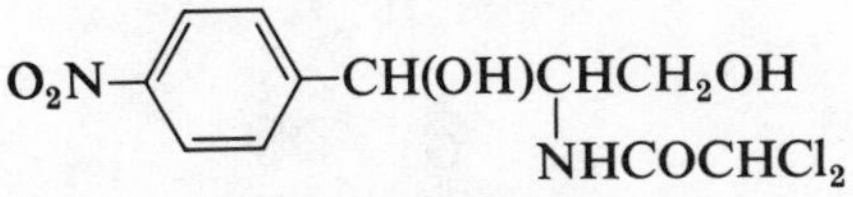

Pyrrolnitrin
Pseudomonas pyrrocinia

$O_2N-C_6H_4-CH(OH)CH(NHCOCHCl_2)CH_2OH$

Chloromycetin
Streptomyces venezuelae

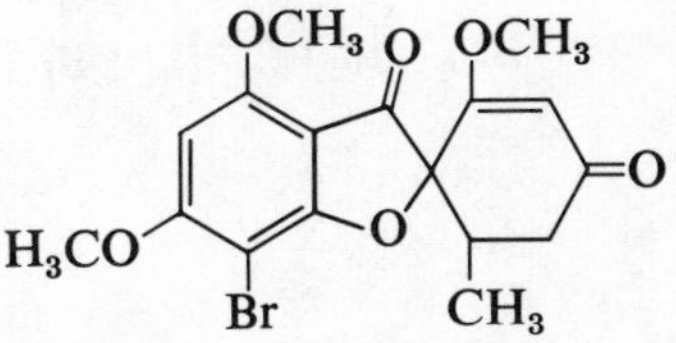

Griseofulvin (Br-analogue)
Penicillium griseofulvum

Fig. 22 Structures of some naturally occurring terrestrial halogen compounds of polyketide origin

simple volatile polyhalogenated hydrocarbons, such as $CHBr_3$, CBr_4, CHBrClI, $BrCH_2COCH_2I$, $Br_2CHCH(OAc)CHI_2$, $ClCH_2COOH$, $BrCH{=}CBrCOCH_2Br$, $Br_2C{=}CH\ COOH$, etc. They were identified by GC–MS analyses and comparison with authentic samples.

No biogenetic studies have been undertaken and little is known about the mechanism of halogenation. The relative concentration of bromine and iodide in sea water is very low in comparison to that of chlorine and contrasts sharply to the frequent occurrence especially of bromine containing metabolites. This is explained by the capability of seaweeds to accumulate bromine and iodine and the low oxidation potential of these halogens. Peroxidases are present in marine organisms and hydrogen peroxide in conjunction with halide ions is an excellent halogenation system. The halogenated acetic acids arise by halogenation of malonic acid or, together with haloforms, by the classical haloform fission of perhalogenated acetone derivatives which are formed by halogenation and decarboxylation of acetoacetic acid. The halogenated acrylic acid could be formed by a Favorsky rearrangement of halogenated acetone (Fig. 23). A route

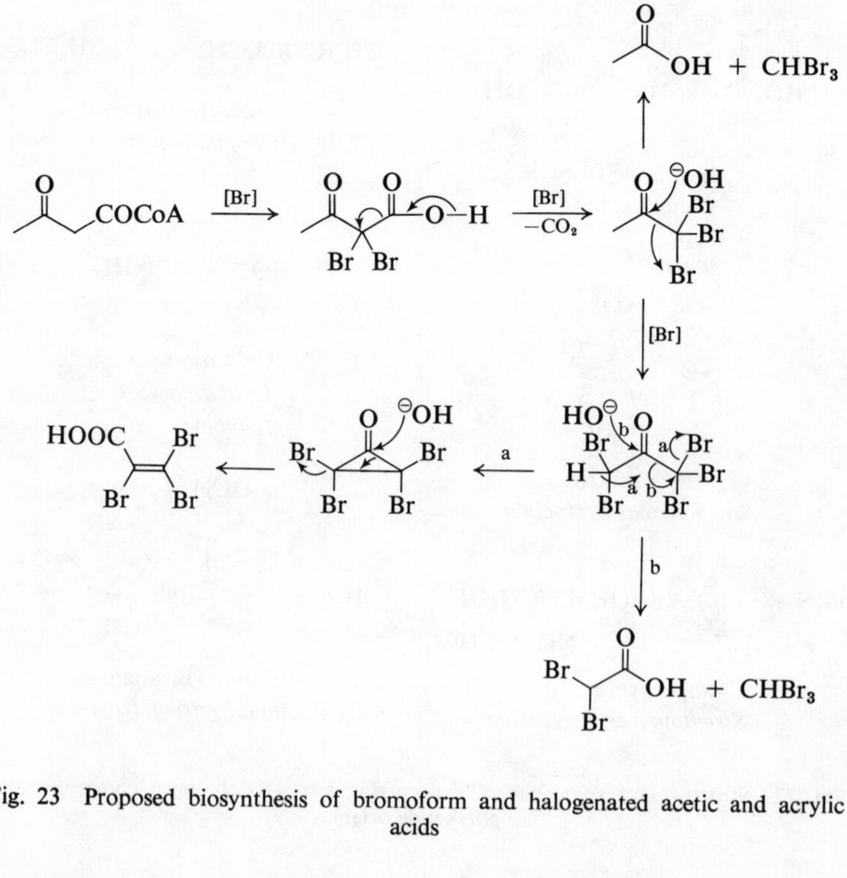

Fig. 23 Proposed biosynthesis of bromoform and halogenated acetic and acrylic acids

to the butenones must be still more speculative. Several chlorine containing acetylenes are known of lipid origin in the *Compositae* family. Epoxides are present as cometabolites and the halides are formed by nucleophilic opening of the epoxide ring with HCl as in a polyyne from the *Gnaphalium* sp. (Fig. 22).[18]

Bromonium ions or their equivalents initiate cyclizations as in the biosynthesis of nidificene from farnesyl phosphate (56).

H
OP
H
$Cl^{\ominus}$
$Br^{\oplus}$ $Br^{\oplus}$

(56)

Cl
H
Br
Br

Nidificene
Laurencia nidifica

4.15 Modification of the carbon skeleton

We have already encountered the insertion of a carbon atom by methionine into the benzenoid ring to form tropolones, oxidative cleavages of the chain as in chartreusin aglycone, and 1,2 shifts as in the flavone–isoflavone rearrangement or the homogentisic acid rearrangement. These processes are comparably simple to survey but rather drastic rearrangements also occur which have challenged the ingenuity of chemists. Oxidative ring cleavage of aromatics often leads to secondary products of unusual structure, the genesis of which is hard to visualize. Patulin, a metabolite of *Penicillium patulum*, was known to originate from 6-methylsalicylic acid. *m*-Cresol, *m*-hydroxybenzyl alcohol, gentisylalcohol, and gentisaldehyde were identified as cometabolites and incorporate well into patulin.[84,85] The route via toluquinol appears to be a side reaction. When 2,4,6-$^{2}H_3$-*m*-cresol was fed to the glucose-deficient culture medium up to 57 per cent incorporation was observed.[86] The fragmentation pattern of patulin in the mass spectrometer is well understood and therefore it was possible to locate the position of the isotopes in the molecule. Fig. 24 depicts the biosynthetic pathway.

Penicillic acid produced by *Penicillium cyclopium* or *P. baarnense* derives from orsellinic acid via oxidative ring fission. 1-^{14}C-acetic acid was incorporated

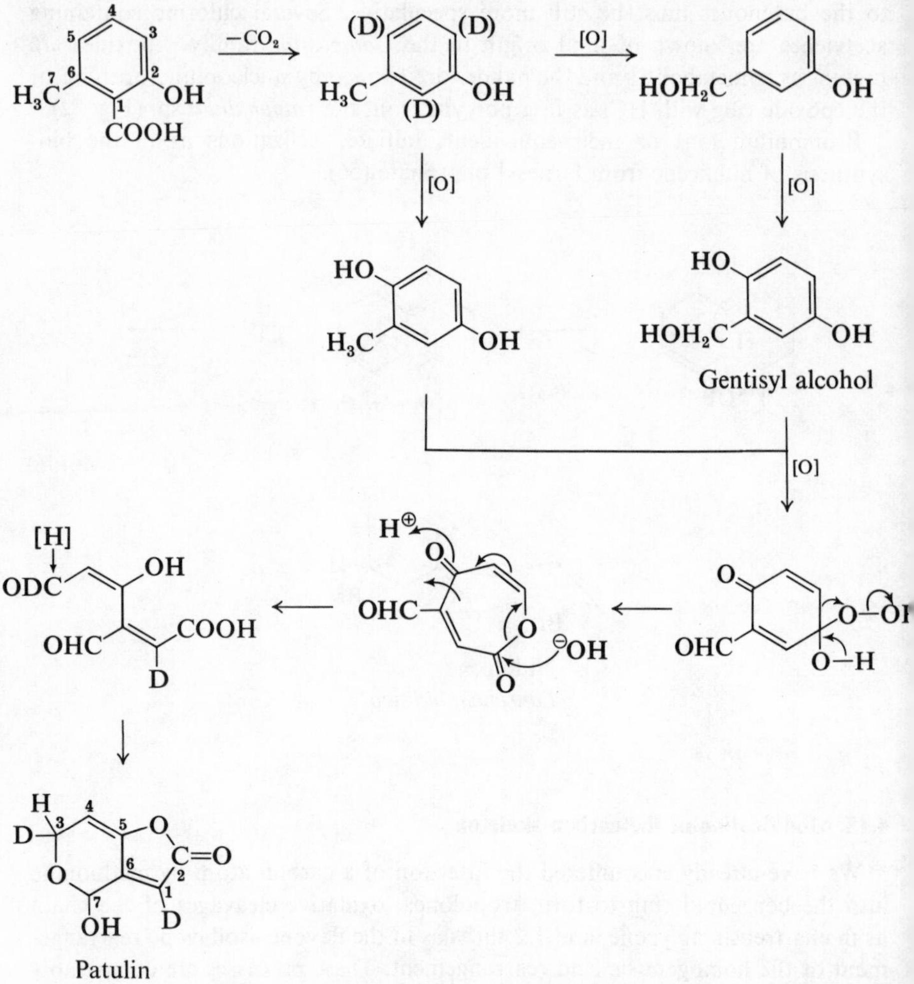

Fig. 24 Biosynthesis of patulin

according to Fig. 25. If 1-^{14}C-malonic acid was administered, the methyl group stayed unlabelled. The ring fission, suggested to occur at $C^{4,5}$,[87] was confirmed by ^{13}C NMR double labelling technique using 90 per cent enriched 1,2–$^{13}C_2$ acetate.[88] Two pairs of $^{1}J_{^{13}C-^{13}C}$ coupling between $C^{2,3}$ ($J = 77$ Hz, two sp_2 carbons, originally $C^{2,3}$) and $C^{5,7}$ ($J = 44$ Hz one sp^3 and one sp^2 carbon, originally $C^{6,7}$) observed. All the ^{13}C shifts were assigned by off-resonance decoupling and by comparison with model compounds. 2,5-Dihydroxy-3-methoxytoluene has been isolated from cultures of *Penicillium baarnense* supporting the biosynthetic scheme.[89]

The biosynthesis of penicillic acid has also been studied by ^{3}H NMR providing further details of the enzymatic processes (Fig. 26).[90] ^{3}H-Acetate labels

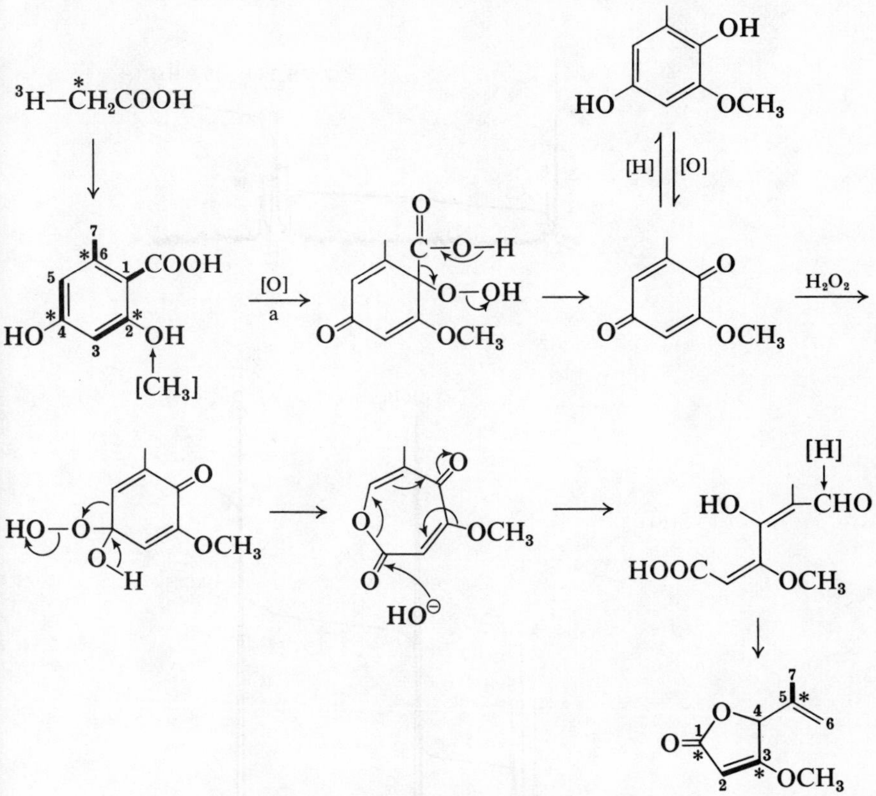

Fig. 25 Biosynthesis of penicillic acid

orsellinic acid at $C^{3,5,7}$ and penicillic acid is therefore expected to be labelled at $C^{2,6,7}$ which proved to be correct. C^3–^3H and C^5–^3H appear as singlets and C^7–^3H as a triplet, coupling geminally with two ^1H (orsellinic acid numbering). The very different intensities of the vinylic C^5–^3H absorptions (C^6H_2) indicate a high degree of stereospecificity for the formation of this function. The C^7–^3H showed the highest relative intensity in accord with C^7 being the starter end of the tetraketide.

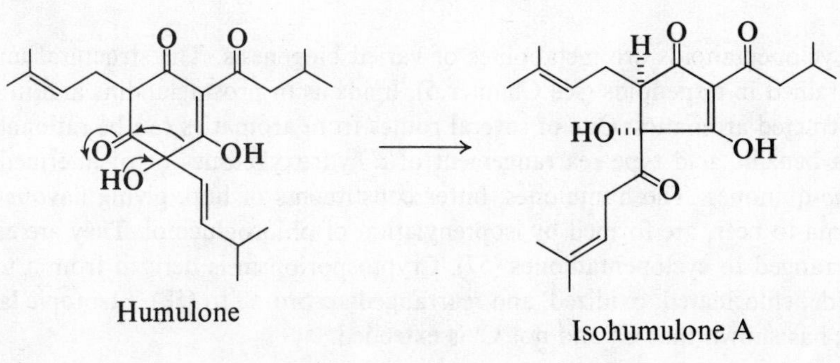

(57)

Humulone

Isohumulone A

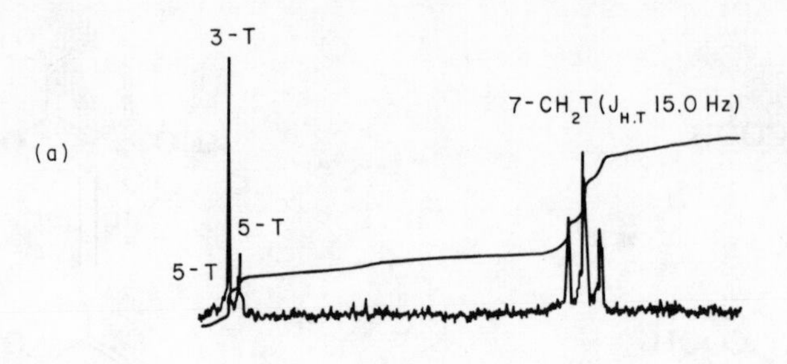

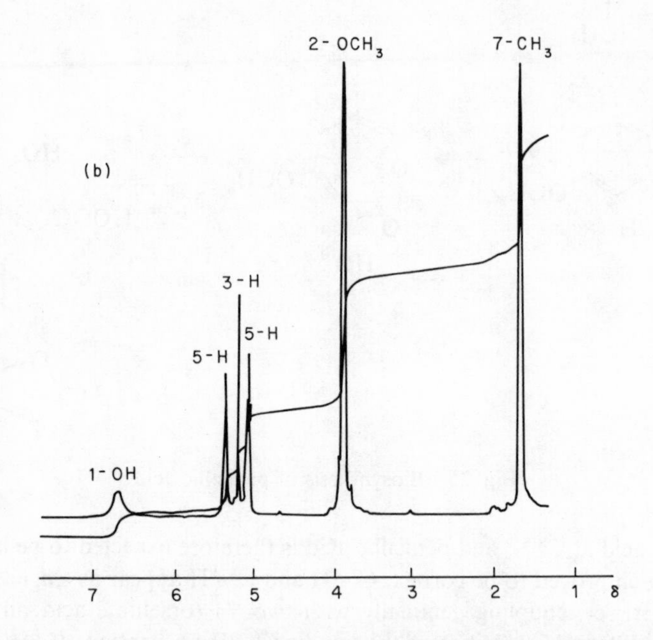

Fig. 26 (a) FT-^{3}H NMR spectrum of penicillic acid in acetone, d_6, 25°C, 1.12×10^4 pulses at 1.75 second intervals. (b) ^{1}H NMR spectrum. The numbering is that of the precursor, orsellinic acid. (Reproduced by permission of The Royal Society of Chemistry from *J. Chem. Soc. Chem. Commun.* **1974,** 220)

Cyclopentanoids are metabolites of varied biogenesis. The structural unit is contained in terpenoids (see Chapter 5), lipids as in prostaglandins and in ring contracted aromatics. One of several routes from aromatics can be rationalized as a benzilic acid type rearrangement of α-hydroxyketones or of intermediary *ortho*-quinones. The humulones, bitter constituents of hop, giving flavour and aroma to beer, are formed by isoprenylation of phloroglucinol. They are easily rearranged to cyclopentadiones (57). Cryptosporiopsin is derived from a tetraketide, chlorinated, oxidized, and rearranged according to (58).[91] Isotopic labelling has shown that C^6 and not C^1 is extruded.

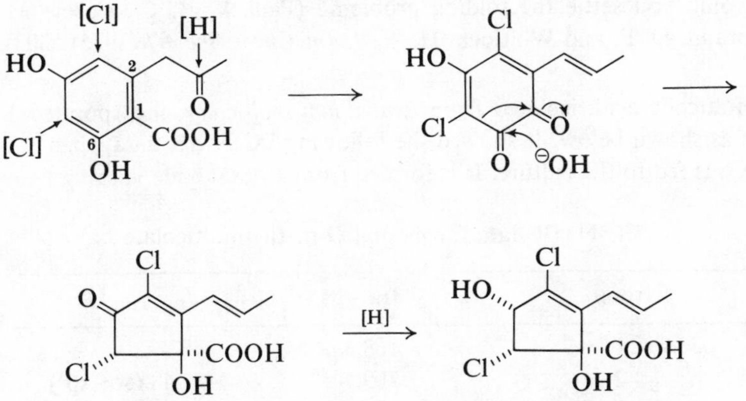

(58)

Cryptosporiopsin

Dihydrocryptosporiopsin

4.16 Problems

4.1 The bacterial monoenoic acid 14:1 (9c) is supposed to be formed by dehydration of 3-hydroxyacyl-CoA to the corresponding *cis*-olefinic acyl-CoA derivative. Subsequent addition of further acetate units to the unsaturated acyl-CoA completes the biosynthesis. What is the structure of this 3-hydroxy acyl group? Formulate the biosynthetic pathway leading to 14:1 (9c) acid. (Scheuerbrandt, G. and Bloch, K. *J. Biol. Chem.* **235** (1960) 337).

4.2 Suggest a mechanism for formation of aurones, e.g. sulphuretin, from the corresponding chalcone. Radical as well as ionic mechanisms can be invoked.

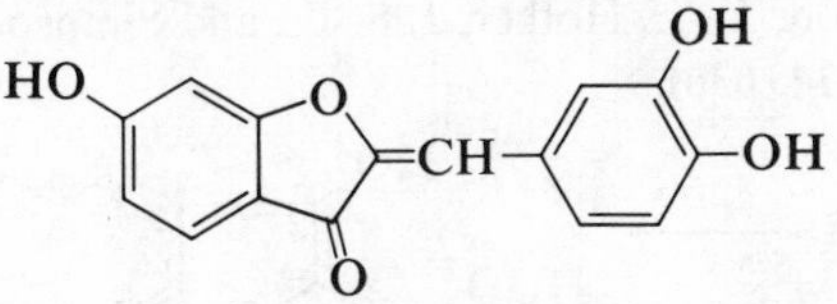

Sulphuretin

4.3 It is suggested that islandicin is assembled from eight acetate units according to equation (33), Chapter 4. However, it can also be formed by folding the octaketide chain in the following manner:

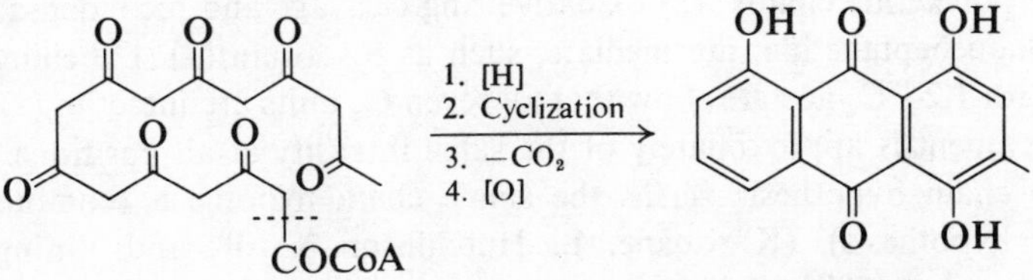

1. [H]
2. Cyclization
3. $-CO_2$
4. [O]

Islandicin

How would you settle the folding problem? (Paulick, R. C., Casey, M. L., Hillenbrand, D. F. and Whitlock, H. W. *J. Am. Chem. Soc.* **97** (1975) 5303).

4.4 Multicolic acid, isolated from *Penicillium multicolor*, incorporated 1-^{13}C-acetate as shown below. It showed the following ^{13}C NMR data when 1,2-^{13}C-acetate was fed to the culture. It is formed from a hexaketide via ring fission of

^{13}C NMR data for methyl *O*-methylmulticolate

Carbon	δ_C	$J_{^{13}C-^{13}C'}$, Hz
1	168.3	—
2	109.7	48 (sp^2–sp^3)
3	160.6	—
4	150.2	90 (sp^2–sp^2)
5	23.4	48
6	29.7	35 (sp^3–sp^3)
7	25.5	35
8	32.2	35
9	62.2	35
10	100.9	90
11	163.8	—
CH_3OOC–	52.0	—
CH_3O–	59.5	—

an aromatic intermediate. Discuss the biosynthetic pathway, the reaction mechanisms involved, and indicate which acetate bonds are still intact in the metabolite. (Gudgeon, J. A., Holker, J. S. E., and Simpson, T. J. *J. Chem. Soc. Chem. Commun.* **1974**, 636).

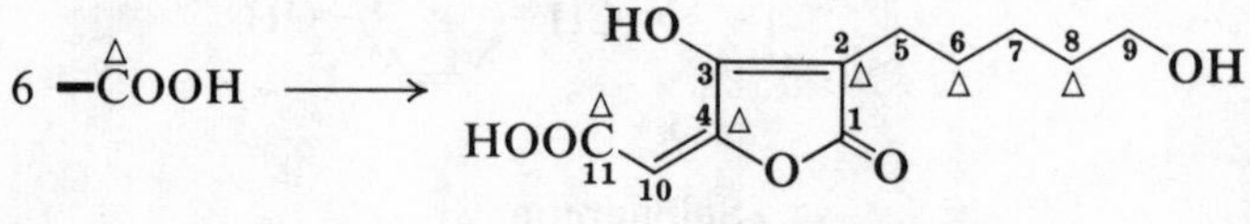

6 —$\overset{\Delta}{C}$OOH ⟶

Multicolic acid

4.5 Citromycetin (1) and fulvic acid (2) are formed from seven acetate units but they cannot be formed by simple linear chain folding. No introduction of C_1 units is observed. Two pathways have been considered: (a) condensation of two separate polyketide chains; (b) oxidative ring cleavage and recondensation of a single chain heptaketide intermediate, such as fusarubin (3). Labelling experiments with 1,2-$^{13}C_2$-acetate show that all seven C_2 units are intact in (1)–(3) and the enrichment is approximately of the same intensity at all positions. Discuss the two chain hypothesis *versus* the single chain hypothesis (common intermediate hypothesis). (Kurobane, I., Hutchinson, C. R. and Vining, L. C. *Tetrahedron Lett*, **1981**, 493).

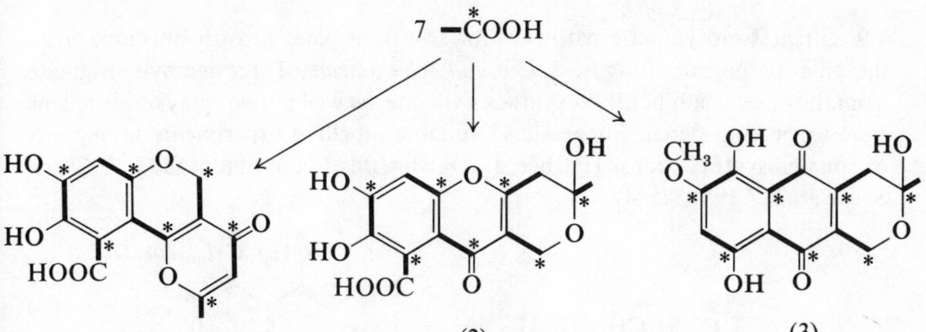

7 —$\overset{*}{C}OOH$

(1) (2) (3)

4.6 The antibiotic isolasalocid A has the following structure:

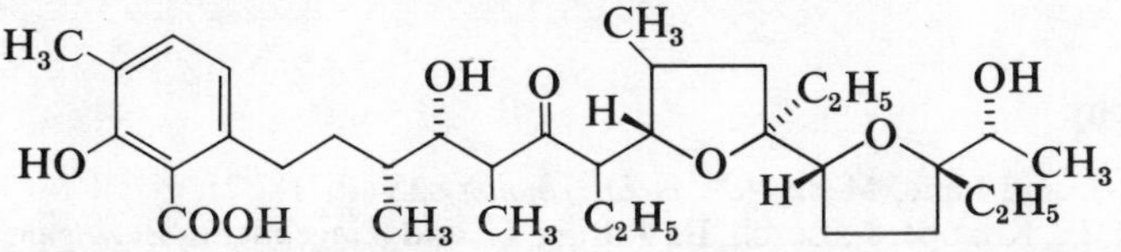

Suggest a reasonable biosynthetic pathway which accounts for the substitution, branching, and oxygenation patterns and the formation of the benzene and tetrahydrofuran rings. (Westley, J. W., Preuss, D. L. and Pitcher, R. G. *J. Chem. Soc. Chem. Commun.* **1972**, 161).

4.7 The phenanthrene quinones, piloquinone and hydroxypiloquinone, R = H or OH, are derived from nine malonate units and a starter. Suggest a plausible biosynthetic route to these metabolites. (Zylber, J., Zissman, E., Polonsky, J. and Lederer, E. *European J. Biochem.* **10** (1969) 278).

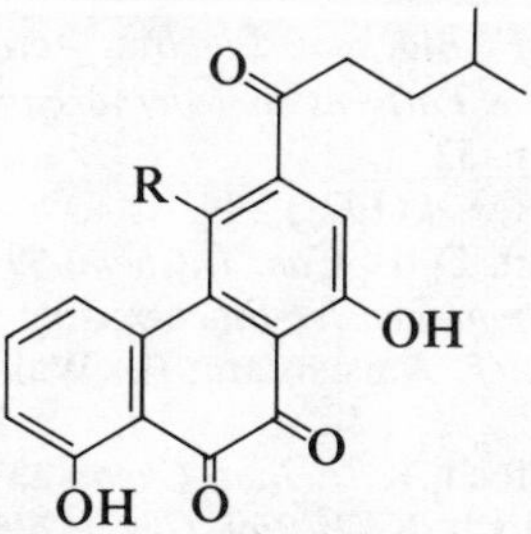

4.8 Suggest a biosynthetic pathway for the fungal metabolite chloromycorrizin A, an antagonist of *Fomes annosus*, which causes great losses in Swedish forestry. The biosynthesis is related to the early stages of the biosynthesis of another fungal metabolite cryptosporiopsin (section 4.15). (Trofast, J. and Wickberg, B. *Tetrahedron* **33** (1977) 875).

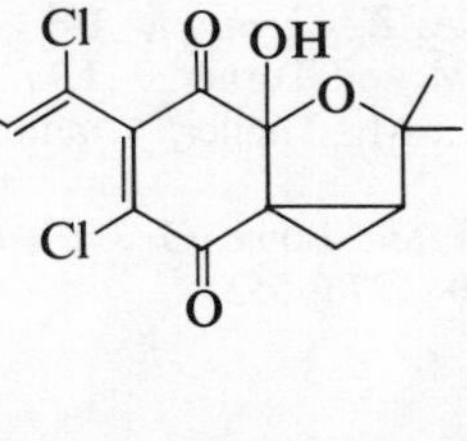

4.9 Suggest biosynthetic pathways for sclerin, a plant growth hormone from the phytopathogenic fungus, *Sclerotinia sclerotiorum.* Three methyls originate from the one carbon pool. Biosynthesis via one, as well as two, polyketide chains have to be considered. Suggest also suitable labelling experiments in support of your biosynthetic ideas (Barber, J., Garson, M. J. and Staunton, J. *J. Chem. Soc. Perkin I* **1981**, 2584).

$$3\ \overset{\blacktriangle}{C} + CH_3COOH \longrightarrow$$

Bibliography

1. Moss, J. and Lane, M. in *Adv. in Enzymology* **35** (1971) 321.
2. Lynen, F., Knappe, J., Lorch, E., Jutting, G. and Ringelmann, E. *Angew. Chem.* **71** (1959) 481.
3. Bonnemere, C., Hamilton, J. A., Steinrauf, L. K. and Knappe, J. *Biochemistry* **4** (1965) 240.
4. Rose, I. A., O'Connel, E. and Solomon, F. *J. Biol. Chem.* **251** (1976) 902.
5. Sedgwick, B., Cornforth, J. W., French, S. J., Grey, R. T., Kelstrup, E. and Willadsen, P. *European J. Biochem.* **75** (1977) 481.
6. Arnstad, K.-I., Schindlbeck, G. and Lynen, F. *European J. Biochem.* **55** (1975) 561.
7. Sedgwick, B., Morris, C. and French, S. J. *J. Chem. Soc. Chem. Commun.* **1978,** 193.
8. Buckner, J. S., Kolattukudy, P. E. and Rogers, L. *Arch. Biochem. Biophys.* **186** (1978) 152.
9. Cramer, U. and Spener, F. *Biochem. Biophys. Acta* **450** (1976) 261.
10. Grisebach, H. *Biosynthetic Patterns in Microorganisms and Higher Plants*, John Wiley, New York, 1967, p. 32.
11. Law, J. H. *Acct. Chem. Res.* **4** (1971) 199.
12. Buist, P. H. and MacLean, D. B. *Can. J. Chem.* **59** (1981) 828.
13. (a) O'Leary, W. M. in *Comprehensive Biochemistry* **18** (1970) 229, Florkin, M. and Stotz, E. H. (Eds.), Elsevier, Amsterdam; (b) Wakil, S. J. and Barnes Jr., E. M. *Ibid.* **18S** (1971) 57.
14. Scheuerbrandt, G. and Bloch, K. *J. Biol. Chem.* **237** (1962) 2064.
15. Bloomfield, D. K. and Bloch, K. *J. Biol. Chem.* **235** (1960) 337.
16. Cherif, A., Dubacq, J. P., Mache, R., Oursel, A. and Tremalieres, A. *Phytochemistry* **14** (1975) 703.
17. Stoffel, W. *Biochem. Biophys. Res. Commun.* **6** (1961) 270.
18. Bohlmann, F., Burkhardt, T. and Zdero, C. *Naturally Occurring Acetylenes*, Academic Press, London, 1973.
19. Bu'Lock, J. D. and Smith, G. N. *Biochem. J.* **1962,** 35.
20. Hatt, H. H. and Szumer, A. Z. *Chem. Ind.* **1954,** 962.
21. Jones, E. R. H., Thaller, V. and Turner, J. L. *J. Chem. Soc. Perkin I* **1975,** 424.
22. Jones, E. R. H., Piggin, C. M., Thaller, V. and Turner, J. L. *J. Chem. Research* **1977** (S) 68: (M) 0744.
23. Hodge, P., Jones, E. R. H. and Lowe, G. *J. Chem. Soc.* (*C*) **1966,** 1216.
24. Money, T. *Chem. Revs.* **70** (1970) 553.

25. Birch, A. J. and Donovan, F. W. *Austr. J. Chem.* **6** (1953) 360; Birch, A. J. *Fortschr. Chem. Org. Naturstoffe* **14** (1957) 186, Springer, Wien.
26. Gatenbeck, S. and Mosbach, K. *Acta Chem. Scand.* **13** (1959) 1561.
27. Bu'Lock, J. D., Smalley, H. M. and Smith, G. N. *J. Biol. Chem.* **237** (1962) 1778.
28. Dimroth, P., Walter, H. and Lynen, F. *European J. Biochem.* **13** (1970) 98.
29. Gatenbeck, S. and Hermodson, S. *Acta Chem. Scand.* **19** (1965) 65.
30. Gatenbeck, S. *Acta Chem. Scand.* **14** (1960) 296.
31. Garson, M. J. and Staunton, J. *Chem. Soc. Rev.* **8** (1979) 539.
32. Wehrli, F. W. and Nishida, T. *Prog. Chem. Org. Nat. Prod.* **36** (1979) 1.
33. Tanabe, M. and Detre, G. *J. Am. Chem. Soc.* **88** (1966) 4515.
34. Sato, Y., Oda, T. and Saito, H. *Tetrahedron Lett.* **1976,** 2695; *J. Chem. Soc. Chem. Commun.* **1978,** 135.
35. Wat, C.-K., McInnes, A. G., Smith, D. G. and Vining, L. C. *Can. J. Biochem.* **50** (1972) 620.
36. Burlingame, A. L., Balogh, B., Welch, J., Lewis, S. and Wilson, D. *J. Chem. Soc. Chem. Commun.* **1972,** 318.
37. Sankawa, U., Shimada, H. and Yamasaki, K. *Tetrahedron Lett.* **1978,** 3375.
38. Canham, P., Vining, L. C., McInnes, A. G., Walter, J. A. and Wright, L. J. C. *J. Chem. Soc. Chem. Commun.* **1976,** 319.
39. Seto, H., Cary, L. and Tanabe, M. *J. Chem. Soc. Chem. Commun.* **1973,** 867.
40. Birch, A. J. and Donovan, F. W. *Austr. J. Chem.* **8** (1955) 529.
41. Birch, A. J. and Donovan, F. W. *Austr. J. Chem.* **6** (1953) 373.
42. Thomson, R. H. *Naturally Occurring Quinones*, 2nd. Edn, Academic Press, London, 1971.
43. Leistner, E. and Zenk, M. H. *Tetrahedron Lett.* **1971,** 1677.
44. Harborne, J. B. *Phytochemistry* **6** (1967), 1415, 1643.
45. Grisebach, H. *Biosynthetic Patterns in Microorganisms and Higher Plants*, J. Wiley, New York, 1967, p. 1.
46. Hrazdina, G. and Creasy, L. L. *Phytochemistry* **18** (1979) 581.
47. Saleh, N. A. M. and Fritsch, H., Kreuzaler, F. and Grisebach, H. *Phytochemistry* **17** (1978) 183.
48. Crombie, L., Dewick, P. M. and Whiting, D. A. *Chem. Commun.* **1970,** 1469; **1971,** 1182, 1183.
49. Dreyer, D. L., Munderloh, K. P. and Thiessen, W. E. *Tetrahedron* **31** (1975) 287.
50. Roitman, J. N. and Jurd, L. *Phytochemistry* **17** (1978) 161.
51. Saitoh, T., Shibata, S. and Sankawa, U. *Tetrahedron Lett.* **1975,** 4463.
52. Ellis, B. E. *Lloydia* **37** (1974) 168.
53. Hösel, W., Frey, G. and Barz, W. *Phytochemistry* **14** (1975) 417.
54. Birch, A. J. and Thompson, D. J. *Austr. J. Chem.* **25** (1972) 2731.
55. Stocker, M. and Pohl, R. *Phytochemistry* **15** (1976) 571.
56. von Rudloff, E. and Jorgensen, E. *Phytochemistry* **2** (1963) 297.
57. Pryce, R. J. *Phytochemistry* **10** (1971) 2679.
58. Rupprich, N. and Kindl, H. *Hoppe-Seyler's Z. Physiol Chem.* **359** (1978) 165.
59. Bentley, R. *J. Biol. Chem.* **238** (1963) 1889, 1895.
60. Scott, A. I. and Wiesner, K. J. *J. Chem. Soc. Chem. Commun.* **1972,** 1075.
61. *Oxidative Coupling of Phenols*. Taylor, W. I. and Battersby, A. R. (Eds.), M. Dekker, New York, 1967.
62. Rhodes, A., Somerfield, G. A. and McGonagle, M. P. *Biochem. J.* **88** (1963) 349.
63. Barton, D. H. R., DeFlorin, A. M. and Edwards, O. E. *J. Chem. Soc.* **1956,** 530.
64. Taguchi, H., Sankawa, U. and Shibata, S. *Chem. Pharm. Bull.* (*Japan*) **17** (1969) 2054.
65. Davidson, T. A. and Scott, A. I. *J. Chem. Soc.* **1961,** 4075.
66. Cameron, D. W. and Lord Todd in *Oxidative Coupling of Phenols*, Taylor, W. I. and Battersby, A. R. (Eds.), M. Dekker, New York, 1967, p. 203.

67. Brockmann, H. *Prog. Chem. Nat. Prods.* **14** (1957) 142.
68. Anderson, J. M. and Murray, J. *Chem. Ind.* **1956,** 376.
69. Allport, D. C. and Bu'Lock, J. D. *J. Chem. Soc.* **1958,** 4090.
70. Howard, D. H. and Raistrick, H. *Biochem. J.* **57** (1954) 212.
71. Shibata, S., Murakami, T., Kitagawa, I. and Kishi, T. *Pharm. Bull.* (*Japan*) **4** (1956) 111.
72. Sultanbawa, M. V. S. *Tetrahedron Reports* **84; *Tetrahedron*** **36** (1980) 1465.
73. Atkinson, J. E., Gupta, P. and Lewis, J. R. *Chem. Commun.* **1968,** 1386.
74. Floss, H. G. and Rettig, A. *Z. Naturforsch.* **19B** (1964) 1103.
75. Gottlieb, O. R. *Phytochemistry* 7 (1968) 411.
76. Scott, A. I. *Quart. Revs.* **19** (1965) 1.
77. Marais, J. S. C. *Ondersteport J. Vet. Sci.* **20** (1944) 67.
78. McEwan, T. *Nature* **201** (1964) 827.
79. Peters, R. A., Hall, R. J., Ward, P. F. W. and Sheppard, N. *Biochem. J.* **77** (1960) 17.
80. Sinda, J. F. and De Bernardis, J. F. *Lloydia* **36** (1973) 107.
81. Moore, R. E. in *Marine Natural Products*, Vol. I, Scheuer, P. J. (Ed.), Academic Press, New York, 1978, p. 44.
82. Martin, J. D. and Darias, J. in *Marine Natural Products*, Vol. I, Scheuer, P. J. (Ed.), Academic Press, New York, 1978, p. 125.
83. Faulkner, D. J. in *Environmental Chemistry*, Vol. IA, Hutzinger, O. (Ed.), Springer, Berlin, 1980, p. 229.
84. Murphy, G., Vogel, G., Krippahl, G. and Lynen, F. *European J. Biochem.* **49** (1974) 443.
85. Scott, A. I., Zamir, L., Phillips, G. T. and Yalpani, M. *Bioorg. Chem.* **2** (1973) 124.
86. Scott, A. I. and Yalpani, M. *Chem. Commun.* **1967,** 945.
87. Mosbach, K. *Acta Chem. Scand.* **14** (1960) 457.
88. Gudgeon, J. A., Holker, J. S. E. and Simpson, T. J. *J. Chem. Soc. Chem. Commun.* **1974,** 636.
89. Better, J. and Gatenbeck, S. *Acta Chem. Scand.* **B30** (1976) 368.
90. Elvidge, J. A., Jaiswal, D. K., Jones, J. R. and Thomas, R. *J. Chem. Soc. Perkin I* **1977,** 1080.
91. Holker, J. S. E. and Young, K. *J. Chem. Soc. Chem. Commun.* **1975,** 525.

Chapter 5

The mevalonic acid pathway
The terpenes

5.1 Introduction

When the number of structurally defined compounds rapidly accumulated in the late nineteenth century, Wallach noted that many compounds, especially the fragrant principles of plants—the essential oils—could be formally dissected into branched C_5 units called isopentenyl or isoprene units. These compounds, which typically possessing the molecular composition $C_{10}H_{16}$, were given the collective name terpenes, etymologically derived from the terebinth tree, *Pistacia terebinthus*, which exudes a resin. Conifers, eucalyptus trees, and citrus fruits are rich in these low molecular weight volatile terpenes. Limonene can, in principle, be synthesized from two moles of isoprene by a Diels–Alder reaction (1) but it

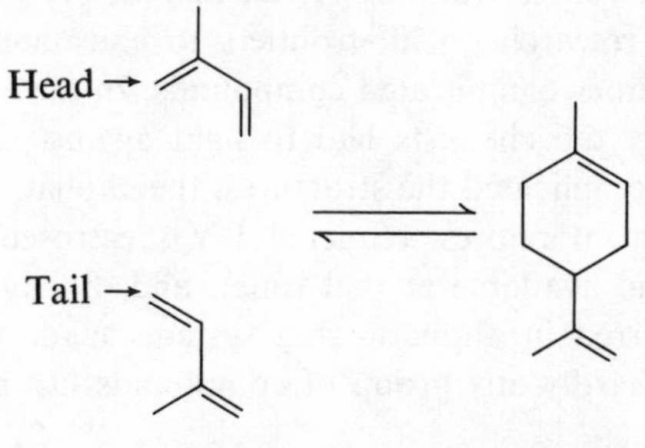

Head →

Tail →

(1)

2 Isoprenes Limonene

was rapildy perceived that isoprene itself could not be the functional unit used by nature. Nevertheless, the isoprene unit was a useful device for rationalizing the structures of many more complex compounds of higher molecular weight (2).

The isoprene rule[1] states that terpenes are multiples of C_5 units linked together head to tail. Several modes of cyclization are conceivable and lead to various skeletons just like the cyclization of polyketide chains. The terpenes are classified

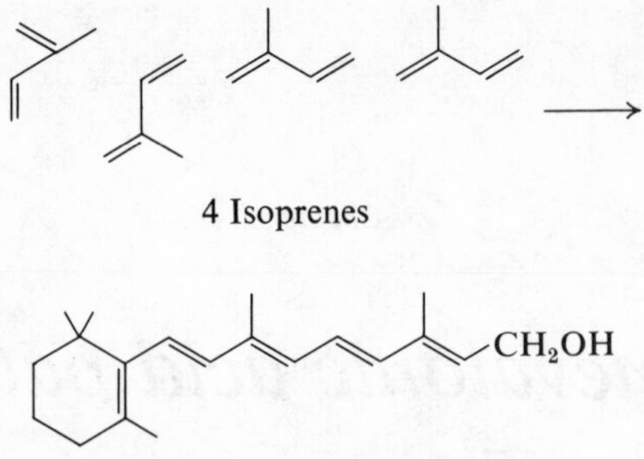

4 Isoprenes

(2)

CH_2OH

Vitamin A

according to the number of C_5 units: monoterpenes, C_{10}; sesquiterpenes, C_{15}; diterpenes, C_{20}; sesterterpenes, C_{25}; triterpenes, C_{30}; and tetraterpenes, C_{40}. It soon appeared that neither steroids, C_{27}, nor several other related compounds obeyed the rule. Degradation had occurred, the head to tail principle was violated, and the skeleton could not be dissected into isoprene units. These changes could be rationalized on the assumption of cleavages and rearrangements of the original skeleton. In 1956 Folkers isolated the easily incorporated mevalonic acid, and subsequently Bloch, Lynen, Cornforth, Eggerer, and Popjak showed how it functioned as a building block. It was then possible to interpret the biosynthesis and secondary modifications of terpenes correctly.

The terpenes house a wealth of significant compounds. The perfume industry is interested in the 'essential' oils, terpentine is used for painting, and most importantly, we find among the terpenes physiologically very active compounds governing the life processes, such as adrenal hormones (cortisone) sex hormones (oestrone and testosterone), vitamins A, D, and E, etc. This has, of course, triggered an intense research on all frontiers to elucidate the structures and properties of these rather complicated compounds. In the early days of terpene and steroid chemistry the chemists had to fight against difficult odds. Many asymmetric centres complicated the structures, the aliphatic nature of the compounds, with few chromophores, rendered UV spectroscopy less helpful (the only spectroscopic aid available at that time), and, finally, the ease by which rearrangements occurred in aliphatic ring systems made chemists frustrated. On the other hand, hardly any group of compounds has repaid the efforts of

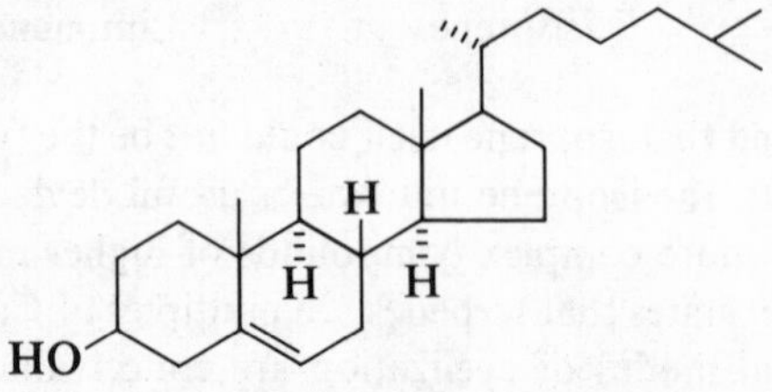

H H H HO

Fig. 1 Structure of cholesterol

chemists so well as the terpenes. The research has contributed immensely to the development of conformational chemistry, understanding of mechanisms, and synthesis. The need for suitable spectroscopic methods was articulated and became gradually available.

The many problems faced are reflected in the rather late date by which the final structural clarification of cholesterol (Fig. 1) was achieved by Wieland and Windaus (1932) and that was with the help of Bernal, an X-ray crystallographer who, by measuring the dimensions of the unit cell of ergosterol, indicated the most likely structure.

A few C_5 compounds occur in nature but not all of them are of isoprenoid origin (Fig. 2).

$(H_3C)_2CHCH_2CHO$

Isovaleraldehyde I

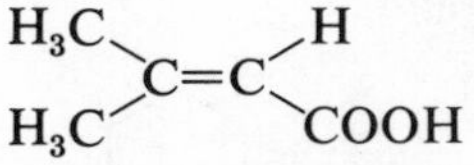

$(H_3C)_2C=CH-COOH$

Senecioic acid II

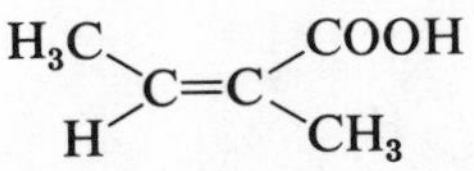

$H_3C(H)C=C(COOH)CH_3$

Angelic acid III

$H_3C(H)CHCH(CH_3)COOH$

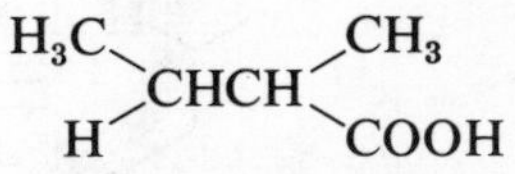

α-Methylbutyric acid IV

Fig. 2 Some naturally occurring C_5 compounds. I comes from mevalonic acid, III and IV from isoleucine, and II can effectively be synthesized from both precursors

Single isoprene units are found in many natural compounds, so-called hemiterpenes, e.g. in furanocoumarins (section 3.5), hop constituents (section 4.15), quinones (sections 3.6 and 4.10), and in a variety of alkaloids (Chapter 7).

The formation of the very large number of terpenes from C_{10}–C_{30} precursors can be rationalized by a few basic reaction types. In the starting reaction a discrete carbonium ion is generated, either by solvolysis of an allylic pyrophosphate, opening of an epoxide or protonation or halogenation (marine organisms) of double bonds. The further reaction proceeds by electrophilic cyclization, by Wagner–Meerwein shifts (1,2 shifts), or less frequently by electrophilic substitution at the aliphatic bond (1,3 1,4 and 1,5 shifts). The reaction is terminated by proton elimination or addition of water (Fig. 3). The 1,2 shifts and 1,3 and higher shifts deserve some comment. The structure of carbonium ions[2] has been object of considerable speculation which undoubtedly has promoted much valuable research in the field. The problem was whether the equilibrating classical carbonium ion with localized charge, or the non-classical carbonium ion, represented the structure of lowest energy. The problem has received an answer in favour of the non-classical structure by recording NMR spectra of various ions at low temperature in SbF_5–SO_2ClF solutions. It was found that the non-classical

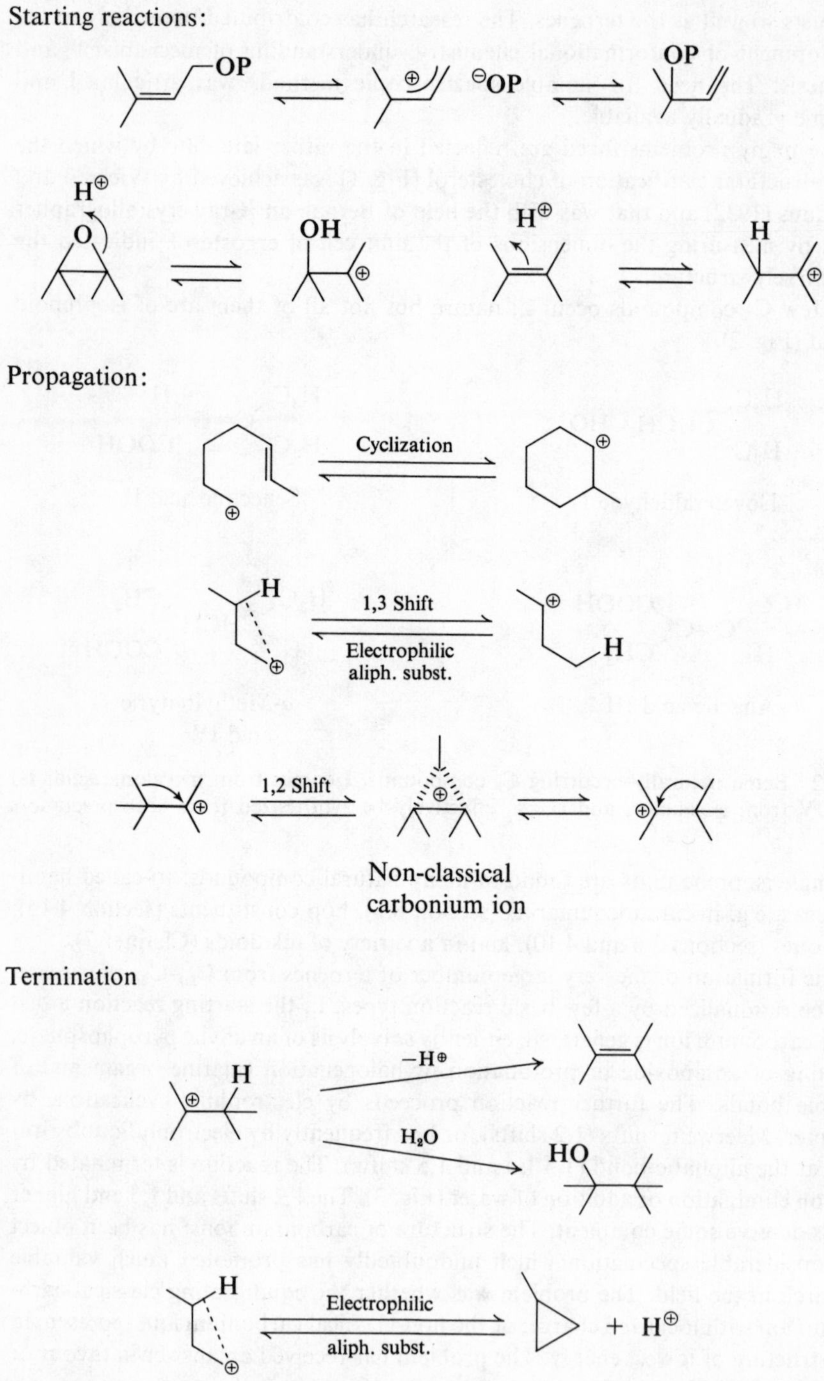

Starting reactions:

OP ⇌ ⊕ ⊖OP ⇌ OP

H⊕ O ⇌ OH ⊕ H⊕ ⇌ H ⊕

Propagation:

⊕ Cyclization ⇌ ⊕

H ⊕ 1,3 Shift ⇌ Electrophilic aliph. subst. ⊕ H

⊕ 1,2 Shift ⇌ ⊕ ⇌ ⊕

Non-classical carbonium ion

Termination

H ⊕ $-H^{\oplus}$ → ; H_2O → HO

H ⊕ Electrophilic aliph. subst. ⇌ + $H^{\oplus}$

Fig. 3 Reaction types in terpene chemistry

structure represents a transition state of slightly higher energy, and that the equilibrium is very rapid on the NMR time scale, $E_a \sim$ 5–10 kcal/mol. Thus, the NMR spectrum of the cyclopentyl carbonium ion shows only one peak corresponding to 9 fast exchanging hydrogens.

1,3 and higher hydrogen shifts and eliminations are synonymous with electrophilic aliphatic substitution, *cf. trans*-annular reactions. The potential surface of a carbonium ion attacking an aliphatic σ-bond has a minimum in the direction perpendicular to the σ-bond. This contrasts with the radical attack which occurs in the direction of the C–H bond (Fig. 4). This implies that electrophilic aliphatic substitution in principle occurs with retention of configuration. The activation energy for the 1,3 hydrogen shift is *ca.* 10 kcal/mol.

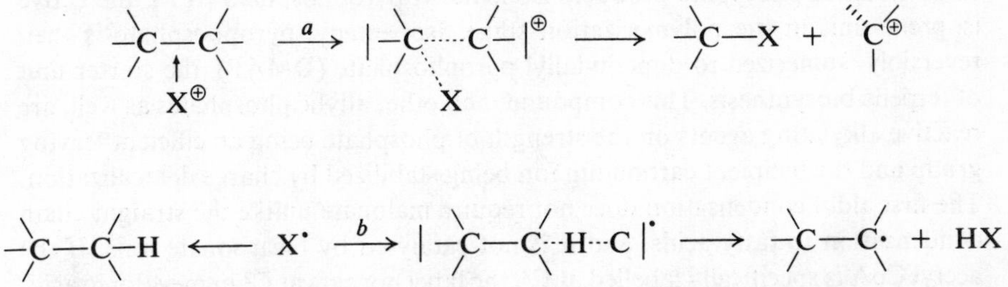

Fig. 4 (a) Preferred direction of attack in ionic reactions and (b) radical reactions

The steric structure of the product is determined by the folding of the polyprenyl chain and the principle of a synchronized antiparallel addition leading directly to the product. The few exceptions to this rule are explained by the stability of an intermediate carbonium ion or the intervention of a 'Y' group from the enzyme that is subsequently expelled by substitution and inversion of configuration (Fig. 5).

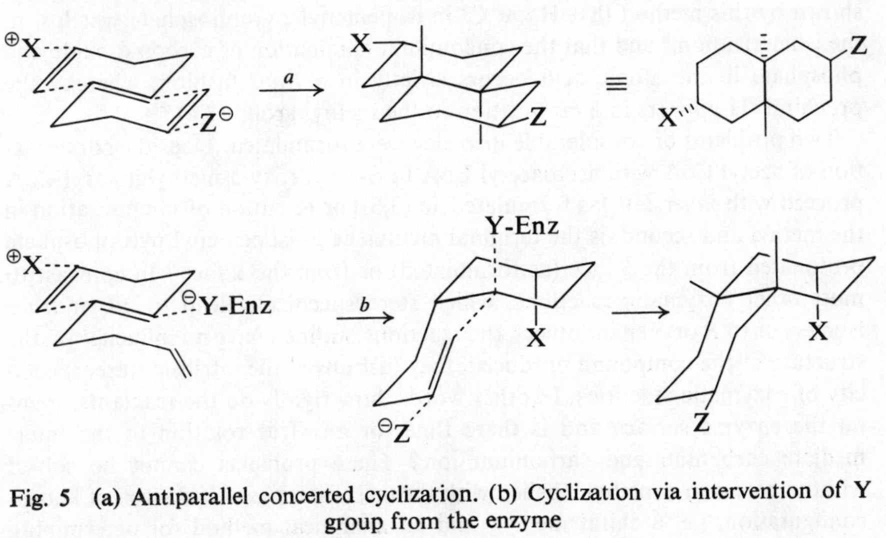

Fig. 5 (a) Antiparallel concerted cyclization. (b) Cyclization via intervention of Y group from the enzyme

5.2 Biosynthesis of mevalonic acid and the active isoprene units. The chiral methyl

Early isotopic studies revealed that the carbon skeleton of terpenoids is acetate based.[3] The hunt for active intermediates continued and a major break-through came with the demonstration of mevalonic acid as a general precursor which could replace acetate as an essential growth factor.[4,5] Its formation in the cell is shown in (3).[6] Acetyl CoA combines with a sulphhydryl group at the active site of the enzyme and condenses with acetoacetyl CoA in a branched fashion to give 3-hydroxy-3-methylglutaryl CoA after hydrolysis. The two-step reduction with NADPH then affords mevalonic acid via mevaldic acid. Only 3*R*-mevalonic acid is biologically active. Phosphorylation, decarboxylation, and elimination of phosphate produces isopentenyl pyrophosphate (IPP), the active isoprene unit in the polymerization stage. Isopentenyl pyrophosphate is then reversibly isomerized to dimethylallyl pyrophosphate (DMAP), the starter unit of terpene biosynthesis. This compound, and other allylic phosphates as well, are reactive alkylating agents on the strength of phosphate being an efficient leaving group and the incipient carbonium ion being stabilized by charge delocalization. The first aldol condensation does not require malonate unlike the straight chain condensation to fatty acids, and it is not catalysed by bicarbonate ions. If the acetyl CoA is specifically labelled at C^2, the label appears at C^2 in mevalonic acid, at C^4 in isopentenyl pyrophosphate, and the labelled methyl group appears eventually in the *trans* position[7] to the chain in dimethylallyl pyrophosphate. The two methylene protons of a CH_2XY centre are stereochemically different as a consequence of a particular orientation of the molecule in a chiral environ-ment. Mevalonic acid contains three such so-called prochiral methylenes at C^2, C^4 and C^5. By chemical and enzymatic procedures deuterium and tritium were stereospecifically introduced in these positions. Isotopic analysis of the meta-bolites gave information about the steric course of the enzymatic steps. It was shown by this method that H_R at C^2 in isopentenyl pyrophosphate was lost in the isomerization,[8] and that the concomitant elimination of carbon dioxide and phosphate in mevalonic acid occurs strictly in a *trans* fashion, whereby the prochiral ^3H appears in a *cis* position to the methyl group (3,4).[9]

Two problems of considerable intricacy were formulated. Does the condensa-tion of acetyl CoA with acetoacetyl CoA to *S*-3-hydroxy-3-methylglutaryl CoA proceed with inversion [as formulated in (3,5)] or retention of configuration in the methyl and second, is the terminal methylene in isopentenyl pyrophosphate protonated from the β-face (as formulated) or from the α:face? In contrast to most other enzymatic reactions, which stereospecifically lead to *cis* or *trans* isomers or to *R* or *S* enantiomers, the reactions outlined have no influence on the structure of the compound produced, they just unveil the intrinsic stereospecifi-city of enzymatic reactions. In other words, how rigidly do the reactants orient on the enzyme surface and is there time for any free rotation of the inter-mediate carbanion and carbonium ion? These problems cannot be solved without access to a methyl labelled with the three isotopes of hydrogen of known configuration, i.e. a chiral methyl, and an analytical method for determining

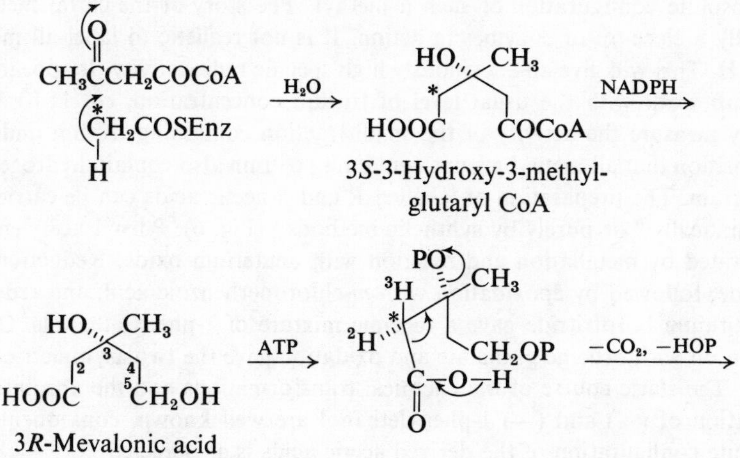

$$CH_3C(=O)CH_2COCoA + HCH_2COSEnz \xrightarrow{H_2O} HOOC\text{-}CH_2\text{-}C^*(HO)(CH_3)\text{-}CH_2COCoA \xrightarrow{NADPH}$$

3*S*-3-Hydroxy-3-methyl-
glutaryl CoA

$$\text{3}R\text{-Mevalonic acid} \xrightarrow{ATP} \xrightarrow{-CO_2,\ -HOP} \tag{3}$$

3*R*-Mevalonic acid

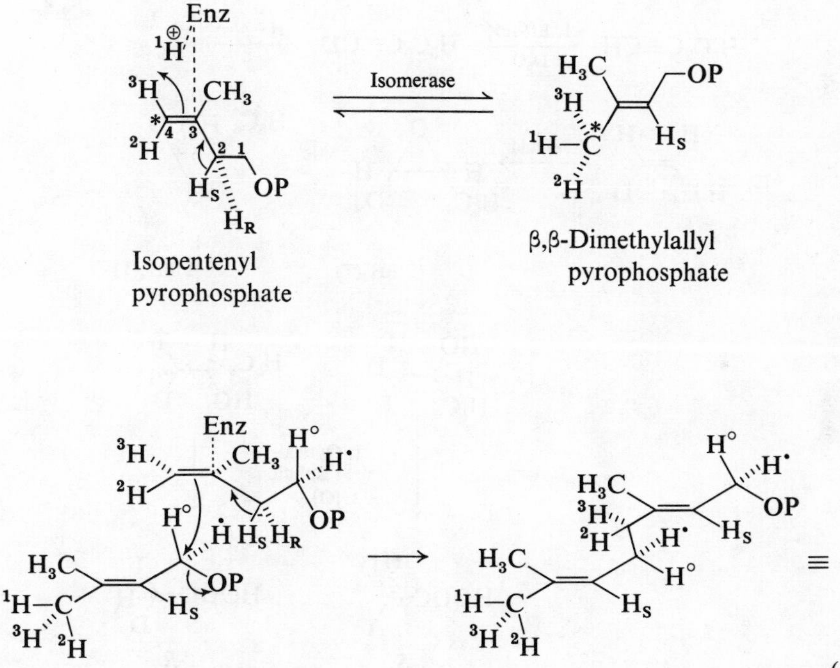

Enz

Isomerase

Isopentenyl
pyrophosphate

β,β-Dimethylallyl
pyrophosphate

Enz

(4)

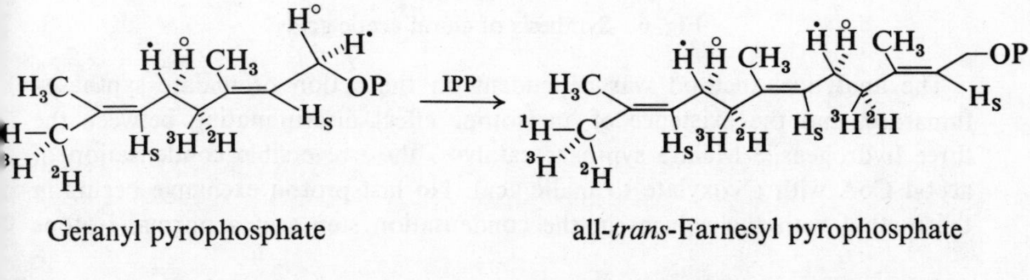

IPP

Geranyl pyrophosphate

all-*trans*-Farnesyl pyrophosphate

the absolute configuration of such a methyl. The story of the chiral methyl is actually a close-up of enzymes in action. It is not realistic to label all methyls with ^3H. This will give an enormously high specific radioactivity. It appears that one can work with the usual level of tritium concentration, *ca.* 1: 10^{-6}, and merely measure the activity of the small fraction containing tritium under the supposition that all methyl groups containing tritium also contain hydrogen and deuterium. The preparation of labelled *R* and *S* acetic acids can be carried out enzymatically[10] or purely by synthetic methods[11] (Fig. 6). Phenyl acetylene was deuterated by metallation and reaction with deuterium oxide. Reduction with diimide, followed by epoxidation with *m*-chloroperbenzoic acid, and reduction with lithium borotritride gave a racemic mixture of 1-phenylethanols. Optical resolution with brucine phthalate and oxidation gave the two asymmetric acetic acids. The steric course of the chemical transformations and the absolute configuration of (+) and (—) 1-phenylethanol are well known, consequently the absolute configuration of the derived acetic acids is as depicted.

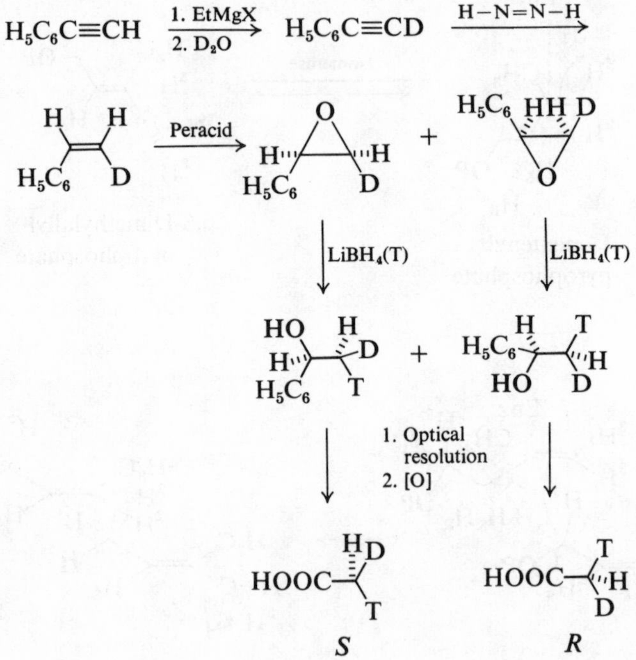

$$H_5C_6C{\equiv}CH \xrightarrow[2.\ D_2O]{1.\ EtMgX} H_5C_6C{\equiv}CD \xrightarrow{H-N=N-H}$$

Peracid

LiBH$_4$(T)

1. Optical resolution
2. [O]

S *R*

Fig. 6 Synthesis of chiral acetic acids

The analytical method was dependent on the action of malate synthase–fumarase, and the existence of an isotope effect discriminating between the three hydrogens.[12] Malate synthase catalyses the irreversible condensation of acetyl CoA with glyoxylate to malic acid. No fast proton exchange occurs in the methyl with the solvent in the condensation step and a normal isotope effect was observed. Fumarase eliminates reversibly and stereospecifically water from malic acid in a *trans* fashion.

Chiral acetyl CoA gives a mixture of tritium labelled malic acid I and II. If we suppose that the condensation is stereospecific, e.g. inversion of configuration as in Fig. 7, then I/II = k_H/k_D. Prolonged equilibration with fumarase will exchange all D in I with protons from the solvent, but tritium will be retained. On the other hand, II will lose all of its tritium. Thus measurement of the activity of malic acid produced from chiral acetic CoA and the activity of malic acid resulting from a sample incubated with fumarase will give us the relative amounts of I and II, and consequently, k_H/k_D. Malate from *R* acetyl CoA was found to retain 69 per cent of its activity. When the *S* acetate was used 31 per cent of the

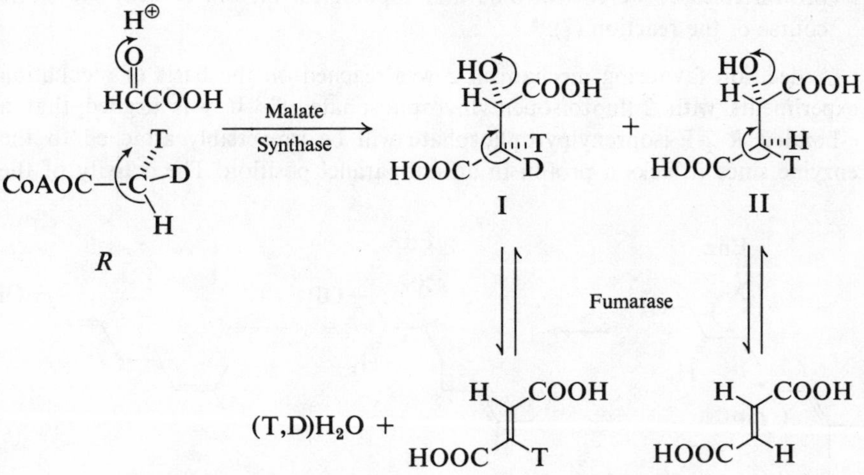

Fig. 7 Reactions of malate synthase and fumarase

activity was retained, e.g. $k_H/k_D = 2.2$. This is a normal isotope effect and the malate condensation proceeds thus with inversion. Without regard to the isotope effect, this series of enzymatic transformations constitutes a method to determine the chirality of a given acetic acid. When dimethylallyl pyrophosphate was degraded and the acetic acid formed was put through the malate synthase–fumarase procedure, it turned out that the steric course followed (3), i.e. protonation occurs from the β-face or the *re* side of IPP.[13] Isoprenylation, on the other hand, proceeds with the opposite configurational outcome (4). The incoming isoprenyl group and the released proton are here located on the same side. By using chiral acetyl CoA it was also demonstrated that the condensation with acetyl CoA to β-hydroxy-3-methylglutaryl CoA proceeded stereospecifically with inversion of configuration and involved a normal hydrogen isotope effect (5).[14]

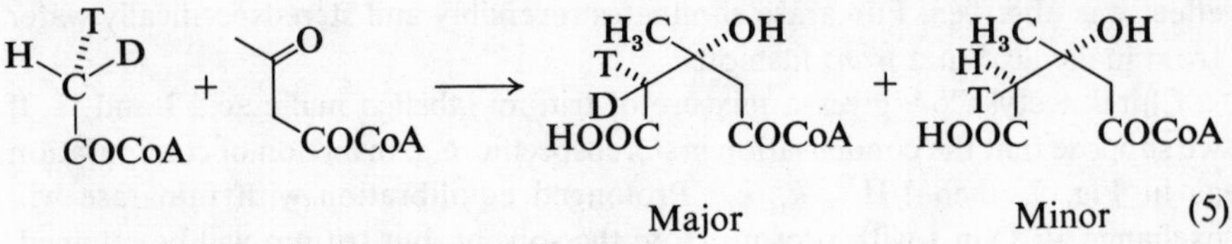

(5) [Reaction scheme: H(T)(D)C–COCoA + acetoacetyl-CoA (O, COCoA) → Major: H_3C, OH, T, D, HOOC, COCoA + Minor: H_3C, OH, H, T, HOOC, COCoA]

Major Minor (5)

Two mechanisms have been advanced to account for the stereochemistry of the prenylation reaction:

1. consecutive *trans*-1,2-addition and *trans*-1,2-elimination involving formation of an intermediate σ-bond with the enzyme (6);[9]
2. ionization of the allylic pyrophosphate synchronized with alkylation and proton release: stereoelectronic and topological factors control the steric course of the reaction (7).[15]

A decision favouring mechanism 2 was reached on the basis of incubation experiments with 2-fluoroisopentylpyrophosphate (8). It was argued that a σ-bonded *R* 2-F-isoprenylpyrophosphate will be irreversibly attached to the enzyme since it lacks a proton in an antiparallel position. The activity of the

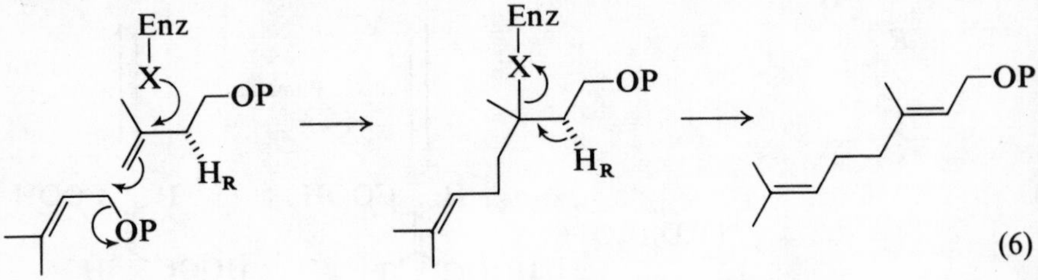

(6) [Scheme: Enz–X, OP, H_R, OP → Enz–X, OP, H_R → OP]

(6)

(7) [Scheme: OP → Ion pair (⊕⊖OP), OP, H_R → OP]

Ion pair

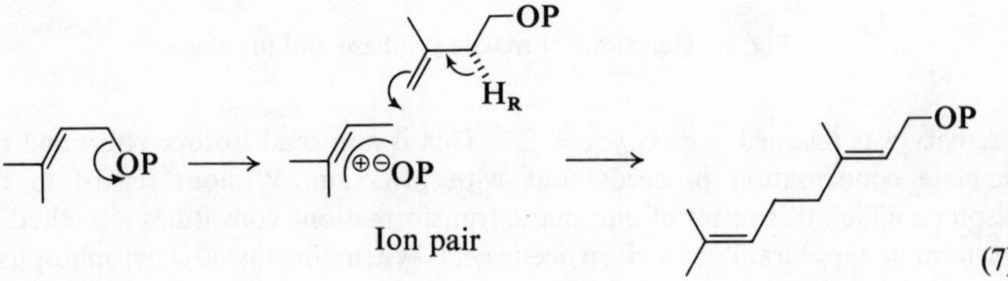

(7)

(8) [Scheme: OP, F, H_R, OP → OP, F]

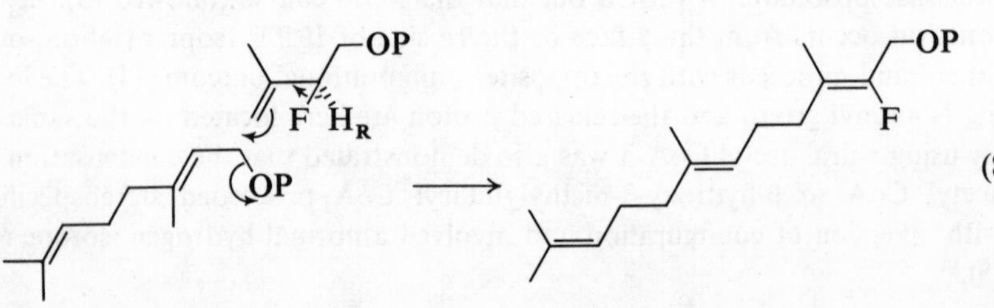

(8)

enzyme was indeed reduced, but restored again after dialysis. This suggests a competitive absorption at the active site but no participation of an X-group of the enzyme. 2-Fluorofarnesylpyrophosphate was identified as a metabolite from incubation of geranyl pyrophosphate and *R,S*-2-fluoroisopentenyl pyrophosphate. In another series of experiments it was found that the relative rate for geranylation and 2-fluorogeranylation of IPP is nearly identical to the relative rate for solvolysis of the geranyl and 2-fluorogeranyl methanesulphonates (*ca.* 10^3). The electron withdrawing effect of fluorine retards both ionization and prenylation but it has practically no effect on the rate of S_N2 displacements at C^1. This implies that the C–OP bond breaking is far advanced before alkylation of the double bond occurs. The steric integrity at C^1 of the allyl carbonium ion is maintained by a rotational barrier of 28 kcal.[16] The experiments also show that the substrate specificity of prenyl transferase is not very stringent.

5.3 Monoterpenes

Geranyl pyrophosphate is the parent compound for the monoterpenes. Labelling experiments in higher plants have shown that mevalonic acid is incorporated preferentially in the part derived from isopentenyl pyrophosphate. The reason for the imbalance is not clear, but it is suggested that leucine participates in the biosynthesis of monoterpenoids by being primarily converted to dimethylallyl pyrophosphate through an alternative route.[17] In contrast to the situation in higher plants the fungus *Ceratocystis monoliformis* produces monoterpenes with symmetrical incorporation of labelled mevalonic acid.[18]

Before cyclization can occur, *trans*-geranyl pyrophosphate must rearrange to *cis*-geranyl pyrophosphate = neryl pyrophosphate (Fig. 8). This can be accomplished in two ways:

1. by a redox reaction via the aldehyde; rotation bout $C^{2,3}$ is achieved by a reversible Michael addition (9); and
2. intermediate rearrangement to the tertiary allylic pyrophosphate, linalyl pyrophosphate, allowing rotation about the $C^{2,3}$ (10).[19]

The same mechanisms also hold for the rearrangement of *trans–trans*-farnesol to *cis–trans*-farnesol. Mechanism 1 implies that one of the C^1–H is removed and possibly lost during oxidation, and analysis of the isomerization products shows that this is the case,[20,21] but some of these results have been opposed.[22,23] There are several experiments supporting mechanism 2, i.e. isomerization without loss of C^1–H.[19] The tertiary allylic pyrophosphate needs not to be a mandatory intermediate. The barrier of rotation about $C^{2,3}$ becomes smaller the tighter the phosphate ion is associated with the tertiary centre in the ion pair. It is also argued in favour of the redox mechanism that a transiently generated NADH could eventually return the same hydrogen to the intermediate aldehyde. However, experiments seem to indicate that the cyclase does not require $NAD^{\oplus}$. It has been demonstrated on several occasions that geranyl, neryl, and linalyl pyrophosphate (and the corresponding *trans–trans*, *cis–trans*-farnesyl and

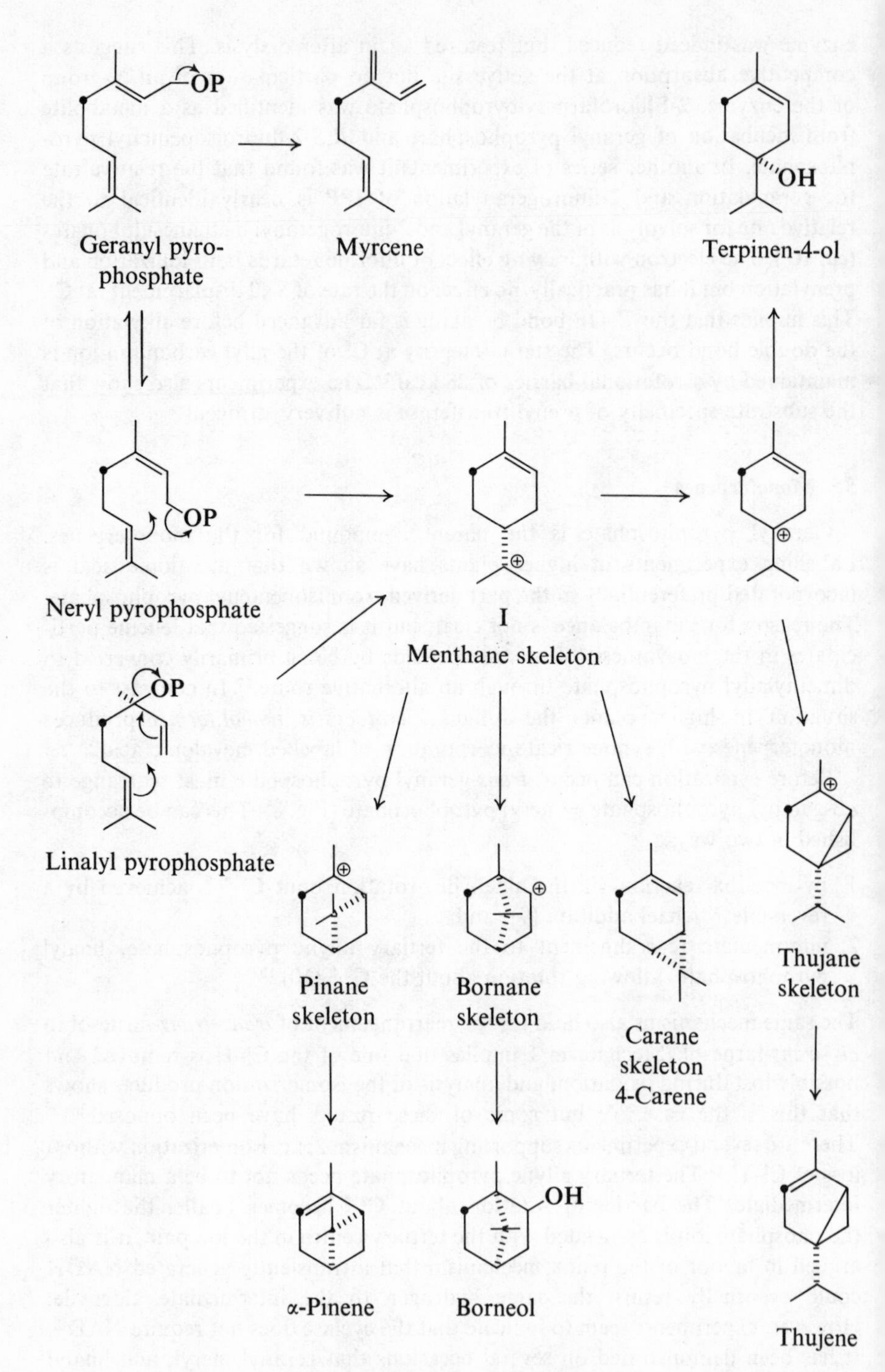

Fig. 8 Biosynthesis of monoterpenes

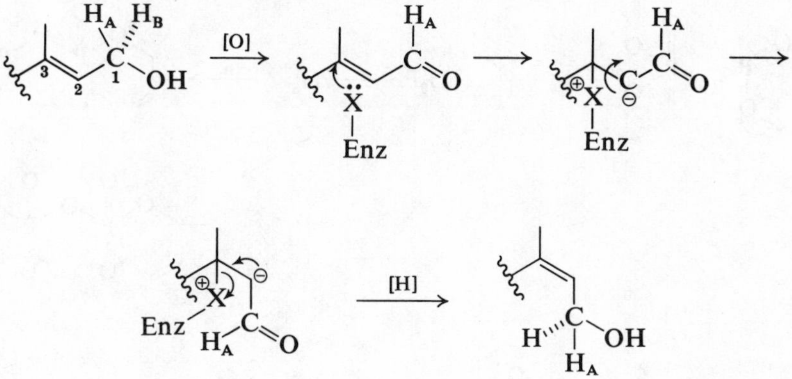

(9)

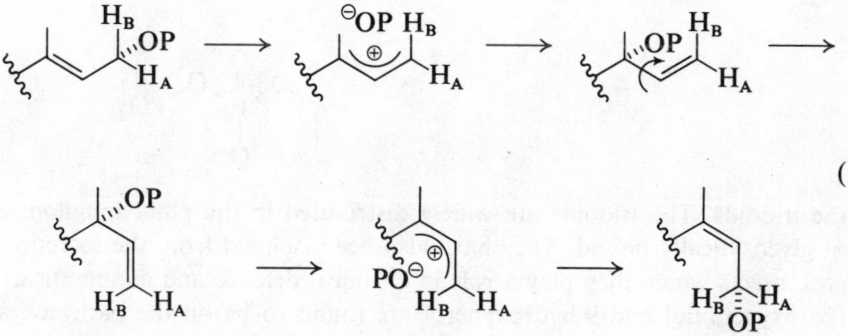

(10)

nerolidyl pyrophosphates in the sesquiterpene series) are cyclized to the same product but no interconversion is observed. This finding can be taken as evidenec for formation of a common asymmetric ion pair which cyclizes faster than it rearranges.

The allylic rearrangement exhibits some interesting features. It is established to proceed with *syn* stereochemistry. C^1–^{18}O labelling of farnesol shows that the allylic tertiary alcohol carries one-third of the activity of the precursor. That means that a carbonium–pyrophosphate ion pair is formed in which there is sufficient time for free rotation (11a) about the P–O–P bond, but no time for tumbling (11b). A 1,3-sigmatropic tight rearrangement requires unchanged activity in the tertiary alcohol (11c) and a concerted six-membered transition state requires loss of activity (11d).[19]

The elimination of pyrophosphate from neryl pyrophosphate, anchimetrically assisted by the double bond, gives the terpinyl cation I which forms a branch point to the various monoterpene skeletons. By using labelled 2-^{14}C-mevalonic acid it was established that the cyclization took the steric course depicted in Fig. 8.

A slightly different folding of geranol gives rise to the cyclopentene skeleton

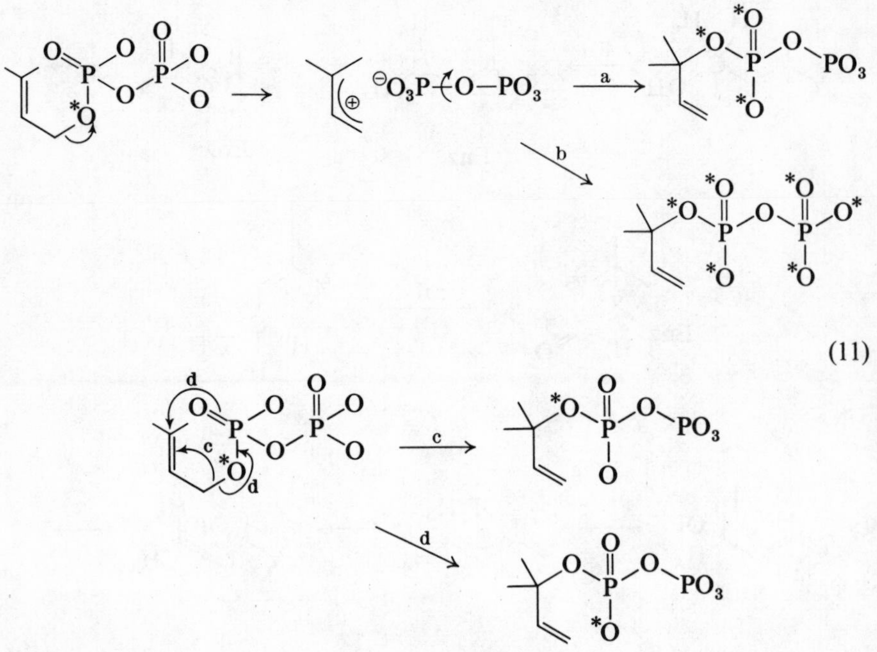

(11)

of the iridoids. The iridoids are widely distributed in the plant kingdom and often glycosidically bound. They have also been isolated from the secretion of several insects where they play a role in chemical defence and communication. 9-Hydroxygeraniol and 9-hydroxynerol are found to be on the pathway, and

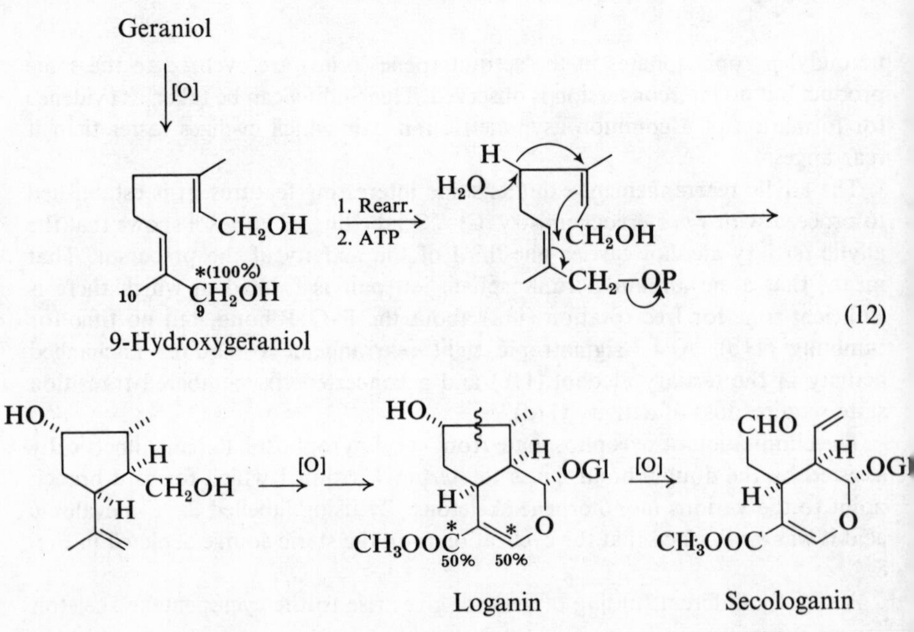

Geraniol

[O]

1. Rearr.
2. ATP

*(100%)

9-Hydroxygeraniol

(12)

[O]

[O]

50% 50%

Loganin

Secologanin

phosphorylation of this hydroxyl group followed by cyclization could lead to the iridoid skeleton (12). At one stage of the subsequent oxidation carbons 9 and 10 of geraniol become equivalent. Loganin is a precursor for the monoterpene and indole alkaloids.

The biosynthesis of some irregular monoterpenes, e.g. artemisia, yomogi, and santolina alcohols and chrysanthemic acid, a component of the efficient pyrethrum insecticides, is started by condensation of two molecules of dimethylallyl pyrophosphate. Nucleophilic aliphatic substitution gives chrysanthemyl pyrophosphate and opening of the ring gives the various alcohols in a reaction initiated by solvolysis of the pyrophosphate (13). Geraniol and nerol are not obligatory precursors for these monoterpenes because extensive scrambling of label was observed when these alcohols were incorporated into artemisia ketone by *Artemisia annua*.[24]

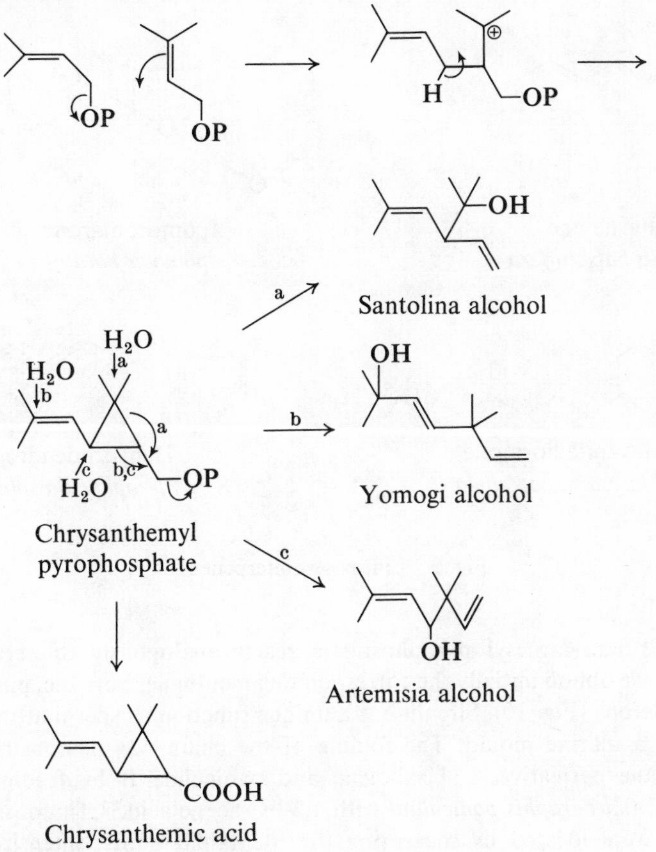

OP OP H OP OH

Santolina alcohol

H_2O a H_2O b a b OH (13) c b,c H_2O OP

Yomogi alcohol

Chrysanthemyl pyrophosphate

c OH

Artemisia alcohol

COOH

Chrysanthemic acid

5.4 Sesquiterpenes

When the chain length increases, the number of conceivable cyclizations and secondary modifications increases dramatically. This is manifested in the large number of compounds isolated from all parts of the plant and animal kingdoms.

The structural variation is impressive, especially within the sesquiterpene and diterpene series. None the less the skeletons can be derived by suitable folding of the chain and by applying the basic reactions outlined in Fig. 3. The success of this semi-theoretical approach and the success of many syntheses inspired by biosynthetic principles lend strength to our conceptions of enzymes at work.

A few acyclic sesquiterpenes are known (Fig. 9). Dehydration of farnesol gives farnesenes. Furanofarnesenes are detected in such different phyla as marine sponges, ants, and sweet potato infected by *Ceratocystis fimbriata*. It is suggested that the ethyl groups in the juvenile hormone comes from propionic acid by the synthesis of the methyl homologue of mevalonic acid.[25] The juvenile hormone prevents the metamorphosis of larvae and hampers the development of insects into adults. Compounds of this type have interest as insecticides.

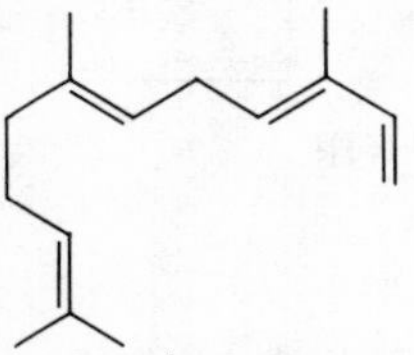

β-Farnesene
Populus balsamifera

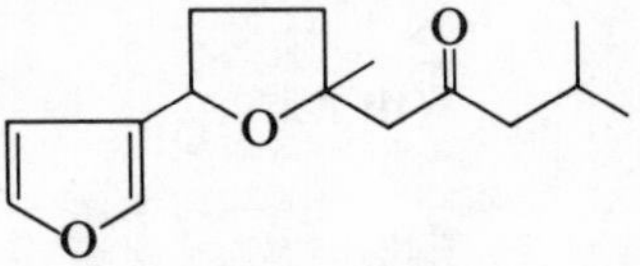

Ipomoeamarone
Ipomoea batatas

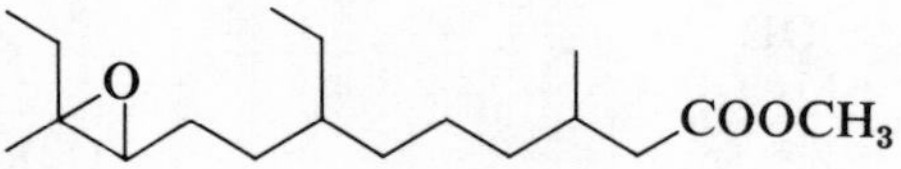

Juvenile hormone
Platysamia cecropia

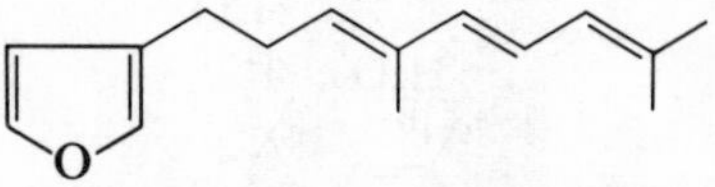

Dehydrodendrolasin
Pleraphycilla spinifera

Fig. 9 Linear sesquiterpenes

When *cis-trans*-farnesyl pyrophosphate reacts analogously to neryl pyrophosphate, we obtain initially the corresponding menthane, bornane, pinane and carane skeletons (Fig. 10). Sirenin has a unique function as sperm attractant in *Allomyces*, a marine mould. The folding of the chain was demonstrated for the menthane derivatives, γ-bisabolene and paniculide B by feeding tissue cultures of *Andrographis paniculata* with 1,2-^{13}C-acetic acid.[26] The positions of the labels were located by measuring the ^{13}C NMR shifts, intensities, and $^{1}J_{^{13}C^{13}C}$, and were found to be in agreement with the biosynthetic scheme (Figs. 10, 11). *cis–trans*-Farnesol, but not *cis–cis*-farnesol, was incorporated.

If the C^{10} double bond of the bisabolyl cation is involved in the second cyclization, we obtain the skeletons of the cuparane–widdrane families (Fig. 12). *trans*-Alkylation of the central double bond with C^1 and C^{11} and 1,5-hydrogen

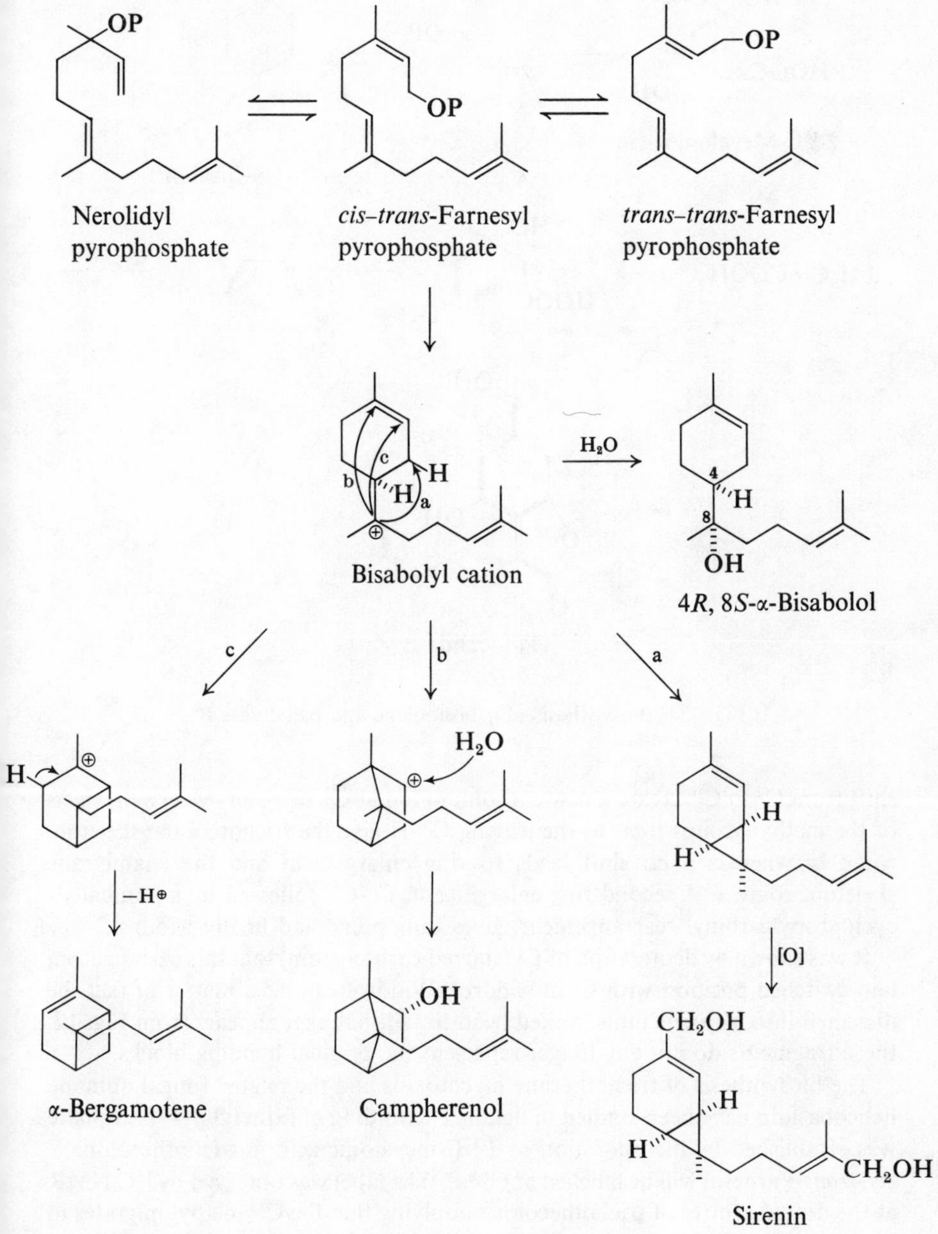

Fig. 10 Generation of menthane, bornane, pinane, and carane skeletons from *cis–trans*-farnesyl pyrophosphate

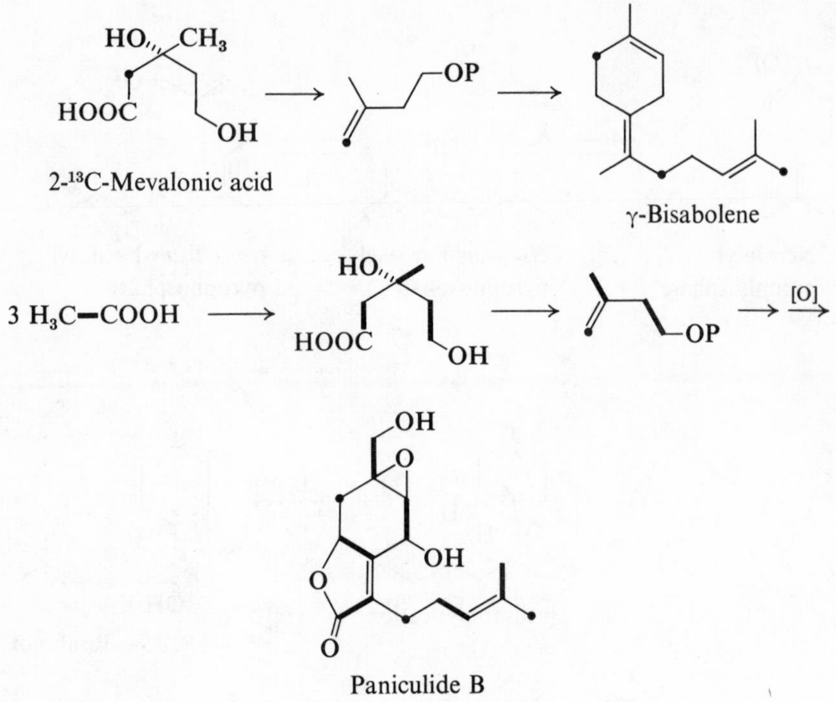

Fig. 11 Biosynthesis of γ-bisabolene and paniculide B

shift from C^6 to C^{10} create a new carbonium ion at C^6. Wagner–Meerwein shifts of the methyl groups *trans* to the leaving C^6–H give the trichothecane skeleton, route b, whereas a *cis* shift leads to ring enlargement and the chamigrane skeleton, route a. A second ring enlargement, C^5–C^7, followed by a homoallyl-cyclopropylcarbinyl rearrangement, gives thujopsene and finally widdrol.

It was shown by deuteration of C^1 (starred carbon atom) that this carbon atom had switched position with C^2 in widdrol. Widdrol can, as a matter of fact, be dissected into isoprene units, linked head to tail, but as it appears from Fig. 12, these fragments do not any longer represent the original building blocks.

The biosynthesis of trichothecane mycotoxins and the related fungal quinone helicobasidin have been studied in detail. The folding of farnesyl pyrophosphate was established by incorporation of 2-^{13}C-mevalonic acid in trichothecolone.[27] *cis–trans*-Farnesol will be labelled at $C^{4,8,12}$. The label was observed by ^{13}C NMR at the dotted centres of trichothecolone implying that the C^{12} methyl migrates in a *cis* fashion to C^7 (Fig. 12). The 1,5-hydrogen shift was established by incorporation of 6-^{3}H-farnesyl pyrophosphate. Tritium was recovered with no loss at C^{10} showing that the subsequent hydroxylation at C^{10} occurs with retention of configuration.[28] It is found that trichodiene, but not γ-bisabolene, is a precursor for trichothecolone.[29]

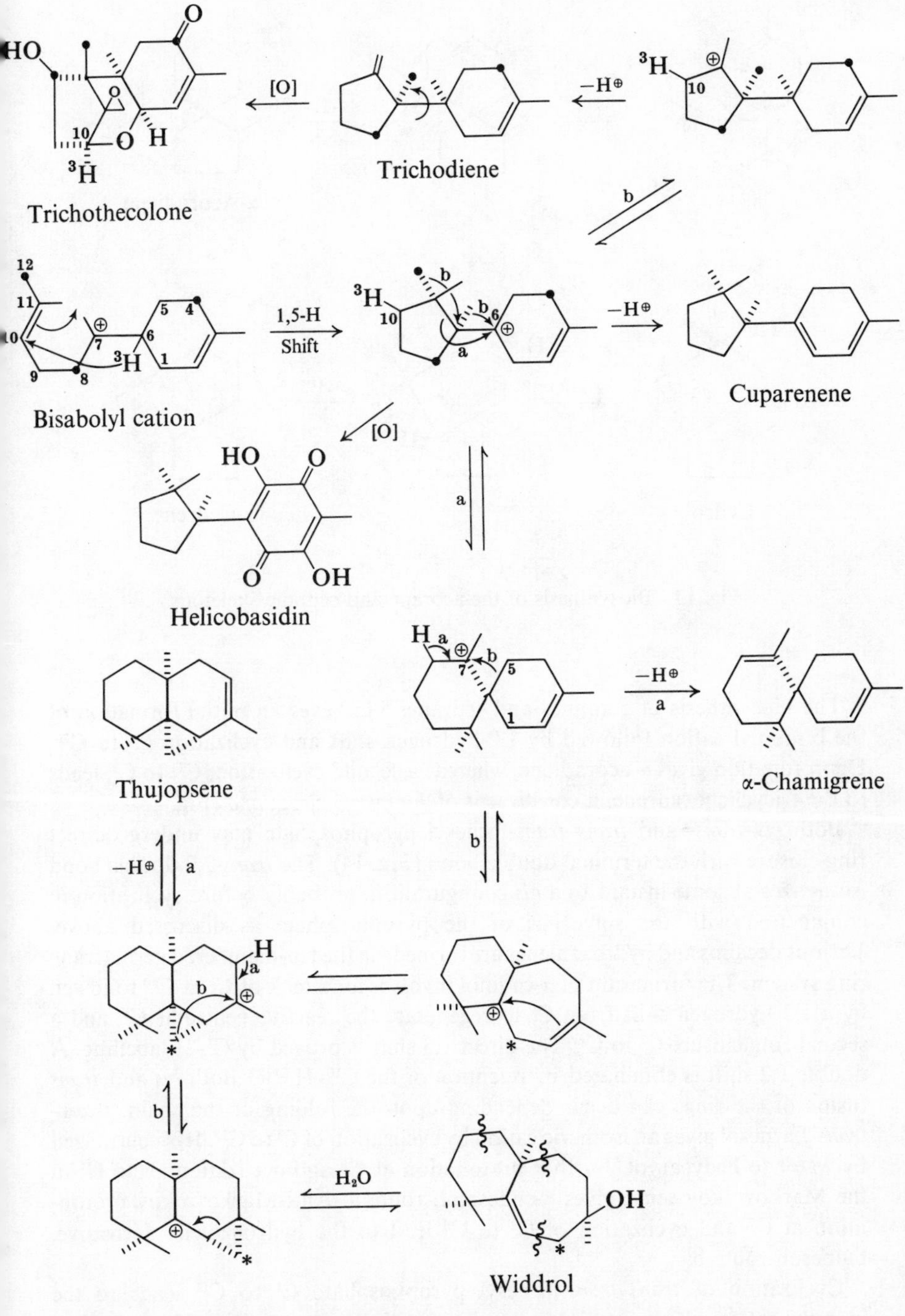

Trichothecolone

Trichodiene

Bisabolyl cation

1,5-H Shift

Cuparenene

[O]

Helicobasidin

Thujopsene

α-Chamigrene

Widdrol

H_2O

Fig. 12 Derivation of some sesquiterpene skeletons

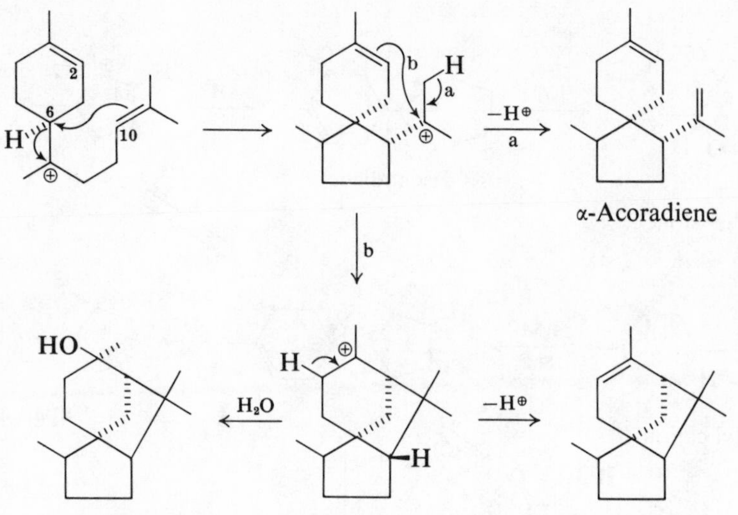

Fig. 13 Biosynthesis of the acorane and cedrane skeletons

The biosynthesis of acoranes and cedranes[30] involves an initial formation of the bisabolyl cation followed by 1,2-hydrogen shift and cyclization C^6 to C^{10}. Deprotonation gives α-acoradiene, whereas a second cyclization, C^{11} to C^2, leads to the tricyclic α-cedrene, a constituent of *Juniperus virginiana* (Fig. 13).

Both *cis–trans-* and *trans–trans*-farnesyl pyrophosphate may undergo direct ring closure with the terminal double bond (Fig. 14). The *trans*-2,3 double bond isomerizes at some instant to a *cis* configuration, probably before cyclization in conjunction with the solvolysis of the pyrophosphate as discussed above. Various decalins and hydroazulenes are formed via the ten-membered germacrane ring system. The formation of α-cadinol involves an attack of C^1 at C^{10} followed by a 1,3 hydrogen shift from C^1 to regenerate the reactive centre at C^1, and a second ring closure C^1 to C^6. The direct 1,3 shift is proved by C^1–H labelling. A double 1,2 shift is eliminated by retention of the C^{10}–H.[31,32] Both *cis* and *trans* fusion of the rings can occur depending upon the folding of the chain. *trans–trans*-Farnesol gives an isomeric cation by cyclization of C^1 to C^{10}. It is neutralized by water to hedycaryol. Further protonation at C^6 and cyclization C^2 to C^7 in the Markovnikov sense gives β-eudesmol, route a. Anti-Markovnikov protonation at C^7 and cyclization of C^2 to C^6 lead to the hydroazulene derivative, bulnesol, route b.

Cyclization of *trans–trans*-farnesyl pyrophosphate C^1 to C^{11} leads to the humulyl cation which gives humulene by proton elimination (Fig. 15, route a) or undergoes a further cyclization to caryophyllene, route b. Cyclization of *cis–trans-*

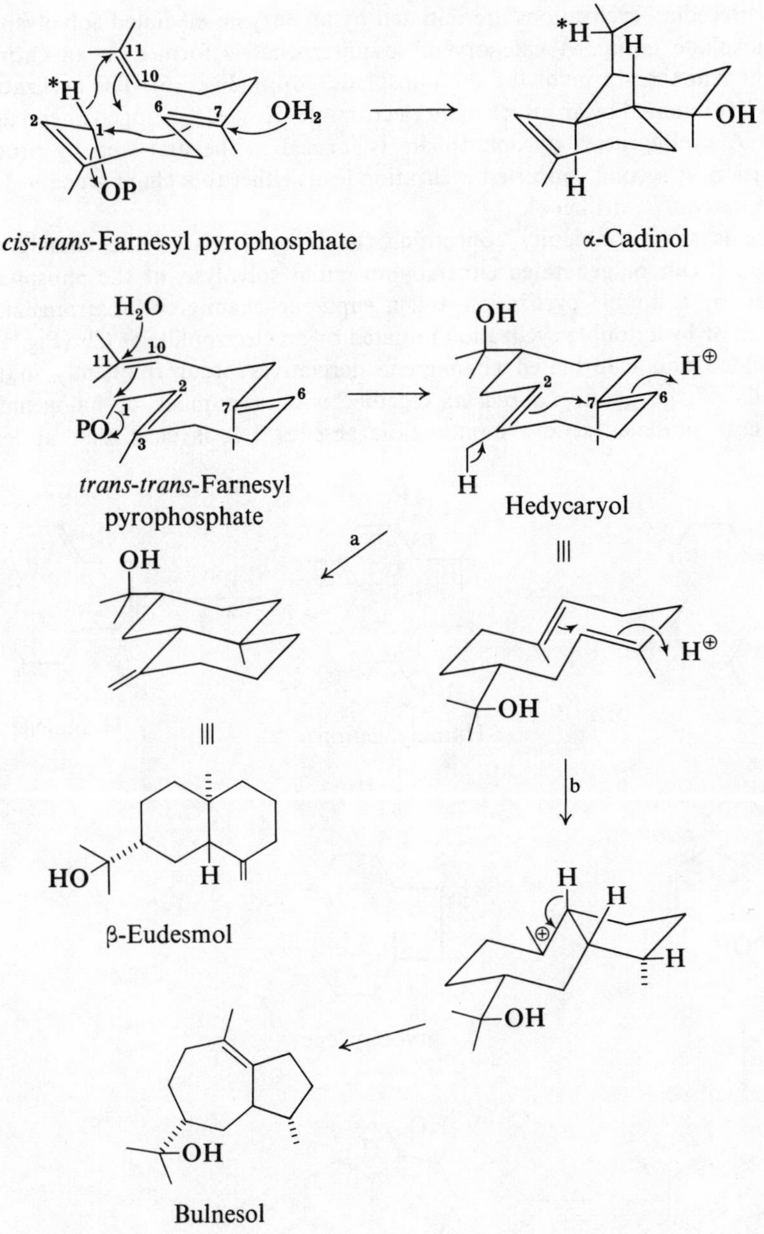

cis-trans-Farnesyl pyrophosphate → α-Cadinol

trans-trans-Farnesyl pyrophosphate → Hedycaryol

β-Eudesmol

Bulnesol

Fig. 14 Biosynthesis of decalin and hydroazulene derivatives

farnesyl pyrophosphate C^1 to C^{11} followed by a 1,3 hydrogen shift from C^1 to C^{10} and cyclization C^1 to C^6 gives himachalol on neutralization with water (Fig. 15).

The preceding cyclizations are initiated by an enzyme-mediated solvolysis of the phosphate group. A category of sesquiterpenes is formed in an entirely different manner not involving the phosphate group (Fig. 16). The cyclization C^{11} to C^6 is initiated by an attack of an electrophile ($X = H^{\oplus}$, halogen or epoxide) at C^{10}. A cyclofarnesyl pyrophosphate is formed in the first step by proton elimination. A second concerted cyclization leads either to a chamigrene or to a decalin derivative, drimenol.

There is some ambiguity concerning the biosynthesis of the chamigrene skeleton. It can be generated either by an initial solvolysis of the phosphate, followed by a double cyclization and a cuparene–chamigrene rearrangement (Fig. 12), or by a double cyclization initiated by an electrophile at C^{10} (Fig. 16). Brominated and chlorinated chamigrene derivatives occur frequently in the seaweeds of the genus *Laurencia*, together with a number of halogenated cuparenes, bisabolenes, and monocyclofarnesenes.[33] It is clear that at least

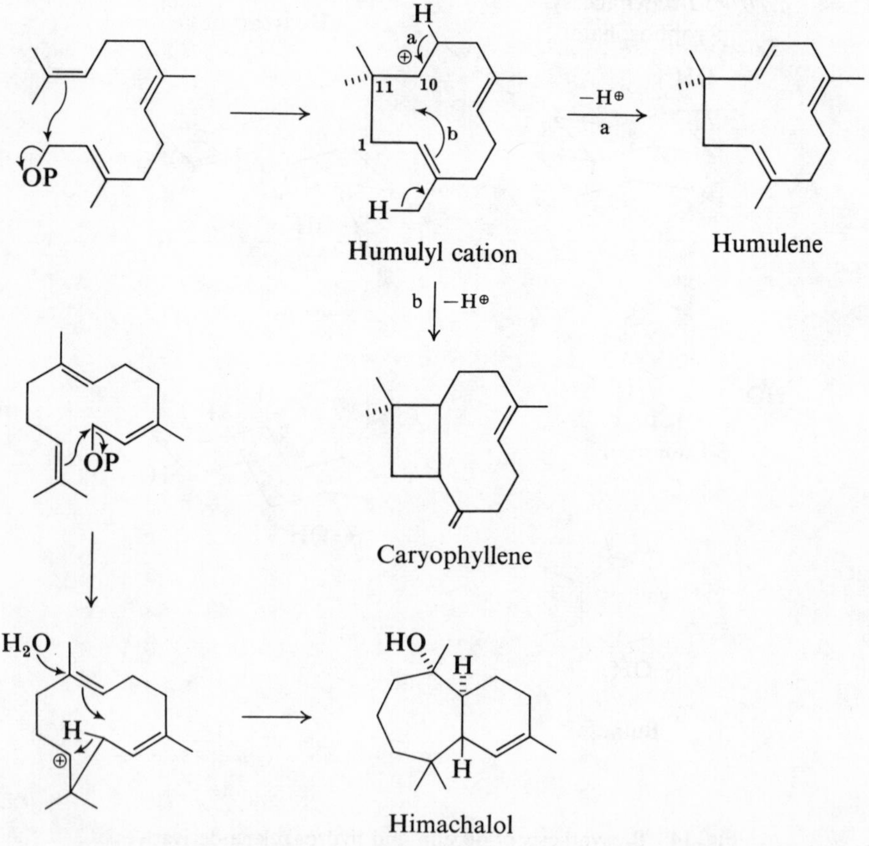

Fig. 15 Biosynthesis of humulanes, caryophyllanes, and himachalanes

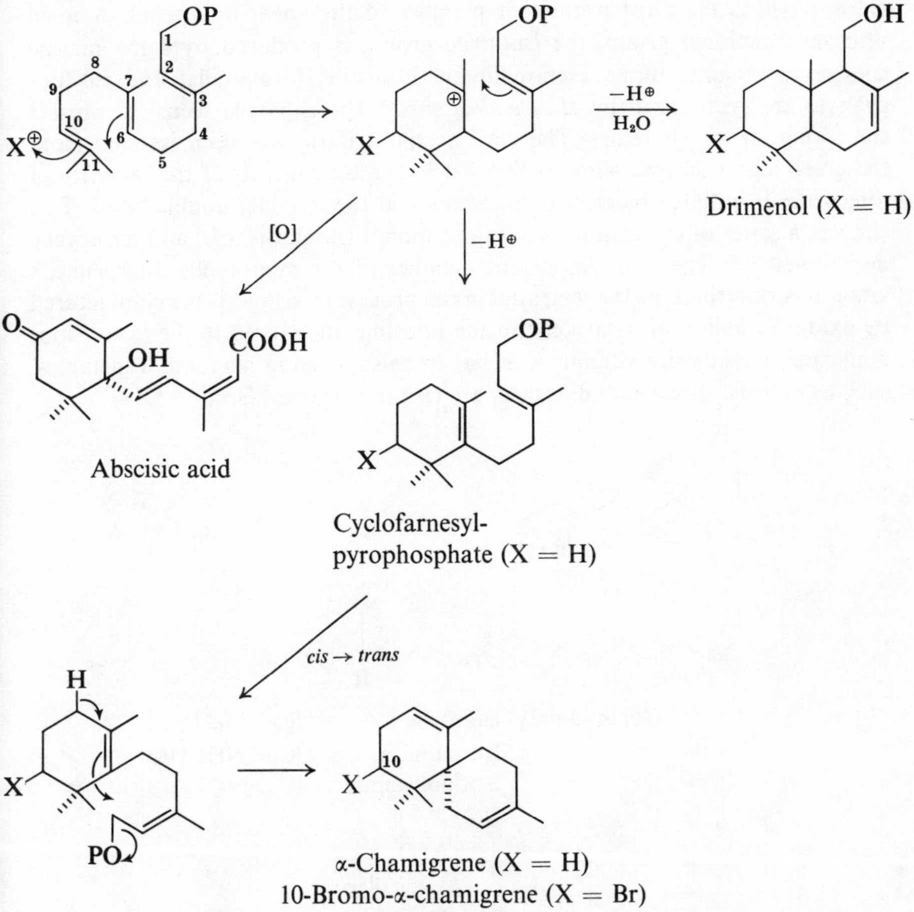

Drimenol (X = H)

Abscisic acid

Cyclofarnesyl-pyrophosphate (X = H)

α-Chamigrene (X = H)
10-Bromo-α-chamigrene (X = Br)

Fig. 16 Sesquiterpene cyclizations initiated by electrophilic attack at C^{10}

in this particular environment the cyclization is initiated by the peroxide–bromide couple at C^{10}. There are also indications that chamigrenes rearrange to cuparenes in *Laurencia* spp. Note that chamigrene biosynthesized according to Fig. 16 contains intact isoprene building blocks in contrast to chamigrene formed according to Fig. 12.

The cyclization leading to decalins is much more common in the di- and tri-terpene series. The cyclofarnesene derivative abscisic acid (Fig. 16) is a plant hormone controlling the shedding of leaves.

5.5 Diterpenes

Geranylgeranyl pyrophosphate is the precursor of the diterpenes. Phytol, 6,7,10,11,14,15-hexahydrogeranylgeraniol, forming the lipophilic side chain of chlorophyll, is the most prominent member of the linear diterpenes. A most unusual functional group, the isonitrile group, is produced by some marine sponges and some fungi. Geranyllinalyl isonitrile, formamide, and isothiocyanate are synthesized by *Halicondria* spp.[34] The isonitrile seems to protect the sponge from predators. The enol acetate trifarin has been isolated from the green alga *Caulerpa trifaria* (Fig. 17).[35] The biosynthesis of the majority of diterpenes is initiated by electrophilic attack at the terminal double bond. This triggers a series of cyclizations leading to mono- (rare), di-, tri-, and tetracyclic derivatives.[36,37] The most important member of the monocyclic diterpenes is vitamin A or retinol, p. 168, essential in the process of vision. It is manufactured by oxidative fission of β-carotene in the intestine and stored in the liver. Since man cannot synthesize vitamin A, it has to be supplied in his food. Vegetables, such as carrots, spinach, and lettuce, are rich in carotenes (section 5.10).

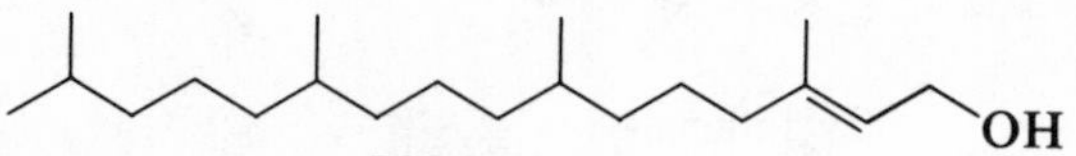

OH

Phytol

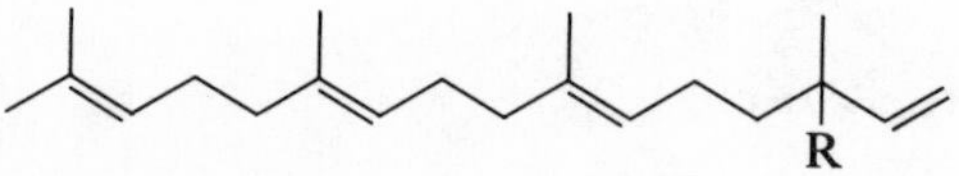

R

Geranyllinalyl isonitrile R = N=C
formamide R = NHCHO
isothiocyanate R = NCS

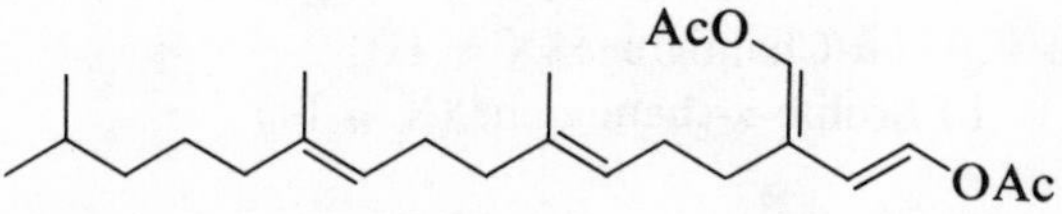

AcO

OAc

Trifarin

Fig. 17 Linear diterpenes

The cyclization normally progresses directly to the bicyclic decalin system which is subsequently discharged by addition of water or proton elimination, often via a series of Wagner–Meerwein shifts. The configuration of $C^{5,8,9,10}$ (steroid numbering) is determined by the folding of the *trans–trans*-geranylgeraniol chain. A chair–chair conformation gives the labdane skeleton (Fig. 18). Hydration of the C^8-carbonium ion and allylic rearrangement in the sidechain give sclareol. Deprotonation to form the C^8 exocyclic double bond produces labdadienol pyrophosphate which is an important intermediate for biosynthesis of tri- and tetracyclic diterpenes. A five-step concerted Wagner–Meerwein shift with inversion at each centre, route a, leads to the skeleton of hardwickiic acid. Modifications in the side chain are less stereospecific. It is characteristic for

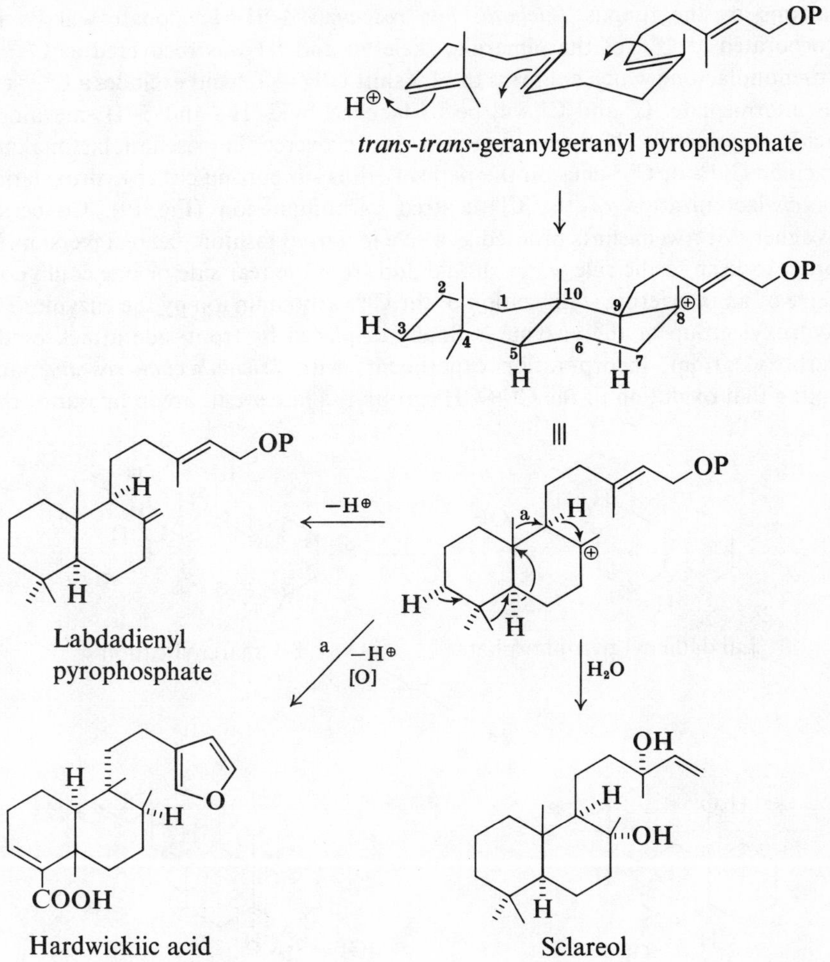

Fig. 18 Cyclization of all-*trans*-geranylgeranyl pyrophosphate in a chair–chair conformation to bicyclic diterpenes

diterpenes that often both enantiomeric forms are produced, occasionally in the same species. The cyclizations have been imitated with great success *in vitro* and consequently the mode of cyclization is not entirely determined by the topology of the enzyme but is stereoelectronically controlled.[38,39]

Solvolysis of labdadienyl pyrophosphate gives, via the 8-pimarenyl cation, pimarane and abietane derivatives, widely distributed in *Coniferae*. Formation of rosenonolactone is rationalized by a double Wagner–Meerwein shift, $C^9 \rightarrow C^8$ and $C^{10}CH_3 \rightarrow C^9$, route a, giving a carbonium ion at C^{10} which finally is neutralized either by water or by direct lactonization. The scheme was verified by incorporation of $4R$-4-^3H, 2-$^3H_2$ and 5-$^3H_2$ mevalonic acid in rosenonolactone by the fungus *Trichothecium roseum*.[40] 4-^{3}H-Mevalonate will be incorporated at $C^{5,9}$ of the pimarane skeleton and ^{3}H was recovered at $C^{5,8}$ in rosenonolactone which confirms the 1,2 shift $C^9H \rightarrow C^8$ and excludes a $C^{5,10}$-ene as intermediate. C^1 and C^6 will be ^{3}H-labelled by 2-$^3H_2$- and 5-$^3H_2$-mevalonic acid, respectively. Both hydrogens were recovered in rosenonolactone and exclude $C^{1,10}$- or $C^{5,6}$-enes on the pathway thus supporting either hydroxylation or *cis*-lactonization of the C^{10} centred carbonium ion (Fig. 19). Concerted Wagner–Meerwein shifts proceed as a rule in a *trans* fashion, i.e. by inversion. In order to keep to the rule water should add from the rear side or one could conceive of an interfering stabilization of the C^{10} carbonium ion by the enzyme. The hydroxyl group or the enzyme is finally displaced by front–side attack of the carboxyl group. Incorporation experiments with *Trichothecium roseum*, indicating that oxidation of the C^4-β-CH_3 group is a late event, are in favour of this

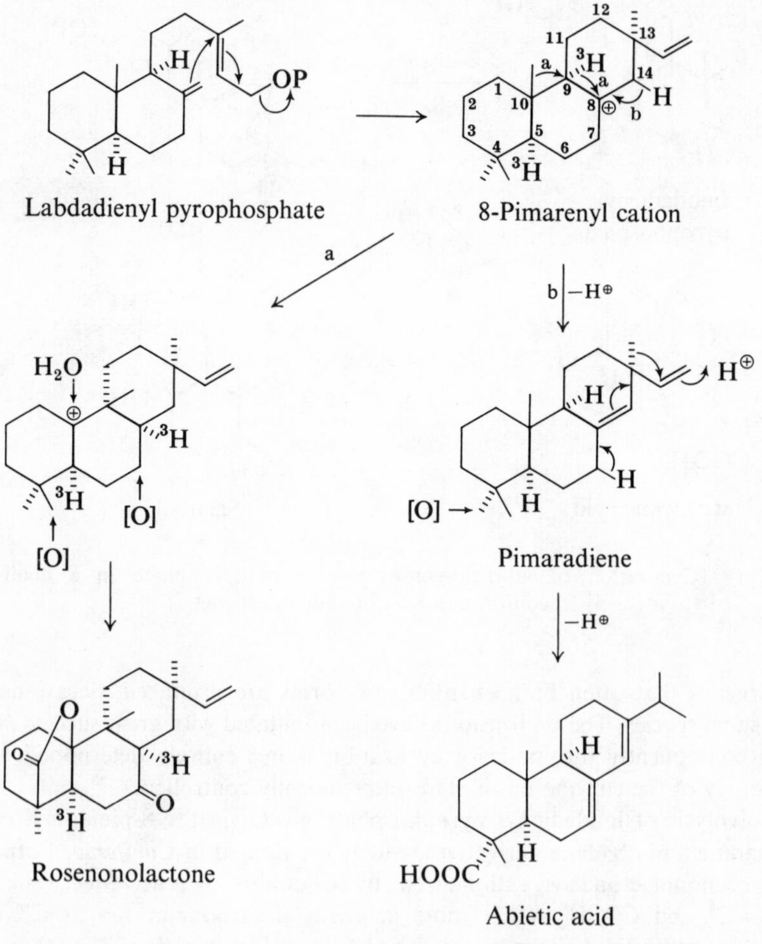

Labdadienyl pyrophosphate

8-Pimarenyl cation

Rosenonolactone

Pimaradiene

Abietic acid

Fig. 19 Biosynthesis of tricyclic diterpenes

mechanism. On the other hand, the direct lactonization, implying that the oxidation of the β-CH_3 group occurs at an early stage before rearrangement, is supported by an *in vitro* synthesis, mimicked after these biogenetic principles.[41]

Oxidation of ring C of the abietanes leads to a number of aromatic diterpenes. C^{14}–O opening of an assumed epoxide at $C^{13,14}$ accounts for the C^{13} to C^{14} migration of the isoprenyl group in totarol (Fig. 20).

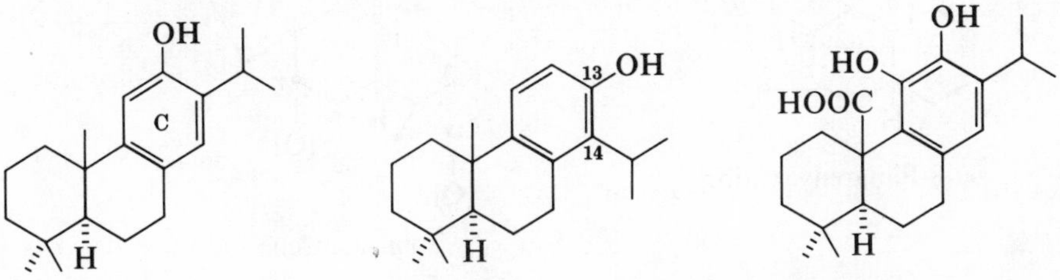

Ferruginol
Podocarpus ferruginea

Totarol
Podocarpus ferruginea

Carnosic acid
Rosmarinus officinalis

Fig. 20 Aromatic diterpenes

The tetracyclic gibberellins are of importance as plant growth hormones. They control cell elongation and were first isolated from the fungus *Gibberella fujikuroi*, a parasite of rice causing reduced straw stiffness as a result of formation of too long and limp straw cells. This feared infection causes great losses of rice crops. It was later found that gibberellins are produced universally by plants in small quantities as natural plant growth hormones. The chemistry and biosynthesis of gibberellins have been studied extensively.[42] The immediate precursor is *ent*-kaurene derived from *ent*-labdadienyl pyrophosphate = copalyl pyrophosphate (Fig. 21). C^{19} of *ent*-kaurene is oxidized to a carboxylic group followed by 7α- and 6α-hydroxylation and ring contraction via a diol-one rearrangement to gibberellin A_{12} aldehyde, the first detectable derivate with a gibbane skeleton. The aldehyde is then oxidized to the dicarboxylic acid.[43–46] The order of the following oxidative events is dependent on the metabolizing organism. There seems to be some consensus that hydroxylation at C^3 occurs first followed by oxidative elimination of $C^{10}CH_3$ as carbon dioxide with concomitant formation of the lactone bridge.[47] There is no loss of oxygen from ^{18}O-labelled C^{19}-carboxylate and no loss of hydrogen atoms from the adjacent centres. These findings suggest a peracid as a plausible intermediate (14). Dehydrogenation of $C^{1,2}$ and hydroxylation of C^{13} are late events.

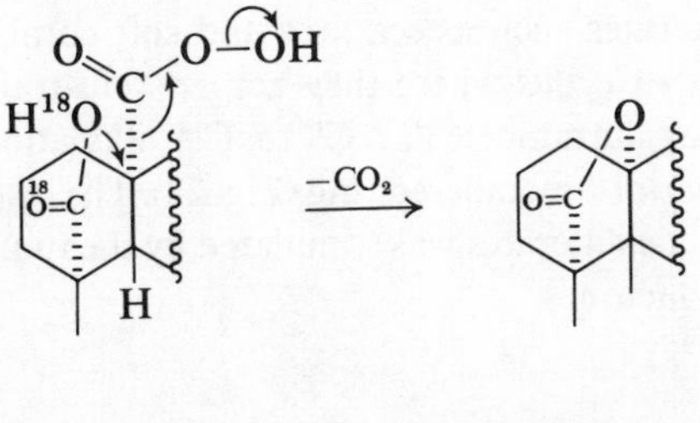

$$\xrightarrow{-CO_2} \qquad (14)$$

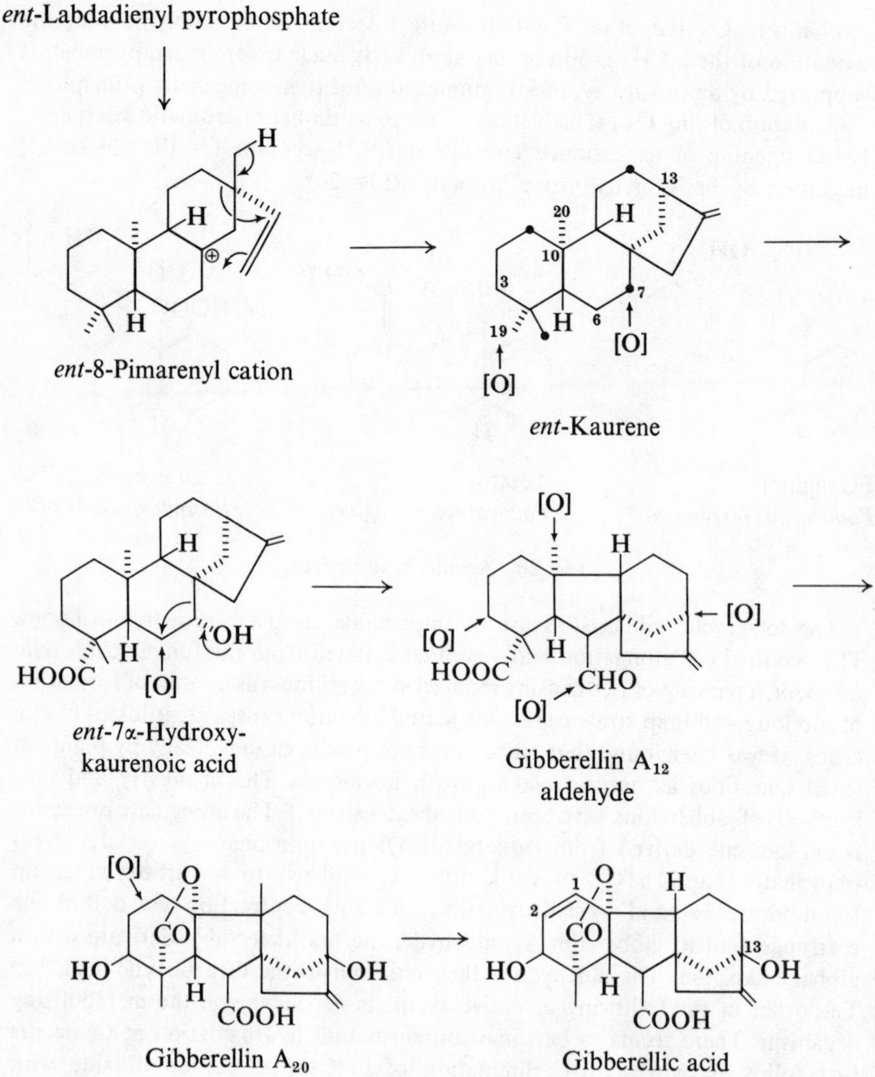

ent-Labdadienyl pyrophosphate

ent-8-Pimarenyl cation

ent-Kaurene

ent-7α-Hydroxy-kaurenoic acid

Gibberellin A_{12} aldehyde

Gibberellin A_{20}

Gibberellic acid

Fig. 21 Biosynthesis of gibberellic acid

Macrocyclic diterpenes,[48] cembranes, have been isolated from *Pinus* spp., from marine coelenterates such as sea fans and soft corals, and also from the termite *Nasutitermes exitiosus* where they act as scent-trail pheromones. They are biosynthesized by intramolecular C^1 to C^{14} alkylation of geranylgeranyl pyrophosphate to give a 14-membered ring (Fig. 22). The research on constituents of the shallow water coelenterates was stimulated by the tumour inhibiting effects of certain cembrane lactones.

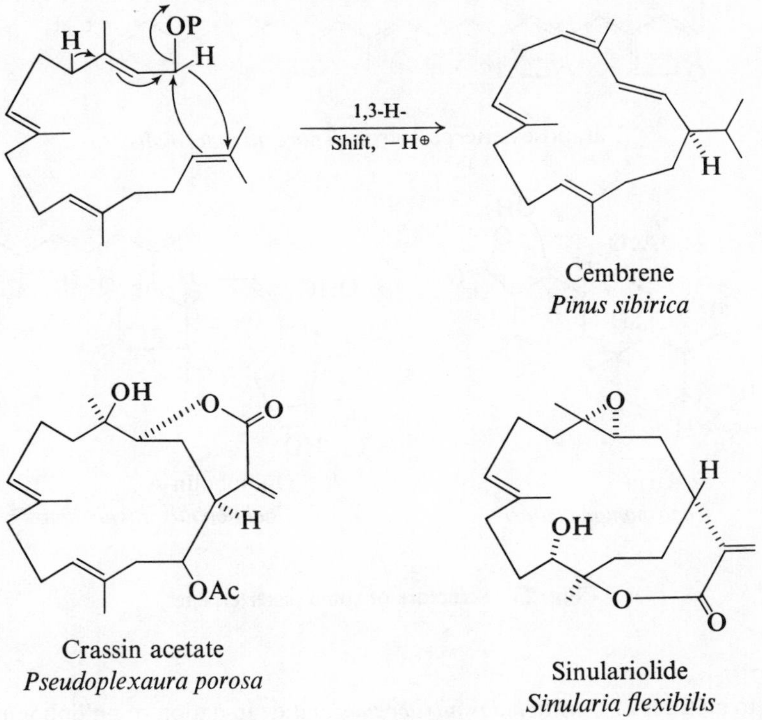

1,3-H-
Shift, $-H^{\oplus}$

Cembrene
Pinus sibirica

Crassin acetate
Pseudoplexaura porosa

Sinulariolide
Sinularia flexibilis

Fig. 22 Macrocyclic diterpenes

5.6 Sesterterpenes [49]

The parent linear compound is geranylfarnesol, isolated from the wax of the insect *Ceroplastes albolineatus*.[50] Until recently members of this group of compounds were rare. Several linear furanosesterterpenes as well as tetracyclic sesterterpenes have now been isolated from marine sponges.[51] A group of tricyclic sesterterpenes, the ophiobolanes, with a 5–8–5 ring-system was isolated from phytopathogenic fungi (Fig. 23). To account for the biosynthesis of ophiobolin A,[52] all-*trans*-geranylfarnesyl pyrophosphate is rearranged to the 2,3-*cis* isomer and folded according to Fig. 24. The cyclization is initiated by alkylation of C^{11} with C^1 phosphate and C^{10} reacts concertedly with C^{14} thus creating a carbonium ion at C^{15}. One of the C^8 hydrogens is transferred in a 1,5 shift to C^{15}. The 5–8–5 ring structure is finally accomplished by $C^{2,6}$ cyclization and *trans* attack by water at C^3. This scheme, which initially leads to ophiobolin F, is supported by the finding that C^3OH derives from water but C^{14}–O in ophiobolin A is derived from atmospheric oxygen.[53] By administration of *R*2–^3H mevalonic

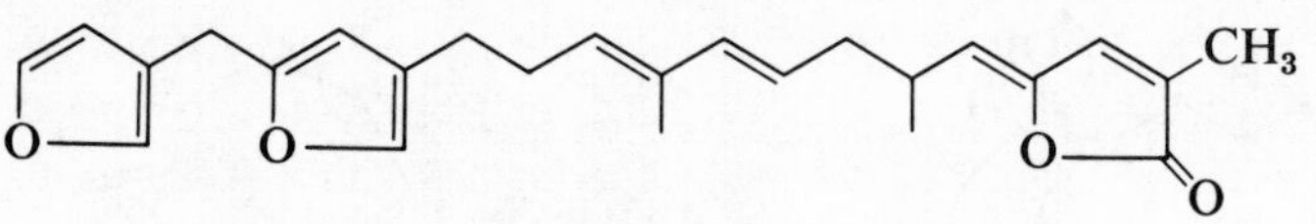

Furanosesterterpene from *Thorecta marginalis*

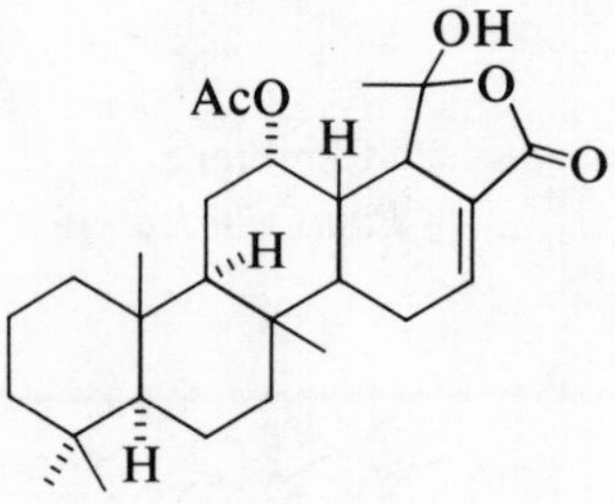

Scalarin
Cacospongia scalaris

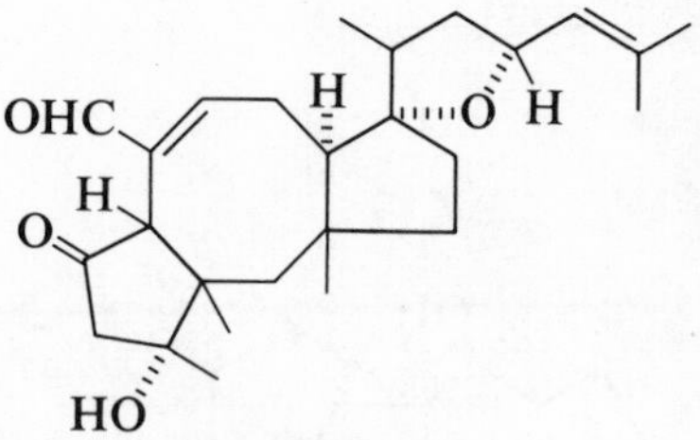

Ophiobolin A
Cochliobolus miyabeanus

Fig. 23 Structure of some sesterterpenes

acid to cultures of *Cochliobolus miyabeanus* and degradation of ophiobolin A, it was demonstrated that the label was retained at C^{15} this proving that the 8-α-H (= $R2^3$H of mevalonic acid) is shifted stereospecifically. Apparently the 8α-H lies closer to the carbonium ion centre at C^{15} than the 8-β hydrogen.[54]

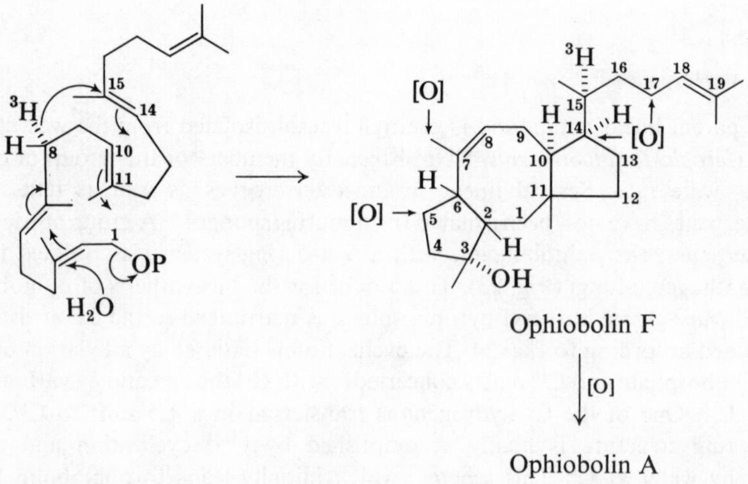

Ophiobolin F

[O]

Ophiobolin A

Fig. 24 Folding of geranylfarnesyl pyrophosphate for the biosynthesis of the ophiobolane skeleton

5.7 Squalene. Triterpenes

Parallel studies on the different aspects of terpene and steroid chemistry gradually focused the interest around a rare C_{30} hydrocarbon, squalene, as a conceivable progenitor of the higher terpenoids. Squalene was first isolated from shark liver, *Squalus* spp, but was later found to be ubiquitously distributed. By folding this compound in certain modes one can construct the basic triterpenoid skeleton with the angular methyls and side chain in correct positions. Experiments soon revealed that squalene was indeed efficiently incorporated into cholesterol.

Squalene consists of two all-*trans* farnesyl groups joined tail to tail. The mechanism of this puzzling coupling remained unsolved for a long period. It was first noted that *one* proton of the central unit was derived from NADPH by using an isotopic labelling technique. The solution of the problem came eventually with the isolation of an intermediate, presqualene pyrophosphate, having a cyclopropane structure.[55,56] The biosynthesis of presqualene pyrophosphate follows initially the same sequence as the synthesis of chrysanthemyl pyrophosphate (section 5.3). The 2,3 double bond of one farnesyl pyrophosphate molecule is alkylated by another farnesyl pyrophosphate with inversion of configuration, and its C^1H_S hydrogen is stereospecifically eliminated thus giving rise to the cyclopropane moiety of presqualene. In absence of NADPH this compound accumulates. It is formed more rapidly than squalene in yeast microsomes, which is consistent with its function as a true intermediate. The cyclopropane rearranges by ring expansion and inversion of configuration at C^4 and collapse of the discrete cyclobutyl carbonium ion to the linear all-*trans*-squalene (Fig. 25).[57]

Squalene can be folded in a number of ways both with regard to ring size, start and terminus of the chain and prechair–preboat conformation of the chain. There exists also the possibility that the naturally occurring all-*trans* form may isomerize at an olefinic centre at some stage of the cyclization. In the sesqui- and diterpene series the enzymes have admirably demonstrated their diligence and aptitude for the construction of a variety of skeletons, but surprisingly, with regard to the long multifunctional squalene chain, there are few skeletal variations in the triterpene series. With few exceptions the A, B, and C rings are six-membered rings and the cyclization is always initiated at the terminal double bond. However, there are numerous secondary modifications such as oxidations, dehydrogenations and Wagner–Meerwein shifts within the framework.

A salient feature of nearly all triterpenes is the equatorial hydroxy group at C^3. The process leading to this functionality was first formulated as an attack at the terminal double bond of $OH^{\oplus}$ or some less clearly defined biochemical equivalent, which triggers the multiple cyclization. This species is highly improbable in a biological environment but nevertheless useful for rationalization of many triterpenoid structures.

A mechanistically more satisfactory process that leads to the same products is the proton catalysed opening of an epoxide. Epoxide formation is a common reaction mediated by an epoxydase that requires NADPH and molecular oxygen. The partaking of cytochrome is questionable since carbon monoxide does not

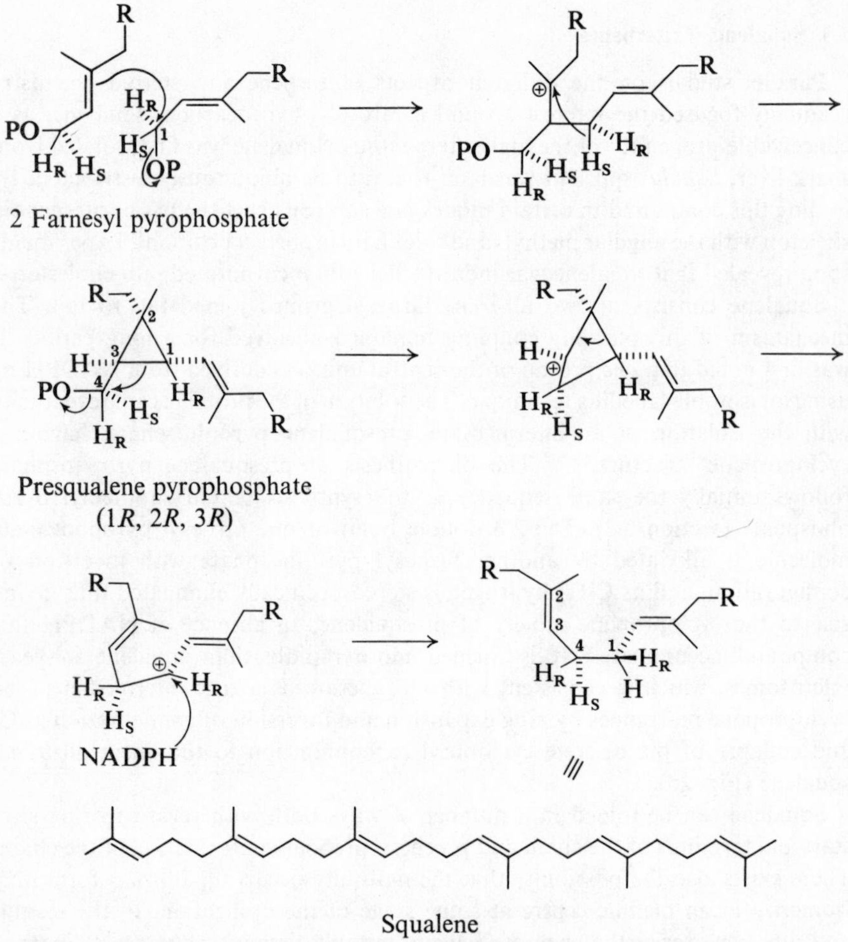

2 Farnesyl pyrophosphate

Presqualene pyrophosphate
(1*R*, 2*R*, 3*R*)

Squalene

Fig. 25 Mechanism of the tail to tail coupling of two farnesyl pyrophosphates to squalene. R = geranyl

affect the epoxidation. The reduction of oxygen to activated hydrogen peroxide can very well be mediated by $FADH_2$ (section 3.3). Lanosterol was indeed formed when squalene-2,3-oxide was incubated with enzyme preparations[58,59] and it was later isolated from a number of plants. Its formation required molecular oxygen and NADPH.[60] Consequently, it is a true intermediate on the path to the cyclic triterpenes. Squalene is epoxidized at either one of the terminal double bonds to the 3*S*-isomer indicating that it is assembled on another enzyme and then released.[61]

One can recognize two main groups of triterpenes, the tetracyclic triterpenes including the sterols, and the pentacyclic triterpenes. The conformation of squalene oxide on the enzyme surface and the extent of backbone rearrangement which follows upon cyclization, give rise to a number of subgroups.[37,62] The widely distributed lanosterol derives from 3*S*-squalene-2,3-oxide in a chair–boat–chair–boat conformation via the hypothetical protosterol carbonium ion I

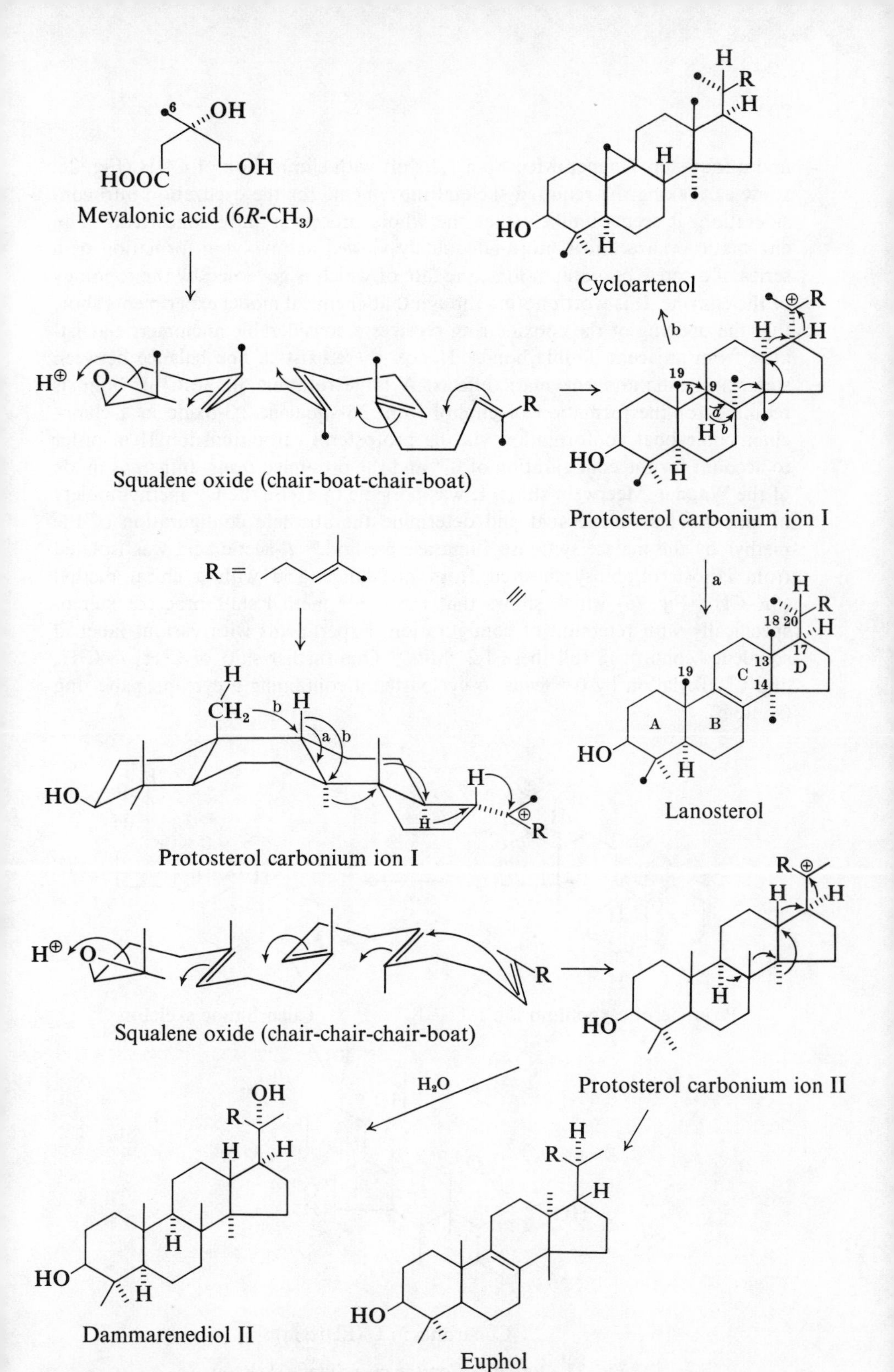

Fig. 26 Biosynthesis of tetracyclic triterpenes. Incorporation of mevalonic acid, labelled in the methyl group, into lanosterol

and a four-step Wagner–Meerwein 1,2 shift with elimination of C^9–H (Fig. 26, route a). Taking the required skeletal movements for the cyclization into consideration, it seems unlikely that the whole process is fully concerted.[63] The enzymatic cyclization is more adequately viewed as involving formation of a series of discrete carbonium ions, the fate of which is governed by the topology of the enzyme. It is worth noting though that chemical model experiments show that the opening of the epoxide ring receives a considerable anchimetric assistance from adjacent double bonds. Hence, there exists a fine balance between stereoelectronic and enzymatic effects. A large rotation of the $C^{17,20}$ bond is required for the formation of euphol from 3*S*-squalene-2,3-oxide in a chair–chair–chair–boat conformation via the protosterol carbonium ion II in order to account for the configuration of C^{20} and the presumed *trans*–anti-*trans* mode of the Wagner–Meerwein shifts. It was possible to excise the C^{18} methyl moiety of lanosterol as acetic acid and determine the absolute configuration of the methyl by the malate synthase–fumarase method.[64] *R*-acetic acid was isolated from lanosterol, biosynthesized from mevalonic acid with a chiral methyl (6*R*–CH_3, Fig. 26) which shows that the $C^{14,13}$ methyl shift proceeds stereospecifically with retention of configuration. Experiments with various labelled squalenes confirm in full these 1,2 shifts.[65] One further shift of $C^9H_\beta \rightarrow C^8H_\beta$ and C^9 alkylation by C^{18} leads to cycloartenol containing a cyclopropane ring (route b).

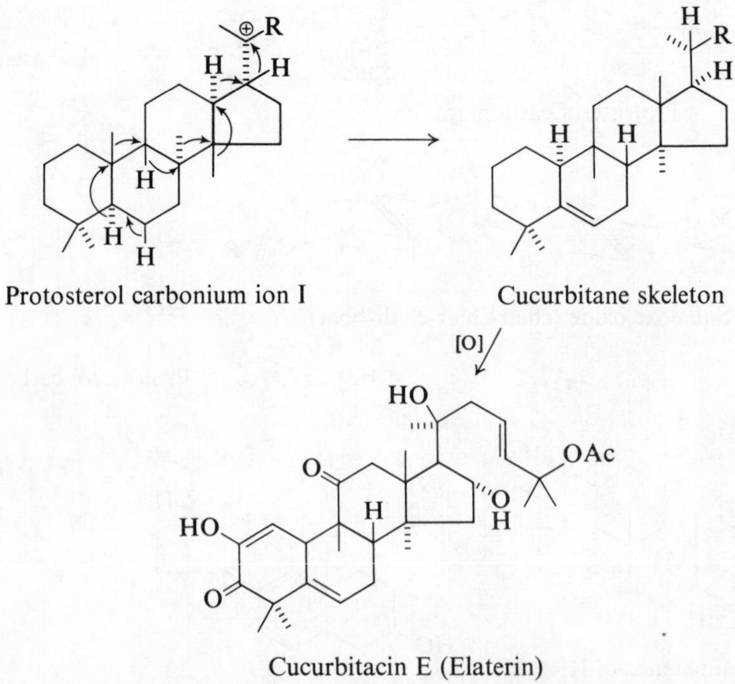

Protosterol carbonium ion I

Cucurbitane skeleton

[O]

Cucurbitacin E (Elaterin)

Fig. 27 Variations on the cucurbitane skeleton

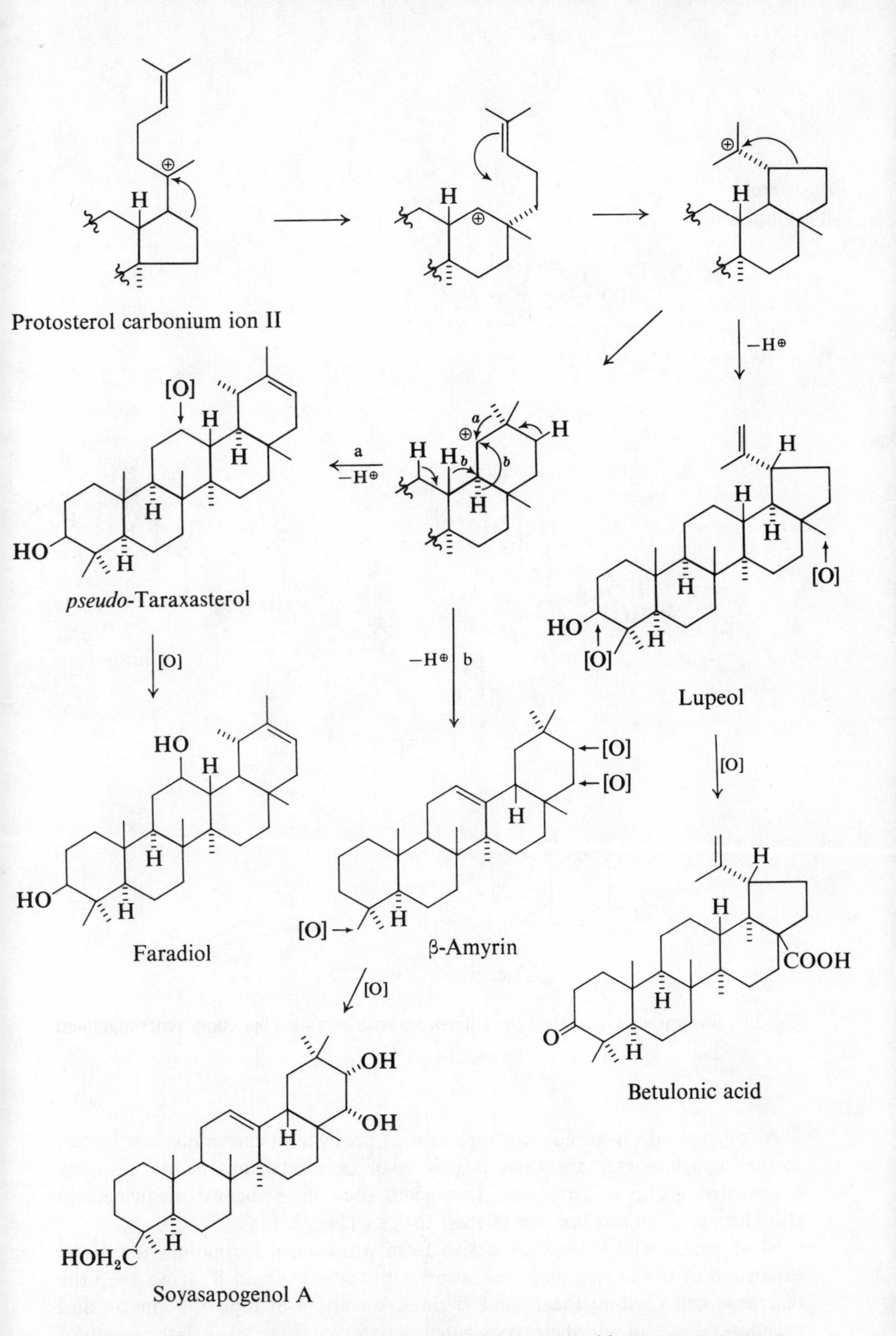

Fig. 28 Biosynthesis of pentacyclic triterpenes with no backbone rearrangements

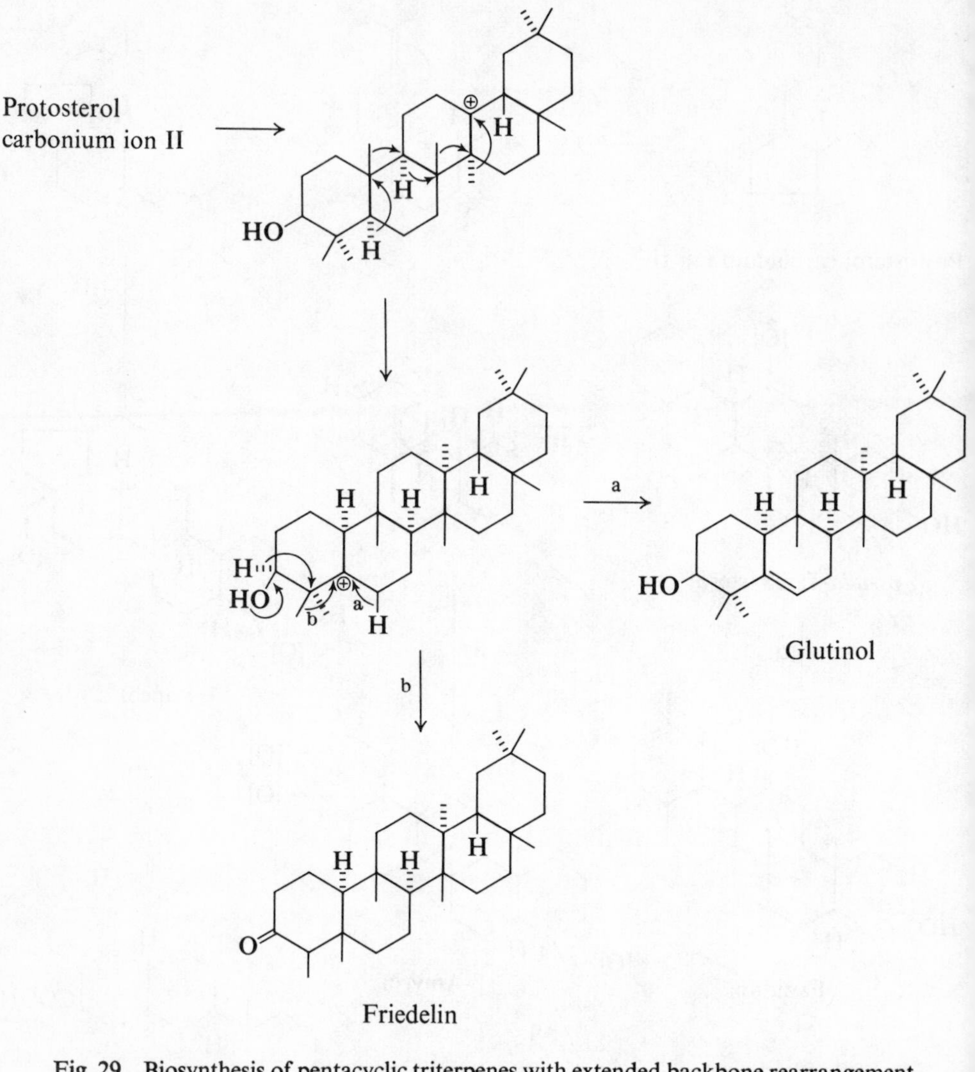

Protosterol carbonium ion II

HO H H H

H H H
H
HO a b H
a

H H H
HO
Glutinol

b

H H H
O
Friedelin

Fig. 29 Biosynthesis of pentacyclic triterpenes with extended backbone rearrangement

A still more deep-seated rearrangement of protosterol carbonium ion I leads to the cucurbitanes,[66] the toxic principles of cucurbitaceous plants, a highly oxygenated group of tetracyclic triterpenes used in mediaeval medicine and still attracting interest because of their toxicity (Fig. 27).

Most pentacyclic triterpenes derive from protosterol carbonium ion II by expansion of the D ring and cyclization with the side chain R. If we keep the rearrangements within the D and E rings, we arrive at lupeol, β-amyrin and pseudotaraxasterol and their oxygenated derivatives (Fig. 28). A large group of

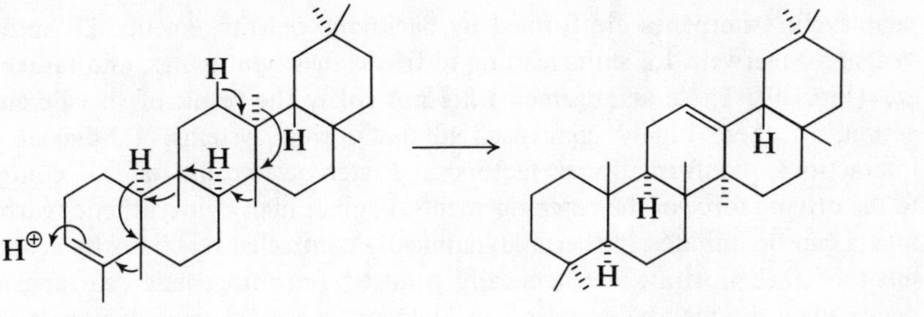

Δ^3-Friedelene Δ^{12}-Oleanene

Fig. 30 **Acid induced retrobiosynthetic backbone rearrangement**

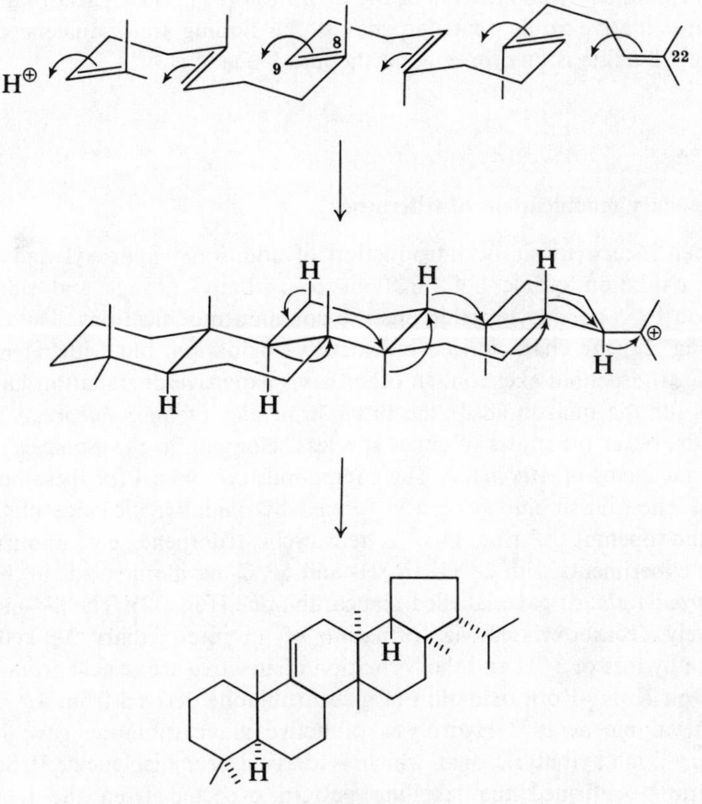

Fern-9-ene

Fig. 31 **Biosynthesis of fern-9-ene**

pentacyclic triterpenes are formed by backbone rearrangements, i.e. multistep Wagner–Meerwein 1,2 shifts leading to friedelanes, glutinanes, and taraxeranes etc. (Fig. 29). These arrangements are not solely the result of specific enzyme action. In these highly condensed aliphatic ring systems, 1,3-diaxial steric interactions, conformational factors and stereoelectronic effects contribute to the driving force of the rearrangements. A great many biosynthetic rearrangements can be imitated in thermodynamically controlled reactions by acid treatment of the substrate.[38] Chemically induced retrobiogenetic rearrangements occur when the metabolite possesses high free energy. This is demonstrated by the concerted retrobackbone rearrangement of Δ^3-friedelene to the more stable Δ_{12}-oleanene[67] (Fig. 30). Cyclizations initiated by epoxide cleavage are rather uncommon in lower terpenoids, whereas this is the rule for the triterpenoids. A small number of triterpenes are derived by an initial protonation, e.g. the filicenes and the fernenes from ferns. The pentacyclic system is generated from squalene in an all-chair conformation by a series of concerted cyclizations generating a carbonium ion at C^{22} followed by a six-step concerted Wagner–Meerwein shift that is terminated by elimination of the C^9 proton (Fig. 31). In accordance with this non-oxidative route to triterpenes is the finding that squalene but not squalene-2,3-oxide is incorporated in the fern-9-ene.[68]

5.8 Secondary modifications of triterpenes

Wagner–Meerwein shifts, introduction of additional hydroxyl and olefinic groups, oxidation of alcohol functions to carbonyl groups and side chain alkylation by *S*-adenosyl methionine are common modifications. The cucurbitanes (Fig. 27) are characterized by extensive oxidation, but with retention of the basic triterpenoid skeleton. In other cases extensive degradation has taken place as for the quassinoids,[69] the bitter principles of *Simaroubaceae*, for the limonoids, bitter principles of citrus species, belonging to the *Rutaceae* family, and the meliacins of *Meliaceae*. The compounds are typical for these botanical families. The quassinoid skeleton is formed by oxidative cleavage of the side chain and opening of ring D of a tetracyclic triterpene, e.g. apotirucallol. Feeding experiments with 2-^{14}C, 4*R*-^{3}H- and 5-^{14}C-mevalonic acids in the seeds of *Simarouba glauca* gave labelled glaucarubinone (Fig. 32). The β-4-methyl is oxidatively decarboxylated via formation of an intermediary 3-β-keto ester as shown by loss of 3-^{3}H and the formation of inactive acetic acid from $C^{4,10,13}$ methyls on Kuhn–Roth oxidation of glaucarubinone derived from 4*R*-^{3}H- and 2-^{14}C mevalonic acids.[70] Hydrolysis of active glaucarubinone gave inactive 2-hydroxy-2-methylbutyric acid which is derived from isoleucine.[71] Selective degradation confirmed the labelling pattern expected from the tetracyclic triterpene precursor. The finding that 9-^{3}H is retained excludes a precursor with a C^8 double bond.

In the limonoids the side chain is transformed to a furane and ring A is

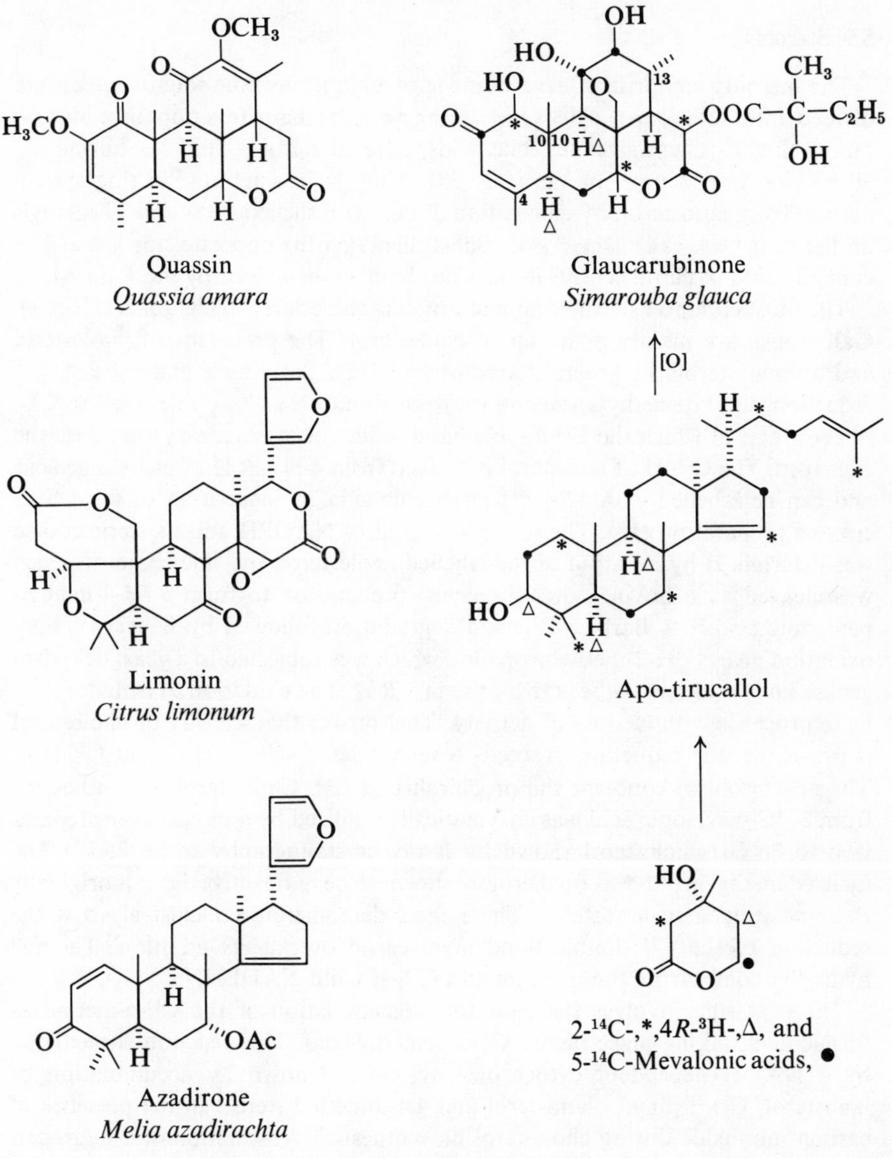

Quassin
Quassia amara

Glaucarubinone
Simarouba glauca

[O]

Apo-tirucallol

Limonin
Citrus limonum

Azadirone
Melia azadirachta

2-^{14}C-,*,4*R*-^{3}H-,Δ, and
5-^{14}C-Mevalonic acids, ●

Fig. 32 Partially degraded triterpenes of *Simaroubaceae*, *Rutaceae* and *Meliaceae*

further oxidized to a lactone. The constituents of *Meliaceae* are closely related to the limonoids but have an intact ring A (Fig. 32).

The fundamental secondary modification that leads to the steroids is selective C_4 and C_{14} demethylation. This reaction is separately discussed in the next section.

5.9 Steroids

The steroids comprise a large number of ubiquitous compounds which are divided into subgroups, chiefly according to side chain functionality: sterols, sapogenins, cardiac aglycones, bile acids, adrenal steroids, and sex hormones, all of vital importance for life (Fig. 33). They have a tetracyclic ring system derived from lanosterol, by convention drawn with the axial 18- and 19-methyls on the near face, called the β-side. Substituents on the opposite side are α. The configuration of chiral centres in the side chain are described by the *R*,*S* system.

The most common sterol of animal origin is cholesterol[72] (section 5.1, Fig. 1). Gall stones are mainly made up of cholesterol. The precursor of cholesterol and animal sterols in general, is lanosterol (Fig. 26) which undergoes $C^{24,25}$ reduction, $C^{4,14}$ demethylation and transposition of the C^8 double bond to C^5.

The stage, at which the C^{24} double bond reduction occurs, may vary with the organism. The C^{24}–H of lanosterol is derived from 4-pro-*R* H of mevalonic acid and can be labelled with 4*R*-4-^{3}H-mevalonic acid by incubation of a rat liver enzyme preparation (15). The reduction requires NADPH and its steric course was determined by isolation of the labelled cholesterol, the side chain of which was cleaved by a bovine adrenal enzyme preparation to form 3-^{3}H-4-methylpentanoic acid.[73] A Barbier–Wieland degradation followed by Baeyer–Villiger oxidation gives 1-^{3}H-2-methylpropanol, which was subjected to a yeast dehydrogenase known to remove selectively the pro-*R* H. The oxidation to isobutyraldehyde proceeds without loss of activity. This proves that C^{24}–^{3}H of cholesterol is pro-*R*, i.e. the reduction proceeds from the back side = *si*-side at C^{24} (15). The next problem concerns the prochirality at C^{25}. Cholesterol biosynthesized from 2-^{14}C-mevalonic acid was enzymatically oxidized by a mouse liver preparation to 26-OH-cholesterol shown by X-ray crystallography to be 25*S*.[74] The hydroxymethyl group was furthermore shown to be radioactive, i.e. it is originally *trans* positioned in lanosterol. These facts demonstrate conclusively that the reduction of the C^{24} double bond involves an overall *cis* addition. The C^{24} hydrogen comes from the medium and C^{25}–H from NADPH.

The next step involves the oxidative demethylation of the C^{14}-α-methyl as formic acid, taking place before C^4 is demethylated. This reaction is mediated by a NADH dependent cytochrome system as shown by accumulation of lanosterol, $C^{24,25}$ dihydrolanosterol and 4,4-dimethyl sterols in the presence of carbon monoxide during cholesterol biosynthesis.[75] Abstraction of a hydrogen from the C^{14} methyl by a hydrogen peroxide radical initiates the oxidation to a C^{14} formyl group which is eliminated by the mechanism suggested in (16).[76] This last oxidative step produces formic acid with incorporation of molecular oxygen. It was observed that the C^{15}-α-hydrogen is lost in a *cis* elimination during cholesterol formation and the $C^{8,14}$ diene is considered to be an intermediate on the pathway. Contrary to the C^{14}-methyl, the C^4-methyls are lost as carbon dioxide[77] starting with a three-stage oxidation of the α-methyl involving oxygen and NAD(P)H. The decarboxylation is mediated by oxidation of 3-OH to a β-keto acid. The oxidative enzyme system has rigid steric requirements.[78]

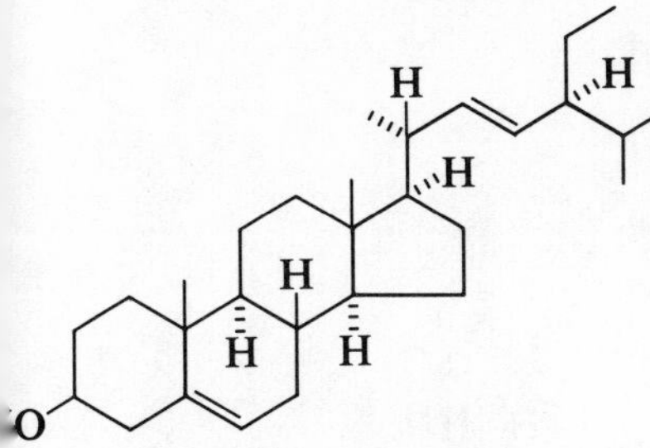

Stigmasterol (Phytosterol)

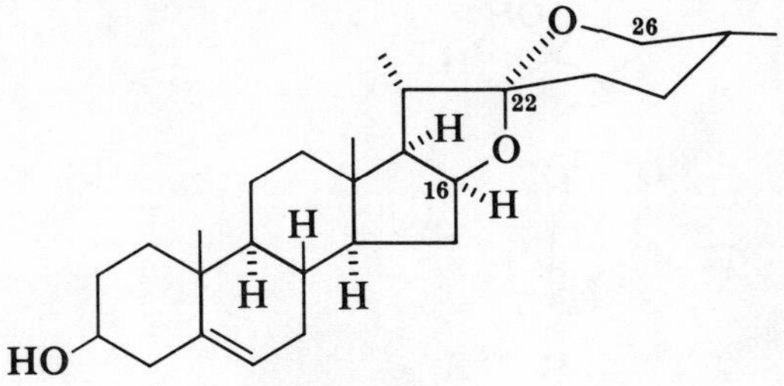

Diosgenin
(Sapogenin)

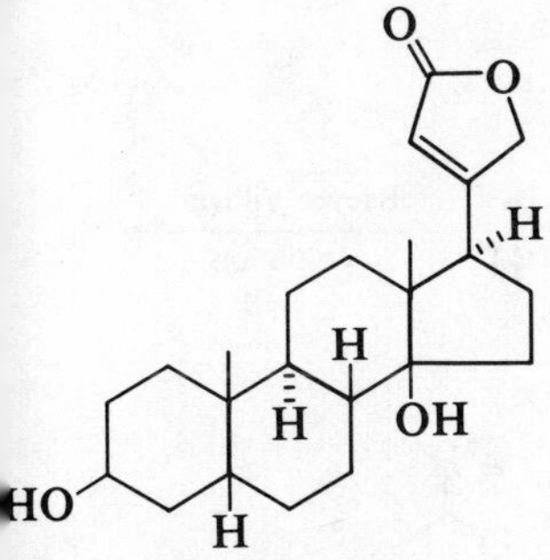

Digitoxigenin
(Cardiac aglycone)

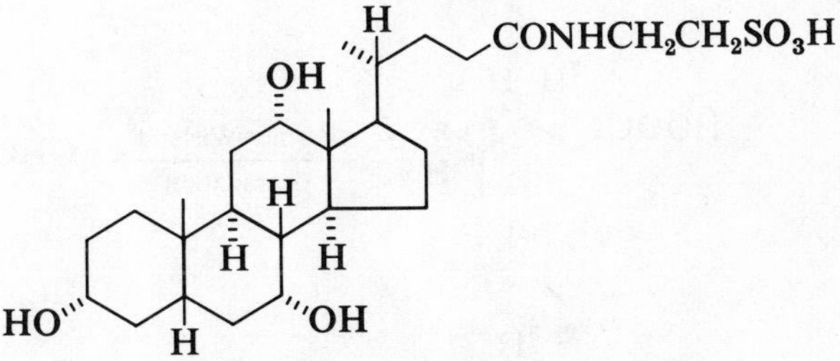

Tauro-cholic acid (Bile acid)

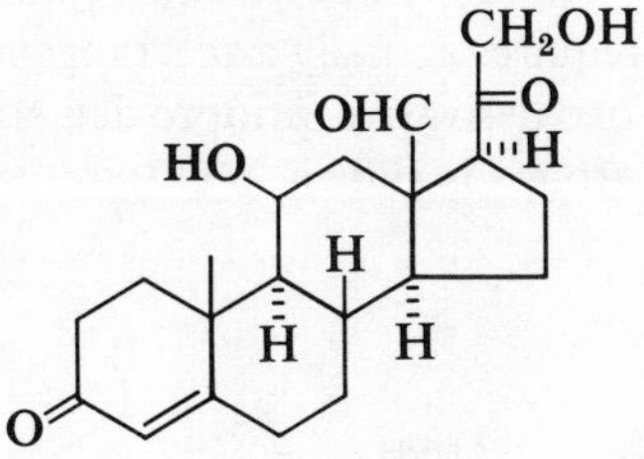

Aldosterone (Adrenal steroid)

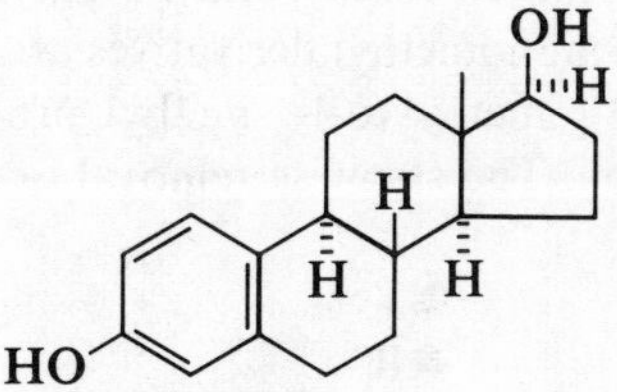

β-Oestradiol
(Female sex hormone)

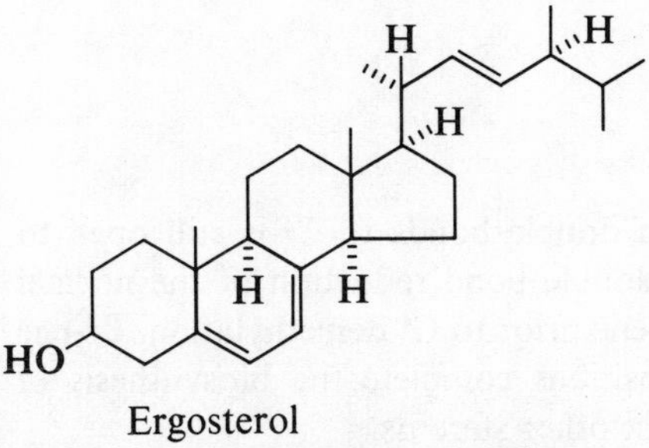

Ergosterol
(Precursor of vitamin D)

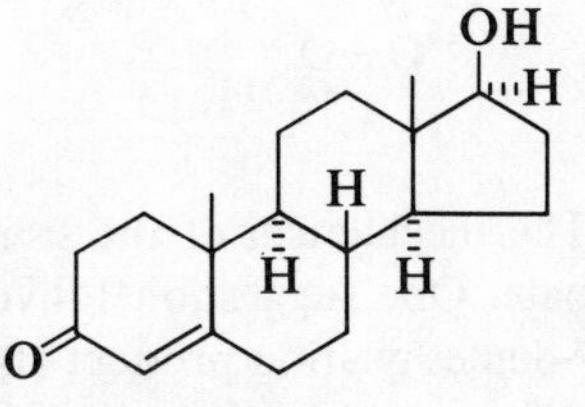

Testosterone
(Male sex hormone)

Fig. 33 Representative steroids

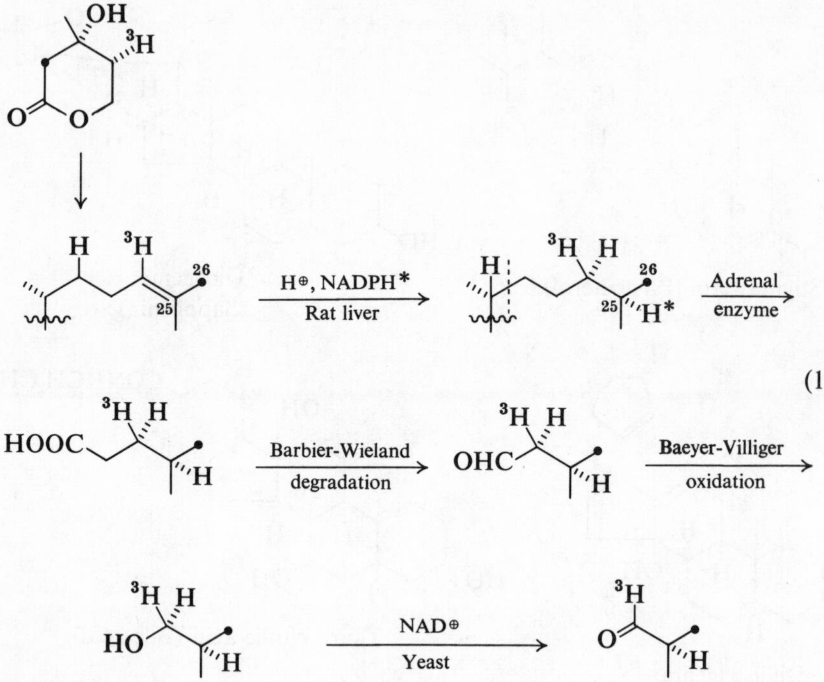

OH ^3H O O

↓

H ^3H 26 25 —H⊕, NADPH*/Rat liver→ H ^3H H 26 25 H* —Adrenal/enzyme→

(15)

^3H H HOOC H —Barbier-Wieland/degradation→ ^3H H OHC H —Baeyer-Villiger/oxidation→

^3H H HO H —NAD⊕/Yeast→ ^3H O H

Thus, the 4-α-ethyl and 4-β-ethyl homologues, 4-β-hydroxymethylisomers and the 4-β-monomethyl derivatives are inactive. This requires actually a rearrangement of 4-β-methyl to 4-α-methyl, probably via the 3-oxo derivative, prior to demethylation. The stepwise removal of the C^4-methyl groups is shown in (17).[79]

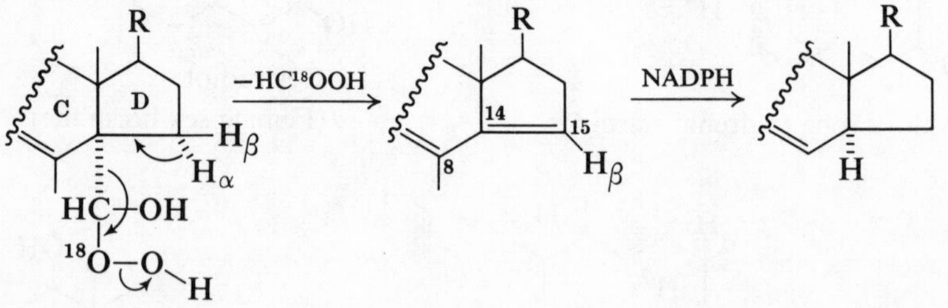

R C D H_β H_α HC—OH ^{18}O—O H —$-HC^{18}OOH$→ R 14 15 8 H_β —NADPH→ R H

(16)

The mechanism of the transposition of double bonds to C^5 is still open to debate. One suggestion[80] involves a C^{14} double bond reduction of the normal C^{14}-demethylation product (16) to the C^8-ene prior to C^4-demethylation. C^8-ene → C^7-ene → $C^{5,7}$-diene → C^5-ene transpositions complete the biosynthesis of cholesterol which is the precursor for most other steroids.

Sapogenins[81] are aglycones of a class of steroid glycosides, the saponins, often containing a spiroketal side chain (Fig. 33). They acquired their name from their

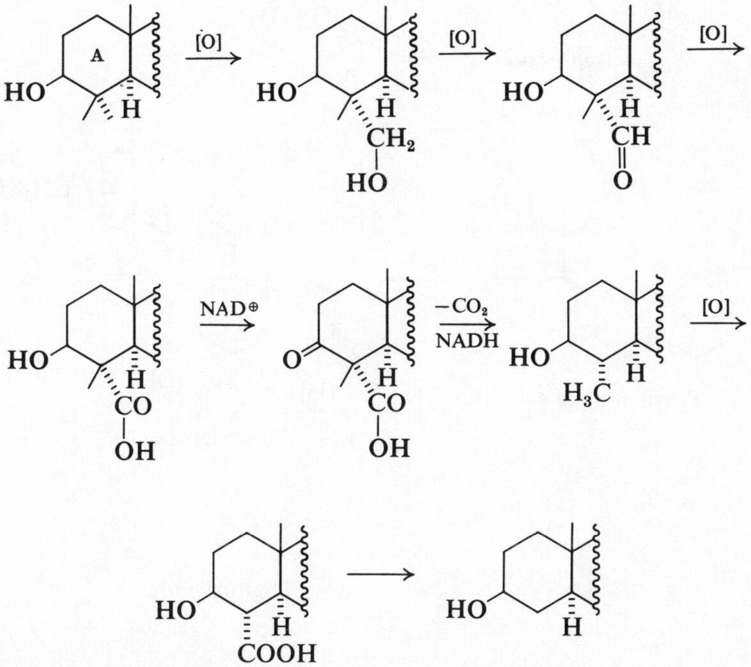

(17)

property for forming soapy emulsions in water, a property they share with the cardiac glycosides and the bile acids. They cause haemolysis by destroying the membranes of the erythrocytes. These compounds contain a hydrophobic and a hydrophilic part which make them surface active. Diosgenin (Fig. 33), a sapogenin from yams, a *Dioscorea* sp., is used as a valuable starting material for steroid hormone synthesis. It has the same skeletal configuration as cholesterol from which it is biosynthesized by oxidation of C^{22} to a carbonyl function and introduction of hydroxyl groups at $C^{16,26}$, which then ketalize the carbonyl group. The bile acids emulsify lipids thereby promoting the absorption through the intestinal wall.

In the cardiac glycosides, e.g. digitoxigenin (Fig. 33), the side chain has been converted to an α,β-unsaturated γ-lactone. It appears that the side chain of cholesterol rather unexpectedly is cleaved first at C^{20}, via a C^{20} peroxide[82] which fragments thus forming pregnenolone, which is subsequently oxidized to progesterone, and then condensed with a C_2 unit (18). Both of these pregnane derivatives are incorporated into digitoxigenin by *Digitalis lanata.*[83] Rings A,B and C,D are *cis* fused in digitoxigenin and C^3–OH is coupled to various sugars in the glycosides, Pregnenolone and progesterone are precursors for the sex hormones and are produced in the ovaries.

The cardiac glycosides have a powerful and specific action on the heart muscle and are most valuable agents in the treatment of heart ailments. These glycosides are also the active principle in African arrow poisons from *Strophanthus* spp.

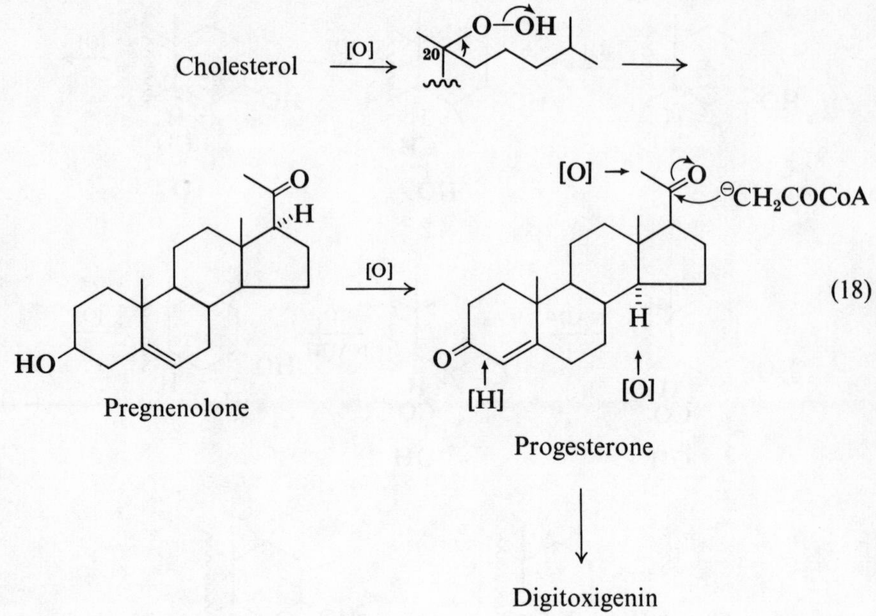

Cholesterol $\xrightarrow{[O]}$ [20-hydroperoxy intermediate: O–OH] $\longrightarrow$

Pregnenolone $\xrightarrow{[O]}$ Progesterone ([O], [H], [O], $^{\ominus}CH_2COCoA$)

(18)

Progesterone $\longrightarrow$ Digitoxigenin

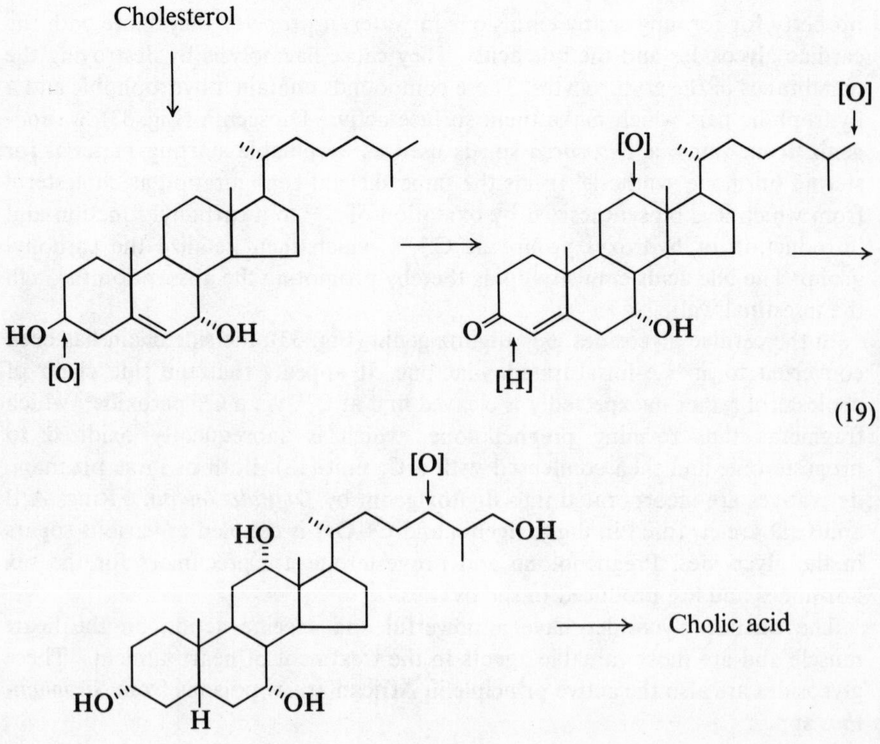

Cholesterol $\longrightarrow$ [O] $\longrightarrow$ [O], [O], [H] $\longrightarrow$ [O] $\longrightarrow$ Cholic acid

(19)

In the bile acids[84] additional hydroxyl groups are introduced in the nucleus and the side chain is shortened to a 24-oic acid. The order of events is not completely settled but nuclear hydroxylation occurs prior to side chain cleavage (19). The bile acids, conjugated with taurine or glycine, occur in the bile.

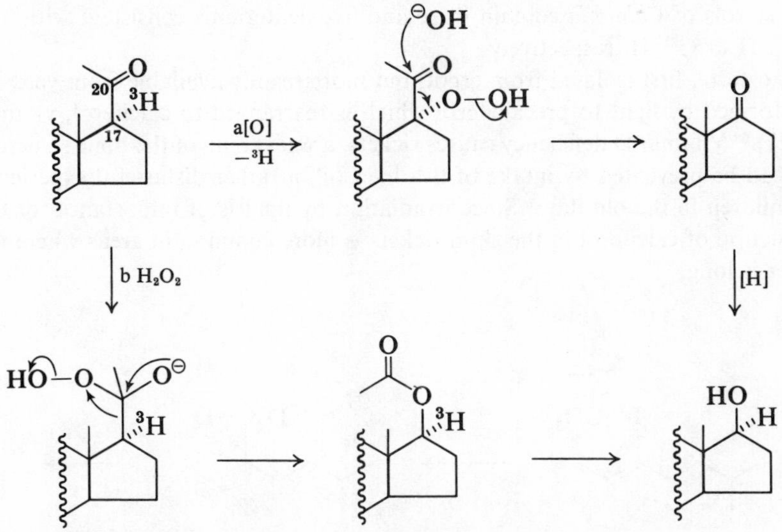

(20)

The adrenal hormones, e.g. aldosterone (Fig. 33), are formed from cholesterol by side chain cleavage at C^{20} followed by C^{17} and C^{21} hydroxylations. These hormones are secreted by two small glands associated with the kidneys. They regulate the metabolism of sugars and proteins, the salt balance in the animal organism, and are used in medicine for the treatment of inflammations and rheumatoid arthritis. Further cleavage of the side chain leads to the sex hormones, the C_{19} steroids (Fig. 33), controlling the development of male and female characteristics, lactation and the menstruation cycle, etc. The microbiological cleavage of the $C^{17,20}$ bond in progesterone to testosterone proceeds via hydroperoxidation of C^{20} (route 20a), rather than by a Baeyer–Villiger rearrangement (20b) because C^{17}–H is lost in the process.[85] Oestradiol is produced in the ovaries and testosterone in the testes. Contraceptives are modified sterols suppressing ovulation.

Cycloartenol (Fig. 26) is considered to represent the first stable tetracyclic product from squalene-2,3-oxide cyclization in plants on the grounds that it is efficiently incorporated into phytosterols and that it is widely distributed in plants as compared to lanosterol.[86,87] The sterols of higher plants, algae, and fungi are characterized by having extra carbons at C^{24} and often a C^{22} double bond. Several marine sterols[88] with methyls at C^{22}, C^{23}, C^{24}, C^{26} and C^{27} are known as well as some containing a cyclopropyl side chain.

Labelling experiments with *Saccharomyces cerevisiae* using deuterated methionine and tritiated mevalonic acid show that in ergosterol only two deuteriums are retained in the C^{24} methyl, and C^{24}–H is shifted to C^{25}. The biosynthesis thus proceeds by way of a C^{24} methylene intermediate and this is also the case for the methyl sterol produced by *Hordeum vulgare* (21).[89] The ethyl sterol contained four deuteriums indicating an ethylidene intermediate. Various mechanisms are operating in the C^{24} alkylation since labelling experiments show that methyl and ethyl sterols of *Chlorella* contain three and five deuteriums consistent with loss of C^{24}–H or C^{25}–H, respectively.

Ergosterol, first isolated from ergot, but more readily available from yeast, is transformed by light to precalciferol which is rearranged to calciferol, vitamin D_2 (22).[90] Vitamin D deficiency causes rickets, a weakening of the bone structure that can be prevented by intake of fish liver oil, a rather distasteful experience for children in the old days. Since irradiation by light is of importance for the production of calciferol in the skin, rickets is more common in areas where the winter is long.

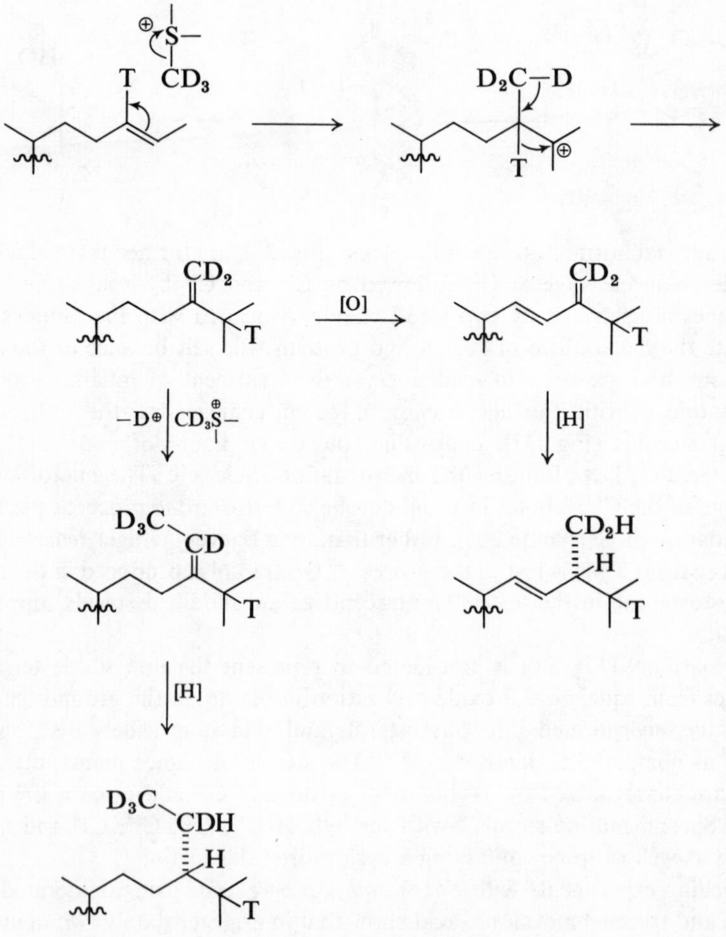

T CD_3 $\overset{\oplus}{S}$ → D_2C—D, T, ⊕ →

CD_2, T —[O]→ CD_2, T

$-D^{\oplus}$ | $CD_3\overset{\oplus}{S}$– ↓ [H] ↓ (21)

D_3C, CD, T; CD_2H, H, T

↓ [H]

D_3C, CDH, H, T

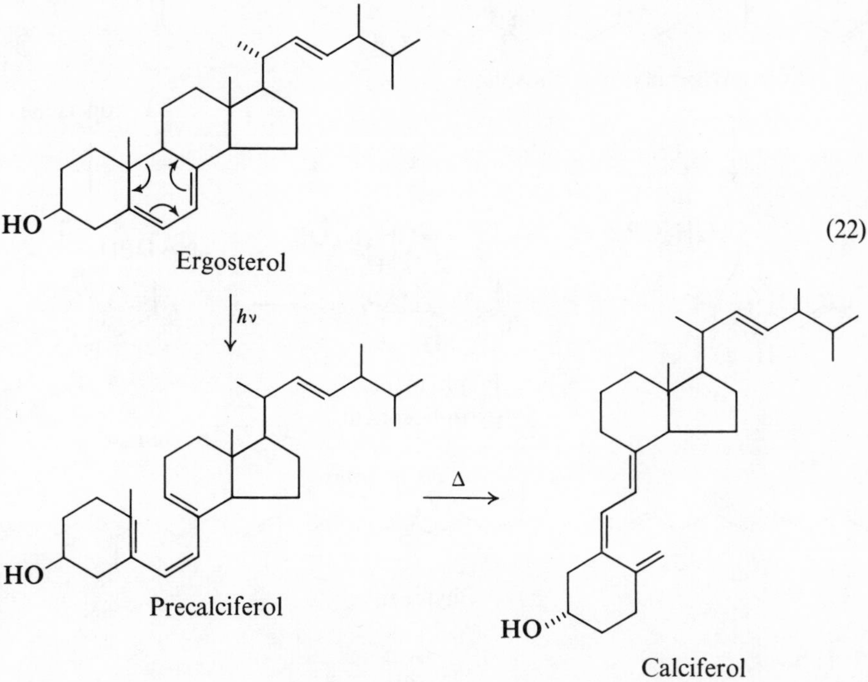

Ergosterol

$\downarrow h\nu$

Precalciferol $\xrightarrow{\Delta}$ Calciferol (22)

Insects do not synthesize steroids *de novo* but are capable of processing suitable materials taken with the diet to provide for hormonal functions. Of special interest is the conversion of cholesterol to the moulting hormones, the ecdysones, essential for insect development. The ecdysones have an opposite effect to that of the juvenile hormones in that they stimulate metamorphosis (Fig. 34). The name derives from ecdysis (Greek, shedding), the entomological term for moulting. Several ecdysones occur also in higher plants, where they presumably play an ecological role in affecting the metabolsim of phytophagous insects.

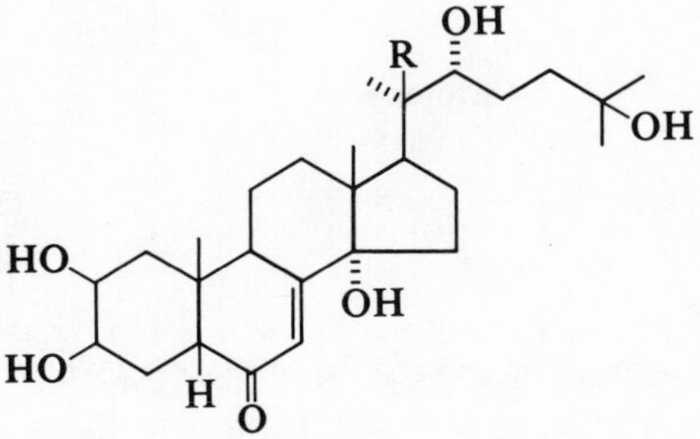

Fig. 34 Structure of insect moulting hormones. R = H, ecdysone; R = OH, ecdysterone

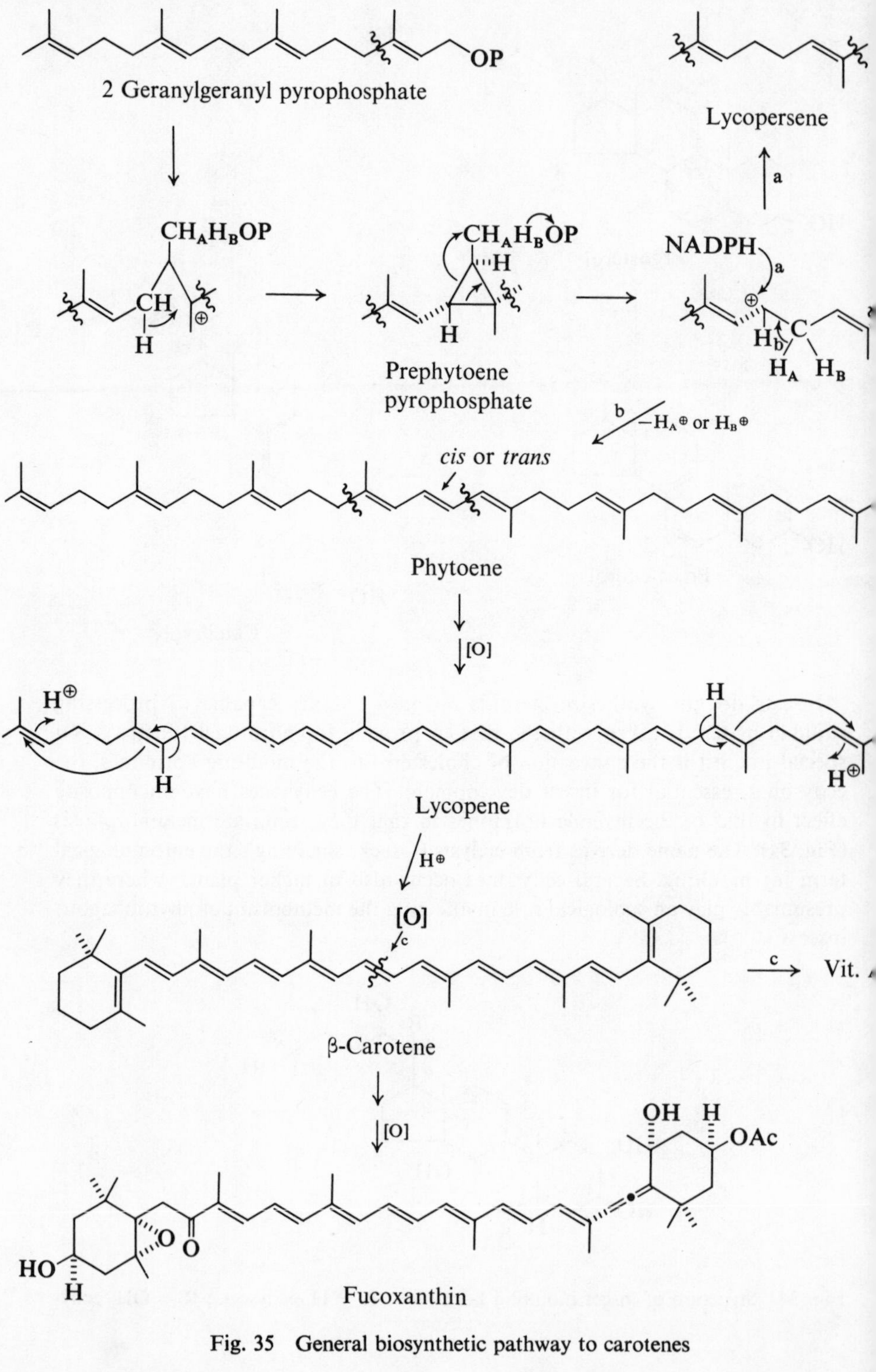

OP

2 Geranylgeranyl pyrophosphate

Lycopersene

a

CH_AH_BOP

CH

H

CH_AH_BOP

H

H

NADPH

a

H_b C

H_A H_B

Prephytoene pyrophosphate

b

$-H_A^{\oplus}$ or $H_B^{\oplus}$

cis or *trans*

Phytoene

[O]

$H^{\oplus}$

H

H

H

$H^{\oplus}$

Lycopene

$H^{\oplus}$

[O]

c

c

Vit. A

β-Carotene

[O]

OH H

OAc

O

O

HO

H

Fucoxanthin

Fig. 35 General biosynthetic pathway to carotenes

5.10 Carotenes. Polymers

Tail-to-tail coupling of geranylgeranyl pyrophosphate gives the C_{40} terpenes or carotenes, an important group of yellow-red conjugated polyene pigments with widespread occurrence in nature.[91,92] They are the chief pigments of egg yolk, yellow corn, carrots, tomatoes, pansy flowers, yellow autumn leaves, algae, etc. The biosynthesis (Fig. 35) parallels that of squalene. A cyclopropanoid prephytoene[93,94] is formed first which rearranges, either assisted by NADPH to lycopersene (route a) or, depending upon which of the prochiral hydrogens, H_A or H_B, is eliminated, to *cis*- or *trans*-phytoene 1(b). Further dehydrogenations and terminal cyclizations lead to a variety of carotenes, such as lycopene, the red pigment in tomatoes, or to β-carotene, a widespread pigment which produces two moles of vitamin A (section 5.1, (2)) in the intestine by symmetrical oxidative fission (c). Fucoxanthin, a common carotene confined to the marine environment in algae, has the unusual allene function.

The carotenoids function as supplementary light receptors in photosynthesis, transmitting their excitation energy to chlorophyll.

Some bacteria produce C_{45} and C_{50} carotenoids. In continuation of the prenylation process some plants produce polymers as a white fluid, latex. Rubber, obtained from the rubber tree, *Hevea brasiliensis*, is a polymer with about 2000 isoprene units. Nearly all the double bonds are *cis*. The *trans* polymer called gutta-percha is obtained from *Palaquium* spp. and is more horny.

5.11 Optical rotatory dispersion and circular dichroism. The octant rule

Chiroptical methods[95,96] have been applied for more than a century for characterization of optically active compounds. It was already recognized by Pasteur, one of the founders of modern stereochemistry, that optical activity was related to asymmetry of the molecule. The rotatory power measured by the polarimeter has its origin in the different refractive indices of the optically active medium for left and right circularly polarized light. It increases with decreasing wavelength of the incident light in a predictable way and this phenomenon is defined as optical rotatory dispersion, ORD. In the early days of research it was tried more or less successfully to relate sign and molecular rotatory power, measured at one wavelength, commonly the sodium D line, to the configuration of the molecule, i.e. chirality. However, it turned out not to be advisable to rely on the sign for assigning the structure. This point is illustrated by the data obtained from the *o*-, *m*-, and *p*-iodophenyl ethers of lactic acid. The *m*- and *p*-derivatives show positive, the *o*-derivative negative rotation at the D line, from which one could be tempted to conclude that the *o*-derivative may possess the opposite configuration at C^2. The positive curvature of ORD, however, demonstrates that all three compounds have the same configuration (Fig. 36). A plain ORD curve can thus occasionally cross the *x*-axis. ORD curves are useful for detection of optical activity in compounds showing very small rotation values

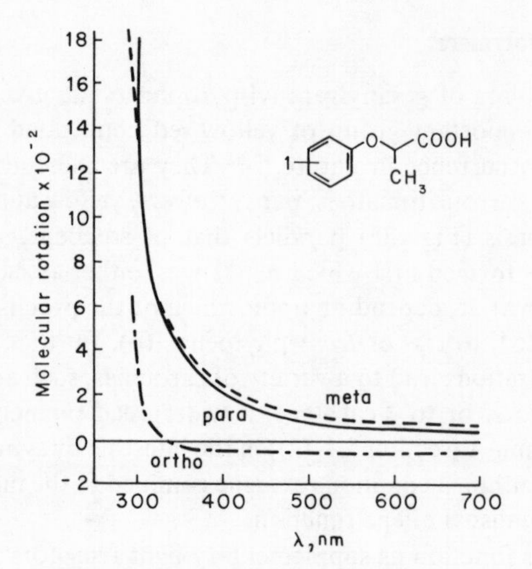

Fig. 36 ORD curves of *o*-, *m*-, and *p*-iodophenyl ethers of lactic acid. (From *Optical Rotatory Dispersion* by C. Djerrassi. Copyright © 1960, McGraw-Hill. Used with the permission of McGraw-Hill Book Company)

at the D line. As a consequence less substance is required for determination of activity. Instrumental improvements, stimulated by the demand from organic chemists for simple spectroscopic methods for determination of absolute configuration and conformation of optically active natural products, has considerably facilitated rapid automatic recordings of rotatory power as a function of wavelength. A novel valuable tool became available to the chemist. It turned out that chiroptical techniques were of immense importance for structural elucidations of terpenoids containing many chiral centres.

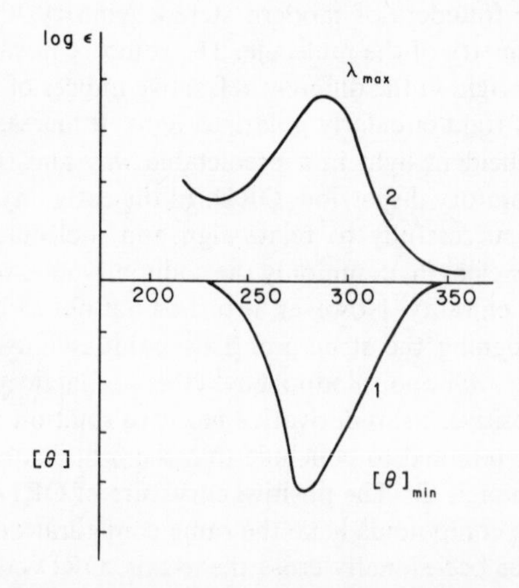

Fig. 37 CD curve, 1, showing a negative Cotton effect, and the UV absorption curve, 2

If the compound has a chromophore which is asymmetrically perturbated by a chiral centre, left and right circularly polarized light is absorbed to different extents. This phenomenon is called circular dichroism, CD. The CD curve graphs the difference in molecular extinction coefficients, $\Delta\varepsilon = \varepsilon_L - \varepsilon_R$, or ellipticity $[\theta] = 3300 \cdot \Delta\varepsilon$, against wavelength. This difference can be positive or negative, depending upon the chirality at the active centre. Its maximum, $\Delta\varepsilon_{max}$, coincides with the maximum, λ_{max}, of the absorption curve (Fig. 37). These conditions are visible as a drastic anomaly in the ORD curve, the Cotton effect. A single positive Cotton effect gives rise to a sharp maximum (peak) which suddenly drops to a minimum (trough) (Fig. 38). If the peak in the ORD curve occurs at shorter wavelength than the trough, the curve is defined as a negative Cotton effect curve. These anomalous curves provide more information than the plain curves and are important for structural determinations. A molecule containing several chromophores gives rise to a complex spectrum with multiple Cotton effects.

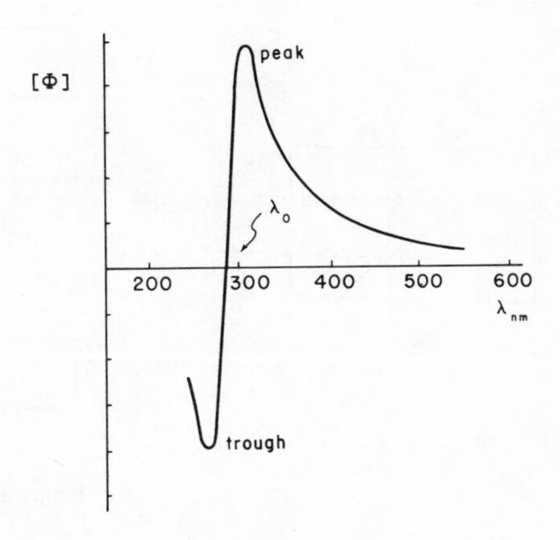

Fig. 38 ORD curve showing positive Cotton effect. λ_0 corresponds closely to λ_{max} of the UV absorption curve

The carbonyl function was found to be most suitable for rotatory dispersion studies. It is a common function in natural products and hydroxyl groups can easily be oxidized to carbonyl groups. It absorbs in a readily accessible spectral region, it has a low extinction coefficient, which does not interfere with ORD measurements, and most importantly it is electronically perturbated by proximate chiral centres. The terpenes offered numerous compounds of known structure with the carbonyl function located in a variety of molecular environments, such as different ring sizes and proximities to chiral centres of different configuration and conformation. In condensed cyclic structures the number of conformers is considerably reduced and they are controlled in a predictable way. Correlations between sign and magnitude of the Cotton effect and the absolute configuration of chiral centres and their disposition relative to the carbonyl group were noted, and eventually these systematic studies culminated in the formulation of the semi-empirical *octant rule*. Cyclohexanone in its chair form, the most stable form, is oriented in a three-dimensional coordinate system so that the *yz* plane is bisecting the ring through the carbonyl group, the *xz* plane passes through the O=C(C)(C) atoms and the *xy* plane bisects the C=O bond (Fig. 39). The *octant rule* states that substituents lying in the different octants contribute to the Cotton effect with the sign shown. Substituents located in the planes and distantly located from the chromophore make a negligible contribution. Only rarely do substituents fall into the front octants.

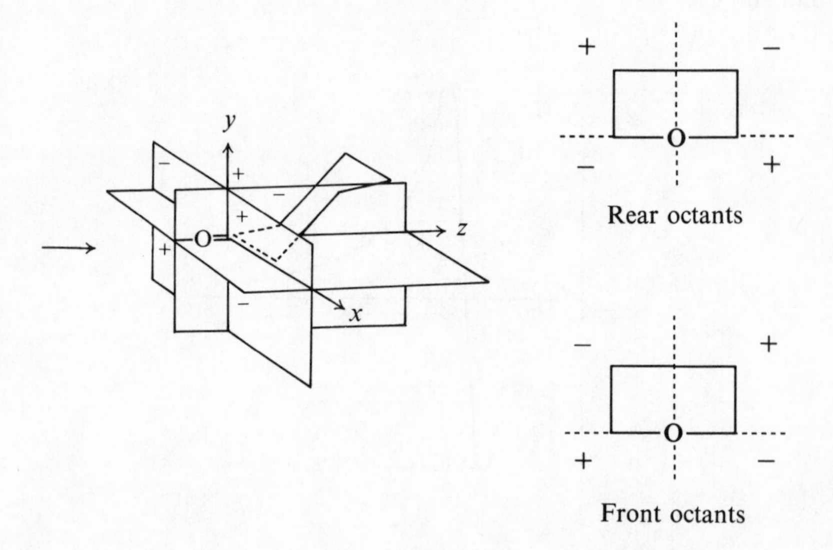

Fig. 39 The octant rule. A projection of cyclohexanone in a three dimensional co-ordinate system and simplified front views along the CO bond (along the arrow)

Two examples will clearly demonstrate the usefulness of the octant rule. (+)-3-Methylcylohexanone has been shown to have *R* configuration and a postive Cotton effect. Consequently the methyl groups must be located in a positive octant and that is only possible if the molecule acquires a conformation with an equatorial methyl group (Fig. 40). It is known from infrared spectroscopic studies that the energy difference between axial and equatorial methylcyclohexane is 1.7 kcal, i.e. *ca.* 95 per cent of the compound has the equatorial conformation. Provided with the information it is possible, on the other hand,

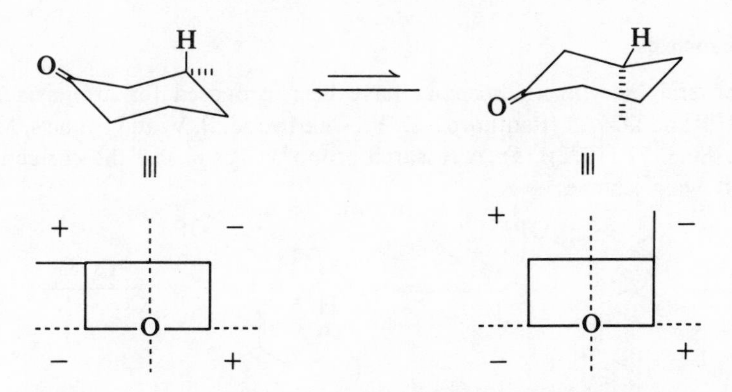

Fig. 40 The preferred conformation of $R(+)$-3-methylcyclohexanone has an equatorial methyl. The Cotton effect is positive

to determine the absolute configuration of $(+)$-3-methylcyclohexanone from the ORD curve.

Reduction of podocarpic acid gave two ketones exhibiting positive and negative Cotton effects. Application of the octant rule shows that the compound exhibiting positive Cotton effect is *trans* fused, whereas the other must be *cis* fused (Fig. 41).

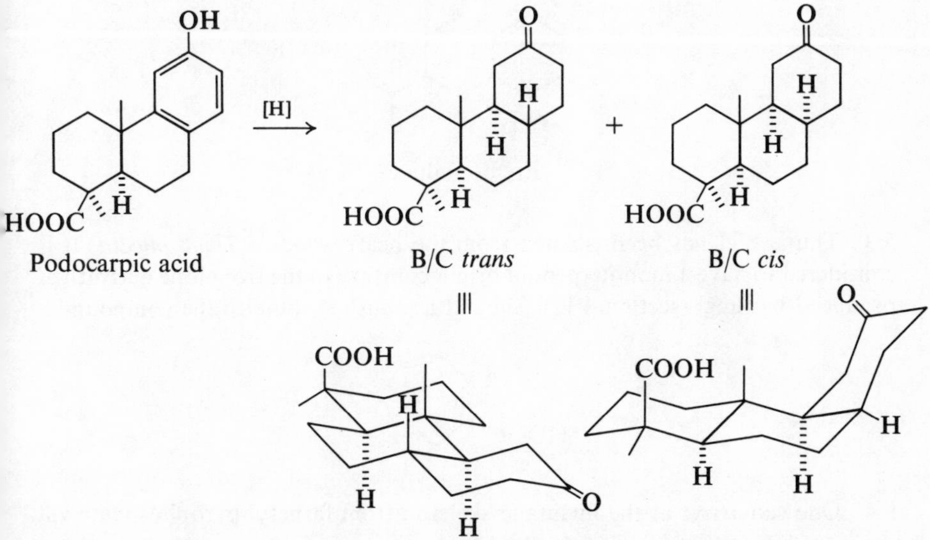

Podocarpic acid [H] → B/C *trans* + B/C *cis*

Fig. 41 Reduction of podocarpic acid. The projections of *trans* and *cis* products fall in positive and negative octants, respectively

5.12 Problems

5.1 Several biosynthetic schemes have been proposed for artemisia ketone from IPP and DMAP (Banthorpe, D. V., Charlwood, B. V. and Francis, M. J. O. *Chem. Revs.* **72** (1972)115). A research group wishes to test the correctness of the following scheme:

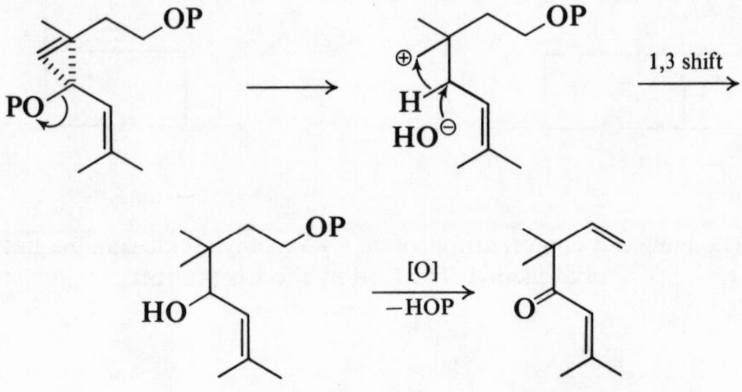

OP PO OP H HO 1,3 shift

OP HO [O] −HOP O

Artemisia ketone

How would you attack the problem of verifying the proposed pathway?.

5.2 The biosynthesis of a large number of sesquiterpenes is initiated by cyclization of *trans–trans*-farnesyl pyrophosphate with the terminal double bond. Suggest a biosynthetic pathway to eremophilone (Coates, R. M. *Prog. Chem. Org. Nat. Prod.* **33** (1976) 74).

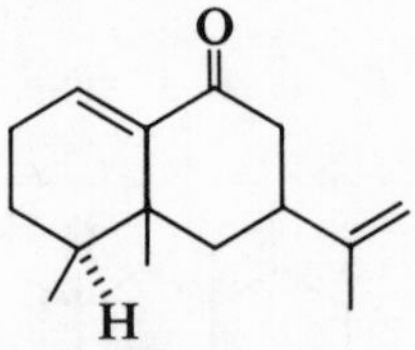

O H

Eremophilone

5.3 Thujic acid has been isolated from the heart wood of *Thuja plicata*. It is considered to have a monoterpenoid origin contrary to the tropolone derivatives produced by fungi (section 4.12). Suggest a plausible route to the compound.

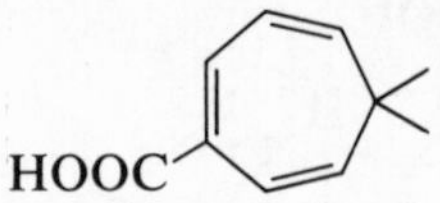

HOOC

5.4 One can arrive at the hirsutane skeleton from farnesyl pyrophosphate via the humulyl cation (section 5.4, Fig. 15) along three possible cyclization routes. The correct pathway was determined by fermentation of *Coriolus consors* in the presence of 1,2-^{13}C-acetate which produced 5-dihydrocoriolin. The $^{1}J_{^{13}C^{13}C}$ of the

^{13}C NMR spectrum of labelled 5-dihydrocoriolin indicated the presence of six intact acetate units in the ring and four intact units in the side chain. Discuss the biosynthetic pathway for the metabolite. The ^{13}C NMR data are summarized in the table (Tanabe, M., Suzuki, K. T. and Jankowski, W. C. *Tetrahedron Lett*, **1974**, 2271).

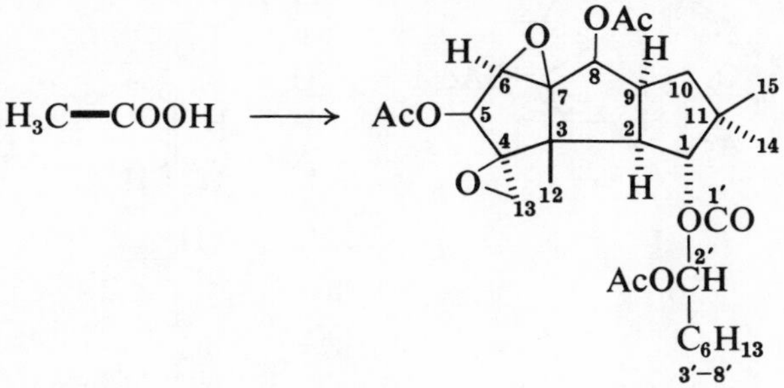

H_3C—COOH ⟶ AcO— [structure: OAc, H, O, 1′ OCO, 2′ AcOCH, C_6H_{13} 3′–8′]

Table 1 ^{13}C NMR data for dihydrocoriolin C (as triacetate)

Position, δ, multiplicity	$^{1}J_{^{13}C^{13}C}$
C^1 (80.0,d)–C^2 (51.0,d)	42
C^3 (47.1,s)–C^{12} (13.1,q)	38
C^4 (64.5,s)–C^{13} (44.9,t)	30
C^5 (71.7,d)	
C^6 (60.9,d)–C^7 (73.5,s)	26
C^8 (72.7,d)	
C^9 (41.3,d)–C^{10} (37.6,t)	33
C^{11} (43.9,s)–C^{15} (21.4,q)	34
C^{14} (26.5)	
$C^{1'}$ (170.0,s)–$C^{2'}$ (72.4,d)	65
$C^{3'}$ (31.3,t)–$C^{4'}$ (25.2,t)	34
$C^{5'}$ (28.8,t)–$C^{6'}$ (31.5,t)	35
$C^{7'}$ (22.5,t)–$C^{8'}$ (14.0,q)	34
CH_3CO (20.5, 20.5, 20.9, 170.0, 170.5, 170.8)	

The multiplicity (s = singlet, d = doublet, t = triplet, q = quartet) refers to off-resonance residual couplings

5.5 The coelenterates (Octocorallia), forming a prominent part of the biomass in the tropical reefs, produce a number of terpenoids of unusual structures. Xenicin, I, was isolated from the Australian soft coral *Xenia elongata* (Alcyonaria) and from the Okinawan soft coral, *Alcyonium* Sp., alcyonolide II was isolated. Suggest plausible biosynthetic pathways for the metabolites. Lead: the highly oxygenated skeletal frameworks could be formed by cleavage of a caryophyllene type precursor (Vanderah, D. J., Steudler, P. A., Ciereszko, L. S., Schmitz, F. J., Ekstrand, J. D. and van der Helm, D. *J. Am. Chem. Soc.* **99** (1977) 5780; Kobayashi, M., Yasuzawa, T., Kobayashi, Y., Kyogoku, Y. and Kitagawa, I. *Tetrahedron Lett.* **1981**, 4445).

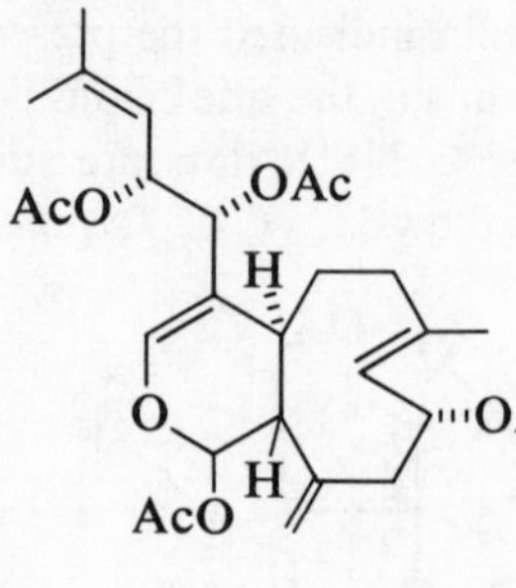

I

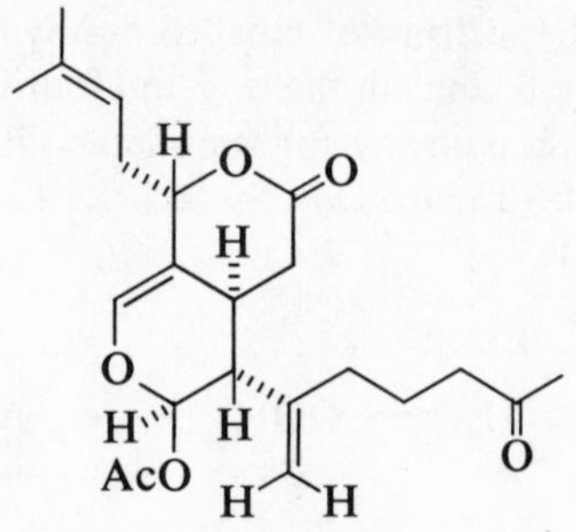

II

5.6 The termite *Nasutitermes rippertii* produces a gluey defence excretion from which 3α-hydroxy-15-rippertene was isolated. Suggest a biosynthetic scheme for this tetracyclic terpenoid (Prestwick, G. D., Spanton, S. G., Lauber, J. W. and Vrkoč, J. *J. Am. Chem. Soc.* **102** (1980) 6825).

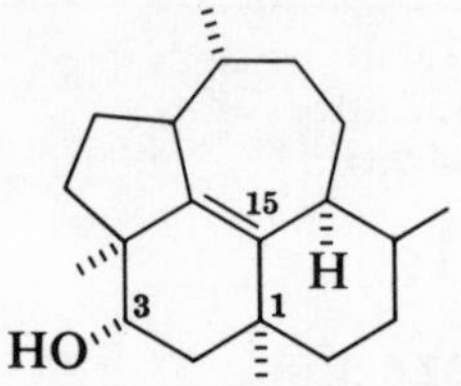

3α-Hydroxy-15-rippertene

Bibliography

1. Ruzicka, L. *Experientia* **9** (1953) 357.
2. McManus, S. P. and Pittman, Jr., C. U. in *Organic Reactive Intermediates*, McManus, S. P. (Ed.), Academic Press, New York, 1973, p. 193.
3. Sonderhoff, R. and Thomas, H. *Liebigs Ann.* **530** (1937) 195.
4. Tavormina, P. A., Gibbs, M. H. and Huff, J. W. *J. Am. Chem. Soc.* **78** (1956) 4498.
5. Wolf, D. E., Hoffman, C. H. Aldrich, P. E., Skeggs, H. R., Wright, L. D. and Folkers, K. *J .Am. Chem. Soc.* **78** (1956) 4499.
6. Lynen, F. and Henning, U. *Angew. Chem.* **72** (1960) 820.
7. Stone, K. J., Roeske, W. R., Clayton, R. B. and van Tamelen, E. E. *Chem. Commun.* **1969,** 530.
8. Cornforth, J. W., Cornforth, R. H., Donninger, C. and Popjak, G. *Proc. Roy. Soc.* (*B*) **163** (1965) 492.
9. Cornforth, J. W., Cornforth, R. H., Popjak, G. and Yengoyan, L. *J. Biol. Chem.* **241** (1966) 3970.
10. Lüthy, J., Rétey, J. and Arigoni, D. *Nature* **221** (1969) 1213.
11. Cornforth. J. W., Redmond, J. W., Eggerer, H., Buckel, W. and Gutschow, C. *Nature* **221** (1969) 1212.
12. Cornforth, J. W. *Chem. Brit* **6** (1970) 431.

13. Clifford, K. H., Cornforth, J. W., Mallaby, R. and Phillips, G. T. *Chem. Commun.* **1971,** 1599.
14. Cornforth, J. W., Phillips, G. T., Messner, B. and Eggerer, H. *Eur. J. Biochem.* **42,** (1974) 591.
15. Poulter, C. D. and Rilling, H. C. *Acct. Chem. Res.* **11** (1978) 307.
16. Allinger, N. L. and Seifert, J. H. *J. Am. Chem. Soc.* **97** (1975) 752.
17. Banthorpe, D. V., Charlwood, B. V. and Francis, M. J. O. *Chem. Revs.* **72** (1972) 115; Tange, K., Okita, H., Nakao, Y., Hirata, T. and Suga, T. *Chem. Letters* **1981,** 777.
18. Lanza, E. and Palmer, J. K. *Phytochemistry* **16** (1977) 1555.
19. Cane, D. E. *Tetrahedron* **36** (1980) 1109.
20. Overton, K. H. and Roberts, F. M. *Biochem. J.* **144** (1974) 585.
21. Banthorpe, D. V., Ekundayo, O. and Rowan, M. G. *Phytochemistry* **17** (1978) 1111.
22. Suga, T., Shishibori, T. and Hirata, T. *Chem. Letters* **1977,** 937.
23. Cane D. E., Swanson, S. and Murthy, P. P. N. *J. Am. Chem. Soc.* **103** (1981) 2136.
24. Banthorpe, D. V., Charlwood, B. V., Greaves, G. M. and Voller, C. M. *Phytochemistry* **16** (1977) 1387.
25. Peter, M. G. and Dahm, K. H. *Helv. Chim. Acta* **58** (1975) 1037.
26. Overton, K. H. and Picken, D. J. *J. Chem. Soc. Chem. Commun.* **1976,** 105.
27. Hanson, J. R., Marten, T. and Siverns, M. *J. Chem. Soc. Perkin I* **1974,** 1033
28. Achilladelis, B. A., Adams, P. M. and Hanson, J. R. *J. Chem. Soc. Perkin I* **1972,** 1425
29. Machida, Y. and Nozoe, S., *Tetrahedron* **28,** (1972) 5113.
30. Marshall, J. A., Brady, S. F. and Andersen, N. H. in *Prog. Chem. Org. Nat. Prod.* **31** (1974) 283.
31. Corbella, A., Gariboldi, P. and Jommi, G. *J. Chem. Soc. Chem. Commun.* **1972,** 600.
32. Arigoni, D. *Pure Appl. Chem.* **41** (1975) 219.
33. Martin, J. D. and Darias, J. in *Marine Natural Products*, Scheuer, P. J. (Ed.) Vol. I, p. 125. Academic Press, New York, 1978.
34. Burreson, B. J., Christophersen, C. and Scheuer, P. J. *Tetrahedron* **31** (1975) 2015.
35. Blackman, A. and Wells, R. J. *Tetrahedron Lett.* **1978,** 3063.
36. Hanson, J. R. in *Prog. Chem. Org. Nat. Prod.* **29** (1971) 395.
37. Coates, R. M. in *Prog. Chem. Org. Nat. Prod.* **33** (1976) 73.
38. Johnson, W. S. *Acct. Chem. Res* **1** (1968) 1.
39. Goldsmith, D. in *Prog. Chem. Org. Nat. Prod.* **29** (1971) 363.
40. Achilladelis, B. and Hanson, J. R. *J. Chem. Soc.*(C) **1969,** 2010.
41. Hancock, W. S., Mander, L. N. and Massy-Westropp. R. A. *J. Org. Chem.* **38** (1973) 4090.
42. MacMillan, J. and Pryce, R. J. in *Phytochemistry*. Vol. III p. 283, Miller, L. P. (Ed.), Van Nostrand Reinhold, New York, 1973.
43. Graebe, J. E., Bowen, D. H. and MacMillan, J. *Planta* **102** (1972) 261.
44. Graebe, J. E., Hedden, P. and MacMillan, J. *J. Chem. Soc. Chem. Commun.* **1975,** 161.
45. Hanson, J. R. and Evans, R. *J. Chem. Soc. Perkin I* **1975,** 663.
46. Bearder, J. R., MacMillan, J. and Phinney, B. O. *J. Chem. Soc. Perkin I*, **1975,** 721.
47. Dockerill, B. and Hanson, J. R. *Phytochemistry* **17** (1978) 701.
48. Weinheimer, A. J., Chang, C. W. J. and Matson, J. A. *Prog. Chem. Org. Nat. Prod.* **36** (1979) 285.
49. Cordell, G. A. *Phytochemistry* **13** (1974) 2343.
50. Rios, T. and Perez, C. S. *Chem Commun.* **1969,** 214.
51. Minale, L. in *Marine Natural Products*, Scheuer, P. J. (Ed.), Vol. I, p. 174. Academic Press, New York, 1978.

52. Nozoe, S., Morisaki, M., Tsuda, K., Iitaki, Y., Takahashi, N., Tamura, S., Ishibashi, K. and Shirasaka, M. *J. Am. Chem. Soc.* **87** (1965) 4968.
53. Nozoe, S., Morisaki, M., Tsuda, K. and Okuda, S. *Tetrahedron Lett.* **1967,** 3365.
54. Canonica, L., Fiecchi, A., Galli Kienle, M., Ranzi, B. M. and Scala, A. *Tetrahedron Lett.* **1967,** 4657.
55. Epstein, W. W. and Rilling, H. C. *J. Biol. Chem.* **245** (1970) 4597.
56. Popjak, G., Edmond, J. and Wong, S.-M. *J. Am. Chem. Soc.* **95** (1973) 2713.
57. Poulter, C. D., *J. Agric. Food Chem.* **22** (1974) 167.
58. Corey, E. J., Russey, W. E. and de Montellano, P. P. O. *J. Am. Chem. Soc.* **88** (1966) 4750.
59. van Tamelen, E. E., Willett, J. D., Clayton, R. B. and Lord, K. E. *J. Am. Chem. Soc.* **88** (1966) 4752.
60. Yamamoto, S. and Bloch, K. *J. Biol. Chem.* **245** (1970) 1670.
61. Ebersole, R. C., Godtfredsen, W. O., Vangedal, S. and Caspi, E. *J. Am. Chem. Soc.* **95** (1973) 8133.
62. Mulheirn, L. J. and Ramm, P. J. *Chem. Soc. Revs.* **1** (1972) 259.
63. van Tamelen, E. E. and James, D. R. *J. Am. Chem. Soc.* **99** (1977) 950.
64. Clifford, K. H. and Phillips, G. T. *European J. Biochem* **61** (1976) 271.
65. Popjak, G., Edmond, J., Anet, F. A. L. and Easton, Jr. N. R. *J. Am. Chem. Soc.* **99** (1977) 931.
66. Lavie, D. and Glotter, F. *Prog. Chem. Org. Nat. Prod.* **29** (1971) 307.
67. Courney, J. L., Gasciogne, R. M. and Szumer, A. Z. *J. Chem. Soc.* **1958,** 881.
68. Barton, D. H. R., Mellows, G. and Widdowson, D. A. *J. Chem. Soc.* (*C*) **1971,** 110
69. Polonsky, J. *Prog. Chem. Org. Nat. Prod.* **30** (1973) 101.
70. Moron, J., Merrien, M.-A., and Polonsky, J. *Phytochemistry* **1971,** 585.
71. Moron, J. and Polonsky, J. *European J. Biochem.* **3** (1968) 488.
72. Sabine, J. R. *Cholesterol*, Dekker, New York, 1977.
73. Greig, J. B.. Varma, K. R. and Caspi, E. *J. Am. Chem. Soc.* **93** (1971) 760.
74. Duchamp, D. J., Chidester, C. G., Wickramasinghe, J. A. F., Caspi, E. and Yagen, B. *J. Am. Chem. Soc.* **93** (1971) 6283.
75. Gibbons, G. F. and Mitropoulos, K. A. *Biochem. J.* **132** (1973) 439.
76. Akhtar, M., Alexander, K., Boar, R. B., McGhie, J. F. and Barton, D. H. R. *Biochem. J.* **169** (1978) 449.
77. Gautschi, F. and Bloch, K. *J. Biol. Chem.* **233** (1958) 1343.
78. Nelson, J. A., Kahn, S., Spencer, T. A., Sharpless, K. B. and Clayton, R. B. *Bioorg. Chem.* **4** (1975) 363.
79. Rahimtula, A. D. and Gaylor, J. L. *J. Biol. Chem.* **247** (1972) 9.
80. Schroeper, Jr., G. J., Lutsky, B. N., Martin, J. A., Huntoon, S., Fourcans, B., Lee, W.-H. and Vervilion, J. *Proc. Roy. Soc.* **B180** (1972) 125.
81. Tschesche, R. and Wulff, G. *Prog. Chem. Org. Nat. Prod.* **30** (1973) 461.
82. van Lier, J. E. and Smith, L. L. *Biochem. Biophys. Res. Commun.* **40** (1970) 510.
83. Bennet, R. D., Sauer, H. H. and Heftmann, E. *Phytochemistry* **7** (1968) 41.
84. *The Bile Acids* Vol. 2, Nair, P. P. and Kritchevsky, D. (Eds.), Plenum Press, New York, 1973.
85. Milewich, L. and Axelrod, L. R. *Arch. Biochem. Biophys.* **153** (1972) 188.
86. Goodwin, T. W. in *Rec. Adv. Phytochemistry*, Vol. 6, p. 97, Runeckles, V. C. and Mabry, T. J. (Eds.), Academic Press, New York, 1973.
87. Goad, L. J. and Goodwin, T. W. in *Progress in Phytochemistry*, Vol. 3, p. 113, Reinhold, L. and Liwschitz, Y. (Eds.), J. Wiley, London, 1972.
88. Goad, L. J. in *Marine Natural Products*, Vol. II, p. 75, Scheuer, P. J. (Ed.), Academic Press, New York, 1978.
89. Goad, L. J., Lenton, J. R., Knapp, F. F. and Goodwin, T. W. *Lipids* **9** (1974) 582.
90. Jones, H. and Rasmusson, G. H. *Prog. Chem. Org. Nat. Prod.* **39** (1980) 63

91. Liaan-Jensen, S. *Prog. Chem. Org. Nat. Prod.* **39** (1980) 123.
92. Goodwin, T. W., in *Carotenoids*, Isler, O. (Ed.), Birkhauser, Basel, 1971, p. 577.
93. Altman, L. J., Ash, L., Kowerski, R. C., Epstein, W. W., Larsen, B. R., Rilling, H. C., Muscio, F. and Gregonis, D. E. *J. Am. Chem. Soc.* **94** (1972) 3257.
94. Qureshi, A. A., Barnes, F. J. and Porter, J. W. *J. Biol. Chem.* **247** (1972) 6730.
95. Crabbé, P. *Optical Rotatory Dispersion and Circular Dichroism in Org. Chemistry*, Holden-Day, San Francisco, 1965.
96. Scopes, P. M., *Prog. Chem. Org. Nat. Prod.* **32** (1975) 167.

Chapter 6

Amino acids, peptides and proteins

6.1 Introduction

The title compounds are typical first order metabolites indispensable for life at all levels. They are the starting materials for an array of secondary products such as simple amines, alkaloids, aromatic *N*-heterocycles, as well as phenylpropanoids, i.e. the C_6C_3 compounds, where the amino group is lost (Chapter 3). The amino acid, leucine, is also a precursor of isopentenyl pyrophosphate. Ever since Hofmeister and Fischer proposed at the turn of the century that amino acids are the structural units of proteins, biochemists have extensively devoted their attention and skill to this domain of chemistry. This research was highlighted by such events as the crystallization of the first enzyme, urease, and the proof that it was a protein by Sumner in 1926, the automatic amino acid analyser introduced by Stein and Moore which made rapid protein analysis possible, the high resolution X-ray analysis of enzymes by Kendrew and Perutz in the late 1950s and the unravelling of the mechanisms of action of genes by Jacob, Monod, Nirenberg, Khorana, Ochoa, and others in the 1960s. The biosynthesis, degradation and reactions of amino acids and proteins are well covered in textbooks of biochemistry. Nevertheless, it is appropriate to include a few common transformations here, simply because in a predictable way they illustrate principles of organic reaction mechanism. The elucidation of the biogenetic pathways of amino acids was a difficult task, but of utmost importance for the development of biochemistry, physiology and medicine. It so happens that the classification of amino acids as essential and non-essential has its counterpart in the different compexities of formation of the two groups. The biosynthesis of the essential amino acids is more complex than the biosynthesis of non-essential amino acids, which in principle are available in a few steps by enzymatic reductive amination of α-keto acids originating from the citric acid cycle or from glycolysis.

6.2 Amino acids. Classification, structure and properties

The amino acids, derived from proteins, are all L-α-amino acids, the configuration of which is defined by the Fischer projection in Fig. 1. L-(—)-Serine is related to L-(—)-glyceraldehyde by reactions of known stereochemistry.

$$\begin{array}{c} COOH \\ H_2N - \!\!\!\!+\!\!\!\!- H \\ R \end{array} \qquad\qquad \begin{array}{c} CHO \\ HO - \!\!\!\!+\!\!\!\!- H \\ CH_2OH \end{array}$$

Fig. 1 Fischer projection of L-α-amino acids and L-(−)-glyceraldehyde. L-(−)-Serine = *S*-serine, R = CH_2OH

The first amino acid isolated was the only symmetrical and hence optically inactive amino acid, glycine, obtained by Braconnot in 1820 from a hydrolysate of gelatine. Besides the 20 common amino acids from proteins (Table 1) there are about 300 non-protein amino acids found in nature. Some of them have D-

Table 1 Structure of the commonest amino acids derived from proteins. Name of amine derived by decarboxylation

Amino acid	Structure	Abbreviated symbol	Amine
Alanine N,1	$CH_3CH(NH_2)COOH$	Ala	Ethylamine
Arginine E,3	$H^{\oplus}{}_2N{=}C(NH_2)NH(CH_2)_3CH(NH_2)COOH$	Arg	Agmaline (4-Guanidobutyl-amine)
Aspartic acid N,4	$^{\ominus}OOCCH_2CH(NH_2)COOH$	Asp	β-Alanine
Asparagine N,2	$NH_2COCH_2CH(NH_2)COOH$	Asn	β-Alanyl amide
Cysteine N,2	$HSCH_2CH(NH_2)COOH$	Cys	2-Mercaptoethyl amine
Glutamic acid N,4	$^{\ominus}OOCCH_2CH_2CH(NH_2)COOH$	Glu	γ-Aminobutyric acid (GABA)
Glutamine N,2	$H_2NCOCH_2CH_2CH(NH_2)COOH$	Gln	γ-Aminobutyr-amide
Glycine N,2	$CH_2(NH_2)COOH$	Gly	Methylamine

Table 1 (*continued*)

Amino acid	Structure	Abbreviated symbol	Amine
Histidine E,3	(imidazol-4-yl)$CH_2CH(NH_2)COOH$	His	Histamine
Isoleucine E,1	$C_2H_5CH(CH_3)CH(NH_2)COOH$	Ile	2-Methylbutyl-amine
Leucine E,1	$(CH_3)_2CHCH_2CH(NH_2)COOH$	Leu	3-Methylbutyl-amine
Lysine E,3	$H_2N(CH_2)_4CH(NH_2)COOH$	Lys	Cadaverine, 1,5-Diamino-pentane
Methionine E,1	$CH_3S(CH_2)_2CH(NH_2)COOH$	Met	3-Methylmercapto-propylamine
Phenylalanine E,1	$C_6H_5CH_2CH(NH_2)COOH$	Phe	Phenylethylamine
Proline N,1	(pyrrolidine ring, NH)–COOH	Pro	Pyrrolidine
Serine N,2	$HOCH_2CH(NH_2)COOH$	Ser	Ethanolamine
Threonine E,2	$CH_3CH(OH)CH(NH_2)COOH$	Thr	2-Hydroxypropyl-amine
Tryptophan E,1	(indol-3-yl)$CH_2CH(NH_2)COOH$	Try	Tryptamine
Tyrosine N,2	$HO-C_6H_4-CH_2CH(NH_2)COOH$	Tyr	Tyramine
Valine E,1	$(CH_3)_2CHCH(NH_2)COOH$	Val	*i*-Butylamine

Requirement by man: E, essential; N, non-essential. Polarity of the chain: 1, non-polar; 2, neutral polar; 3, positively charged; 4, negatively charged

configuration. D-Alanine was found free in insects[1] and also in some plants, e.g. *Pisum sativum.*[2] It occurs as the dipeptide, D-alanyl-D-alanine, in the leaves of *Nicotiana tabacum.*[3] D-Glutamic acid is a constituent of the glycopeptides in the cell wall of many bacteria.[4] A few are β-, γ- or δ-amino acids. β-Alanine, which arises by decarboxylation of aspartic acid, is a building block of coenzyme A (section 1.6). γ-Aminobutyric acid, which is formed by decarboxylation of glutamic acid, acts as a relay compound for transmission of nerve impulses.[5] Ornithine, α,γ-diaminovaleric acid, was first isolated by Jaffe in 1877 in excretion products from birds fed with benzoic acid. Birds and reptiles excrete benzoic acid as the dibenzoate of ornithine, ornithuric acid, in contrast to man, who excretes benzoic acid as benzoyl glycine, hippuric acid. If 3 g of benzoic acid is given orally to a man, hippuric acid can readily be extracted in substantial amounts from his urine after 12 h. Conjugation of toxic substances with glycine, glutamine, or cysteine, thereby rendering them more water soluble, is a frequently encountered detoxification mechanism. Sulphatization of hydroxy compounds and acetylation are other detoxification procedures used by the body. Ornithine, a carrier in the urea cycle, is the precursor of the pyrrolidine and tropane alkaloids. It decarboxylates to evil smelling 1,4-diaminobutane, putrescine. Citrulline, or carboxamido-ornithine, homocysteine, homoserine, and β-cyanoalanine, are other non-protein amino acids serving as precursors and intermediates in metabolism.

Bacteria produce several peptide antibiotics,[6,7] e.g. the actinomycins from *Streptomyces* spp. and gramicidin S from *Bacillus brevis*, the latter containing the non-protein amino acids D-phenylalanine and L-ornithine (Fig. 2). Gramicidin S acts—like the crown ethers—by its property to form fat-soluble salts.

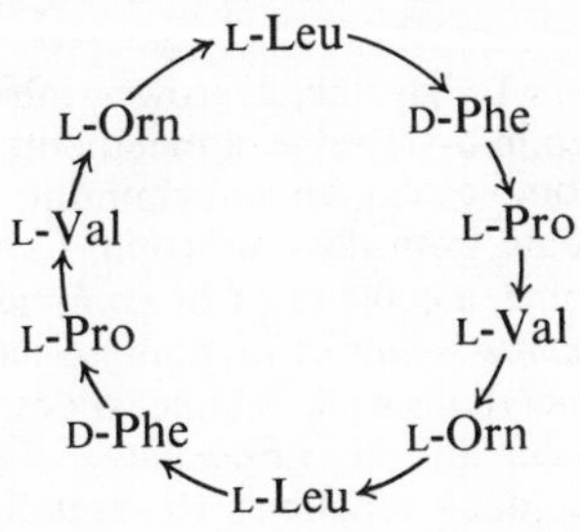

L-Leu → D-Phe → L-Pro → L-Val → L-Orn → L-Leu → D-Phe → L-Pro → L-Val → L-Orn → (L-Leu)

1

OC—L-Me-Val → Sar → L-Pro → D-Val → L-Thr—O → CO

L-Me-Val—CO → Sar → L-Pro → D-Val → L-Thr—O → CO

N, O, NH_2, O, CH_3, CH_3

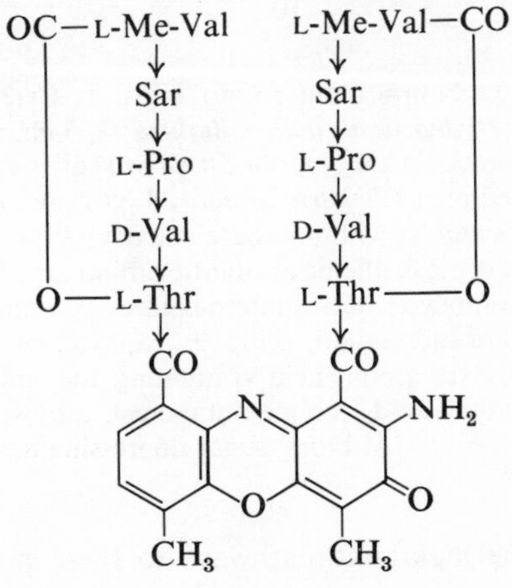

2

Fig. 2 1, Structure of the antibiotic gramicidin S, a cyclic decapeptide containing non-protein D-phenylalanine and L-ornithine. The arrows indicate the N→C direction of the peptide bond. 2, Structure of actinomycin D from *Streptomyces*, a lactone peptide acting as a DNA inhibitor. It contains the non-protein D-valine and *N*-methylated L-valine and *N*-methylated glycine = Sarcosine = Sar. The carbonyl group of L-Me-Val is lactonized with the OH group of L-Thr. The two cyclic peptides are bound to the phenoxazine ring via amide bonds

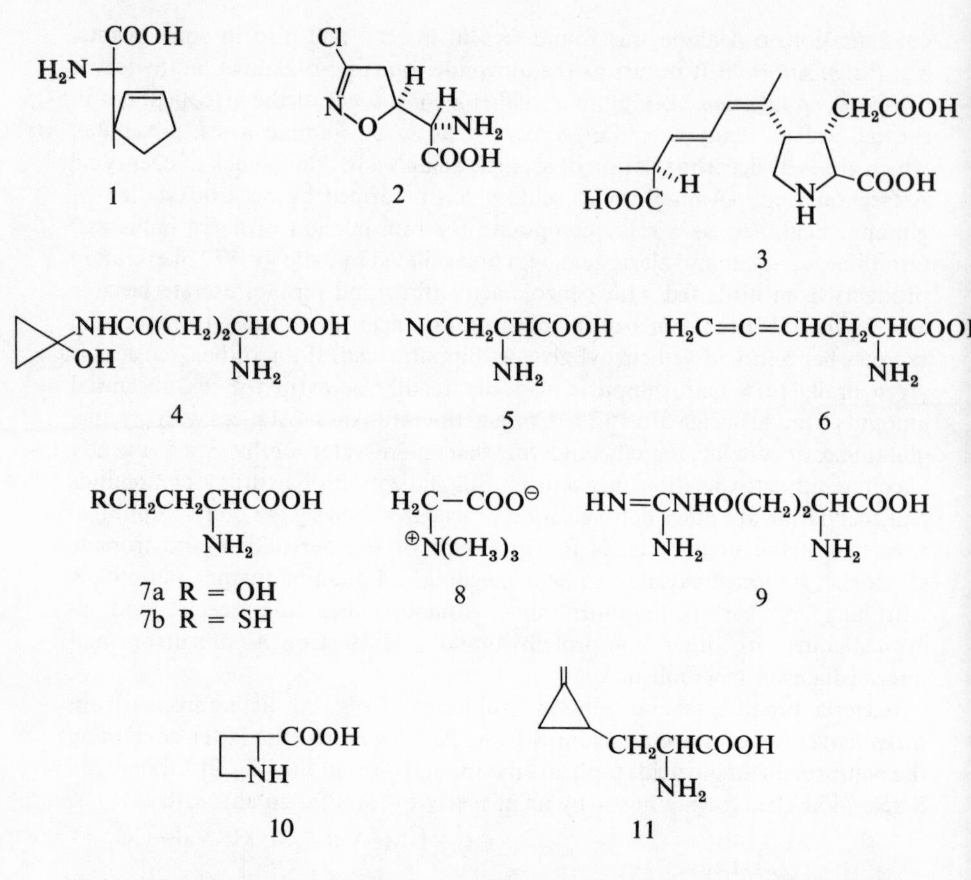

Fig. 3 Non-protein amino acids. 1, Cyclopent-2-en-1-ylglycine, a growth inhibitor from *Hydnocarpus anthelminthica*; 2, 3-chloroisoxazolin-5-ylglycine, a metabolite with antitumour activity from *Streptomyces sviceus*; 3, domoic acid, an anthelminthic from the red alga *Chondria armata*; 4, coprine, a metabolite from the mushroom *Coprinus atramentarius* with antabuse activity; 5, β-cyanoalanine, a constituent of *Lathyrus* and *Vicea* spp.; 6, allenic antibiotic amino acid from *Amanita solitaria*; 7a, homoserine and 7b, homocysteine are intermediates in amino acid metabolism; 8, betaine, widespread zwitterionic amino acid; 9, canavanine, constituent of the *Leguminosae* family, phytotoxic amino acid containing the unique guanidoxy function; 10, azetidine-2-carboxylic acid, proline antagonist, almost ubiquitously distributed; 11, hypoglycine A, blood sugar depressing amino acid from *Blighia sapida*

The metabolic pathways to these non-protein amino acids are not known with certainty yet, but in several cases it is rather obvious that they are constructed in a straightforward manner from mainstream metabolic units discussed in Chapters 3, 4 and 5. The biosynthesis of cyclopent-2-en-1-ylglycine,[8] 1 (Fig. 3) is not known but is related to the biosynthesis of chaulmoogric acid, which is present in the same plant. The isoxazoline,[9] 2, most probably derives from glutamine by oxidation of C^3. Domoic acid,[10] 3, can be dissected into glutamic acid and a monoterpene. It is probably synthesized by *N*-alkylation of glutamic acid with geranyl phosphate followed by ring closure. The exact timing of the formation of the carboxyl group in the terpenoid side chain is difficult to assess. Coprine,[11] 4, is an adduct of cyclopropanone, the origin of which is 1-aminocyclopropane carboxylic acid and glutamic acid. It is shown by radioactive labelling experiments that the biosynthesis of cyanoalanine, 5, takes place from L-cysteine by elimination of hydrogen sulphide and Michael addition of cyanide to the intermediate α-aminoacrylic acid.[12] β-Cyanoalanine is hydrolysed to asparagine,[13] decarboxylated to β-aminopropionitrile and reduced to 2,4-diaminobutyric acid (Fig. 4).

$$HS{-}CH_2{-}CH(NH_2){-}COOH \xrightarrow{PLP} CH_2{=}C(COOH){-}N{=}CH{-}PLP \xrightarrow[H_2O]{CN^{\ominus}}$$

Cysteine

$$NCCH_2CH(NH_2)COOH \xrightarrow{H_2O} NH_2COCH_2CH(NH_2)COOH$$

β-Cyanoalanine → Asparagine

$$NCCH_2CH(NH_2)COOH \xrightarrow{-CO_2} NCCH_2CH_2NH_2$$

β-Aminopropionitrile

$$NCCH_2CH(NH_2)COOH \xrightarrow{[H]} CH_2(NH_2)CH_2CH(NH_2)COOH$$

2,4-Diaminobutyric acid

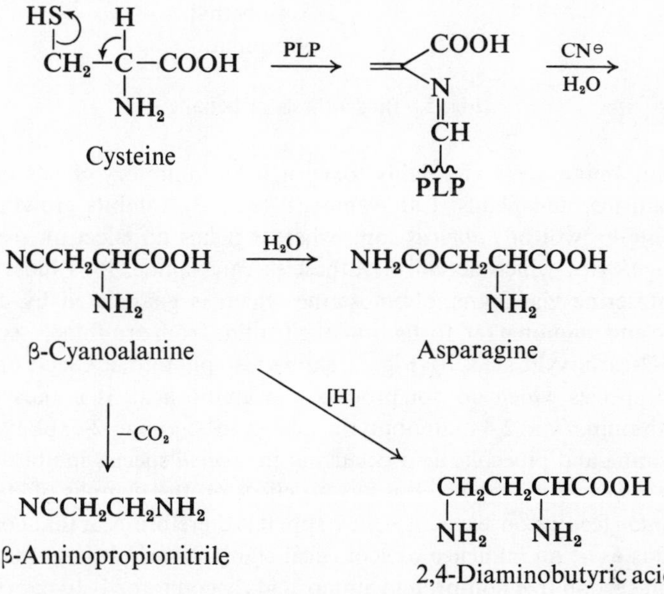

Fig. 4 Biosynthesis and metabolism of β-cyanoalanine. PLP, pyridoxal phosphate

Several unsaturated amino acids of the type 6 (Fig. 3) are known, acetylenic as well as olefinic. The amino group of amino acids has been found to be alkylated to various degrees, methylation being especially common. Glycine gives via sarcosine and dimethylglycine the fully methylated and widely distributed betaine 8 (Fig. 3), which also gave its name to this class of dipolar compounds, the betaines. Another biosynthetic route to betaine starts with serine, which is phosphorylated, decarboxylated, methylated, and finally oxidized (Fig. 5). It was isolated by Huseman in 1863 from *Lycium barbarum* and by Scheibler from the sugar beet, *Beta vulgaris*, in 1869, containing up to 5 per cent in the leaves, dry weight.

The functions of most of these secondary amino acids are unknown but circumstantial evidence indicates that their presence is not meaningless. Several

$$\underset{\text{Glycine}}{CH_2(NH_2)COOH} \xrightarrow{[CH_3]} \underset{\text{Sarcosine}}{CH_2(NHCH_3)COOH} \xrightarrow{[CH_3]} \underset{\text{Dimethylglycine}}{CH_2(N(CH_3)_2)COOH} \xrightarrow{[CH_3]} \underset{\text{Betaine}}{CH_2({}^{\oplus}N(CH_3)_3)COO^{\ominus}}$$

$$\underset{\text{Serine}}{CH_2(OH)CH(NH_2)COOH} \xrightarrow[2.\ -CO_2]{1.\ ATP} \underset{\beta\text{-Aminoethyl phosphate}}{CH_2(OP)CH_2(NH_2)} \xrightarrow[2.\ H_2O]{1.\ 3[CH_3]} \underset{\text{Choline}}{CH_2(OH)CH_2({}^{\oplus}N(CH_3)_3)} \xrightarrow{[O]} \text{Betaine}$$

Fig. 5 Biosynthesis of betaine

non-protein amino acids are highly toxic or have inhibitory effects on insects, microorganisms, and plants. Canavanine, 9 (Fig. 3), inhibits growth of yeast and seedling growth of *Lathyrus* spp., whereas it has no effect on seedlings of *Vicea bengalensis*, which actually synthesizes canavanine. It is biosynthesized from homoserine via *O*-aminohomoserine which is guanidated by carbamylphosphate and ammonia (*cf.* formation of citrulline from ornithine, (section 6.5). Azetidine-2-carboxylic acid, 10 (Fig. 3), shows also phytotoxic effects on seedling growth of species which do not produce the amino acid. It is biosynthesized from methionine via 2,4-aminobutyric acid and 4-amino-2-oxobutyric acid. Homoarginine and pipecolic acid occurring in *Acacia* species inhibit feeding in the locust, *Locusta migratoria*, but has no effect on the nymphs of *Anacridium melanorhodon* feeding on leaves of *Acacia* spp. It is therefore clear that non-protein amino acids exert an influence of ecological significance.[14] It is not uncommon to find that certain free non-protein amino acids account for 1–10 per cent of the dry weight of the tissue. Seeds of *Dioclea megacarpa* contain up to 13 per cent of canavanine and leaves of *Convallaria majalis* contain 3 per cent of azetidine-2-carboxylic acid. An accumulation of this order suggests that they have a storage role. Several amino acids occurring in fodder plants are toxic and could cause death. The unusual cyclopropanoid amino acid hypoglycine A, 11 (Fig. 3), occurs in the fruits of the akee tree, *Blighia sapida*. It causes hypoglycaemia and occasionally death among people in the West Indies. Mimosine (see Fig. 11), a metabolite of the tropical legume *Leucaena glauca*, causes loss of hair and is of great concern to sheep breeders in Australia. Certain plants when growing on selenium rich soil can metabolize selenium and substitute sulphur by selenium. The $SeCH_3$ analogue of cysteine, produced by *Astragalus bisulcatus* in the Western United States, the so-called locoweeds, causes madness among grazing livestock.

Amino acids are classified in various ways. Since the polarity of the R group (Table 1) controls the stability of various conformations of the free acid or still more important controls the conformation and coiling of the peptide chain with consequences for the topology of the enzyme surface, it is meaningful to base a classification on the polarity of the R group:

1. non-polar,
2. neutral polar,
3. positively charged and
4. negatively charged amino acids (see Table 1).

Another classification is based on the nutritional requirements of amino acids. *E. coli* can utilize ammonia alone for the biosynthesis of all of its amino acids and derivatives, whereas most vertebrates including man require some amino acids in their diet. Of the twenty amino acids required for the synthesis of proteins man can manufacture ten, the other ten are called the essential amino acids.

Amino acids are bifunctional compounds having an acid carboxyl group and a basic amino group. Consequently they are better represented by their dipolar ionic or zwitterionic structure (1). They act either as acids or bases and as such

$$\underset{\oplus NH_3}{RCHCOOH} \underset{}{\overset{-H^{\oplus}}{\rightleftharpoons}} \underset{NH_2}{RCHCOOH} \rightleftharpoons \underset{\oplus NH_3}{RCHCOO^{\ominus}} \overset{OH^{\ominus}}{\rightleftharpoons} \underset{NH_2}{RCHCOO^{\ominus}} \quad (1)$$

are called amphoteric compounds. Because of their zwitterionic nature they are high-melting solids, soluble in water and insoluble in non-polar solvents. Having an asymmetric centre α to the carboxyl group they racemize in strong acidic or alkaline media. Actually even at room temperature and under neutral conditions an extremely slow racemization takes place that has been utilized for dating fossils.[15] Different amino acids have different half-lives, and so it has been possible to cover times from *ca.* 100 to 100 000 years. Investigations of proteins from the human lens and tooth enamel show that racemization is related to age and provided that the notorious old Caucasians have not lost their teeth, it should be possible to check their claim of longevity.[16] The conclusions rest on the assumption that conversion of some trapped proteins is low in the organism. It has been proposed that ageing is coupled with racemization of chiral centres in proteins thus making them malfunction. The absolute configuration of amino acids not possessing interfering chromophores can be deduced from their CD curves at the 210 nm n–π^* transition. The L-stereoisomers give a positive Cotton effect,[17] but care must be exercised because conformational factors affect the sign. A sector rule based on the octant rule is proposed on the basic assumption that the N–C^{α}–COO atoms are coplanar in solution.[18]

Amino acids are detected by a sensitive reaction with ninhydrin which gives a bluish-red colour (2). This reaction is closely related to the Strecker degradation of amino acids and the pyridoxal mediated enzymatic decarboxylation of amino acids. The reaction has been used in forensic medicine for the detection of fingerprints on paper. The minute amount of sweat exuded from the pores of a

Ninhydrin

Resonance stabilized ion

Hydrolysis

(2)

Ninhydrin

bluish-red

fingertip contains enough amino acids to give a coloured reproduction of the fingerprints by spraying the paper with ninhydrin and heating it to 100°C.

Few groups of organic compounds have been the object of so much qualitative and quantitative analytical research as the amino acids and peptides. Different kinds of chromatography based on either the principles of partition or absorption have been worked out for complete separation of complicated mixtures, e.g. a hydrolysed protein. These include one- or two-dimensional paper chromatography, silica gel thin layer chromatography (TLC), electrophoresis on different carriers, ion-exchange chromatography and column chromatography for larger quantities. The countercurrent technique developed by Craig is based solely on the principle of partition of a solute between immiscible liquids and chromatographic separations on carriers are to varying extents based on the same principle. On the surface of e.g. cellulose fibres of filter paper, a film of solvent and hydrated polysaccharides is formed, different from the bulk of solvent, and the amino acids are partitioned between these two media. Absorption phenomena have a substantial influence on the partition which can be varied widely by change in solvent composition. The spots or bands are usually developed by ninhydrin. By making amino acids more volatile, by methylation or silylation, it is possible to apply gas chromatographic procedures combined with highly sensitive mass spectrometric recordings. The analytical procedures have continuously been refined and are now automated and carried out with considerable speed and precision.[19]

6.3 Prebiotic formation of amino acids

It caused considerable interest and activity among scientists when Urey and Miller in the 1950s demonstrated that electrical discharge in a primitive mixture of methane, carbon dioxide, water, ammonia and nitrogen gave rise to a complicated mixture of amino acids.[20] These conditions are thought to correspond to the situation on earth at the time when the first organic matter appeared and it was the event that conceivably could have started the evolution of life. The presence of amino acids is the prerequisite for formation of peptides which by organized folding could give rise to the simplest form of replicating life, i.e. the viruses. Glycine, alanine, norvaline, serine, aspartic acid, and other unidentified products can be obtained in a yield of *ca.* 1 per cent. It is also demonstrated that carbon vapour (C_1 or C_2) as only carbon source reacts with ammonia at —196°C and low pressure to give a mixture of simple amino acids, a reaction that may play a role in the interstellar formation of amino acids, detected in meteorites and lunar samples[21] (see section 8.1 for prebiotic formation of pyrimidines etc).

The experimental conditions have been varied with respect to the composition of the prebiotic brew, temperature, pressure, and irradiation. Addition of hydrogen sulphide produced amino acids containing sulphur. Equations (3)–(6) describe a reasonable route to methionine.[22] The individual reactions are known to occur.

$$CH_4 + H_2O \xrightarrow{\text{Discharge}} CH_2 = CHCHO \tag{3}$$

$$CH_4 + H_2S \xrightarrow[h\nu]{\text{Discharge}} CH_3SH \tag{4}$$

$$CH_3SH + CH_2 = CHCHO \xrightarrow{\text{Addition}} CH_3SCH_2CH_2CHO \tag{5}$$

$$CH_3SCH_2CH_2CHO + HCN + NH_3 \xrightarrow{\text{Hydrolysis}} \underset{\text{Methionine}}{CH_3SCH_2CH_2\underset{\displaystyle NH_2}{\underset{|}{C}}HOOH} \tag{6}$$

It ought to be noted that amino acids can be formed in a low yield in a still more primitive system containing water, nitrogen and carbonate as principal components. By the action of light and transition metal catalysts water is split into hydrogen and oxygen and nitrogen is reduced to ammonia.

6.4 Reactions of amino acids promoted by pyridoxal phosphate

Glutamic acid and glutamine occupy a central position in group transfer metabolism for amino acids. Glutamine is the carrier of ammonia in an unreactive form, and it is released from the amido function by enzymatic hydrolysis. Glutamic acid serves as donor of the amino group in transamination. The α-amino group of glutamic acid is introduced by reductive amination of α-keto glutaric acid via the imine, a reversible reaction catalysed by glutamate dehydrogenase which uses NADH or NADPH as reductant (7). It was shown by tracer experiments that the prochiral H_S of NADH was transferred stereospecifically.

$$NH_3 + HOOCCH_2CH_2COCOOH \underset{}{\overset{-H_2O}{\rightleftharpoons}} HOOCCH_2CH_2C(=NH)COOH \underset{NAD^{\oplus}}{\overset{NADH}{\rightleftharpoons}} HOOCCH_2CH_2CH(NH_2)COOH \quad (7)$$

L-Glutamic acid

This represents one of the major pathways to amino acids directly from ammonia and α-keto acids. The other is transamination with pyridoxal phosphate, PLP, as a coenzyme.[23] PLP is anchored as a Schiff base at the active site by a lysine residue as shown by sodium borohydride reduction of the enzyme followed by hydrolysis (8). Glutamic acid displaces the lysine residue forming a new aldimine

Lys-Enz
|
N
‖
CH
PO, OH, N, CH_3

$\xrightarrow[2.\ H^{\oplus}]{1.\ NaBH_4}$

$HN-(CH_2)_4CHCOOH$, CH_2, NH_2
PO, OH, N, CH_3 (8)

PLP-Enz-aldimine
(Schiff base)

in which H^{α} is doubly activated and can be revoved by a strategically located basic function on the enzyme (Fig. 6). The aldimine rearranges to the ketimine which on hydrolysis gives α-ketoglutarate and pyridoxamine phosphate. This constitutes the first half of the transamination, the mechanism of which originally was proposed by Snell and Braunshtein. The $H^{\oplus}$ transfer takes places intramolecularly to some extent. The second half is a reversal of the first part with another keto acid generating a new amino acid and pyridoxal phosphate etc. The two steps can be summarized as in (9). The mechanism of action of PLP dependent racemization is readily envisioned according to the scheme above.

$$\text{Amino acid}^1 + \text{Keto acid}^2 \rightleftharpoons \text{Keto acid}^1 + \text{Amino acid}^2 \quad (9)$$

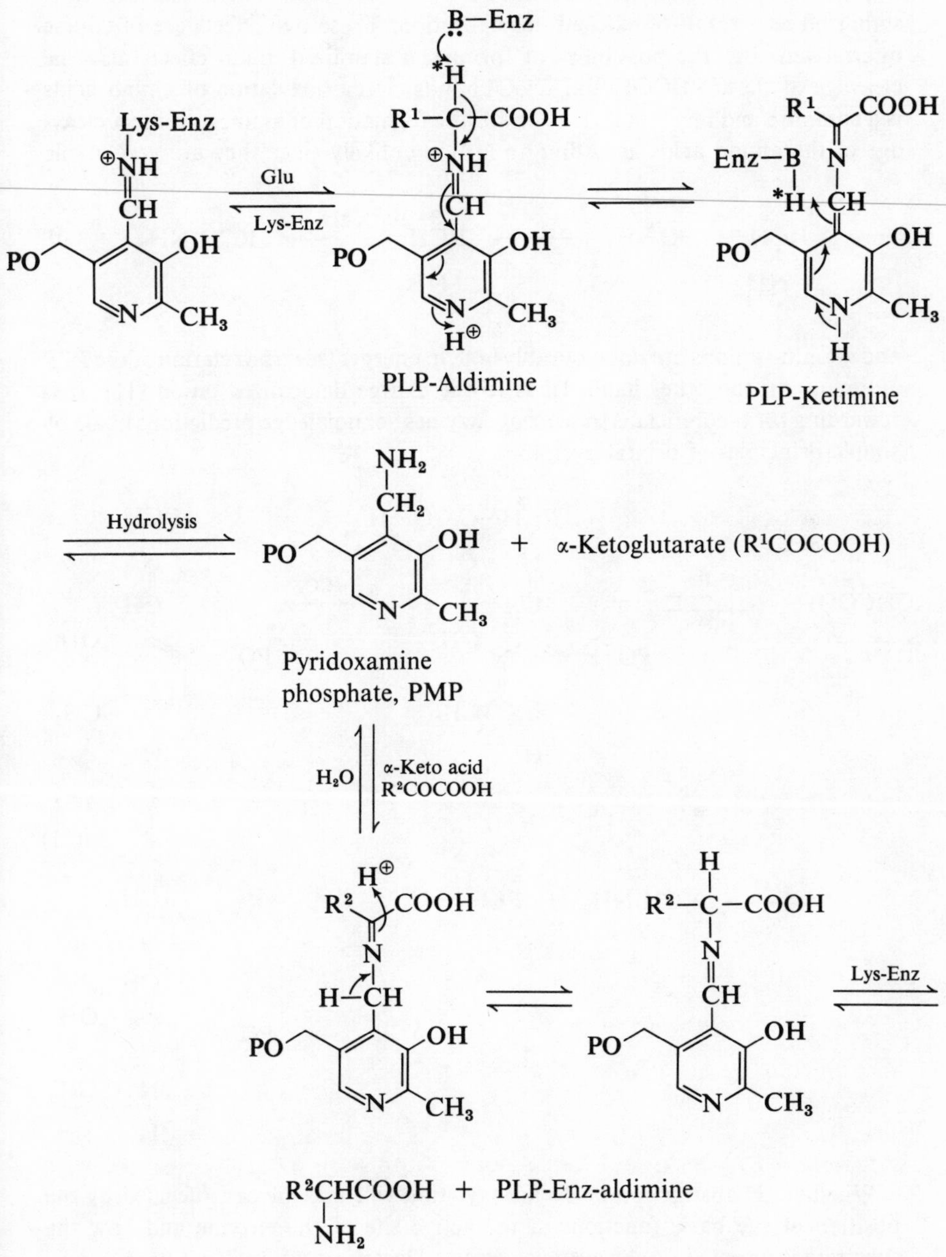

Lys-Enz

Glu

Lys-Enz

B—Enz

R^1—C—COOH

PLP-Aldimine

R^1 COOH

Enz—B

PLP-Ketimine

Hydrolysis

Pyridoxamine phosphate, PMP + α-Ketoglutarate ($R^1COCOOH$)

H_2O

α-Keto acid $R^2COCOOH$

R^2 COOH

R^2—C—COOH

Lys-Enz

$R^2CHCOOH$ (NH_2) + PLP-Enz-aldimine

Fig. 6 Mechanism of transamination with glutamic acid as donor of the amino group

The purpose of the coenzyme is thus to activate the C^α–H bond to such a degree that the proton can be removed by a weak base. The anion formed is stabilized as a result of efficient delocalization. These two effects are of course interrelated. But the possibility of forming a stabilized anion effects also the cleavage of the C^α–COOH and C^β–C^α bonds. Decarboxylation of amino acids is a common and important reaction for the formation of amines. A direct cleavage of the amino acids according to (10) is unlikely since they are very stable

$$\underset{NH_2}{RCH}-\underset{O}{\overset{\|}{C}}-O-H \longrightarrow \underset{NH_2}{R\overset{\ominus}{C}H} \xrightarrow{H^\oplus} RCH_2NH_2 \qquad (10)$$

and α-amino anions are unfavourably high in energy. Decarboxylation of the PLP aldimine, on the other hand, leads to the charge delocalized anion (11). It is rewarding for theoreticians to see that enzymes acknowledge predictions made by simple principles of orbital overlap.

$$\underset{NH_2}{RCHCOOH} \underset{H_2O}{\overset{PLP}{\rightleftharpoons}} RCH(-N{=}CH-\text{Py})-C(=O)-O-H \xrightarrow{-CO_2} R\overset{\ominus}{C}H-N{=}CH-\text{Py} \leftrightarrow RCH{=}N-CH{=}\text{Py} \qquad (11)$$

(Py = pyridinium ring bearing PO–CH₂, OH, CH₃ and N–H substituents: PO, OH, $\overset{\oplus}{N}$–H, CH_3)

$$\xrightarrow{H_2O} RCH_2NH_2 + PLP$$

Whether H^α abstraction or decarboxylation will take place is decided by the position of the basic functions at the active site of the enzyme and how the substrate is oriented on the enzyme surface. Decarboxylation is not preceded by any H^α exchange. The product from the decarboxylation of tyrosine contains one deuterium when the decarboxylation is run in D_2O and it was found that it proceeds with retention of configuration.[24]

Generation of the stabilized C^α-anion of PLP-glycine by cleavage of a C^β–C^α

$$\text{CH}_2\text{-CHCOOH(OH)-N=CH-PLP} \rightleftharpoons \text{CH}_2\text{O} + {}^{\ominus}\text{CHCOOH-N=CH-PLP} \quad (12)$$

bond occurs in the retro aldol cleavage of serine and threonine catalysed by serine hydroxymethylase (12). Formaldehyde is not set free but scavenged by tetrahydrofolate for later usage. The presence of tetrahydrofolate is found to increase the rate of cleavage as a result of formaldehyde being removed from the equilibrium. Experiments with chirally labelled glycine[25] show that the enzyme exclusively abstracts the H_S and that substitution takes place with retention of configuration at C^2 of glycine, e.g. reversal of (12). The mechanism depicted in (12) is analogous to that for decarboxylation of amino acids. No C^{α}–H exchange occurs with the solvent implying that dehydroalanine is not on the pathway and tetrahydrofolate does not directly participate in the displacement as formulated in (13). Consistent with the transient formation of formaldehyde are results with serine, chirally labelled with ^3H at C^3. Partial loss of chirality was observed in the CHT fragment of N^5,N^{10}-methylene tetrahydrofolate in accordance with the intermediary formation of formaldehyde, free long enough to rotate.[26] The cleavage of threonine is independent of tetrahydrofolate.[27] N^5,N^{10}-methylene tetrahydrofolate is one of the two most important donors of the

$$\text{HO-CH}_2\text{-C(H)COOH-N=CH-PLP} \xrightleftharpoons{\text{THF}} \text{CH}_2\text{=CCOOH-N=CH-PLP} + N^5\text{H}, N^{10}\text{H (THF)} \rightleftharpoons \quad (13)$$

$$\text{HOOCC(H)(N=CH-PLP)-CH}_2\text{-}N^5 \rightleftharpoons \text{H}_2\text{C=}N^{5\oplus} \; :N^{10}\text{H} \rightleftharpoons \text{H}_2\text{C-N (ring)} + \text{H}^{\oplus}$$

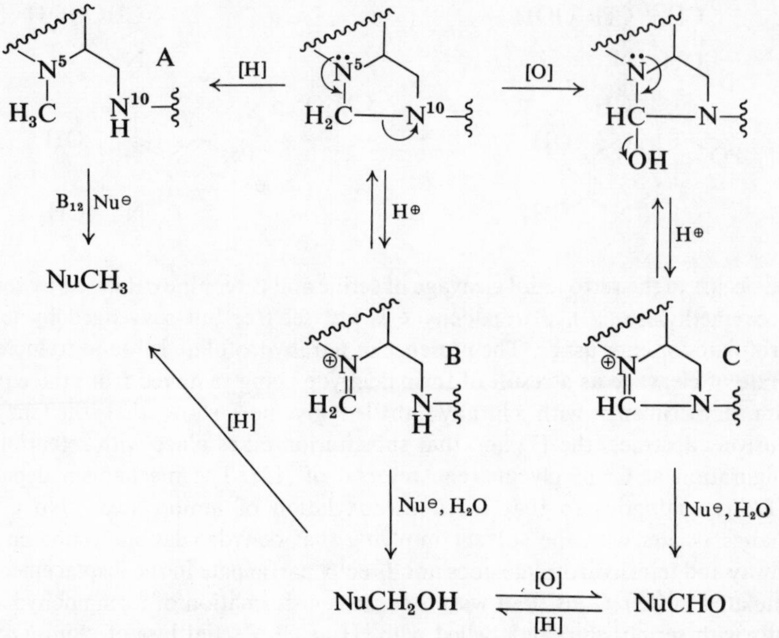

$$\mathrm{H_3C{-}N^5\cdots N^{10}H}\ (\mathrm{A}) \xleftarrow{[H]} \mathrm{H_2C{-}N^{10}} \xrightarrow{[O]} \mathrm{HC(OH){-}N}$$

$\mathrm{A} \xrightarrow{B_{12},\ Nu^\ominus} \mathrm{NuCH_3}$; $\mathrm{H^\oplus}$ equilibria to B ($\oplus N{=}CH_2$, NH) and C ($\oplus N{=}CH{-}N$); $\mathrm{B} \xrightarrow{[H]} \mathrm{A}$; $\mathrm{B} \xrightarrow{Nu^\ominus,\ H_2O} \mathrm{NuCH_2OH}$; $\mathrm{C} \xrightarrow{Nu^\ominus,\ H_2O} \mathrm{NuCHO}$

$$\mathrm{NuCH_2OH} \underset{[H]}{\overset{[O]}{\rightleftharpoons}} \mathrm{NuCHO}$$

Fig. 7 Structures of active C_1 transferring tetrahydrofolate at different oxidation levels

one carbon fragment; the other one is *S*-adenosylmethionine which actually receives its methyl from N^5,N^{10}-methylene tetrahydrofolate. The C_1unit can be used directly at the formaldehyde oxidation level or be oxidized or reduced to active formic acid or methyl, respectively. The active species in formylation is the amidinium ion C and in hydroxymethylation the immonium ion B (Fig. 7). It is more difficult to reconcile the methylating ability of the N^5–CH_3 derivative A with the known sluggishness of trialkylamines as alkylating agents in regular organic reactions. Hence, it seems more plausible that B is the active species in combination with a reductant. It ought to be mentioned that the methylation seems to be mediated by vitamin B_{12} and it is suggested that Co–CH_3 bonding is involved.[28] The reaction pattern of organocobalt compounds is not well understood and does not follow the traditional mechanistic principles organic chemists are used to handle. An example of an unusual skeletal rearrangement is the methylmalonyl-CoA–succinyl-CoA rearrangement which also is mediated by vitamin B_{12} (14). There are indications that the rearrangement is radical in nature[29] but the mechanism of $-CH_2-{}^*H$ homolysis and $^*H$ transfer is not clear. The action of vitamin B_{12} will be discussed in greater detail in section 8.4.

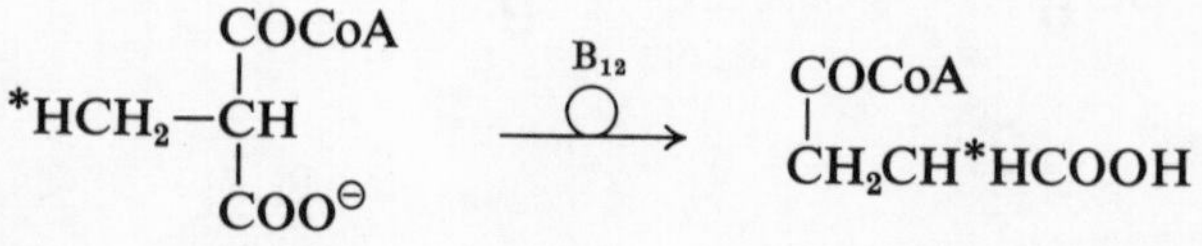

$$\mathrm{{}^*HCH_2{-}CH(COCoA)(COO^\ominus)} \xrightarrow{B_{12}} \mathrm{CH_2(COCoA)CH^*HCOOH} \qquad (14)$$

The first part of the mechanism in (15) represents an α,β-elimination which also is observed for serine and other amino acids with good leaving groups at C^3, e.g. 3-chloroalanine and cysteine. The intermediate α-iminoacrylate gives either pyruvic acid on hydrolysis (15a), or if it finds a good nucleophile in the vicinity of the active site of the enzyme, it undergoes β-substitution (15b). We have already encountered such cases in the biosynthesis of tryptophan (section 3.2) and cyanoalanine $Nu^{\ominus} = CN^{\ominus}$

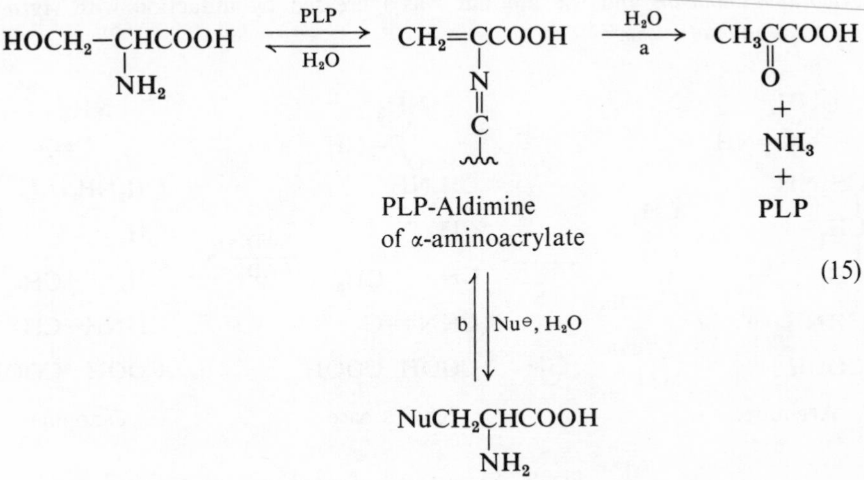

$$HOCH_2-CH(NH_2)COOH \underset{H_2O}{\overset{PLP}{\rightleftharpoons}} CH_2{=}C(COOH)-N{=}C\sim \xrightarrow[a]{H_2O} CH_3C(=O)COOH + NH_3 + PLP \tag{15}$$

PLP-Aldimine
of α-aminoacrylate

$$\downarrow b \quad Nu^{\ominus}, H_2O$$

$$NuCH_2CH(NH_2)COOH$$

PLP also activates reactions at γ-positions as exemplified by its role as cofactor for cleavage of cystathionine to cysteine and α-ketobutyric acid, the last step in the biosynthesis of cysteine (Fig. 8). In support for the depicted β,γ-elimination mechanism which presupposes β-ionization, are experiments carried out in D_2O showing incorporation of one $3S$-^2H in α-ketobutyrate.[30]

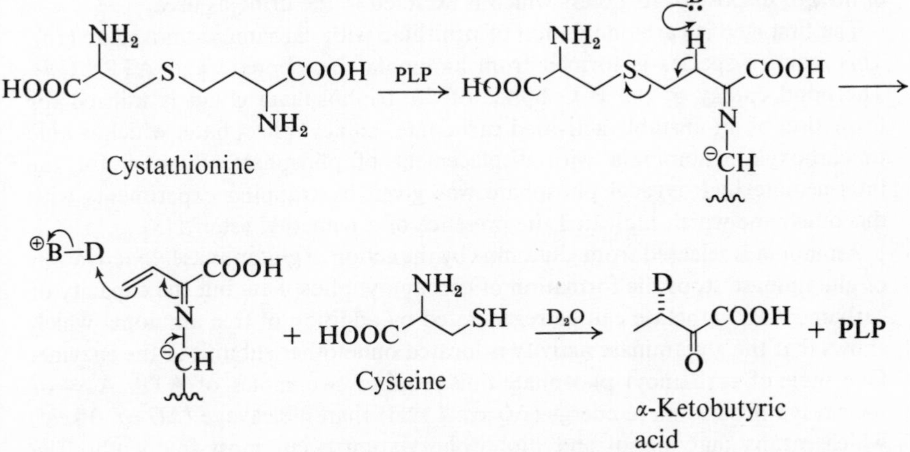

Cystathionine $\xrightarrow{PLP}$ HOOC–CH(NH$_2$)–CH$_2$–S–CH$_2$–CH(H···B:)–C(COOH)=N–$^{\ominus}$CH$\sim$ ⟶

$^{\oplus}$B–D + CH$_2$=CH–C(COOH)=N–$^{\ominus}$CH$\sim$ + HOOC–CH(NH$_2$)–CH$_2$SH (Cysteine) $\xrightarrow{D_2O}$ CH$_3$–CHD–C(=O)–COOH (α-Ketobutyric acid) + PLP

Fig. 8 Cleavage of cystathionine

6.5 The guanidino function

The guanidino function appears in several compounds from widely different sources not the least in marine metabolites, e.g. in octopine from *Octopus* spp. which, for structural and mechanistic reasons, is supposed to be formed from arginine and pyruvic acid[31] (Fig. 9). It is known that pyruvic acid acts in the role of cofactor in transaminations. Reduction of the Schiff base with NADPH completes the biosynthesis. Octopine was also detected in tissue cultures of *Nicotiana tabacum* and the amount was increased by induction with *Agrobacterium tumefaciens.*[32]

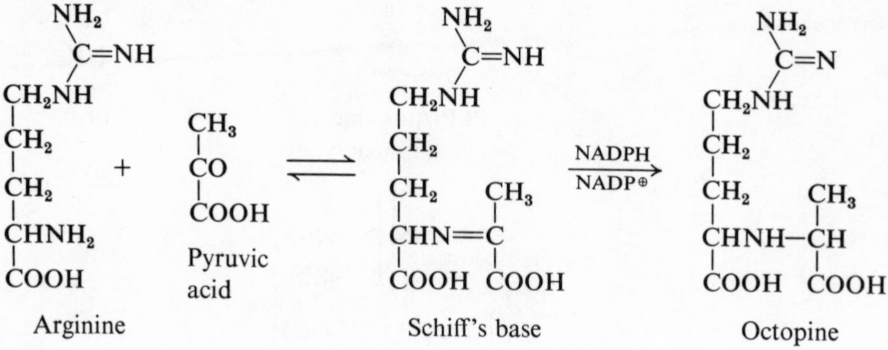

Arginine: NH_2–C(=NH)–NHCH_2–CH_2–CH_2–$CHNH_2$–COOH

\+ Pyruvic acid: CH_3–CO–COOH

⇌ Schiff's base: NH_2–C(=NH)–NHCH_2–CH_2–CH_2–CH(COOH)–N=C(CH_3)–COOH

NADPH → NADP⊕

Octopine: NH_2–C(=N)–NHCH_2–CH_2–CH_2–CH(COOH)–NH–CH(CH_3)–COOH

Fig. 9 Biosynthesis of octopine

The formation of the guanidino function is mechanistically interesting as an illustration of the action of ATP and utilization of its energy. This is discussed in conjunction with a presentation of the urea cycle uncovered by Krebs and Henseleit in 1932 (Fig. 10). The urea cycle solves for the organism the problem of how to dispose of its excess which is excreted in the urine as urea.

The first step is carbamoylation of ornithine with carbamoyl phosphate (16). This reactive species is formed from ammonia, bicarbonate and ATP[33] (17). The bond energy of the P–O bonds of the triphosphate chain is utilized for formation of an unstable activated carbonate, an acyl phosphate, which is able to carboxylate ammonia with displacement of phosphate. Support for the intermediate hydroxyacyl phosphate was given by trapping experiments with diazomethane which indicated the presence of a trimethyl ester[34](18).

Ammonia is released from glutamine by the action of glutaminase. Inactivation of glutaminase stops the formation of carbamoyl phosphate but the capacity of carbamoyl-P synthetase can be regenerated by addition of free ammonia which shows that the glutaminase activity is located on another subunit of the enzyme. One mole of carbamoyl phosphate thus requires two moles of ATP. A γ- or β-cleavage releases more energy (ΔG *ca.* 7 kcal) than α-cleavage (ΔG *ca.* 3 kcal) which means that mono- and diphosphorylation occur most frequently. The acyl phosphate has two reactive sites. If the nucleophile, as in this case, attacks

Ornithine

Citrulline

Argininosuccinic acid

Arginine

Fumaric acid

(16)

Hydrolysis 4

Ornithine

$+ CO(NH_2)_2$

Urea

the polarized carbon, acylation occurs. The nucleophile could also attack phosphorus and the cleavage on the other side of the oxygen leads to phosphorylation. These mixed anhydrides are both good acylating and phosphorylating agents. The mode of action depends on how the enzyme orients the reactants.

The second step in the urea cycle is formation of argininosuccinic acid from citrulline and aspartic acid. This reaction requires another mole of ATP for activation of the carbamoyl group of citrulline. In the third step the synthesis of the guanidine group is completed by elimination of fumaric acid. The last step is a hydrolysis to urea and ornithine by the action of arginase, an enzyme requiring $Mn^{2\oplus}$ as cofactor.

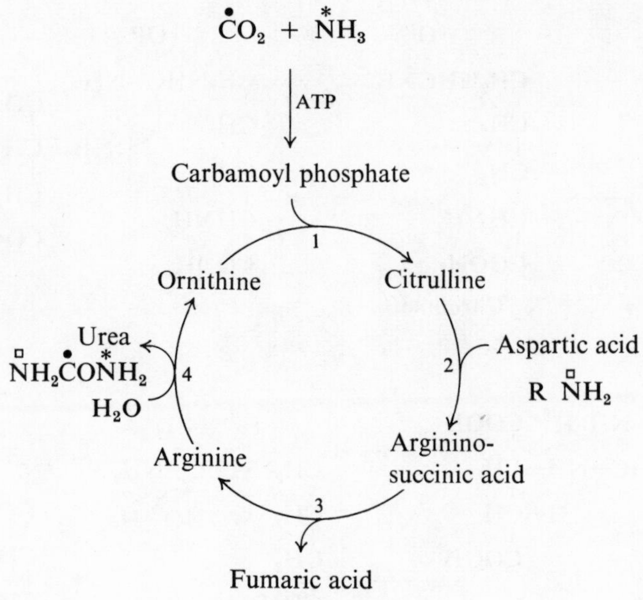

$\overset{\bullet}{C}O_2 + \overset{*}{N}H_3$

ATP

Carbamoyl phosphate

1

Ornithine

Citrulline

Urea

$\overset{\square}{N}H_2\overset{\bullet}{C}O\overset{*}{N}H_2$

4

H_2O

Aspartic acid

2

$R\ \overset{\square}{N}H_2$

Arginine

Arginino-succinic acid

3

Fumaric acid

Fig. 10 The urea cycle

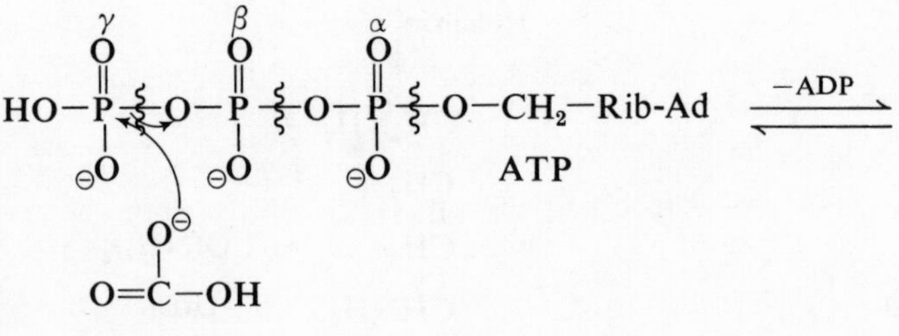

$$HO-\overset{\gamma}{\underset{O^{\ominus}}{\overset{O}{\overset{\|}{P}}}}\sim O-\overset{\beta}{\underset{O^{\ominus}}{\overset{O}{\overset{\|}{P}}}}\sim O-\overset{\alpha}{\underset{O^{\ominus}}{\overset{O}{\overset{\|}{P}}}}\sim O-CH_2-\text{Rib-Ad} \xrightleftharpoons{-ADP}$$

ATP

$$O=\overset{O^{\ominus}}{\overset{|}{C}}-OH \qquad (17)$$

$$HO-\overset{NH_3}{\underset{O}{\underset{\|}{C}}}-O-\underset{O^{\ominus}}{\overset{O}{\overset{\|}{P}}}-OH \xrightarrow{NH_3} H_2N-\overset{O}{\overset{\|}{C}}-O^{\ominus} \xrightleftharpoons{ATP} H_2N-\overset{O}{\overset{\|}{C}}-OP + ADP$$

Acyl phosphate

Carbamoyl phosphate

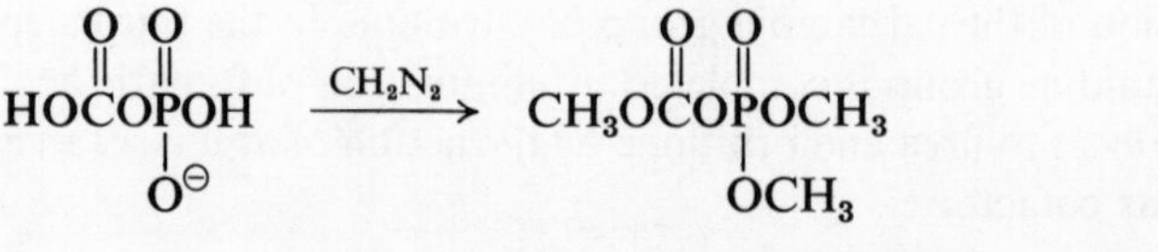

$$HO\overset{O}{\overset{\|}{C}}O\underset{O^{\ominus}}{\underset{|}{\overset{O}{\overset{\|}{P}}}}OH \xrightarrow{CH_2N_2} CH_3O\overset{O}{\overset{\|}{C}}O\underset{OCH_3}{\underset{|}{\overset{O}{\overset{\|}{P}}}}OCH_3 \qquad (18)$$

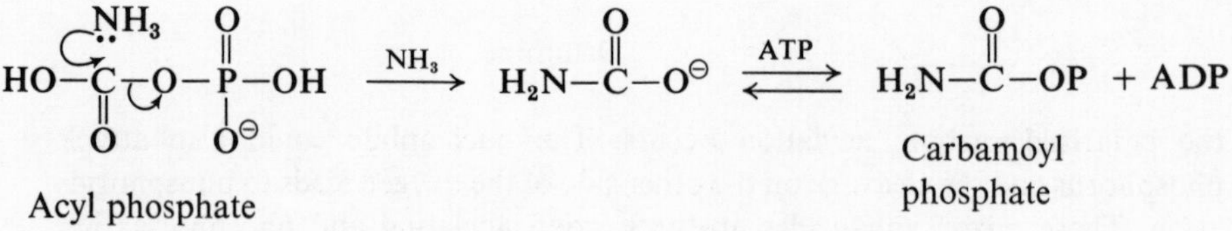

6.6 Secondary products from serine and cysteine

The biosynthetic versatility of serine was pointed out earlier in connection with the reactions of pyridoxal phosphate. Serine is located at a branching point in the secondary metabolism of amino acids. Michael addition of various N-, S- and C-nucleophiles to the derived PLP aldimine leads to β-substituted alanines[35] (Fig. 11). The phenylalanine derivative is especially interesting because from structural reasons it may be concluded that it is derived from shikimic acid as other phenylalanines. Tracer experiments indicate that the aromatic nucleus is derived from orsellinic acid and the side chain from serine,[36] *cf.* the biosynthesis of thyroxine (section 4.13).

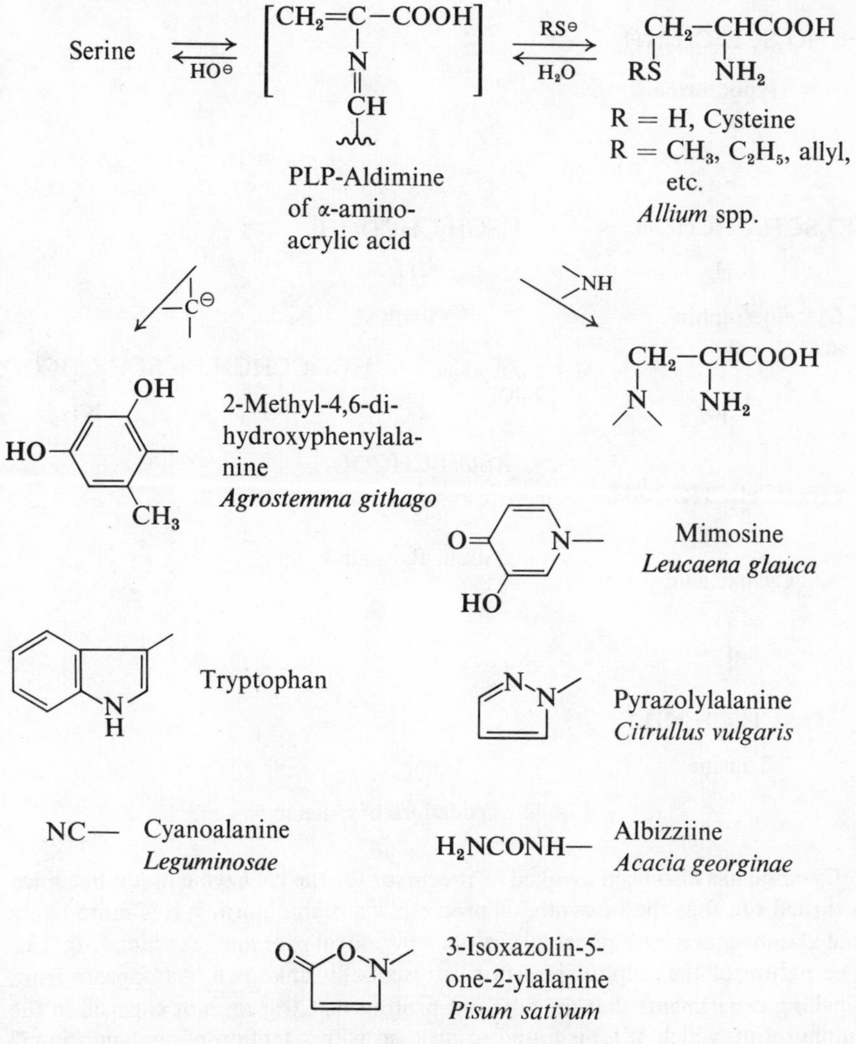

Serine ⇌ ($HO^{\ominus}$) [$CH_2{=}C{-}COOH$, N=CH] ⇌ ($RS^{\ominus}$ / H_2O) $CH_2{-}CHCOOH$ (RS, NH_2)

PLP-Aldimine of α-amino-acrylic acid

R = H, Cysteine
R = CH_3, C_2H_5, allyl, etc.
Allium spp.

$-C^{\ominus}$ >NH → $CH_2{-}CHCOOH$ (N<, NH_2)

OH, HO, CH_3 — 2-Methyl-4,6-dihydroxyphenylalanine
Agrostemma githago

O=, N—, HO — Mimosine
Leucaena glauca

N, H — Tryptophan

N, N — Pyrazolylalanine
Citrullus vulgaris

NC— Cyanoalanine
Leguminosae

H_2NCONH— Albizziine
Acacia georginae

O, O, N — 3-Isoxazolin-5-one-2-ylalanine
Pisum sativum

Fig. 11 Secondary derivatives of serine

It is actually *O*-acetylserine which is the key intermediate in the synthesis of cysteine and other β-substituted alanine derivatives. Acetylation makes the hydroxyl group easier to eliminate. The corresponding *O*-phosphoserine does not seem to have any metabolic function. As other thiols, cysteine is sensitive to oxidation. It dimerizes easily to the disulphide cystine, a reaction of great consequences for the folding of peptide chains. The thiol can also be oxidized to sulphinic and sulphonic acids[37] (Fig. 12). Decarboxylation of the sulphinic acid gives taurine, which is widespread in nature. Oxidation of alkylated cysteines gives sulphoxides, e.g. alliin, common in the *Liliaceae*, *Cruciferae* and *Mimosaceae* families. *S-trans*-propenyl-L-cysteine functions as the progenitor of the characteristic lachrymatory principle in onions.

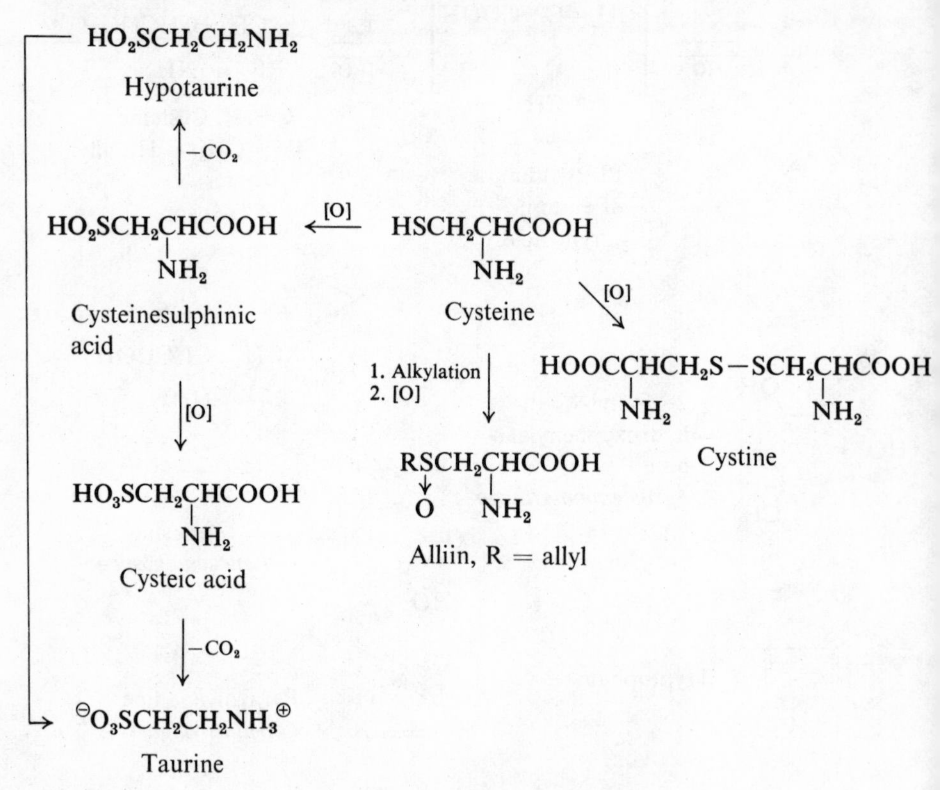

Fig. 12 Oxidations of cysteine

Cysteine has also been invoked as precursor for the coenzyme biotin but since it turned out that the biosynthesis proceeds via dethiobiotin,[38] it is more likely that alanine reacts with pimeloyl CoA in a pyridoxal promoted reaction (Fig. 13). The nature of the sulphur insertion is essentially unknown. As appears from labelling experiments that the adjacent protons at $C^{2,3,5}$ are not engaged in the sulphuration which at C^4 is found to proceed with retention of configuration.[39] Evidently sulphur is not introduced by a Michael addition.

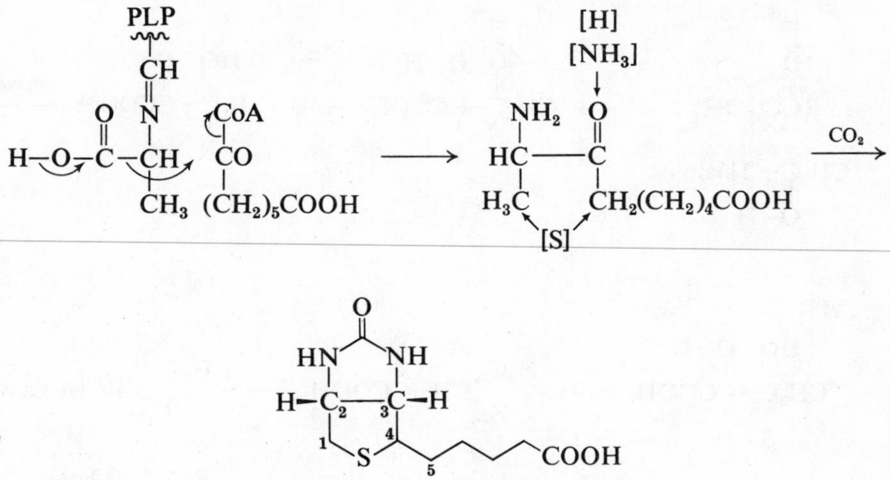

Fig. 13 Biosynthesis of biotin

6.7 Secondary products from valine, leucine and isoleucine

The biosynthesis of these three essential amino acids is similar. It embraces some general mechanistic principles and is therefore discussed here.

We have first the thiamine catalysed decarboxylation of pyruvic acid leading to a pole reversal of acetaldehyde (section 2.3 (14)) which condenses with an α-keto acid (Fig. 14). Pinacolone type rearrangement and NADPH reduction lead to the β-branched α-keto acid which either is directly aminated to valine and isoleucine or undergoes chain lengthening to leucine ($R = CH_3$). This sequence, which has several analogies, starts with a condensation with acetyl CoA to form an α-substituted malic acid, *cf.* the biosynthesis of mevalonic acid. Elimination and readdition of water gives β-substituted malic acid, *cf.* the citric acid cycle. Oxidation of the hydroxy group and decarboxylation of the β-keto acid formed leads, eventually, to the homologous α-keto acid and, after transamination with glutamic acid, to leucine.

Pantoic acid, one of the building blocks in coenzyme A, is synthesized from the intermediary α-ketoisovaleric acid by condensation with N^5,N^{10}-methylene-tetrafolic acid followed by reduction (Fig. 14). Erythroskyrin and tenuazonic acid are formed by mixed biogenesis from valine and isoleucine, respectively, and a polyketide (Fig. 15).

The pyrrolizidine alkaloids are esters of aminoalcohols, necines, and branched carboxylic acids, necic acids, e.g. echimidinic acid and senecic acid, originally thought to be derived from acetate. Tracer experiments have shown, however, that they originate from simple amino acids. Echimidinic [40] acid originates from valine, senecioic acid[41] from leucine or mevalonic acid and senecic acid [42] by decarboxylation and dimerization of two isoleucine units (Fig. 16).

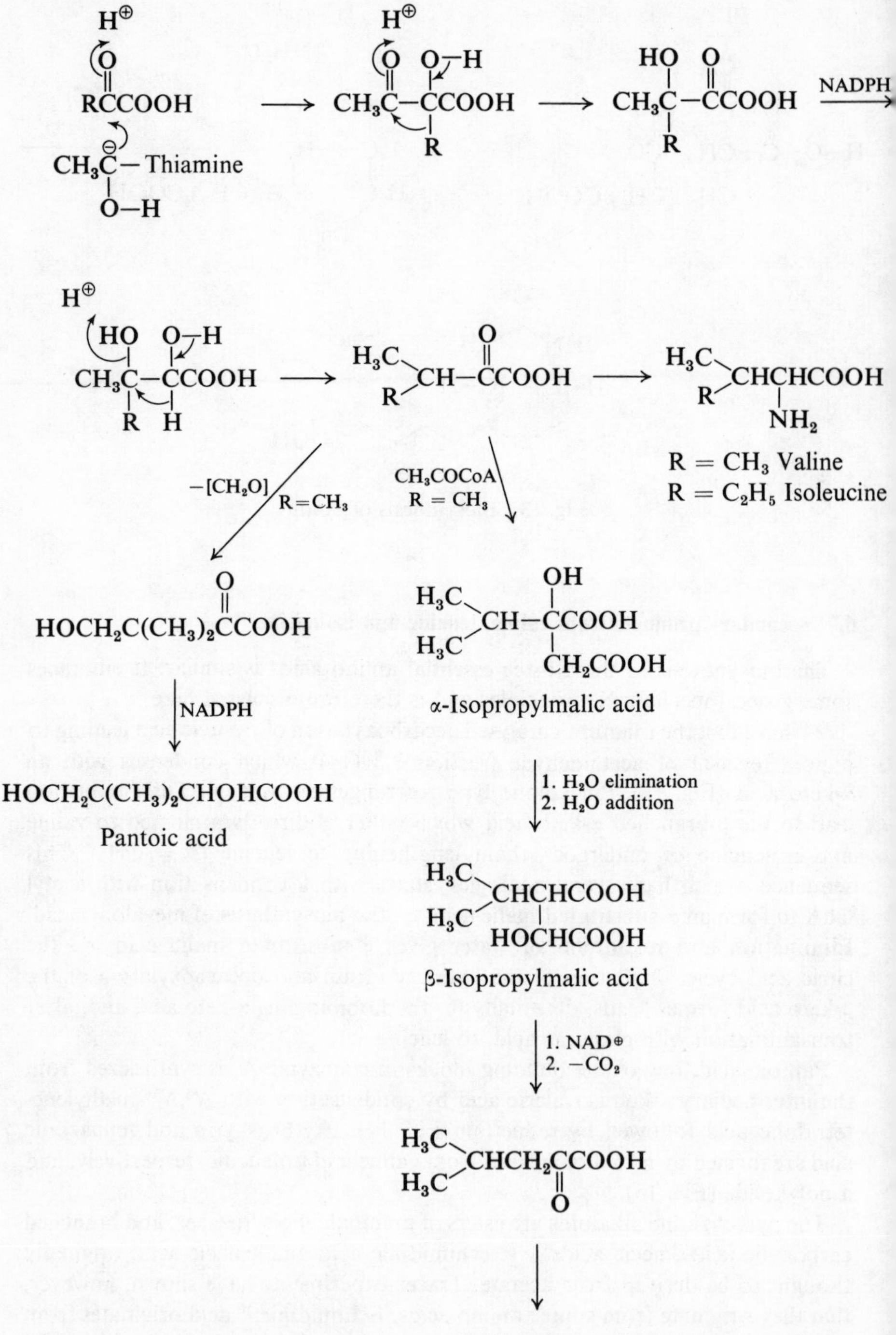

$H^{\oplus}$ O ‖ RCCOOH + $CH_3\overset{\ominus}{C}$—Thiamine (O—H) ⟶ $H^{\oplus}$ $CH_3C(=O)—C(OH)(R)COOH$ ⟶ HO O $CH_3C—CCOOH$ (R) $\xrightarrow{NADPH}$

$H^{\oplus}$ HO O—H $CH_3C—CCOOH$ (R H) ⟶ H_3C, R >CH—C(=O)COOH ⟶ H_3C, R >CHCHCOOH (NH_2)

R = CH_3 Valine
R = C_2H_5 Isoleucine

−[CH_2O], R=CH_3 ⟶ O ‖ $HOCH_2C(CH_3)_2CCOOH$ $\xrightarrow{NADPH}$ $HOCH_2C(CH_3)_2CHOHCOOH$

Pantoic acid

CH_3COCoA, R = CH_3 ⟶ H_3C, H_3C >CH—C(OH)(COOH)CH_2COOH

α-Isopropylmalic acid

1. H_2O elimination
2. H_2O addition

H_3C, H_3C >CHCHCOOH, HOCHCOOH

β-Isopropylmalic acid

1. $NAD^{\oplus}$
2. $-CO_2$

H_3C, H_3C >$CHCH_2C(=O)COOH$

Leucine

Fig. 14 Biosynthesis of valine, isoleucine, leucine and pantoic acid

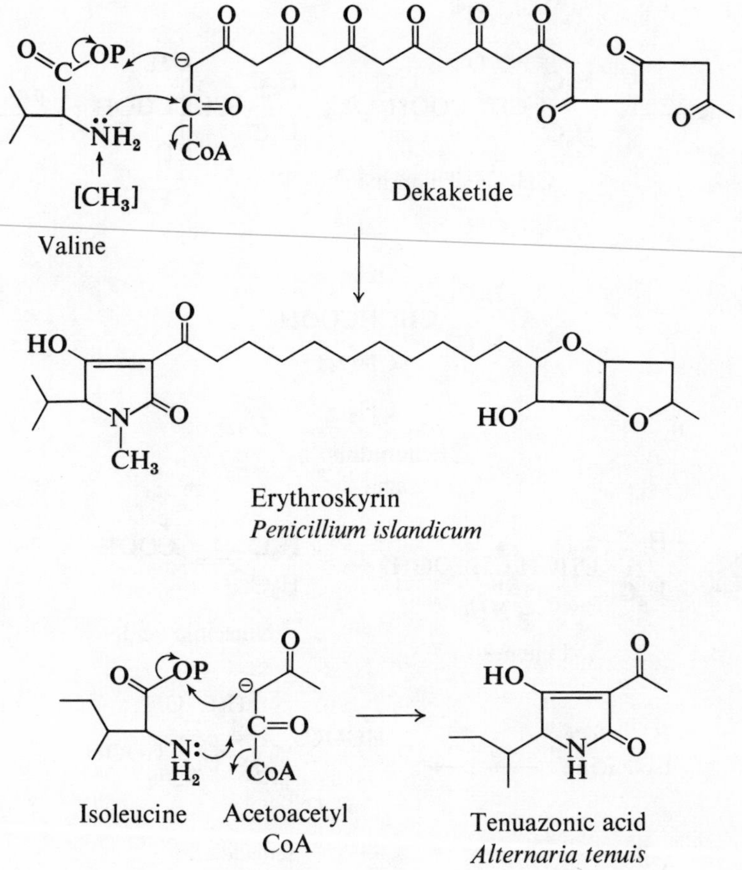

Valine

Dekaketide

Erythroskyrin
Penicillium islandicum

Isoleucine

Acetoacetyl CoA

Tenuazonic acid
Alternaria tenuis

Fig. 15 Secondary products from valine and isoleucine

6.8 Cyanogenic glycosides and glucosinolates

The cyanogenic glycosides are the principal precursors for hydrocyanic acids in plants. Strangely enough some plants are able to metabolize the highly toxic cyanide ions known to effectively block the action of ferroporphyrins by stable ligand formation. In higher plants cyanide reacts with cysteine and serine (section 6.5), forming cyanoalanine which can be further hydrolysed to asparagine. This reaction is believed to serve as a detoxication mechanism for cyanide ions. It is well established that the cyanogenic glycosides are derived from amino acids. Experiments with labelled precursors show that the carboxyl group is lost as carbon dioxide, the cyano group is formed from C^2 and the amino group and C^3 becomes hydroxylated (19). There is evidence for the production of isobutyraldoxime and isobutyronitrile from valine in flax seedlings as well as for conversion of these intermediates to glycosides that suggests a Beckmann type

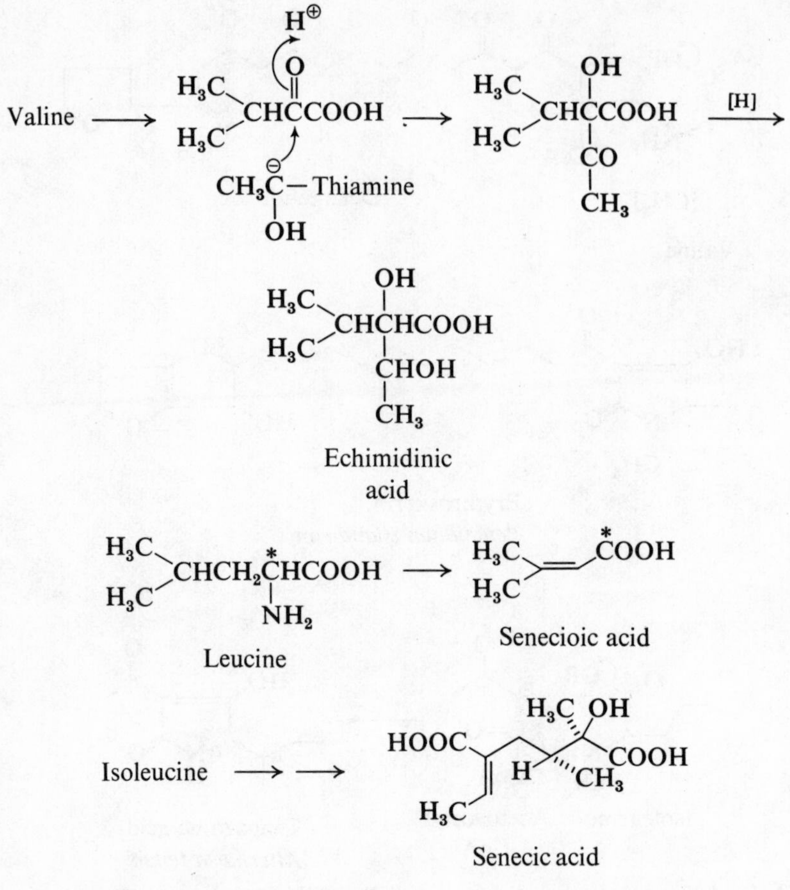

$$\text{Valine} \longrightarrow \underset{\substack{\uparrow \\ \overset{\ominus}{CH_3C}-\text{Thiamine} \\ | \\ OH}}{(H_3C)_2CH\overset{\overset{H^{\oplus}}{\curvearrowright O}}{\overset{\|}{C}}COOH} \longrightarrow (H_3C)_2CH\overset{OH}{\overset{|}{C}}(COCH_3)COOH \xrightarrow{[H]}$$

$$(H_3C)_2CH\overset{OH}{\overset{|}{C}}HCOOH(CHOHCH_3)$$

Echimidinic acid

$$(H_3C)_2CHCH_2\overset{*}{C}H(NH_2)COOH \longrightarrow (H_3C)_2C{=}CH\overset{*}{C}OOH$$

Leucine — Senecioic acid

Isoleucine ⟶ ⟶ Senecic acid ($HOOC$, H_3C, H_3C OH, $COOH$, H, CH_3)

Fig. 16 Pathway for echimidinic, senecioic and senecic acids

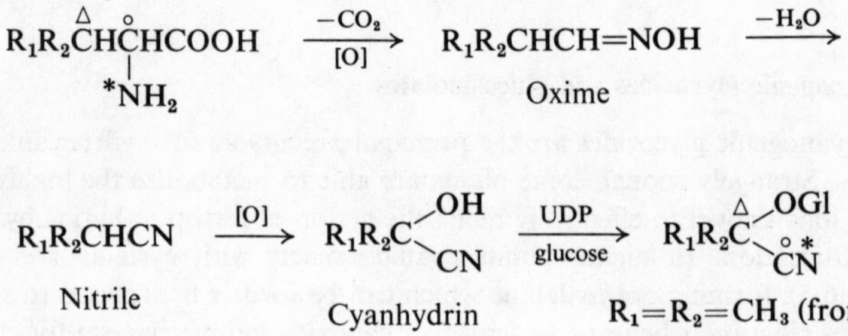

$$R_1R_2\overset{\triangle}{C}H\overset{\circ}{C}HCOOH(^{*}NH_2) \xrightarrow[\text{[O]}]{-CO_2} R_1R_2CHCH{=}NOH \xrightarrow{-H_2O}$$

Oxime

$$R_1R_2CHCN \xrightarrow{\text{[O]}} R_1R_2C(OH)CN \xrightarrow[\text{glucose}]{\text{UDP}} R_1R_2\overset{\triangle}{C}(OGl)\overset{\circ}{C}\overset{*}{N} \qquad (19)$$

Nitrile — Cyanhydrin

$R_1{=}R_2{=}CH_3$ (from valine)
Linamarin
Linum usitatissimum
$R_1{=}H$; $R_2{=}p$-OH-phenyl (from tyrosine)
Dhurrin
Sorghum vulgare

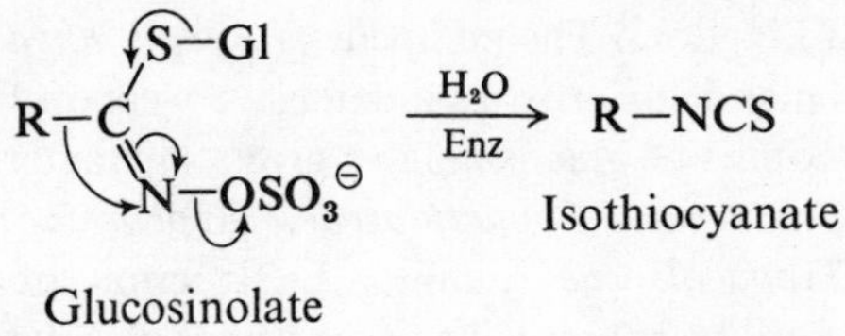

$$R-C(-S-Gl)=N-OSO_3^{\ominus} \xrightarrow[Enz]{H_2O} R-NCS \quad (20)$$

Glucosinolate — Isothiocyanate

dehydration and α-hydroxylation to cyanohydrins as the most likely pathway.[43,44] Molecular oxygen is introduced by peroxidase into aldoximes and cyanides[45] and a β-glucosyltransferase has been partially purified from *Sorghum* seedlings exhibiting activity for cyanohydrins.[46] The occurrence of cyanohydrins is not

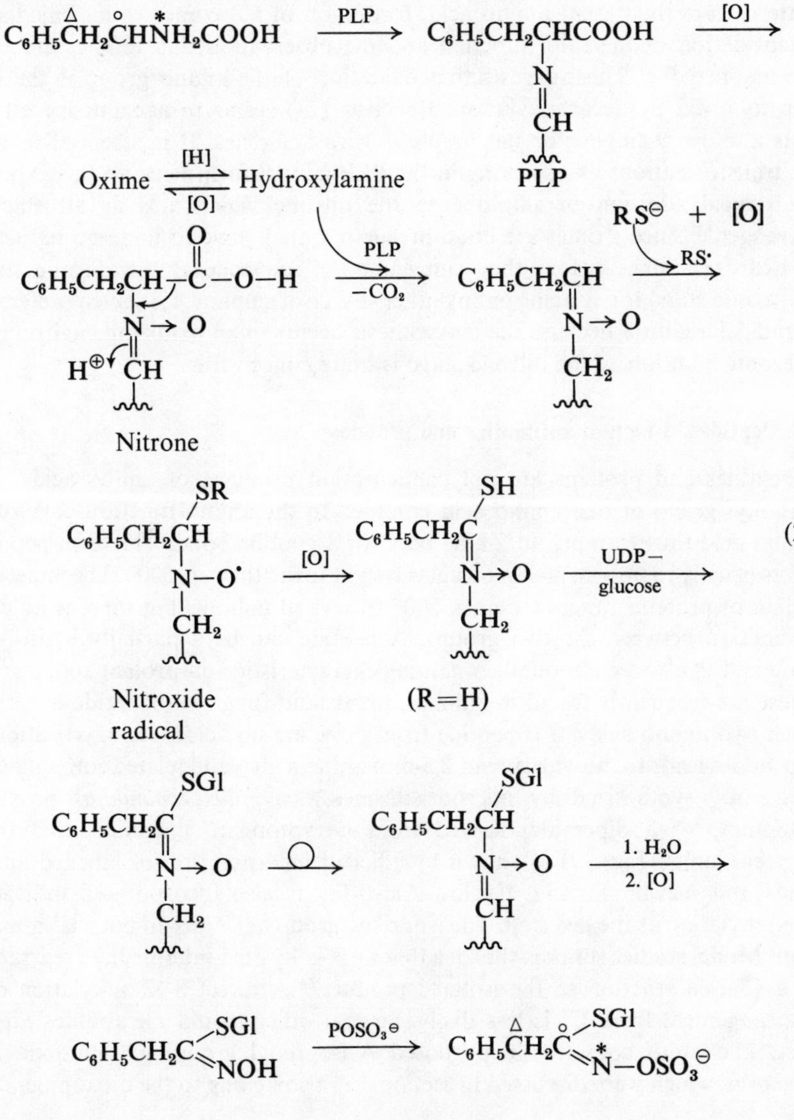

$$C_6H_5\overset{\Delta}{C}H_2\overset{\circ}{C}H\overset{*}{N}H_2COOH \xrightarrow{PLP} C_6H_5CH_2CH(COOH)-N=CH-PLP \xrightarrow{[O]}$$

$$\text{Oxime} \underset{[O]}{\overset{[H]}{\rightleftharpoons}} \text{Hydroxylamine}$$

$$C_6H_5CH_2CH(-C(=O)-O-H)-N(\rightarrow O)=CH\ (H^{\oplus}) \xrightarrow[-CO_2]{PLP} C_6H_5CH_2CH=N(\rightarrow O)-CH_2 \xrightarrow[RS^{\cdot}]{RS^{\ominus} + [O]}$$

Nitrone

$$C_6H_5CH_2CH(SR)-N(-O^{\cdot})-CH_2 \xrightarrow{[O]} C_6H_5CH_2C(SH)=N(\rightarrow O)-CH_2 \xrightarrow{UDP-glucose} \quad (21)$$

Nitroxide radical — (R=H)

$$C_6H_5CH_2C(SGl)=N(\rightarrow O)-CH_2 \xrightarrow{\curvearrowright} C_6H_5CH_2CH(SGl)-N(\rightarrow O)=CH \xrightarrow[2.\ [O]]{1.\ H_2O}$$

$$C_6H_5CH_2C(SGl)=NOH \xrightarrow{POSO_3^{\ominus}} C_6H_5\overset{\Delta}{C}H_2\overset{\circ}{C}(SGl)=\overset{*}{N}-OSO_3^{\ominus}$$

limited to the plant kingdom. The millipede *Harpaphe haydeniana* uses hydrogen cyanide, stored as mandelonitrile, as a defensive weapon.[47]

Degradation products of glucosinolates give a characteristic smell to various plants of the *Cruciferae*, *Capparidaceae*, *Euphorbiaceae*, *Phytolaccaceae*, *Resedaceae* and *Tropaeolaceae* families. Destruction of the compartmental organization of the cell by rubbing the tissue brings glycolytic enzymes in contact with glucosinolates. Isothiocyanates or mustard oils (20) are formed by a Lossen type rearrangement. Glucosinolates derive from amino acids. A common pathway to the aldoximes has been proposed for cyanogenic glycosides and glucosinolates but the details of this transformation are obscure. Characteristic features of the biosynthesis of glucosinolates are preservation of the labelling pattern from the parent amino acid, formation of aldoximes as intermediates, glucosidation of the thiol function and the observation that amines are poor precursors.[48,49,50] This suggests that oxidation of the amino group is the first step followed by decarboxylation. Reaction (21) seems to account for all the facts and every single step has ample *in vitro* analogies. It is essential to view the transformations as occurring in the PLP adduct form because this explains the radical addition of sulphide to the nitrone which acts as an efficient scavenger.[51] Since oximes are good precursors, they have to undergo reduction to hydroxylamines before they can enter the sequence at the nitrone stage. $RS^{\ominus}$ could stand for cysteine or any other source of sulphur. It is given preference to radical addition because the biosynthesis occurs in an oxidizing environment but ionic addition at the nitrone stage is quite conceivable.

6.9 Peptides, β-lactam antibiotics and proteins

Peptides and proteins are polycondensation products of amino acids. The carbonyl group of one amino acid is joined to the amino function of another amino acid thus forming an amide bond or a peptide bond. The term peptides refers usually to polymers of molecular weight lower than *ca*. 5000. The molecular weight of proteins ranges from *ca*. 5000 to several millions but there is no clear distinction between the two groups. A peptide can be a partially hydrolysed protein, but also a compound containing characteristic non-protein amino acids. These are frequently found in animals, plants and fungi. A dipeptide is derived from two amino acids, a tripeptide from three amino acids, etc. Cyclization of dipeptides leads to the widespread 2,5-dioxopiperazines and related compounds,[52] frequently synthesized by microorganisms. *Aspergillus echinulatus* produces echinulin,[53,54] a dipeptide derived from L-tryptophan, L-alanine, and three isoprene units (Fig. 17) as shown by efficient incorporation of labelled amino acids and mevalonic acid. Cyclo-L-Ala-L-Try is also incorporated indicating isoprenylation as the last step. The isoprenyl group at C^2 has an unusual orientation. Model studies support the idea that an *N*-alkylated intermediate rearranges in a Claisen reaction to the isolated product.[55] A direct S_N2' alkylation or a rearrangement from C^3 is less likely. Several other mould metabolites are, in fact, known to contain the proposed *N*-isoprenyl grouping. Gliotoxin and aranotin, which were discussed in section 3.7, also belong to the dioxopiperazine

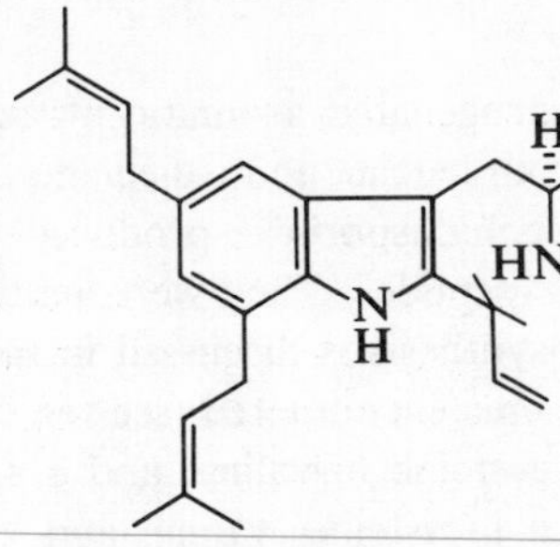

Echinulin
Aspergillus echinulatus

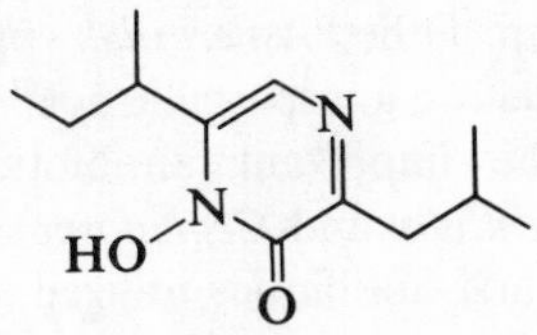

Aspergillic acid
Aspergillus flavus

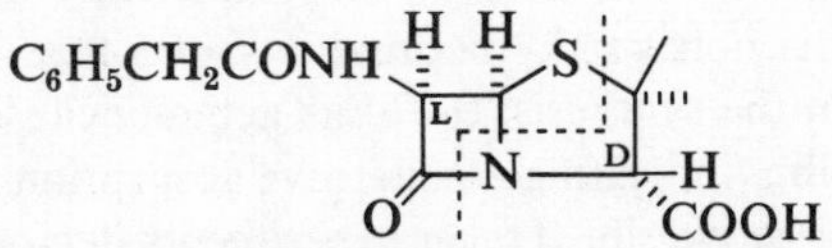

Benzylpenicillin
Penicillin G
Penicillium chrysogenum

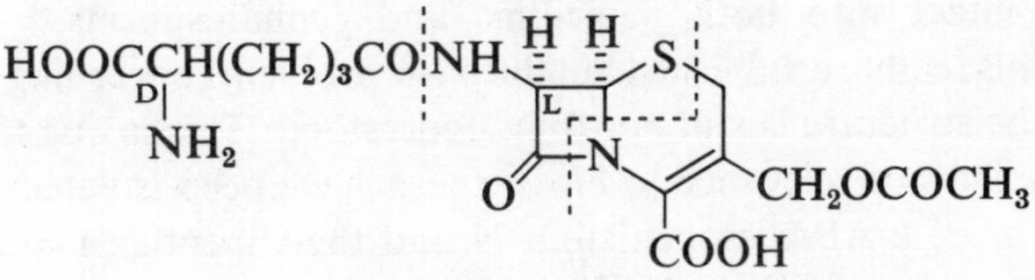

Cephalosporin C
Cephalosporium acremonium

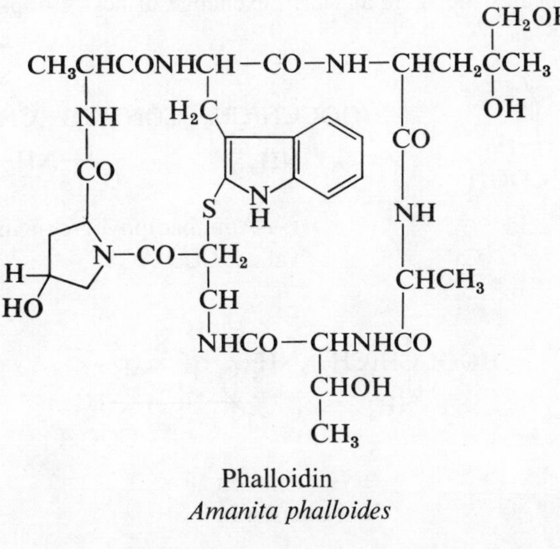

Phalloidin
Amanita phalloides

Fig. 17 Structure of naturally occurring peptides

group. Other *Aspergillus* spp. produce the dehydrogenated aromatic pyrazine nucleus, e.g. aspergillic acid[56] which is derived from leucine and isoleucine.

The important antibiotics penicillin and cephalosporin[57] produced by *Penicillium* and *Cephalosporium* spp. are cyclic tripeptides. They were actually the first antibiotics utilized in medicine. The biosynthesis is discussed in some detail in order to illustrate the mechanistic problems encountered (see ref. 58). The penicillins can visually be dissected into L-cysteine, D-valine, and a substituted acetic acid, and the cephalosporins into L-cysteine, valine, and D-α-aminoadipic acid. The configurational identity of valine in cephalosporins is lost by double bond formation.

The L-amino acids were found to be more efficiently incorporated than the D-amino acids. This is noteworthy because it shows that inversion takes place at some later stage in the synthesis. D-Valine actually caused inhibition[59] of the production of penicillin, indicating competitive absorption, but poor conversion of this isomer, at the active site. Tracer experiments demonstrated phenylacetic acid as the immediate precursor of the benzylpenicillin and this side chain can be varied by adding other appropriate acids to the culture.[60] α-Aminoadipic acid derives from α-ketoglutaric acid by chain elongation (section 6.7) and transamination. By multiple labelling it was shown that the amino acids are incorporated intact into both penicillins and cephalosporins.[61] The building blocks are thus readily established but the basic problem concerning the formation of the bicyclic structure is still not fully understood. This is due to the fact that few intermediates of relevance to biosynthesis have been isolated, e.g. 6-aminopenicillanic acid, 6-APA, isopenicillin N and the tripeptide L-α-aminoadipoyl-L-cysteinyl-D-valine, LLD-ACV (Fig. 18). On the other hand, structures have been proposed for several conceivable intermediates. 6-APA appears at the end of the biosynthesis and cannot shed much light on the earlier steps. Its appearance indicates the stage at which exchange of acyl groups occurs in the side chain.

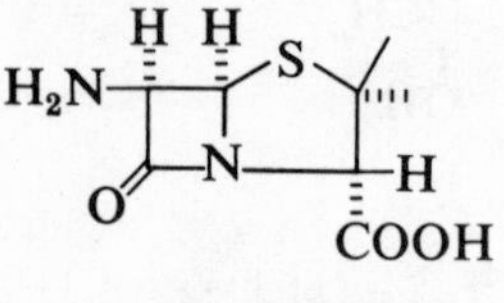

H_2N H H S O N H COOH

Penicillanic acid
6-APA

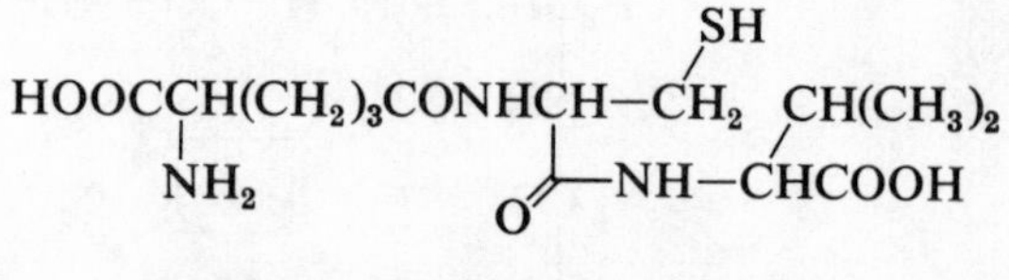

$HOOCCH(CH_2)_3CONHCH{-}CH_2\ \ CH(CH_3)_2$
NH_2 O NH−CHCOOH
SH

L-α-Aminoadipoyl-L-cysteinyl-D-valine, LLD-ACV

$HOOC\overset{*}{C}H(CH_2)_3NH$ H H S 6 5 1 2 7 4 3 N H O COOH
NH_2

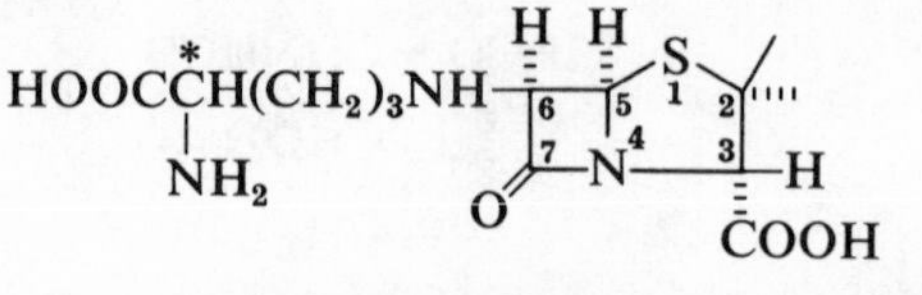

*L Isopenicillin N
*D Penicillin N

Fig. 18 Intermediates in the biosynthesis of β-lactam antibiotics

Enzymatic experiments point to isopenicillin N as the most likely precursor of penicillins with non-polar side chains (22) but it is conceivable that 6-APA appears as a transient in this reaction.[62] It accumulates in the fermentation broth of *P. chrysogenum* in the absence of other side chain precursors. The tripeptide, LLD-ACV, is more interesting and a tracer experiment shows that only this stereoisomer is converted to β-lactam antibiotics by *C. acremonium*.[63] This finding shows, in combination with the uptake studies mentioned earlier, that only the L-amino acids are metabolized and that the valine moiety is inverted at an early stage of the biosynthesis, whereas the α-aminoadipoyl moiety first becomes inverted after the ring condensations. It was also found that the tritium label at C^2 in the valine moiety of LLD-ACV was retained in penicillin N.

$$\text{Isopenicillin N} \xrightarrow[\text{Acyltransferase}]{C_6H_5CH_2COOH} \text{Penicillin G} \qquad (22)$$

The 2,3-dehydrovaline moiety was considered as a likely intermediate for the thiazolidine ring closure by thiol addition but is eliminated by this result unless it is assumed that the very same C^2–H abstracted in the dehydrogenation step is returned in the addition step (23). A final proof that LLD-ACV is a true

SH H N *H COOH Enz [O] SH 4 3 N 2 COOH *H Enz S N H* COOH (23)

intermediate on the road to penicillins was provided by observation of ^{13}C and ^{1}H NMR spectra directly in the NMR probe of the tripeptide incubated with a cell free extract of *Cephalosporium acremonium*. The characteristic peaks originating from the D-valine moiety diminish with time and corresponding signals characteristic for isopenicillin N increase. By studying the NOE effect of H^5 and C^3-α-CH_3 it was also confirmed that S-insertion at C^3 proceeded with retention of configuration.[64] The tripeptide hypothesis caused difficulties in finding an acceptable mechanism for the ring closure. Involvement of 3,4-dehydrovaline was eliminated by a mass spectrometric investigation of phenoxymethylpenicillin derived from D,L-valine-methyl, d_6. All six deuterium atoms were retained in the product.[65] We are finally left with the proposal that the LLD-ACV peptide undergoes oxidase catalysed hydroxylation at C^3 of the valine moiety. Substitution of the hydroxyl group by the thiol could conceivably generate the thiazolidine ring.[58] A similar hydroxylation in the cephalosporin series is the last step in the biosynthesis of cephalosporin C from deacetoxycephalosporin C. However, it cannot be excluded that thiazolidine formation occurs by S-insertion similar to that taking place in the biosynthesis of biotin from dethiobiotin. This reaction proceeds with retention of configuration at C^4 of dethiobiotin.[39] S-Insertion can be formulated as oxidation of thiolate ion to sulphenium ion, which attacks the C–H bond according to an electrophilic aliphatic substitution mechanism and the reaction supposedly occurs with retention (24). ΔH for this process is negative.

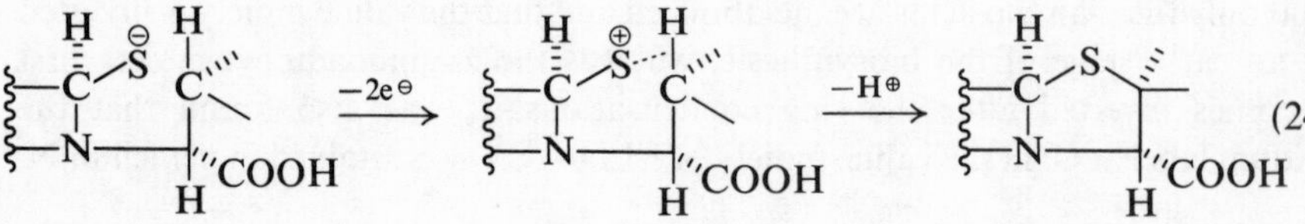

(24)

The stereochemical course of thiazolidine ring formation was determined by synthesis of chiral 2*R*,3*R*-4-^{13}C-valine. This amino acid was converted to phenoxymethylpenicillin by *P. chrysogenum* and to cephalosporin C by *C. acremonium*. The chemical shifts of the geminal methyls in penicillin was known from earlier ^{13}C NMR studies and the labelling pattern from incorporated valine showed inhanced intensity of only one signal consistent with complete retention of configuration at C^3 of valine. Cephalosporin C became stereospecifically labelled in the SCH_2 group[66] (Fig. 19).

The mechanism of formation of the β-lactam ring has been investigated extensively. We know from experiments with 2-^{3}H-cysteine and 3*R*-^{3}H-cysteine that the labels are retained in benzylpenicillin. These results imply that 2,3-dehydrocysteine does not appear on the pathway and that cyclization occurs with retention of configuration at C^3. A Michael type cyclization is therefore

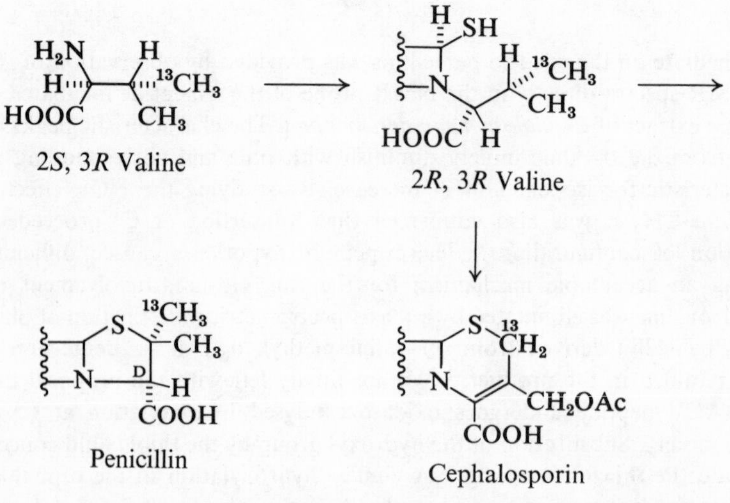

2*S*, 3*R* Valine

2*R*, 3*R* Valine

Penicillin

Cephalosporin

Fig. 19 Stereochemical course of valine incorporation

excluded. The current mechanism assumes oxidation of the thiol to thioaldehyde followed by attack of the amide despite the fact that thioaldehydes have no precedence in natural product chemistry (25). Furthermore, the thioaldehyde is

RCOHN–CH(–CH₂SH)–C(=O)–NH–CH(COOH)– ⟶[O] RCOHN–CH(–CH=S)–C(=O)–NH–CH(COOH)– ⟶ RCOHN–(β-lactam, SH)–N–CH(COOH)

(25)

expected to enolize with ease and consequently also to undergo proton exchange at C². Formation of the cyclic sulphide first, followed by oxidation to sulphoxide and a Pummerer type addition has been suggested as an attractive route (26).

LLD-ACV ⟶ RCOHN–(cyclic sulphide)–NH–CH(COOH) ⟶[O]

(26)

RCOHN–(sulphoxide, $O^{\ominus}$ $H^{\oplus}$)–NH–CH(COOH) ⟶ RCOHN–(bicyclic, S, N)–CH(COOH)

However, the postulated intermediates are not metabolized to penicillins by cell free enzyme preparations.[64] The mode of formation of the bicyclic system must still be regarded as an open question and is as such a matter of much speculation.

One of the toxic principles of the mushroom *Amanita phalloides* was shown to be an unusual bicyclic heptapeptide, phalloidin, derived from cysteine, *allo*hydroxyproline, threonine, oxindolylalanine, γ,δ-dihydroxyleucine and two alanines, all with L-configuration[67] (Fig. 17). The toxic principles in the venoms from snakes and bees consist of single polypeptide chains containing cysteine cross-linkages. The neurotoxin of the spectacled Indian cobra, *Naja naja*, contains 71 residues, the sequence of which has been established.[68]

The peptide bond of protein is synthesized by genetically controlled reactions on ribosomes the details of which are best presented in textbooks of biochemistry.

$$\underset{\displaystyle NH_2}{RCHCOOH} + ATP \rightleftharpoons \underset{\displaystyle NH_2}{RCHCOAMP} + P_2 \tag{27}$$

Small peptides and non-protein peptides are produced enzymatically by activation of the amino acid with ATP according to (27) and by formation of a thioester (28) with the enzyme. Next to this site another amino acid is coming in which fits the topology of the enzyme, forms a thioester and reacts with the first amino acid (29). Subsequent reactions lead to polypeptides.

$$\text{Aminoacyl-AMP} + \text{Enz—SH} \rightleftharpoons \text{Enz—S—}\overset{\displaystyle O}{\overset{\|}{C}}\text{—}\underset{\displaystyle NH_2}{CHR} \tag{28}$$

$$\text{NH}_2\text{—CHR}^1\text{—C(=O)—S—Enz} + \text{H}_2\ddot{\text{N}}\text{CHR}^2\text{—C(=O)—S—Enz} \longrightarrow \text{HS—Enz} + \text{R}^1\text{CH(NH}_2\text{)CONHCHR}^2\text{—C(=O)—S—Enz} \tag{29}$$

The solid phase synthesis[69] invented by Merrifield resembles the polymerizing procedure used by the living cell. The support or carrier consists of a synthetic resin with a functional group that can react with the carboxyl group of the amino acid, e.g. a benzyl chloride residue (Fig. 20). Having anchored the C-terminal amino acid to the solid phase a second amino acid with a protected amino group is added. Its carboxyl group is allowed to react with the free amino group of the anchored residue with the help of the condensing agent dicyclohexyl carbodiimide. The dipeptide formed is bound to the insoluble support that can be washed and freed from all reagents. The N-protecting group is removed and on repetition of the sequence a tripeptide can be prepared, etc. This technique has been automated and in a week's time it is possible to carry out the *ca.* 5000 operations needed for assembling the 51 amino acid residues of insulin, and the yields are throughout very high (> 90 per cent per cycle).

6.10 Problems

6.1 Aspartic acid undergoes β-decarboxylation with formation of alanine. The reaction is catalysed by a PLP-dependent enzyme. It was found that the α-H is exchanged in the process. Occasionally pyruvic acid is formed instead of alanine. Suggest a mechanism explaining the conversions.

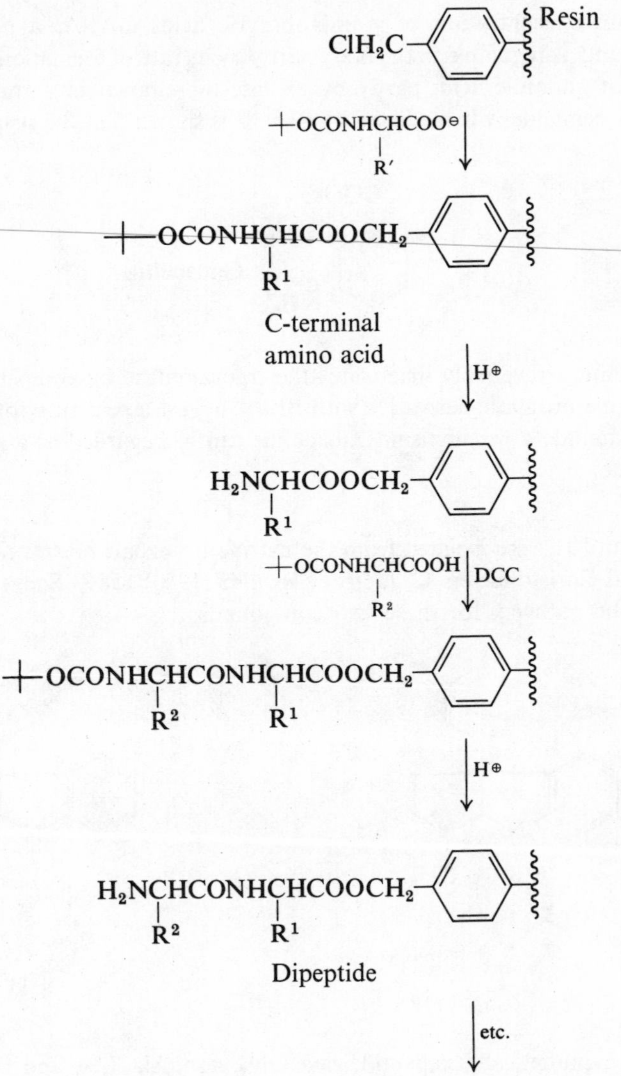

$$ClH_2C-C_6H_4-\text{Resin}$$

$$\downarrow +OCONHCH(R')COO^{\ominus}$$

$$+OCONHCH(R^1)COOCH_2-C_6H_4-\text{resin}$$

C-terminal amino acid

$$\downarrow H^{\oplus}$$

$$H_2NCH(R^1)COOCH_2-C_6H_4-\text{resin}$$

$$\downarrow +OCONHCH(R^2)COOH,\ DCC$$

$$+OCONHCH(R^2)CONHCH(R^1)COOCH_2-C_6H_4-\text{resin}$$

$$\downarrow H^{\oplus}$$

$$H_2NCH(R^2)CONHCH(R^1)COOCH_2-C_6H_4-\text{resin}$$

Dipeptide

$$\downarrow \text{etc.}$$

Fig. 20 Merrifield solid phase synthesis

6.2 It has been suggested that glucosinolates are formed from the corresponding amino acids by oxidation of the amino group to a nitro group as the first step. Rearrangement of the nitro group assisted by ATP, cysteine, and uridine diphosphoglucose (UDPG) as reagents could then lead to the natural product. How would you account for the rearrangements? (Ettlinger, M. and Kjaer, A., in *Recent Advs. in Phytochemistry*, Mabry, T. J., Alston, R. E., and Runeckles, V. C. (Eds.), Appleton-Century-Crofts, New York **1** (1968) 58).

6.3 The concentration of γ-aminobutyric acid, GABA, a neurotransmitter compound, is regulated in the body partly by its rate of formation via decarboxylation of glutamic acid, partly by its rate of removal in a transamination as succinic semialdehyde mediated by PLP. It is shown that the structurally related

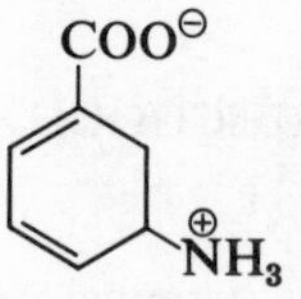

$COO^{\ominus}$

$\overset{\oplus}{N}H_3$

Gabaculine

gabaculine, irreversibly inactivates the transaminase by competitive formation of a stable aromatic derivative with PLP. Suggest a structure for this derivative and formulate a mechanism. Gabaculine can be regarded as a kind of suicidal substrate.

6.4 I and II were isolated from the marine bryozoan *Flustra foliaceae* (Carlé, J. S. and Christoffersen, C. *J. Org. Chem.* **45** (1980) 1586). Suggest a reasonable metabolic pathway for these two compounds.

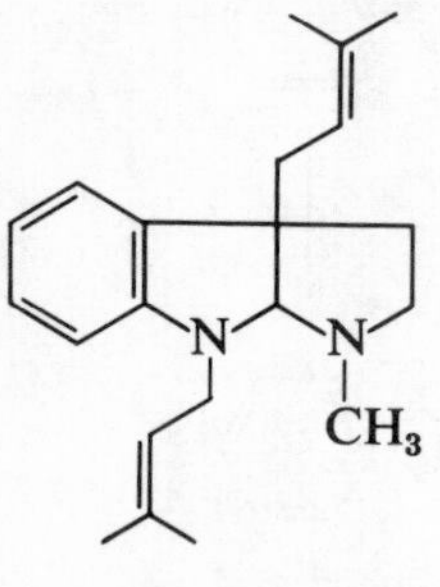

N N CH_3

I

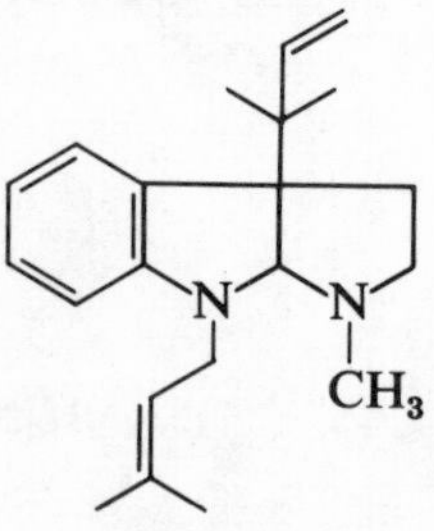

N N CH_3

II

6.5 A hypothetical hexapeptide gave Gly_2, Asn, Ala, Leu, and Tyr on complete hydrolysis. N-Terminal analysis of the peptide with 2,4-dinitrofluorobenzene followed by hydrolysis gave 2,4-dinitrophenylalanate and treatment with carboxypepdidase released first glycine. Partial hydrolysis gave, among others, Ala-Gly and Tyr-Leu. Give structures consistent with the data.

6.6 Fusaric acid is formed by mixed biogenesis from an amino acid and a polyketide. Suggest a biosynthesis route.

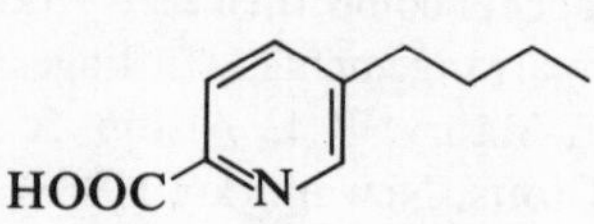

HOOC N

Fusaric acid

6.7 Holomycin is an antibiotic isolated from *Streptomyces griseus* (Ettlinger, L., Gäumann, E., Hütter, R., Keller-Schierlein, W., Kradolfer, F., Neipp, L., Prelog, V. and Zähner, H. *Helv. Chim. Acta* **42** (1959) 563). Suggest precursors for this unusual structure. How would you design experiments to prove your working hypothesis?

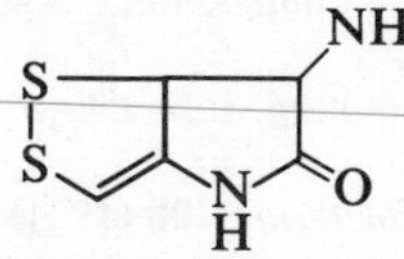

$NHCOCH_3$

6.8 *Streptomyces refuineus* produces the antibiotic anthramycin (Hurley, L. H., Zmijewski, M., and Chang, C-J. *J. Am. Chem. Soc.* **97** (1975) 4372). ^{14}C-labelled methionine was incorporated at the starred positions and 1-^{14}C-labelled tyrosine at the dotted position. Dopa was metabolized with the same efficiency as tyrosine. It was shown that the aromatic ring of anthramycin originated from tryptophan. When 3,5-$^{3}H_2$ tyrosine was fed to the culture, one ^{3}H was retained in the side chain. Suggest a metabolic pathway.

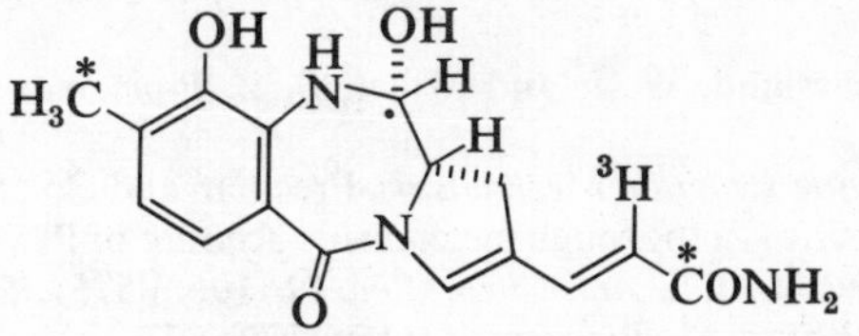

6.9 Suggest plausible precursors for the zoanthoxanthins isolated from the marine colonial organism *Parazoanthus axinellae* related to corals (Cariello, L., Crescenzi, S., Prota, G. and Zanetti, L. *Tetrahedron* **30** (1974) 4191).

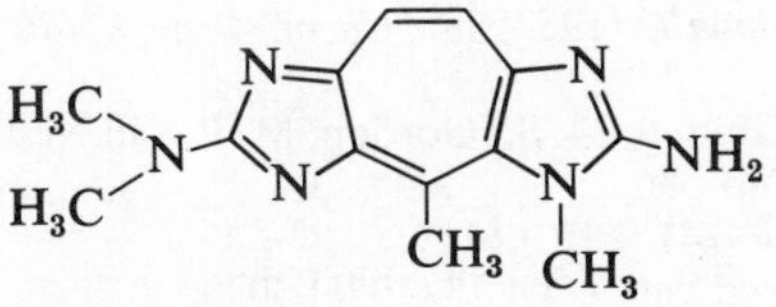

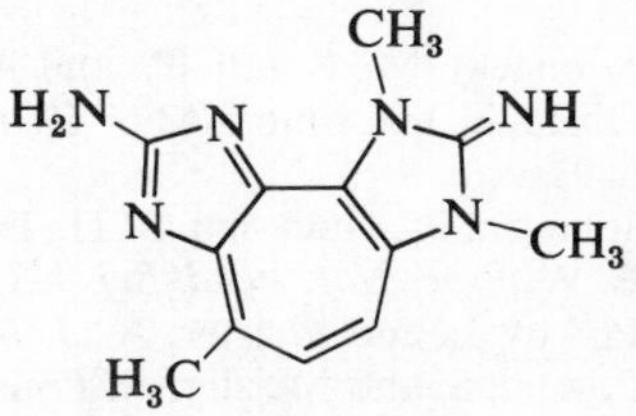

Bibliography

1. Auclair, J. L. and Patton, R. L. *Rev. Can. Biol.* **9** (1950) 3.
2. Ogawa, T., Fukuda, M. and Sasaoka, K. *Biochim. Biophys. Acta* **297** (1973) 60.
3. Noma, M., Noguchi, M. and Tamaki, F. *Agric. Biol. Chem. Japan* **37** (1973) 2439.
4. Marshall, R. D. *Ann. Rev. Biochem.* **41** (1972) 673.
5. Fonnum, F. (Ed.) *Amino acids as Chemical Transmitters*, Plenum, New York, 1978.

6. Walter, R. and Meienhofer, J. (Eds.) *Peptides. Chemistry, Structure and Biology*, Ann Arbor Science Publishers, Ann Arbor, 1975.
7. Ovchinnikov, Yu. A. and Ivanov, V. T. in *Heterodetic Peptides, Int. Review of Science, Organic Chemistry, Series Two* **6** (1976) 183.
8. Cramer, U. and Spener, F. *European J. Biochem.* **74** (1977) 495.
9. Martin, D. G., Duchamp, D. J. and Chidester, C. G. *Tetrahedron Lett.* **1973,** 2549.
10. Takemoto, T., Daigo, K., Kondo, Y. and Kondo, K. *Yakugaku Zasshi* **86** (1966) 874. *Chem. Abstr.* **66** (1967) 28604.
11. Lindberg, P., Bergman, R. and Wickberg, B. *J. Chem. Soc. Chem. Commun.* **1975,** 946.
12. Castric, P. A. and Conn, E. E. *J. Bacteriol.* **108** (1971) 132.
13. Castric, P. A., Farnden, K. J. F. and Conn, E. E. *Arch. Biochem. Biophys.* **152** (1972) 62.
14. Bell, E. A. *Endeavour* **4** (1980) 102.
15. Bada, J. L. *Earth Planet Sci. Letters* **15** (1972) 273.
16. Helfman, P. M. and Bada, J. L. *Proc. Nat. Acad. Sci. USA* **72** (1975) 2891; Barret, G. C. in *Amino acids, Peptides and Proteins*, Specialist Periodical Reports, Chemical Soc., London **8** (1976) 21.
17. Fowden, L., Scopes, P. M. and Thomas, R. N. *J. Chem. Soc.* (*C*) **1971,** 833.
18. Jorgensen, E. G. *Tetrahedron Lett*, **1971** 863.
19. Spackman, D. H., Stein, W. H., and Moore, S. *Anal. Chem.* **30** (1958) 1190.
20. Miller, S. L., Urey, H. C. and Oro, J. *J. Mol. Evol.* **9** (1976) 59.
21. Shevlin, P. B., McPherson, D. W. and Melius, P. *J. Am. Chem. Soc.*, **103** (1981) 7007.
22. Wolman, Y., Haverland, W. J. and Miller, S. L. *Proc. Nat. Acad. Sci. USA* **69** (1972) 809.
23. Walsh, C. *Enzymatic Reaction Mechanisms*. Freeman and Co., San Francisco, 1979, and refs. therein. Gives a thorough mechanistic account of PLP mediated reactions.
24. Belleau, B. and Burba, J. *J. Am. Chem. Soc.* **82** (1960) 5751, 5752.
25. Jordan, P. and Akhtar, M. *Biochem. J.* **116** (1970) 277.
26. Tatum, C. M., Benkovic, P. A., Benkovic, S. J., Potts, R., Schleicher, E. and Floss, H. G. *Biochemistry* **16** (1977) 1093.
27. Schirch, L. and Gross, T. *J. Biol. Chem.* **243** (1968) 5651.
28. Taylor, W. and Weissbach, H. in *The Enzymes*, 3rd. edn. Boyer, P. (Ed.) Academic Press, New York **9** (1973) 121.
29. Tada, M., Miura, K., Okabe, M., Seki, S. and Mizukami, H. *Chem. Letters* **1981** 33.
30. Krongelb, M., Smith, T.. and Abeles, R. *Biochem. Biophys. Acta* **167** (1968) 473.
31. Obata, Y. and Iimori, M. *J. Chem. Soc. Japan* **73** (1952) 832. *Chem. Abstr.* **47** (1953) 6093a.
32. Johnson, R., Guderian, R. H., Eden, F., Chilton, M. S., Gordon, M. P. and Nester, E. W. *Proc. Nat. Acad. Sci. USA* **71** (1974) 536.
33. Pinkus, L. and Meister, A., *J. Biol. Chem.* **247** (1972) 6119.
34. Powers, S. and Meister, A. *Proc. Nat. Acad. Sci. USA* **73** (1976) 3020.
35. Kjaer, A. and Larsen, P. O. in *Biosynthesis*, Specialist Periodical Reports, Chemical Society, London **3** (1973) 71.
36. Schütte, H. R. and Müller, P. *Biochem. Physiol. Pflanz.* **163** (1972) 528.
37. Maw, G. A. in *Sulfur in Organic and Inorganic Chemistry*, Senning, A. (Ed.) M. Dekker, New York **2** (1972) 113.
38. Okumura, S., Tsugawa, T., Tsunanda, T., and Motosaki, S. *J Agric. Chem. Soc. Japan* **36** (1962) 599, 602.
39. Trainor, D. A., Parry, R. J. and Gitterman, A., *J. Am. Chem. Soc.* **102** (1980) 1467, Parry, R. J. and Naidu, M. V. *Tetrahedron Lett.* **1980,** 4783.
40. Crout, D. H. G. *J. Chem. Soc.* **1966,** 1968.

41. O'Donovan, D. G. and Long, D. J. *Proc. Roy. Irish Acad.* **75B** (1975) 465.
42. Cahill, R., Crout, D. G. H., Mitchell, M. B. and Müller, U. S. *J. Chem. Soc. Chem. Commun.* **1980,** 419.
43. Tapper, B. A. and Butler, G. W. *Phytochemistry* **11** (1972) 1041.
44. Conn, E. E. *Naturwissenschaften* **66** (1979) 28.
45. Zilg, H., Tapper, B. A. and Conn, E. E. *J. Biol. Chem.* **247** (1972) 2384.
46. Reay, P. F. and Conn, E. E. *J. Biol. Chem.* **249** (1974) 5826.
47. Duffey, S. S., Underhill, E. W. and Towers, G. H. N. *Comp. Biochem. Physiol.* **47B** (1974) 753.
48. Kindl, H. and Underhill, E. W. *Phytochemistry* **7** (1968) 745.
49. Matsuo, M. and Underhill, E. W. *Biochem. Biophys, Res. Commun.* **36** (1969) 18.
50. Ettlinger, M. and Kjaer, A. *Rec. Advs. Phytochemistry* Mabry, T. J., Alston, R. E., and Runeckles, V. C. (Eds.), Appleton-Century-Crofts, New York, **1** (1968) 59.
51. Janzen, E. G. *Acct. Chem. Res.* **4** (1971) 31.
52. Sammes, P. G. in *Prog. Chem. Org. Nat. Prod.* **32** (1975) 51.
53. Quilico, A. *Res. Prog. Org. Med. Chem.* **1** (1964) 225.
54. Birch, A. J., Blance, G. E., David, S. and Smith, H. *J. Chem. Soc.* **1961,** 3128.
55. Casnati, G. and Pochini, A. *J. Chem. Soc. Chem. Commun.* **1970,** 1328.
56. Newbold, G. T., Sharp, W. and Spring, F. S. *J. Chem. Soc.* **1951,** 2679.
57. Lemke, P. A. and Brannon, D. R. in *Cephalosporins and Penicillins,* Flynn, E. H. (Ed.) Academic Press, New York, 1972.
58. Aberhart, D. J. *Tetrahedron Reports* **30,** *Tetrahedron* **33** (1977) 1545.
59. Warren, S. C., Newton, G. G. F. and Abraham, E. P. *Biochem. J.* **103** (1967) 902.
60. Behrens, O. K. in *The Chemistry of Penicillin,* Clarke, H. T., Johnson, J. R., and Robinson, R. (Eds.) Princeton University Press, Princeton, New Jersey, 1949.
61. Arnstein, H. R. V. and Grant, P. T. *Biochem. J.* **57** (1954) 353, 360.
62. Fawcett, P. A., Usher, J. J. and Abraham, E. P. *Biochem. J.* **151** (1975) 741.
63. Fawcett, P. A., Loder, P. B., Duncan, M. J., Beesley, T. J. and Abraham, E. P. *J. Gen. Microbiol.* **79** (1973) 293.
64. Bahadur, G., Baldwin, J. E., Wan, T., Jung, M., Abraham, E. P., Huddleston, J. A. and White, R. L. *J. Chem. Soc. Chem. Commun.* **1981,** 1146.
65. Aberhart, D. J., Chu, J. Y. R., Neuss, N., Nash, C. H., Occolowitz, J., Huckstep, L. L. and De La Higuera, N. *Chem. Commun.* **1974,** 564.
66. Neuss, N., Nash, C. H., Baldwin, J. E., Lemke, P. A. and Grutzner, J. B. *J. Am. Chem. Soc.* **95** (1973) 3797.
67. Wieland, T. in *Prog. Chem. Org. Nat. Prod.* **25** (1967) 214.
68. Nakai, K., Sasaki, T. and Hayashi, K. *Biochem. Biophys. Res. Commun.* **44** (1971) 893.
69. Neckers, D. C. *J. Chem. Educ.* **52** (1975) 695.

Chapter 7

The alkaloids

7.1 Introduction

The term alkaloids or alkali-like compounds was coined by Meissner, an apothecary in Halle, in 1819. It loosely defines today naturally occurring basic compounds apart from simple amines derived from amino acids and the N-heterocycles of the pyrrole, pyrimidine, purine type, etc. There are exceptions, of course; alkaloids, which are amides, are not basic e.g. colchicine. The classification is fortuitous and chemically artificial because there is no real need to separate the simple amines from alkaloids since both originate from amino acids and so do, in fact, also the *N*-heteroaromatics. The traditional classification is, nevertheless, kept for some practical reasons. The function of the overwhelming majority of alkaloids is yet unknown, though we know that pyrimidine, purine, and pterin derivatives play an important role in the life processes. All alkaloids can be constructed of building blocks from the shikimic acid polyketide, or mevalonic acid pathways in combination with an amino acid, a circumstance that automatically allows a consistent systemization of the numerous and structurally highly diverse compounds. In short, the amino acid component determines the character of the alkaloid and this classification harmonizes well with the classical system based on morphology. Alkaloids are therefore used as evolutionary or biogenetic markers.[1] Comparatively few amino acids are involved in the biosynthesis of alkaloids, e.g. glycine (in the *N*-heterocycles mentioned), glutamic acid, ornithine, lysine, phenylalanine, tyrosine, tryptophan, and anthranilic acid. The majority of alkaloids is found in the plant kingdom, from higher plants down to microorganisms. Few are found in the animal kingdom, and curiously enough, alkaloids are represented extremely sparsely in the marine environment.

Since ancient times few compounds have ever been wrapped in so much mystery as the alkaloids. We find amongst them deadly poisons such as e.g. strychnine. It has been used as a rodenticide and vermin killer for centuries and it is as such responsible for the accidental death of many beloved pets. Since strychnine is very stable it has been detected in exhumed bodies several years after death; consequently not the ideal homicidal agent. Coniine in *Conium maculatum* was used by the ancient Greeks for state executions, and here Socrates is certainly the most famous victim. Several hallucinogens are recognized as alkaloids such as the opium group in *Papaver somniferum*, lysergic acid derivatives in *Claviceps purpurea*, the parasitic fungus on grain, causing convulsive 'St. Anthony's fire' in mediaeval times, mescaline in the Indian Peyote cactus *Lophophora williamsii*, and psilocybin from the Mexican mushroom, *Psilocybe mexicana*, used by the upper priesthood of the Mayas to gain transcendental spiritual contact with their ancient gods. Several alkaloids are used as valuable drugs in medicine, e.g. morphine as a pain reliever, reserpine in psychiatry as a tranquillizer, curare alkaloids in general anaesthesia—also utilized by South American indians as an arrow poison—atropine in eye surgery, ergonovine to induce or make childbirth easier, and quinine as an antimalarial.

7.2 Alkaloids derived from ornithine and lysine. The pyrrolidine and piperidine alkaloids

Ornithine is the immediate precursor for the pyrrolidine alkaloids. The biosynthesis of one of these, hyoscyamine, has been studied extensively. When *Datura stramonium* was fed with 2-^{14}C-labelled ornithine, the hyoscyamine produced was labelled at one bridgehead which demonstrates asymmetric incorporation[2] of ornithine (Fig. 1). Therefore, the symmetric putrescine does not appear free on the pathway, even though feeding experiments show that it can serve as precursor. δ-*N*-Methylated, but not α-*N*-methylated ornithine is efficiently incorporated into hyoscyamine, and ^{15}N-labelling experiments demonstrate further that only the δ-N of ornithine is retained in the pyrrolidine ring.[3] We know that decarboxylation requires PLP and participation of this coenzyme in the biosynthesis of the pyrrolidine nucleus gives a rationale of all the data available. The asymmetric incorporation of ornithine is explained by the assumption that the hydrolysis of the Schiff's base to putrescine is slower than the cyclization and methylation. At the same time putrescine becomes acceptable as a potential alternative precursor. The Schiff's base blocks the α-N and this is specifically eliminated by cyclization and subsequent alkylation. The three carbon bridge originates from acetoacetate as shown by ^{14}C-labelling. The Schiff's base is a more likely intermediate than the often postulated γ-aminobutyraldehyde. However, asymmetric incorporation does not always occur. In this case the Schiff's base is in rapid equilibrium with the free diamine. A considerable body of evidence proves that the pyrrolidine ring of nicotine is formed via a symmetric intermediate,[4] and this is also true for the pyrrolizidine bases[5] (Fig. 2) which are formed from two molecules of ornithine. In nature these bases occur esterified with the necic acids (section 6.7). See problem 7.9 for the biosynthesis of the pyrrolizidine skeleton.

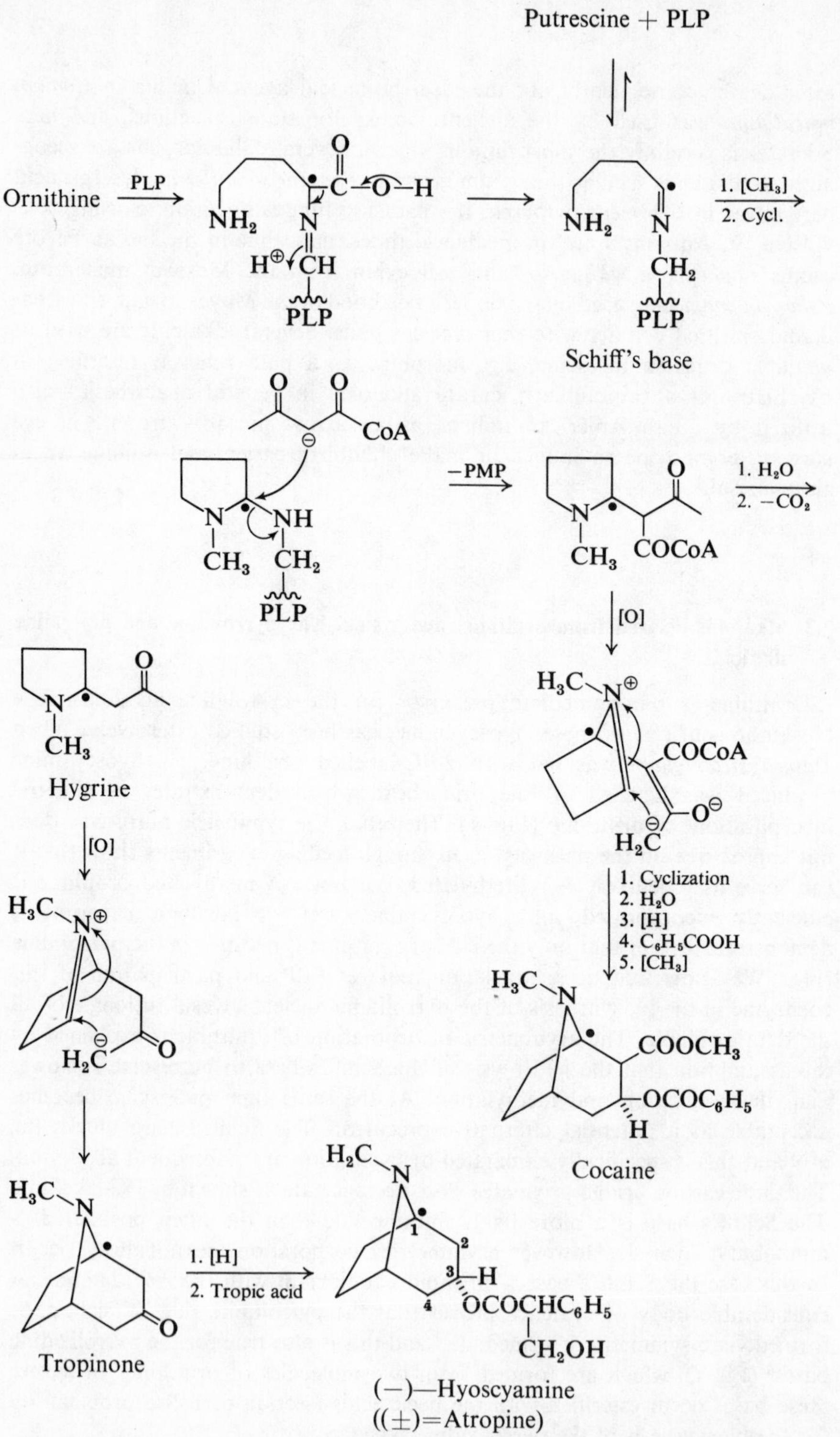

Putrescine + PLP

Ornithine $\xrightarrow{PLP}$ (NH_2, C(=O)–O–H, N=CH, $H^{\oplus}$, PLP) $\longrightarrow$ (NH_2, N, CH_2, PLP) $\xrightarrow[\text{2. Cycl.}]{\text{1. [}CH_3\text{]}}$

Schiff's base

(O, O, CoA, $\ominus$; N, CH_3, NH, CH_2, PLP) $\xrightarrow{-PMP}$ (N, CH_3, O, COCoA) $\xrightarrow[\text{2. }-CO_2]{\text{1. }H_2O}$

[O]

(H_3C–$N^{\oplus}$, COCoA, $O^{\ominus}$, $H_2\overset{\ominus}{C}$)

1. Cyclization
2. H_2O
3. [H]
4. C_6H_5COOH
5. [CH_3]

(H_3C–N, $COOCH_3$, $OCOC_6H_5$, H)

Cocaine

(N, CH_3, O)

Hygrine

[O]

(H_3C–$N^{\oplus}$, $H_2\overset{\ominus}{C}$, O)

(H_3C–N, O)

Tropinone

$\xrightarrow[\text{2. Tropic acid}]{\text{1. [H]}}$

(H_3C–N, 1, 2, 3, 4, H, $OCOCHC_6H_5$, CH_2OH)

(−)—Hyoscyamine
((±)=Atropine)

Fig. 1 Biosynthesis of tropane alkaloids

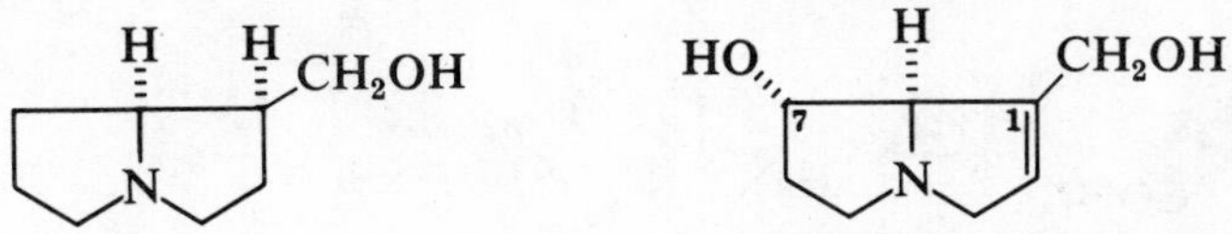

Laburnine

Heliotridine (7α)
Retronecine (7β)

Fig. 2 Structure of some pyrrolizidine bases

In cocaine, the chief alkaloid of the South American coca bush, the carboxyl group is retained as a methyl ester. It has anaesthetic properties but is nowadays replaced in medicine by other drugs since it causes addiction.

Tropinone (Fig. 1) was elegantly synthesized by Robinson under virtually physiological conditions from succinaldehyde, methylamine and acetonedicarboxylic acid (1) and it was once thought—incorrectly—that this mild reaction mimicked the cellular reaction.

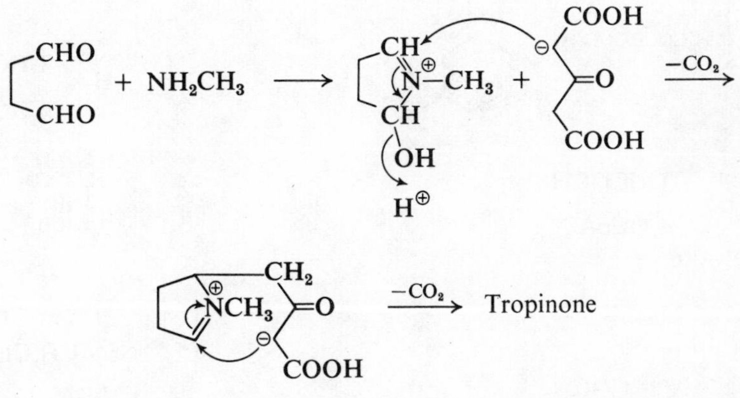

$$\text{succinaldehyde} + NH_2CH_3 \longrightarrow \ldots \xrightarrow{-CO_2} \ldots \xrightarrow{-CO_2} \text{Tropinone} \qquad (1)$$

Lysine undergoes a series of similar reactions and gives rise to the six-membered piperidine ring alkaloids. Isotopic experiments with ψ-pelletierine, *N*-methylpelletierine and sedamine (Fig. 3) show that ε-N, C^6–H_2, and C^2–H of lysine are retained, and that lysine is incorporated unsymmetrically.[6–8] These results exclude α-amino-δ-formylpentanoic acid, α-keto-ε-aminohexanoic acid and free cadaverine as intermediates. The side chain of pelletierine is generated from acetate, whereas that of sedamine originates from phenylalanine. The data are compatible with the scheme presented in Fig. 3. The asymmetry is maintained by the rather stable Schiff's base as is the case for the analogous reaction in Fig. 1. Cadaverine, which acts as a precursor, can enter the system at the step of the Schiff's base formation. Another mechanistic detail becomes evident from the observation that C^2–H is retained. The α-imino group must be generated by decarboxylation. But there are cases, where exchange of C^2–H occurs. The formation of pipecolic acid demands formation of an α-ketocarboxylic acid equivalent, since the C^2–H^* is lost (2).

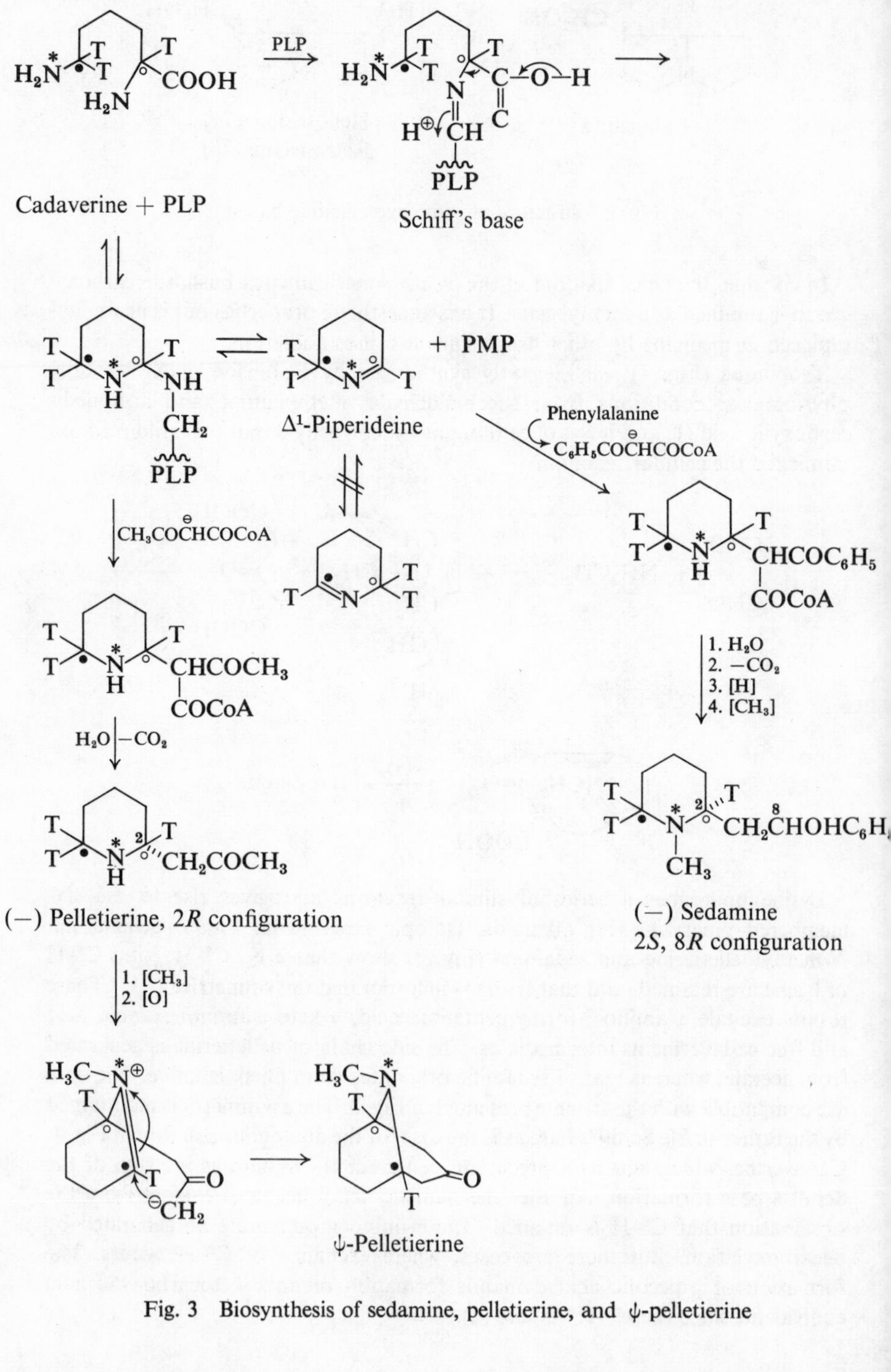

$H_2\overset{*}{N}$ T T T COOH H_2N

PLP

$H_2\overset{*}{N}$ T T T N C—O—H C H⊕ CH PLP

Cadaverine + PLP

Schiff's base

T T $\overset{*}{N}$ H T NH CH_2 PLP

T T $\overset{*}{N}$ T + PMP

Δ^1-Piperideine

T N T T

Phenylalanine

$C_6H_5CO\overset{\ominus}{C}HCOCoA$

T T $\overset{*}{N}$ H T $CHCOC_6H_5$ COCoA

$CH_3CO\overset{\ominus}{C}HCOCoA$

T T $\overset{*}{N}$ H T $CHCOCH_3$ COCoA

H_2O | $-CO_2$

1. H_2O
2. $-CO_2$
3. [H]
4. $[CH_3]$

T T $\overset{*}{N}$ H 2 T CH_2COCH_3

(−) Pelletierine, 2*R* configuration

T T $\overset{*}{N}$ 2 T 8 $CH_2CHOHC_6H_5$ CH_3

(−) Sedamine
2*S*, 8*R* configuration

1. $[CH_3]$
2. [O]

$H_3C—\overset{*}{N}^{\oplus}$ T T O $^{\ominus}CH_2$

$H_3C—\overset{*}{N}$ T T O

ψ-Pelletierine

Fig. 3 Biosynthesis of sedamine, pelletierine, and ψ-pelletierine

$$H_2\dot{N}\text{-}(CH_2)_4\text{-}C^{*}H(NH_2)COOH \xrightarrow{PLP} H_2\dot{N}\text{-}(CH_2)_4\text{-}C(=O)COOH \longrightarrow$$

$$\Delta^1\text{-piperideine-2-COOH} \longrightarrow \text{L-pipecolic acid} \qquad (2)$$

L-pipecolic acid

A final point worth mentioning is the absolute stereostructure at C^2 in (—)pelletierine and (—)sedamine which are of opposite configuration. That means that different faces of the piperideine ring are oriented towards the polyketide on the enzyme surface in the alkylation step in the two cases.

The asymmetric incorporation of lysine is not upheld for the lupine alkaloids,[9] i.e. the hydrolysis of the Schiff's base to symmetric cadaverine and PLP is comparatively rapid. Several hypothetical intermediates such as glutardialdehyde and 5-aminopentanal or its cyclic equivalent Δ^1-piperideine have been proposed, but not identified. Tracer experiments[10] with 2-^{14}C- and 6-^{14}C-Δ^1-piperideine administered by the wick method to *Lupinus angustifolius* give specific incorporation and thus support this compound or its biosynthetic equivalent, i.e. the PMP adduct as true intermediates (Fig. 4). It is assumed that reversal to cadaverine is slow here. Approximately one-third of the total label was recovered at the anticipated sites on degradation in consort with the idea that three Δ^1-piperideines are involved.

The piperidine alkaloids give us another instructive example showing that the biosynthetic origin cannot be deduced simply by analogy and structural considerations. Pinidine in *Pinus* spp. and coniine in *Conium maculatum* resemble stongly pelletierine in *Punica granatum* (Fig. 5), but surprisingly tracer studies demonstrated their acetate origin. As to coniine octanoic acid, 5-keto-octanoic acid, the corresponding aldehyde, and coniceine act as specific precursors.[11,12] These results suggest the sequence presented in Fig. 6. In analogy with coniine, pinidine is labelled by 1-^{14}C-acetate at alternate carbon atoms, $C^{2,4,6,9}$, suggesting the dioxodecanoic acids in Fig. 7 or derivatives as intermediates. The starter end was determined by feeding 1-^{14}C-malonic acid together with inactive acetic acid to *Pinus jeffreyi*. The activity of the derived pinidine was three times higher at C^2, than at C^9, from which it follows that $C^{2,7}$ is the starter.[13]

The biosynthetic path followed by the *Nicotiana* alkaloids turned out to be full of surprises. The aromatic pyridine ring is not derived from lysine by cyclization and dehydrogenation as one intuitively may think, nor is it derived from acetate. Its immediate precursor is nicotinic acid, the important provitamin, trivially called niacin, the amide of which is a constituent of NADH. Deficiency of nicotinic acid leads to pellagra in humans and black tongue in dogs. It was

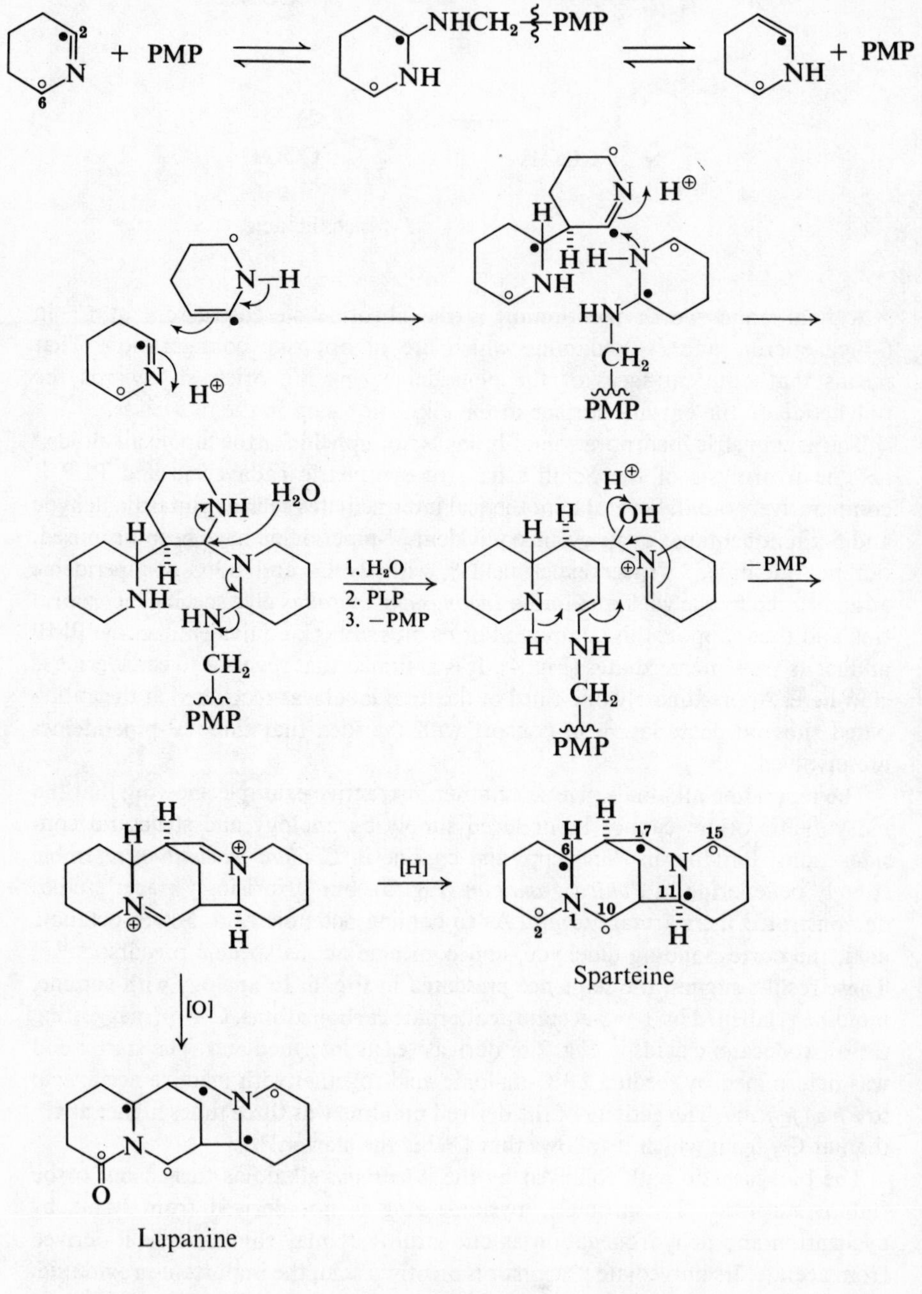

Fig. 4 Biosynthesis of lupine alkaloids

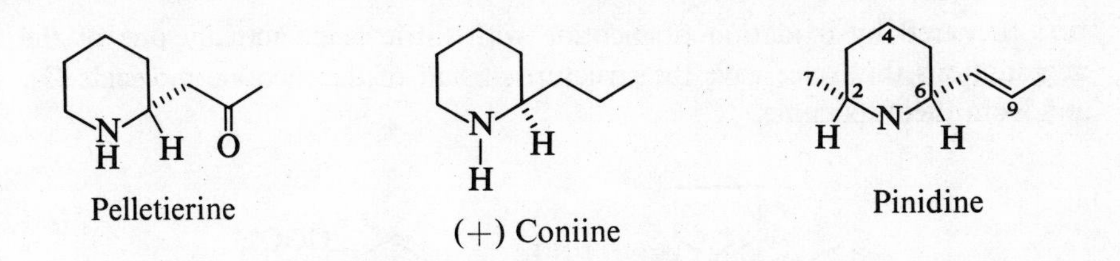

Fig. 5 Structure of piperidine alkaloids of different biogenesis. Pelletierine comes from lysine, whereas coniine and pinidine come from acetate

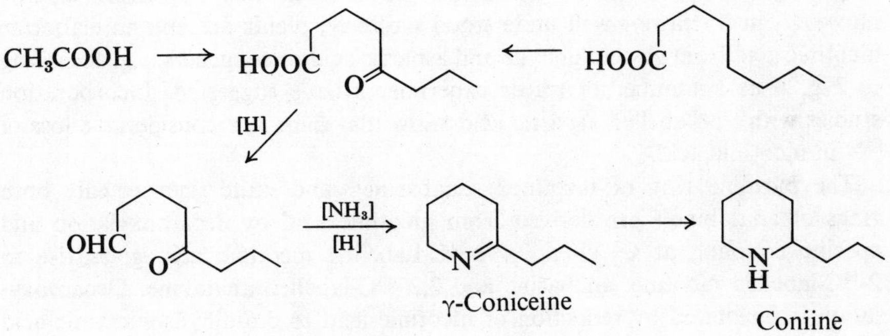

CH_3COOH ⟶ HOOC ... O ⟵ HOOC

[H]

OHC ... O $\xrightarrow[\text{[H]}]{[NH_3]}$ γ-Coniceine ⟶ Coniine

Fig. 6 Biosynthesis of coniine

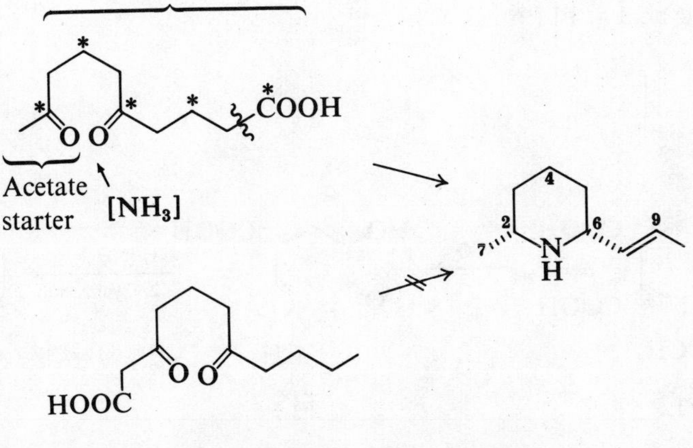

Malonates

*COOH

Acetate starter

$[NH_3]$

Fig. 7 Biosynthesis of pinidine

first prepared by oxidation of nicotine with nitric acid, actually one of the degradations that once gave the structure of half of the nicotine molecule (3), and from there its name.

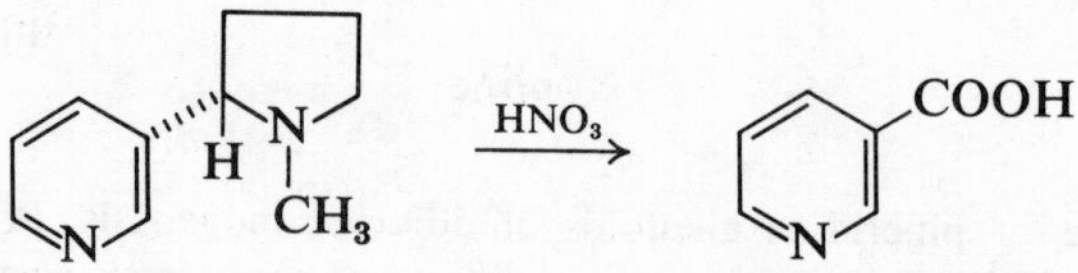

$$\text{Nicotine} \xrightarrow{HNO_3} \text{Nicotinic acid} \quad (3)$$

Nicotinic acid

Most animals can form nicotinic acid catabolically from tryptophan via 3-hydroxyanthranilic acid, a route that is not followed by *Nicotiana* spp. However, in a still not well understood pathway, plants are able to elaborate nicotinic acid from glyceraldehyde and aspartic acid or congeners, e.g. according to Fig. 8 as a number of tracer experiments have suggested. Incorporation studies with ^{15}N-labelled aspartic acid show that there is a considerable loss of ^{15}N in nicotinic acid.

The pyridine ring of nicotine,[14] anabasine[15] and quite unexpectedly both rings of anatabine[16] are derived from nicotinic acid by decarboxylation and specific coupling at C^3 (Fig. 9). 2-^{14}C-Labelled nicotinic acid gives rise to 2-^{14}C-labelled nicotine, anabasine and 2,2′-^{14}C-labelled anatabine. Decarboxylation is facilitated by reduction of nicotinic acid to 3,6-dihydronicotinic acid which contains the β-imino structure necessary for stabilizing an anion. In support for the 3,6-dihydronicotinic acid is an observation of a considerable loss of ^3H label in nicotine from 6-^3H-nicotinic acid as precursor compared with those from 2-^3H-, 4-^3H-, or 5-^3H-nicotinic acid. It is not definitely settled at which stage the carboxyl group of 3,6-dihydronicotinic acid is lost. The fact that the label of 2-^{14}C-nicotinic acid is retained at C^2 of nicotine, speaks in favour of a

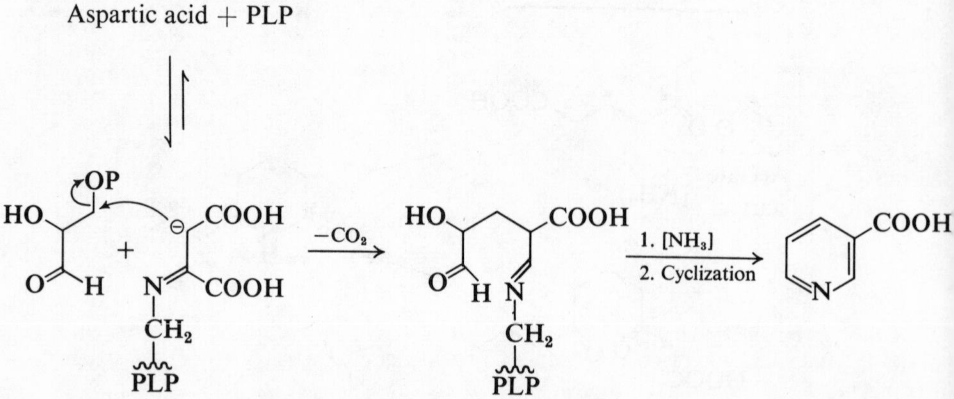

Aspartic acid + PLP

$$\xrightarrow{-CO_2} \quad \xrightarrow[\text{2. Cyclization}]{\text{1. }[NH_3]}$$

Fig. 8 Suggested biosynthesis of nicotinic acid

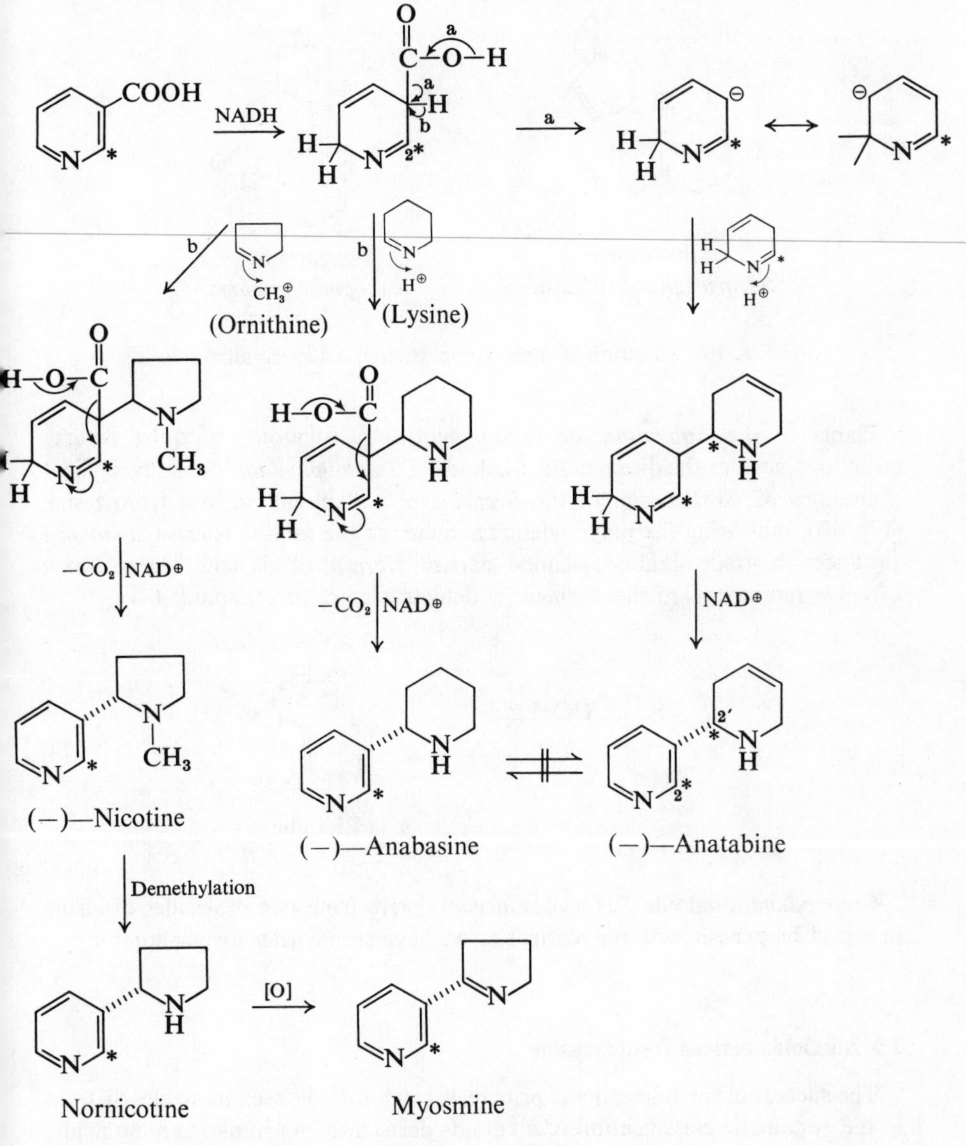

Fig. 9 Biosynthesis of *Nicotiana* alkaloids

coupling with Δ′-pyrroline or Δ′-piperideine preceding or possibly concerted with the decarboxylation (Fig. 9, route b). A delocalized carbanion may give rise to 2,5-^{14}C-labelled nicotine. Lysine serves as a precursor for the piperidine ring in anabasine and it is incorporated unsymmetrically. No interconversion between anabasine and anatabine has been observed.

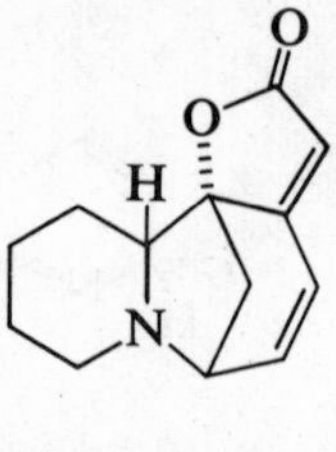

Securinine
Securinega suffruticosa

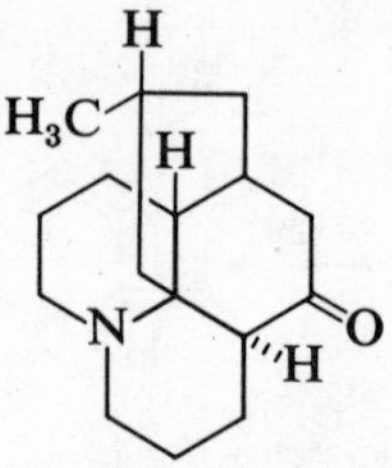

Lycopodine
Lycopodium selago

Fig. 10 Structure of more complex, lysine derived alkaloids

Plants of the *Euphorbiaceae* family synthesize alkaloids of quite diverse structures, such as the diterpenoid alkaloids of *Daphniphyllum*[17a] spp., benzylisoquinolines of *Croton* spp. or the *Securinega*[17b] alkaloids derived from lysine (Fig. 10), indicating the polyphyletic character of the family. *Ricinus communis* produces the toxic alkaloid, ricinine, derived from nicotinic acid. The carboxyl group is retained as nitrile, formed by dehydration of nicotinamide (4).

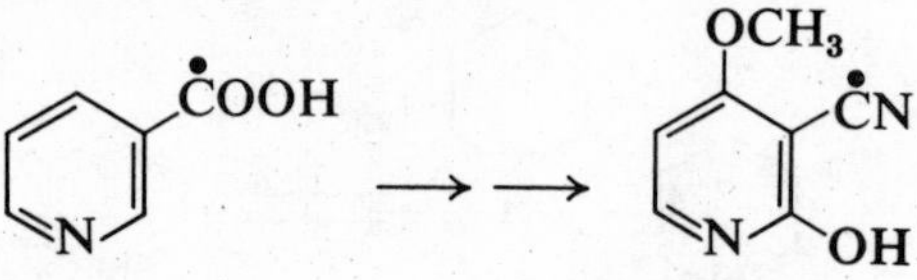

Ricinine (4)

Lycopodium alkaloids[17c] (*Lycopodiaceae*) derive from two molecules of lysine in mixed biogenesis with polyketides as we have seen earlier for pelletierine.

7.3 Alkaloids derived from tyrosine

The success of the biosynthetic principles can hardly be seen more clearly than in the systematic classification of alkaloids derived from aromatic amino acids, the largest group. Yet, in retrospect, with all respect to their success, these principles could not have been discovered without our experience derived at from classical aromatic electrophilic substitution and from radical reactions. The theories of Robinson and Barton *et al.* and the extensive tracer work by Battersby and others gradually gave us a coherent biosynthetic picture. The various skeletons of e.g. simple isoquinolines, opium alkaloids, *Amaryllidaceae* alkaloids, *Ipecacuanha* alkaloids, *Erythrina* alkaloids, benzophenanthridine alkaloids, phenethylisoquinoline alkaloids and of the related indole alkaloids can be constructed by application of a few fundamental principles:

1. aromatic hydroxylation, decarboxylation, and *O*-methylation of the amino acid to form hydroxylated β-arylethylamines;
2. Pictet–Spengler condensation of β-arylethylamine with the appropriate carbonyl compound;
3. a phenol coupling.

Occasionally ring fission and recyclization occur, as for the benzophenanthridine and indole alkaloids.

Most tyrosine-based alkaloids possess additional hydroxyl groups in the aromatic ring. Some of them are methylated or have become part of a methylenedioxy ring. The latter is formed by further oxidation of a methoxy group located *ortho* to a hydroxyl followed by ring closure (5) rather than by direct condensa-

$$\text{HO},\ CH_3O\text{–Ar–R} \xrightarrow{[O]} \text{H–O},\ H_2C{=}\overset{\oplus}{O}\text{–Ar–R} \longrightarrow \text{O–CH}_2\text{–O–Ar–R} \qquad (5)$$

tion of *ortho*-substituted hydroxyls with a formaldehyde equivalent. It appears from feeding experiments in various plants that hydroxylation of tyrosine precedes decarboxylation, whereas methylation is often a later event. This is understandable in the light of the electronic requirements of the electrophilic Pictet–Spengler condensation and the phenolic coupling which may follow. Free hydroxyls promote a much higher electron density in *ortho* and *para* positions than methoxyls as a result of dissociation, and the phenoxy radicals are formed by oxidation of the phenolate ion. Thus, at least one free hydroxyl is required *ortho* or *para* to the position at which ring closure occurs. In salsoline, 1-methyl-6-hydroxy-7-methoxytetrahydroisoquinoline, *O*-methylation could very well occur at any stage in the biosynthesis, but in lophocerine, the 6-methoxy isomer (Fig. 11) methylation presumably takes place at a late stage.

The isopropyl group of lophocerine arises independently, either from mevalonic acid or leucine[18,19] (Fig. 11). Two points in the scheme merit discussion. β-Phenylethylamine could conceivably first be alkylated by isopentenyl phosphate. Oxidation of the secondary amine to an imine followed by cyclization could then give the alkaloid. However, it is known from several other similar cases that the secondary amines act as poor precursors which makes this route unlikely. The carbonyl function is thus a prerequisite in the biosynthesis. The second point concerns the timing of the leucine decarboxylation. Does leucine or the derived α-keto acid form isopentanal as an intermediate or, alternatively, does the α-keto acid condense directly with the amine to a new amino acid which then decarboxylates? This possibility has not been examined in this particular case but it has to be considered because this pathway is followed in several other cases, e.g. in the biosynthesis of the pivotal benzylisoquinoline reticuline (Fig. 12).

The 1-benzylisoquinoline skeleton arises from two molecules of tyrosine one of which is oxidized and decarboxylated to give 3,4-dihydroxyphenylethylamine, dopamine. For a long time it was taken more or less for granted

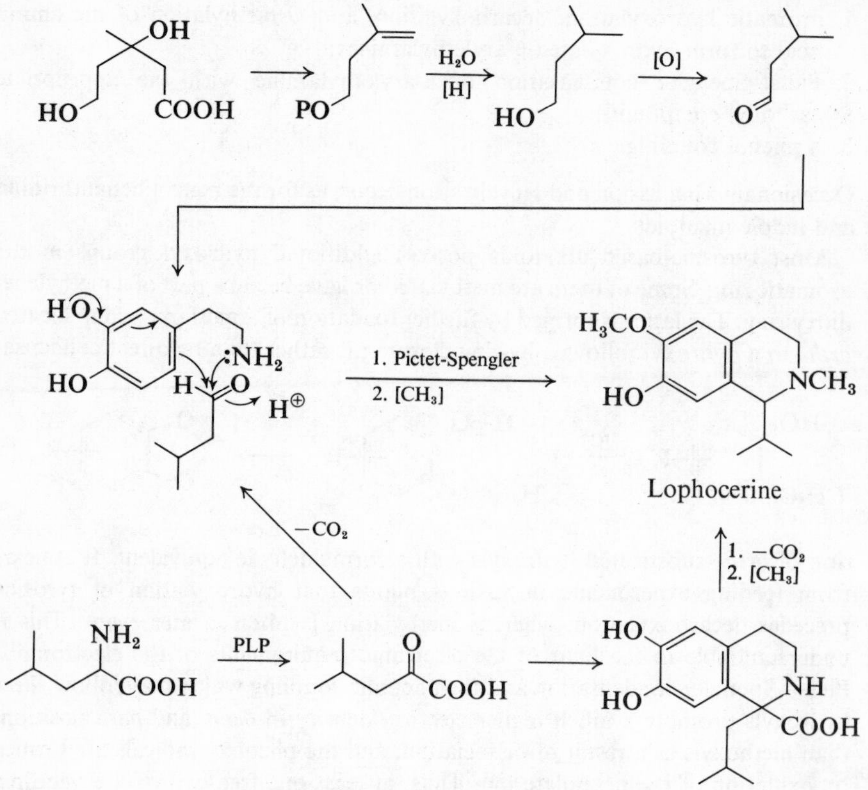

Fig. 11 Possible pathways to lophocerine

that the other tyrosine molecule gave 3,4-dihydroxyphenylacetaldehyde via regular PLP and thiamine mediated oxidation and decarboxylation, respectively, and that this aldehyde then condensed with the amine to form benzylisoquinoline. Extensive tracer work has now provided several important details of the biosynthesis,[20,21] (Fig. 12). 4-Hydroxyphenylpyruvic acid and 3,4-dihydroxyphenylpyruvic acid are, like tyrosine, incorporated into both parts of the molecule, whereas dopa, curiously enough, is incorporated only into the isoquinoline nucleus. 3-^{14}C-Tyrosine labels reticuline equally in the starred positions. The rationale came with the isolation of a new amino acid formed by condensation of 3,4-dihydroxyphenylpyruvic acid with dopamine which in the following step decarboxylated to norlaudanosoline. The specific incorporation of dopa in ring A of reticuline is explained by its inability to undergo rapid transamination to 3,4-dihydroxyphenylpyruvic acid and consequently also to form 3,4-dihydroxyphenylacetaldehyde capable of Pictet–Spengler condensation. 3,4-Dihydroxyphenylpyruvic acid can on the other hand be aminated and thus enter both rings. By examination of the efficiency of incorporation of differently methylated

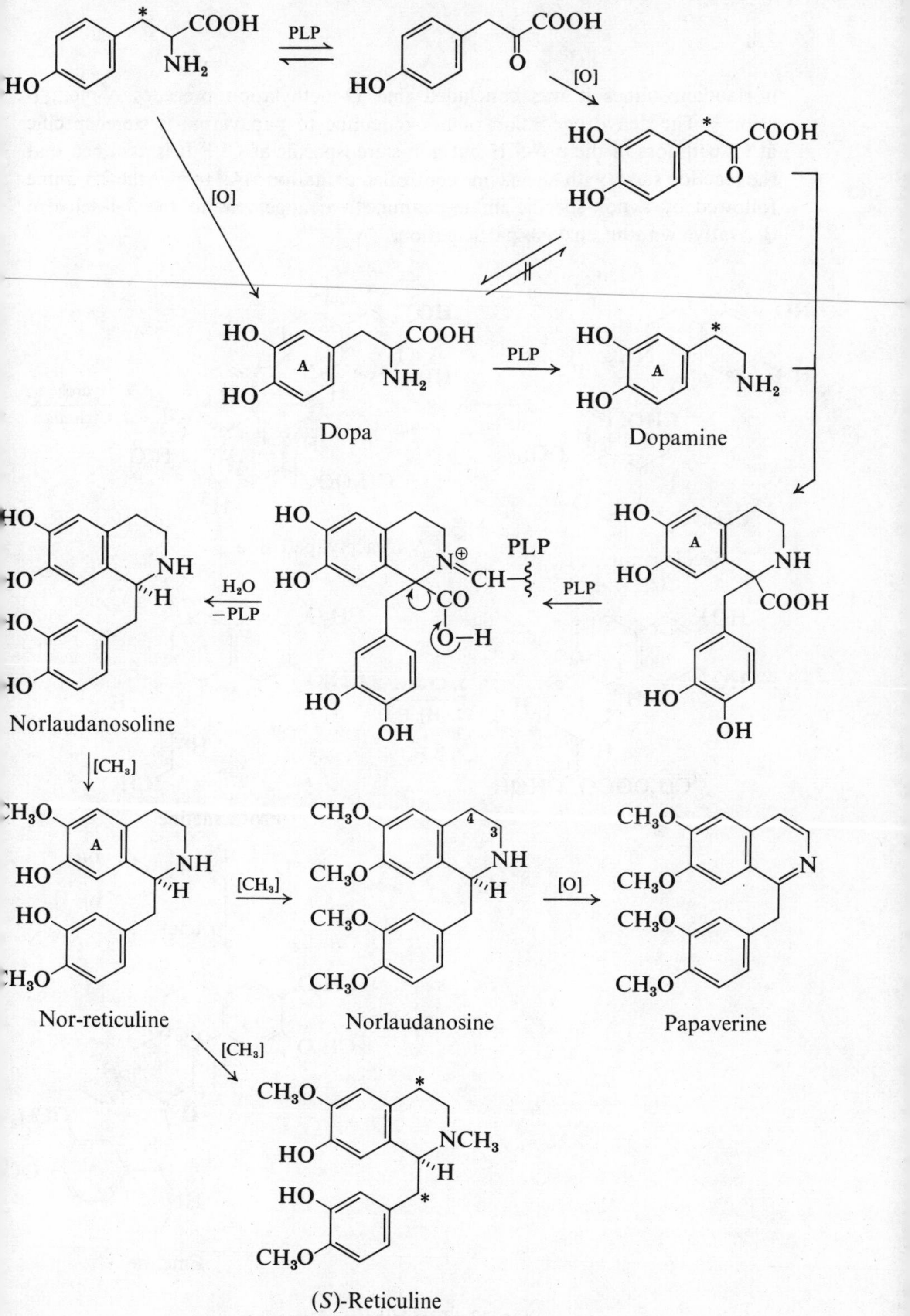

Fig. 12 Biosynthesis of reticuline and papaverine

norlaudanosolines it was concluded that *O*-methylation precedes *N*-methylation.[22] The dehydrogenation of nor-reticuline to papaverine is stereospecific at C^3 with loss of the pro-*S* H but non-stereospecific at C^4.[23] It is assumed that the reaction starts with an enzyme controlled oxidation at C^3 to give the 2,3-imine followed by a non-specific imine–enamine rearrangement to the 3,4-dehydro derivative without enzyme participation.

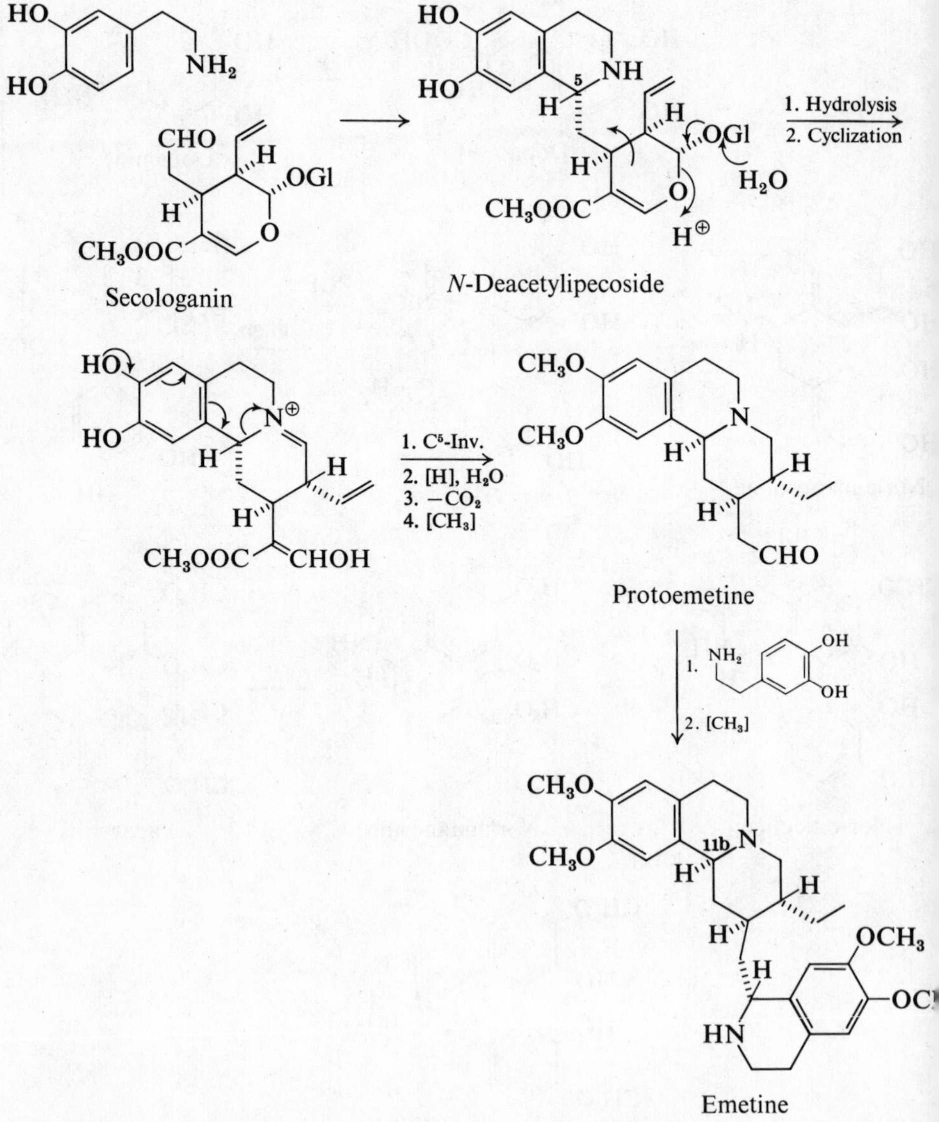

HO HO NH_2 CHO H OGl H CH_3OOC O

Secologanin

HO HO 5 NH H H H OGl CH_3OOC O $H^{\oplus}$ H_2O

1. Hydrolysis
2. Cyclization

N-Deacetylipecoside

HO HO $N^{\oplus}$ H H H CH_3OOC CHOH

1. C^5-Inv.
2. [H], H_2O
3. $-CO_2$
4. $[CH_3]$

CH_3O CH_3O N H H H CHO

Protoemetine

1. NH_2 OH OH
2. $[CH_3]$

CH_3O CH_3O 11b N H H H H OCH_3 OC HN

Emetine

Fig. 13 Biosynthesis of emetine

The important anti-amoebic but rather toxic drug, emetine, contains a more complicated aliphatic part which derives from the monoterpene, loganin (section 5.3). Loganin and its oxidative fission product, secologanin, play a central role in the biosynthesis of the indole alkaloids. It was actually biogenetic considerations based on results gained from the monoterpenoid field that guided the work on emetine towards the elucidation of the correct structure. Condensation of secologanin with dopamine gives (—)-*N*-deacetylipecoside (Fig. 13). Hydrolysis of the glycosidic bond liberates another aldehyde function that cyclizes with the amine to an immonium ion. Reduction, decarboxylation and a new Pictet–Spengler condensation completes the biosynthesis. Rather surprisingly, it turned out that the precursor *N*-deacetylipecoside had the opposite configuration to emetine at C^5. Tracer work showed that deacetylipecoside incorporated well in emetine but its C^5 epimer, *N*-deacetylisoipecoside, with the same configuration as emetine not at all.[24] This implies that only ipecoside fits the enzyme surface adequately for the glycolysis and the cyclization and at some stage an inversion must occur. It was demonstrated by ^3H-labelling of the aldehyde proton in secologanin that it was retained at C^{11b} in emetine. This finding excludes an equilibration of the immonium ion into conjugation with the aromatic ring. But the immonium structure is well suited for another rearrangement which could lead to the wanted inversion. We have a 'push–pull' situation with an electron donating hydroxyl *para* to the carbon carrying the immonium function which can bring about a fission of the C–N bond and rearrangement to the epimer (Fig. 13).

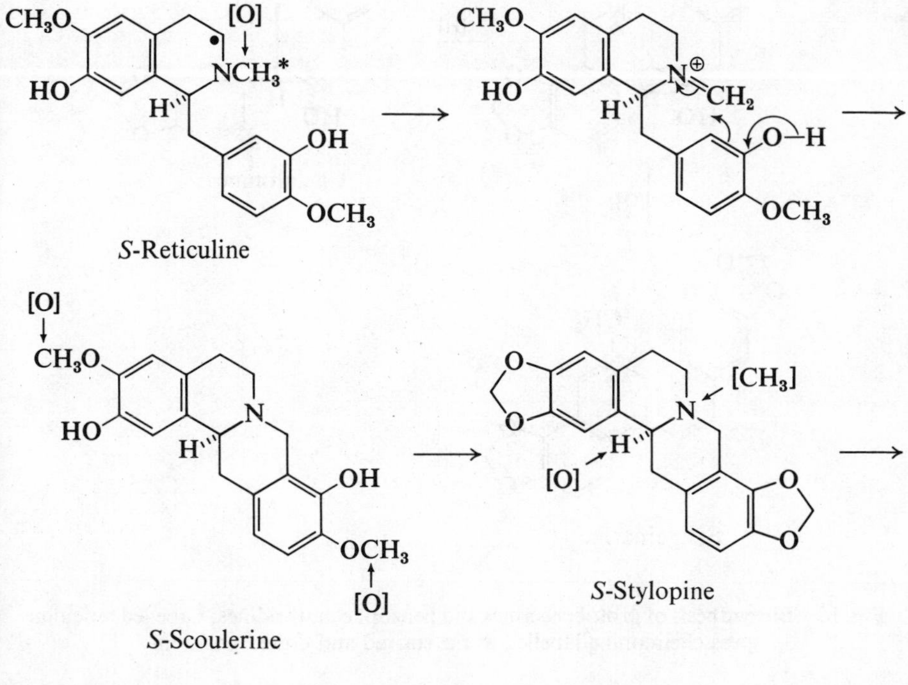

S-Reticuline

S-Scoulerine

S-Stylopine

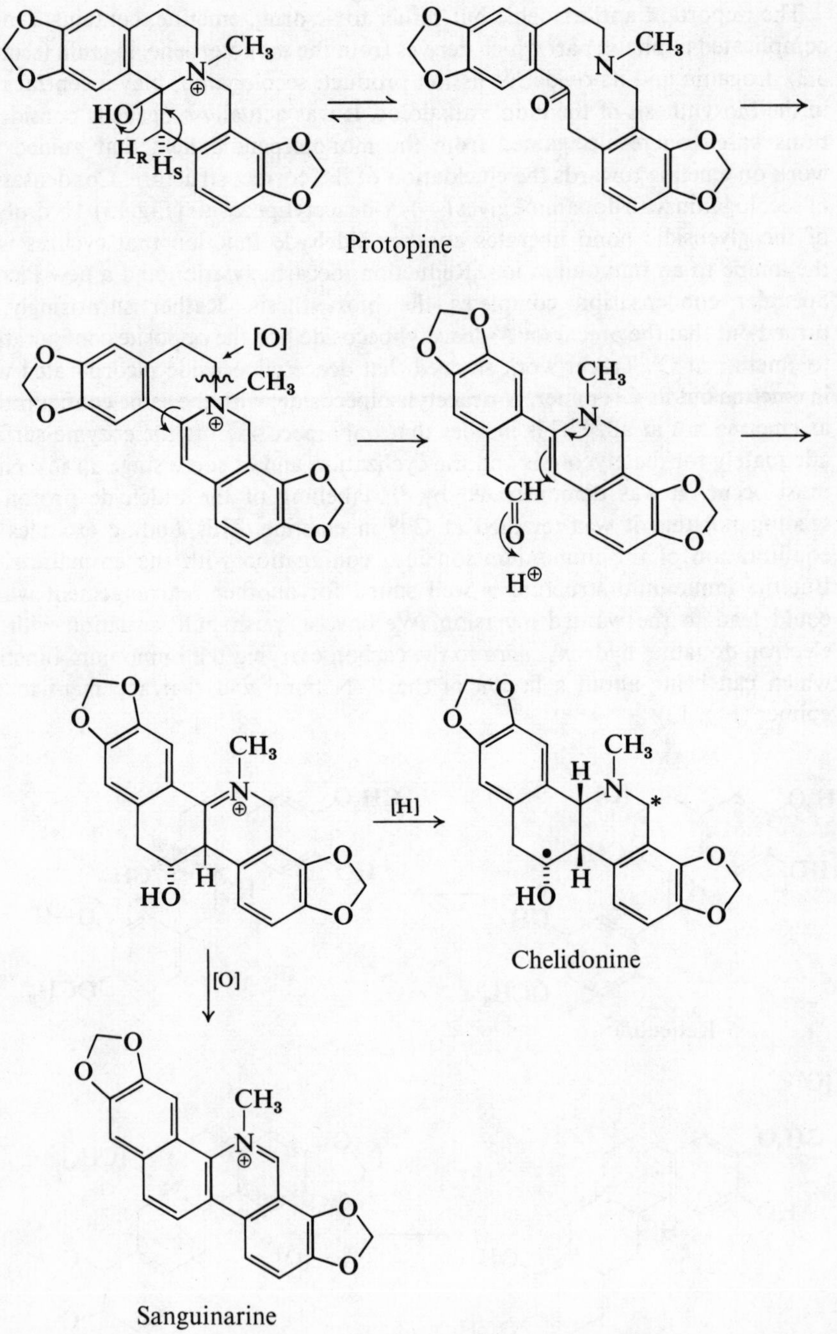

Fig. 14 Biosynthesis of protoberberines and benzophenanthridines. Labelled reticuline gives chelidonine labelled at the starred and dotted positions

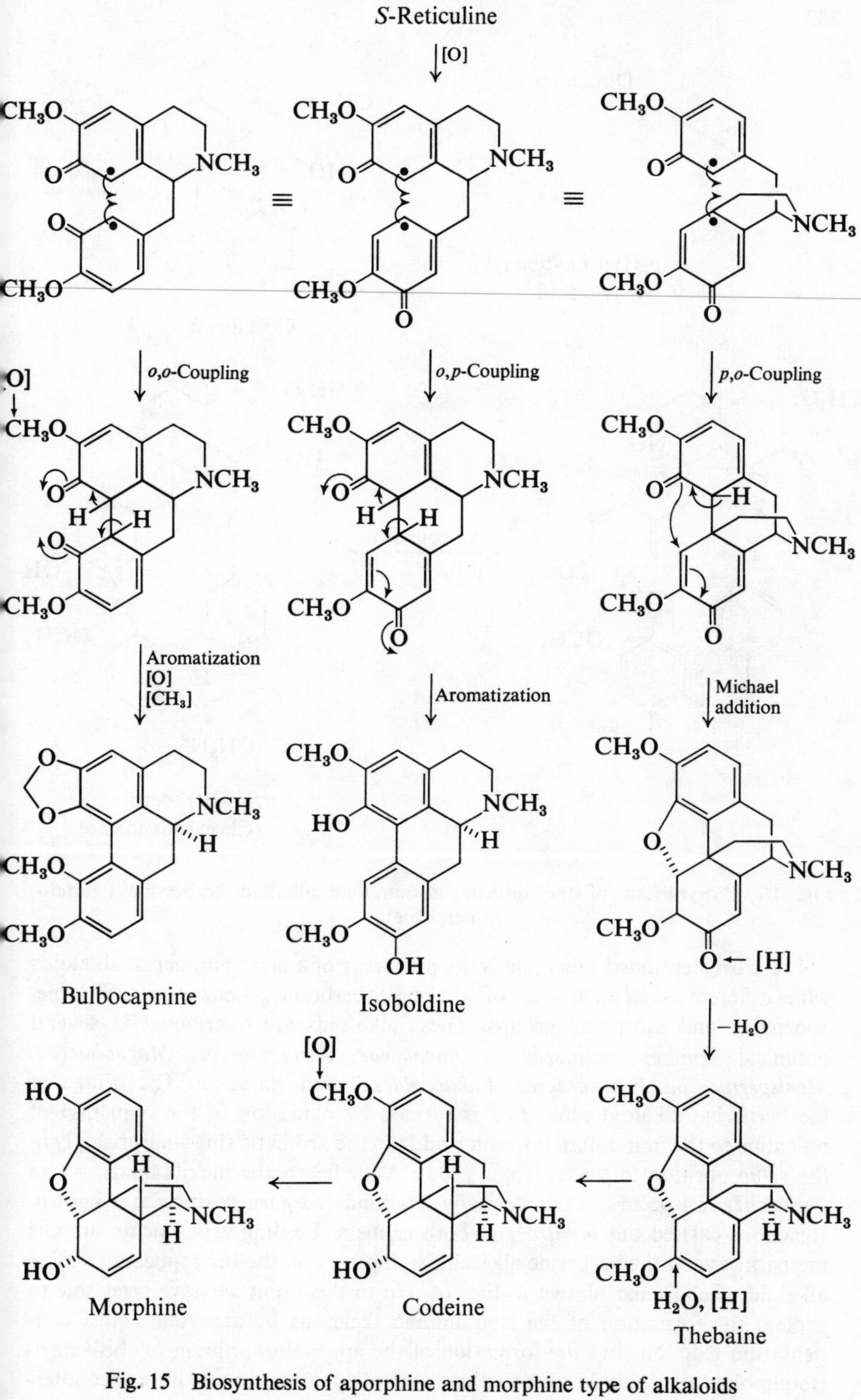

S-Reticuline

[O]

o,o-Coupling

o,p-Coupling

p,o-Coupling

Aromatization
[O]
[CH3]

Aromatization

Michael addition

Bulbocapnine

Isoboldine

[O]

$-H_2O$

H_2O, [H]

Morphine

Codeine

Thebaine

Fig. 15 Biosynthesis of aporphine and morphine type of alkaloids

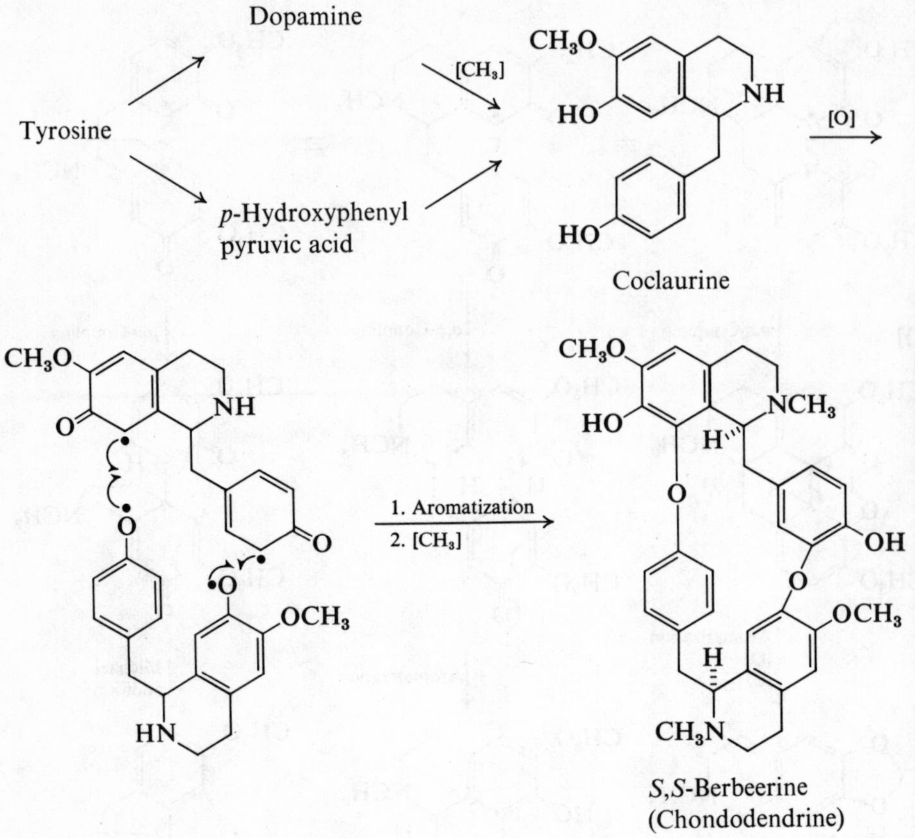

Dopamine

Tyrosine

p-Hydroxyphenyl pyruvic acid

$[CH_3]$

CH_3O HO NH HO

Coclaurine

[O]

CH_3O O NH O O OCH_3 HN O

1. Aromatization
2. $[CH_3]$

CH_3O HO NCH_3 H O OH O OCH_3 H CH_3N

S,S-Berbeerine (Chondodendrine)

Fig. 16 Biosynthesis of the bisbenzylisoquinoline alkaloid berbeerine (chondodendrine)

The aforementioned reticuline is the precursor of a large number of alkaloids with different skeletons, e.g. of the protoberberine, benzophenanthridine, aporphine and morphine groups. These alkaloids are distributed in several botanical families; primarily in *Anonaceae*, *Berberidaceae*, *Magnoliaceae*, *Menispermaceae*, *Papaveraceae*, *Ranunculaceae* and *Rutaceae*. The bridge of the berberine alkaloid scoulerine is formed by oxidation of the *N*-methyl of reticuline to the immonium ion which adds to the aromatic ring—surprisingly in the *ortho* position to the hydroxyl group. Very few berberine alkaloids, where the cyclization occurs in the sterically less hindered *para* position, are known. Reactions carried out *in vitro* give both isomers. Feeding experiments support the pathways to the berberine alkaloid, stylopine, and the benzophenanthridine alkaloid, chelidonine, shown in Fig. 14. Up to this point we have been able to explain the formation of the isoquinoline skeletons by applying simple condensation reactions but the formation of the aporphines, opium or bisbenzylisoquinoline alkaloids demands a phenol coupling, a reaction we have encountered earlier, e.g. in lignin chemistry. Fig. 15 shows how the reticuline biradical by appropriate twisting gives aporphines and the bridged morphine skeleton by intramolecular C–C coupling in *ortho–ortho* and *ortho–para* positions to the free hydroxyl. The comparatively complex morphine skeleton can thus be deduced in an amazingly simple way with the correct oxygenation pattern. Intermolecular C–O couplings give the dimeric bisbenzylisoquinoline alkaloids as exemplified by the formation of berbeerine from coclaurine (Fig. 16). Berbeerine (chondo-

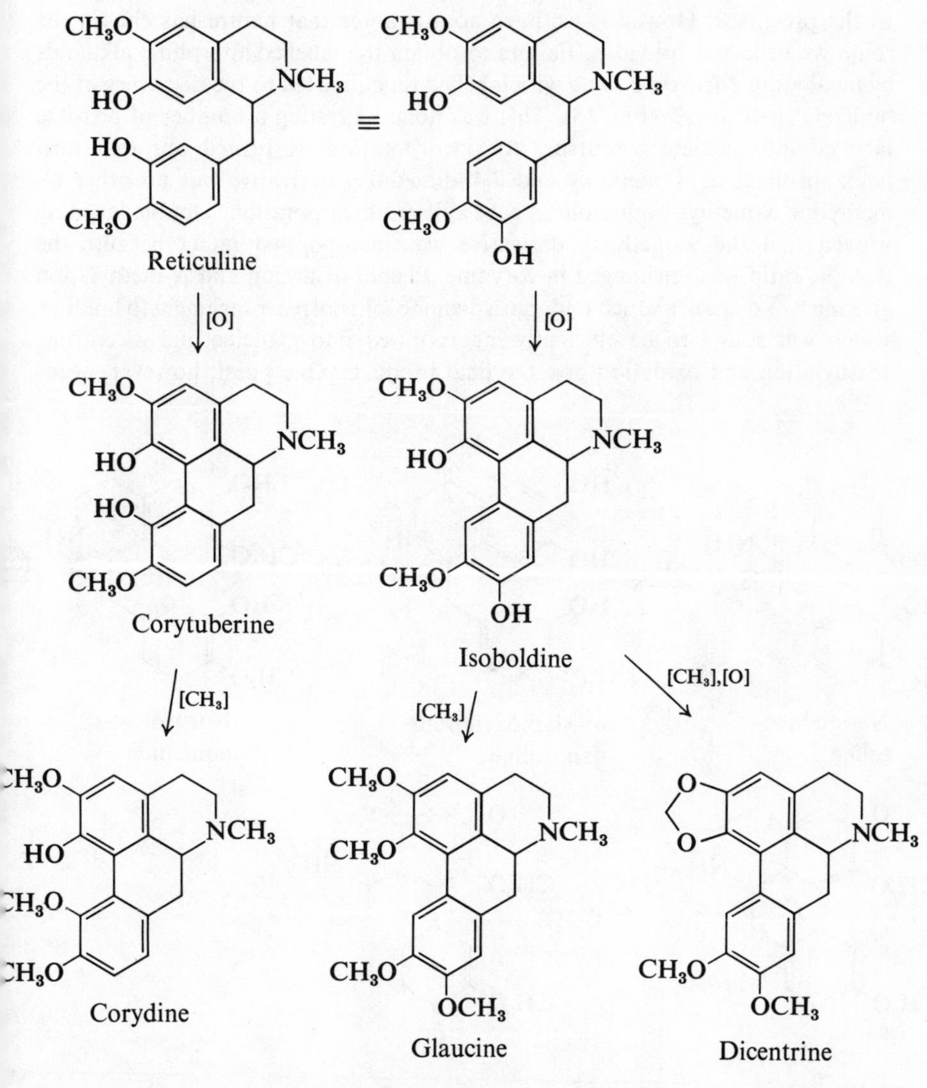

Reticuline

Corytuberine

Isoboldine

Corydine

Glaucine

Dicentrine

Fig. 17 Disproved routes to corydine, glaucine, and dicentrine in *Dicentra eximia*.

dendrine, *S,S* structure) occurs together with curine (*R,R*-isomer) and tubocurarine, the *N,N'*-dimethylammonium salt (*R,S*-structure) in South American *Chondodendron* spp. Tubocurarine which has strong curare-like effects, is the highly active component of the arrow poison prepared by natives of the Amazon.

We must never fall into the habit of routine thinking just because our scheme works beautifully in one case. The straightforward phenol oxidation in Fig. 15 was neatly proved by feeding experiments. Consequently the most 'obvious' routes to the closely related aporphine alkaloids, corydine, glaucine, and dicentrine, should, by analogy, be the ones depicted in Fig. 17 with reticuline as the precursor. However, we have no guarantee that nature has chosen the route we believe is 'obvious. Failure to obtain the labelled aporphine alkaloids by incubating *Dicentra eximia* with labelled reticuline led to the discovery of the 'unlikely' pathways[25] (Fig. 18). This was done by testing a number of possible labelled intermediate precursors for incorporation. It turned out that norlaudanosoline, its 4'-methoxy and 7,4'-dimethoxy derivative but no other *O*-methyl or *N*-methyl isoquinolines gave efficient incorporation. Double labelling proved that the 4'-methoxy derivative was incorporated intact because the isotopic ratio was unchanged in corydine. Phenol oxidation and *N*-methylation give the two dienones which undergo a dienone–phenol rearrangement to boldine, which was found to be efficiently incorporated into glaucine and dicentrine. Methylation and oxidation give the final products. One point, however, needs

Norlaudanosoline → $[CH_3]$ → 4'-Methoxynorlaudanosoline → $[CH_3]$ → Norprotosinomenine → [O]

p,o-Coupling $[CH]_3$

p,p-Coupling $[CH_3]$

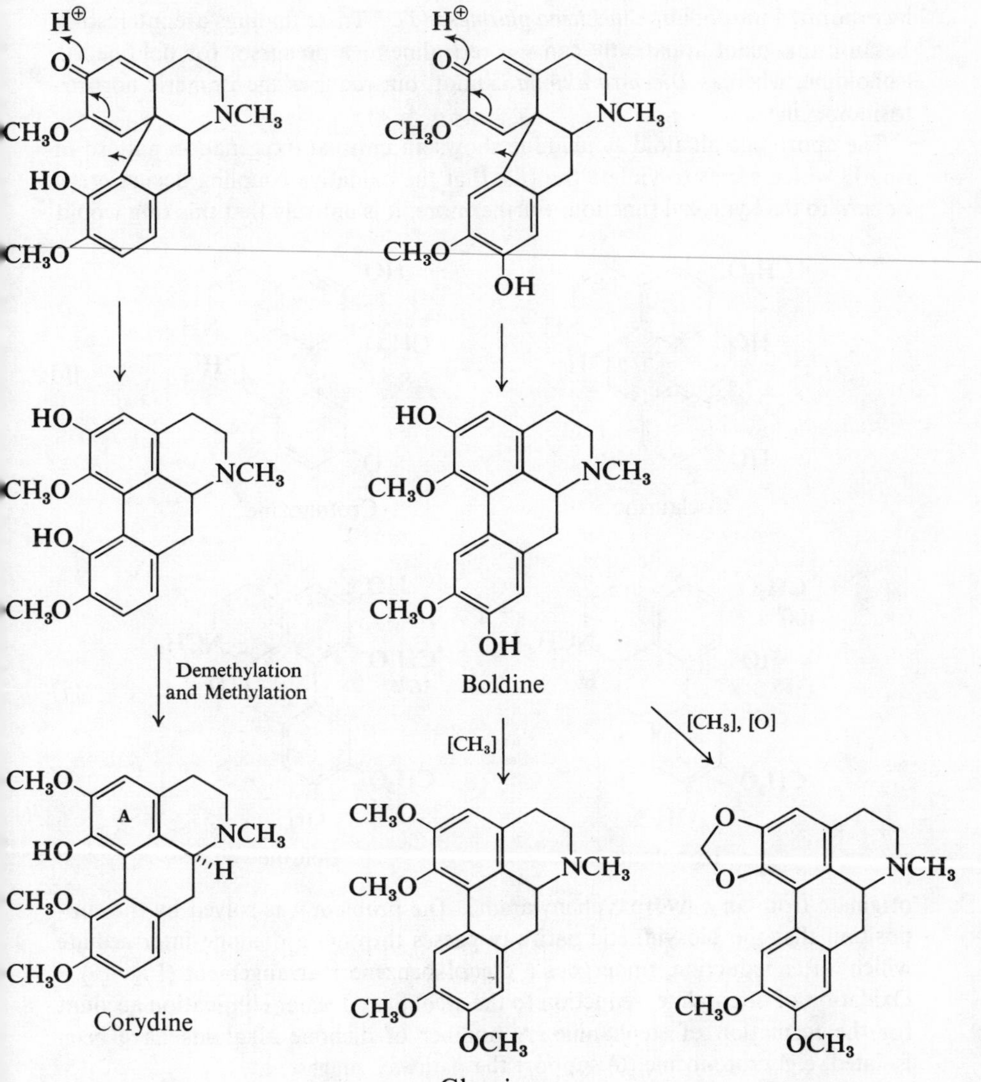

Fig. 18 Biosynthesis of corydine, glaucine and dicentrine

further clarification. Ring A of corydine has the opposite methylation pattern to the postulated aporphine intermediate. It is not yet proved that this is a true intermediate, but if so, how is the methylation pattern changed? From analogous cases it may be concluded that a demethylating–methylating sequence is more likely than a direct methyl transfer. Coclaurine labelled with ^{14}C in the *O*-methyl group lost part of its label during its transformation into crotonosine[26] in *Croton linearis* (6) and (±) 6-O $^{14}CH_3$-reticuline lost 64 per cent of its label when incorporated into boldine in *Litsea glutinosa* (7).[27] These findings are interesting because this plant apparently can use reticuline as a precursor for boldine via isoboldine, whereas *Dicentra eximia* cannot, but requires the isomeric norprotosinomenine.

The aporphine alkaloid stephanine shows an unusual oxygenation pattern in ring D which seems to violate the rule that the oxidative coupling occurs *ortho* or *para* to the hydroxyl function. Furthermore, it is unlikely that this ring would

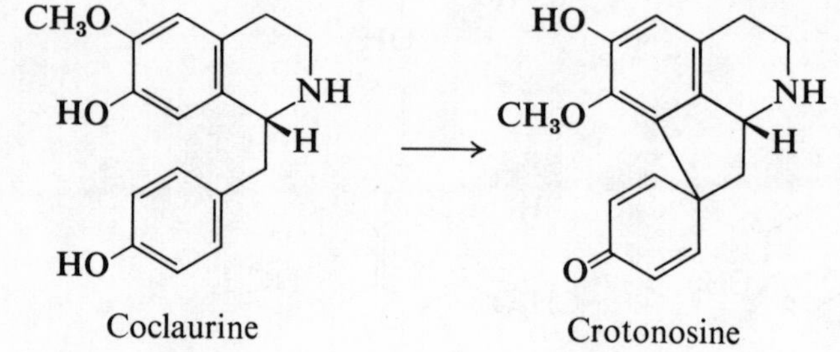

Coclaurine → Crotonosine (6)

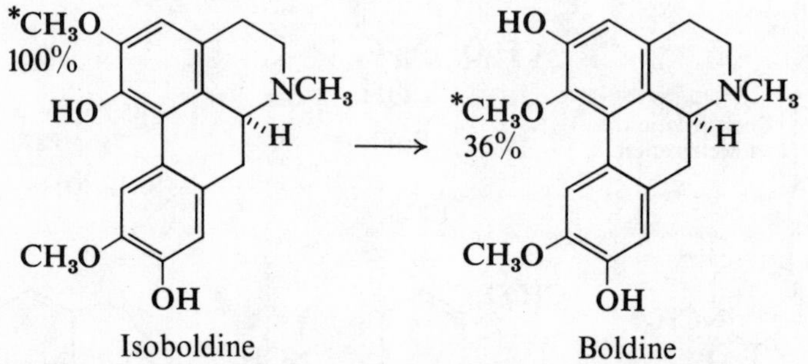

Isoboldine → Boldine (7)

originate from an *o*-hydroxyphenylamine. The problem was solved by the supposition that the biosynthetic pathway passes through a dienone intermediate which, after reduction, undergoes a dienol–benzene rearrangement (Fig. 19).[28] Oxidation of orientaline, reduction to the alcohol and water elimination account for the formation of stephanine. A number of dienone alkaloids have been isolated, e.g. crotonosine (6) support the pathway suggested.

Several of the phenol couplings discussed can be mimicked in the laboratory by oxidation of phenolic tetrahydroisoquinolines with potassium ferricyanide. An early attempt to synthesize the aporphine skeleton by oxidation of laudanosoline did not give the wanted product, but instead a compound having the dibenzopyrrocoline skeleton (8). At that time no alkaloids were known of that type, but several years later alkaloids were isolated from the bark of *Cryptocarya bowiei* which were shown to have this structure. If the amino function is blocked by quaternization, the reaction leads to the aporphine skeleton in good yield.

The occurrence of the *Erythrina* alkaloids is, with few exceptions, limited to the genus *Erythrina* of the *Leguminosae* family. They are physiologically distinguished by a strong curare-like effect. Approximately 1 mg/kg of the most

HO, HO, NCH₃, OH, OH → [O] → HO, HO, NCH₃, O, O

1. *N*-methylation
2. [O]

(8)

HO, HO, $\overset{\oplus}{N}(CH_3)_2$, HO, OH

HO, HO, $\overset{\oplus}{N}CH_3$, OH, OH

potent alkaloids injected intravenously in a frog is sufficient to cause total paralysis. The skeleton of the *Erythrina* alkaloids is another artful variation of the phenol coupling theme. Erythraline and isococculidine (Fig. 20) stem from *S*-norprotosinomenine—only the *S*-isomer is incorporated—by *para–para* coupling followed by cleavage of the C^{Ar}–C^1 bond, supported by interaction of the electron lone pair at the nitrogen. Reduction of the intermediate imine and oxidation of the *p,p*-dihydroxybiphenyl to a diphenoquinone give, after cyclization with the amino function, the *Erythrina* alkaloid skeleton and by conventional reactions finally erythraline. Isococculidine lacks the oxygen function at C^{16}. This can in principle be removed by reductive elimination either at the first asymmetric quinonoid step or at the second quinonoid step, where asymmetry is lost. It was found that 7-$^{14}CH_3O$-norprotosinomenine diverted its label into both parts of isococculidine,[29] i.e. the reduction occurs at the diphenoquinone stage.

All the tyrosine-derived alkaloids we have encountered so far, can be dissected into two C_6–C_2 groups. The *Amaryllidaceae* alkaloids are characterized by having one C_6–C_2 and one C_6–C_1 unit originating from tyrosine and phenylalanine, respectively (Fig. 21). Although tyrosine and phenylalanine are metabolically closely related and the units in the alkaloids are highly oxygenated, very little randomization occurs due to selective enzymatic and compartmental effects. Phenylalanine is degraded to protocatechuic aldehyde (section 3.3) by the sequence phenylalanine → cinnamic acid → coumaric acid → caffeic acid → protocatechuic aldehyde. It was established by ^3H-labelling that phenylalanine

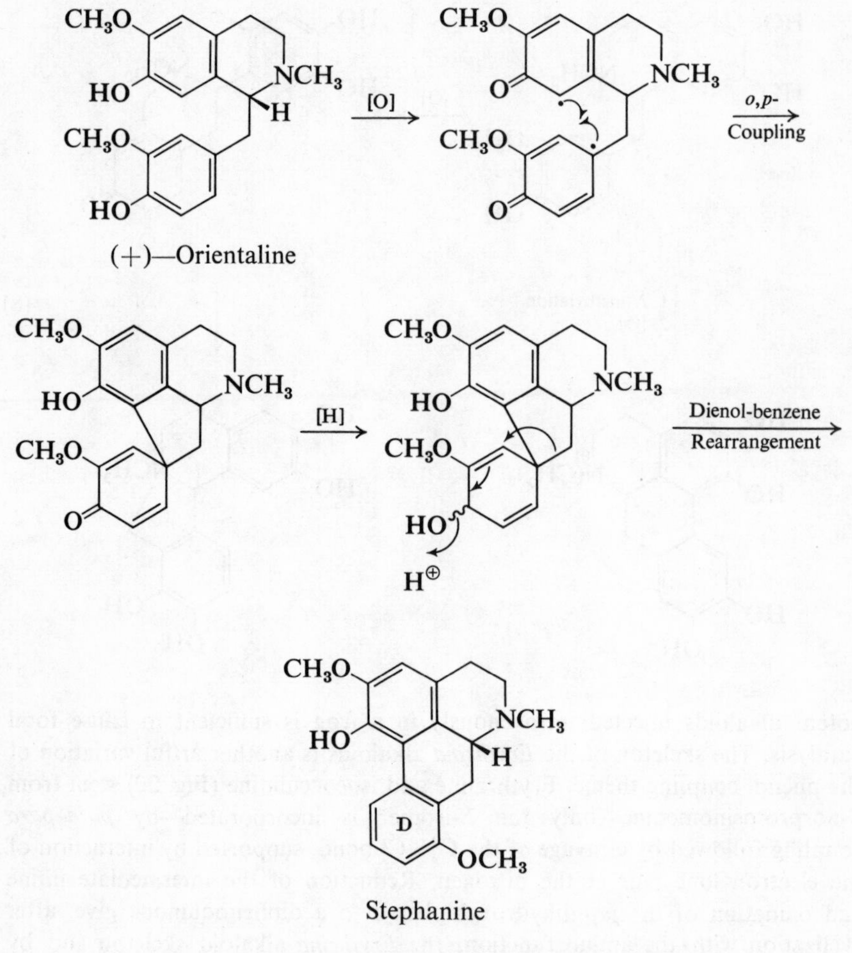

Fig. 19 Biosynthesis of stephanine

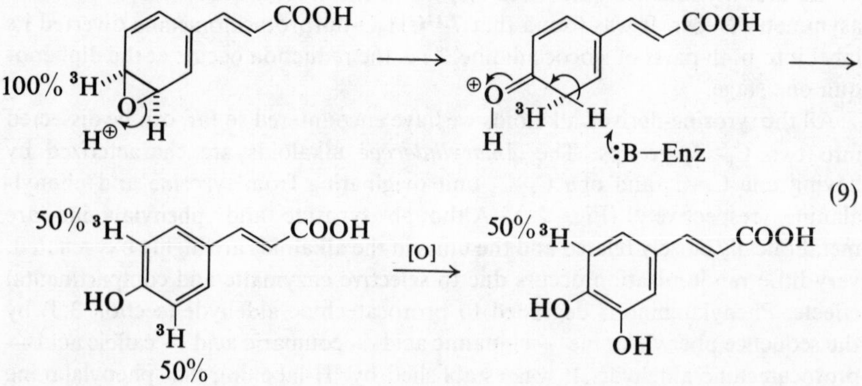

(9)

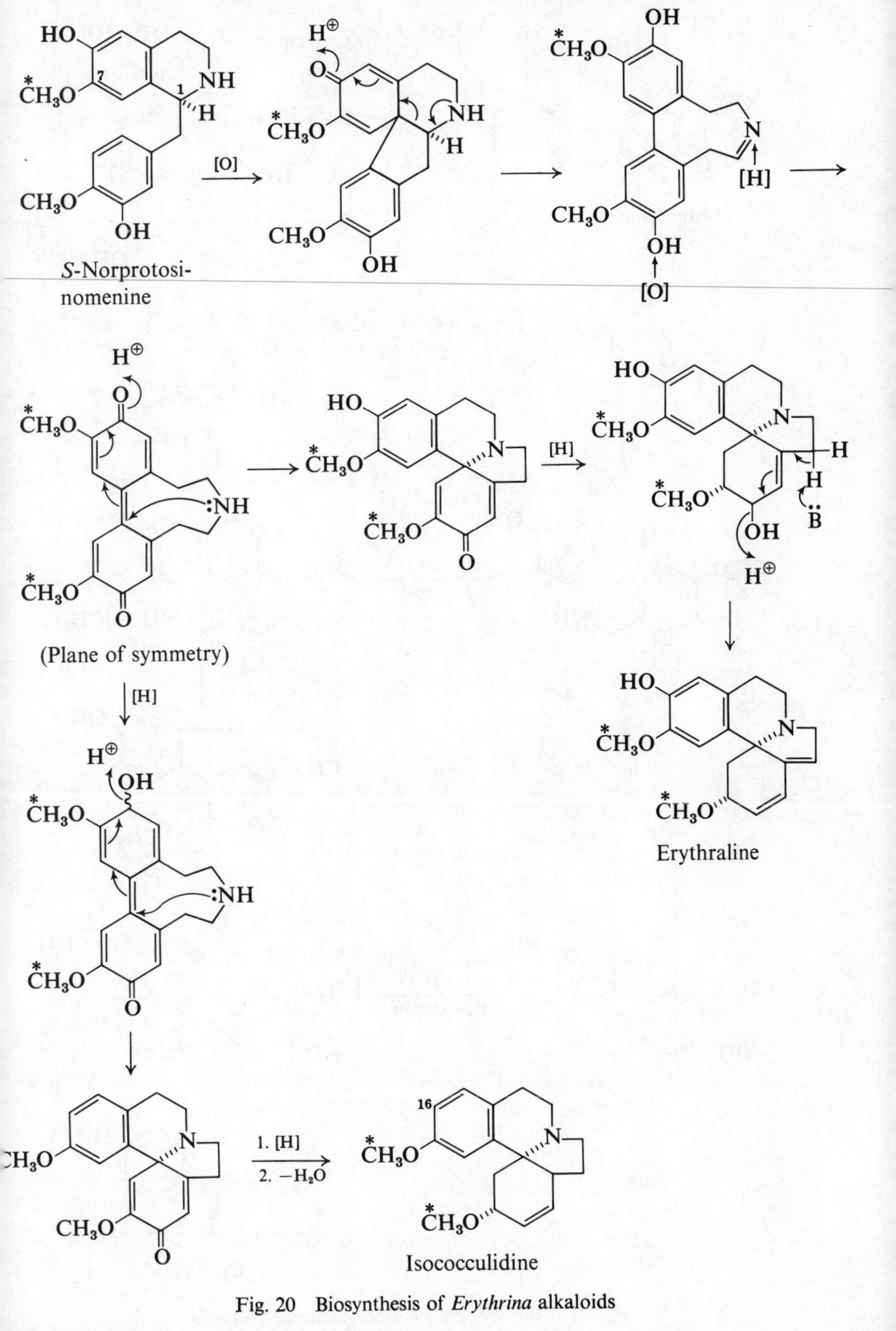

S-Norprotosinomenine

(Plane of symmetry)

Erythraline

Isococculidine

Fig. 20 Biosynthesis of *Erythrina* alkaloids

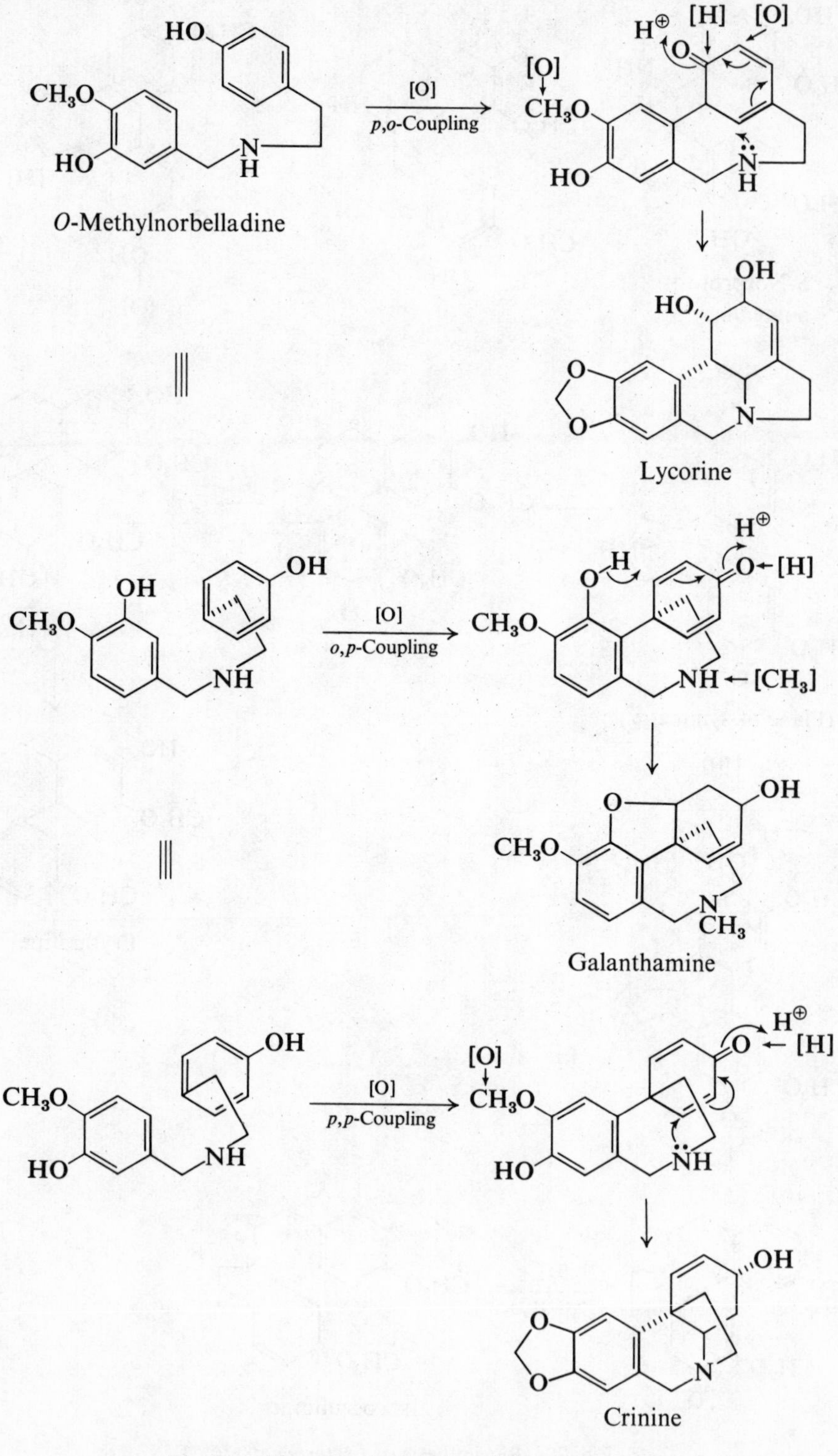

HO

CH_3O

HO

N H

O-Methylnorbelladine

[O]

p,o-Coupling

[O]

CH_3O

HO

$H^{\oplus}$ [H] [O]

O

N H

HO

OH

O

O

N

Lycorine

OH

CH_3O

OH

NH

[O]

o,p-Coupling

$H^{\oplus}$

H

O

O ← [H]

CH_3O

NH ← $[CH_3]$

O

OH

CH_3O

N

CH_3

Galanthamine

OH

CH_3O

HO

NH

[O]

p,p-Coupling

[O]

CH_3O

HO

$H^{\oplus}$

O ← [H]

NH

OH

O

O

N

Crinine

Fig. 21 Biosynthesis of Amaryllidaceae alkaloids

at some stage lost both hydrogens at C^3, which suggests that the fragmentation of the cinnamic acid cannot be formulated simply as a β-hydroxylation followed by a retroaldol condensation. At some stage of the biosynthesis of the C_6–C_1 unit, C_3 of the amino acid is oxidized to the ketone level. The immediate precursor of protocatechuic aldehyde is probably protocatechuoyl-CoA. The biological hydroxylation of 4-^{3}H-cinnamic acid to *p*-coumaric acid involved complete migration and retention of tritium (NIH shift, section 3.7). The introduction of the *m*-hydroxy group proceeded with loss of 50 per cent of the tritium. The first complete retention could conceivably be explained by a large ^{3}H isotope effect but since incorporation of 3,5-$^2H_2$-4-^{3}H-cinnamic acid gave the same result, it is suggested that the stereospecific proton elimination is enzymatically controlled [30,31] (9). Tyrosine is incorporated via tyramine but dopa is not metabolized. Protocatechuic aldehyde condenses with tyramine and gives on reduction the pivotal norbelladine which, depending upon the folding, gives rise to the basic *Amaryllidaceae* skeletons[28] (Fig. 21).

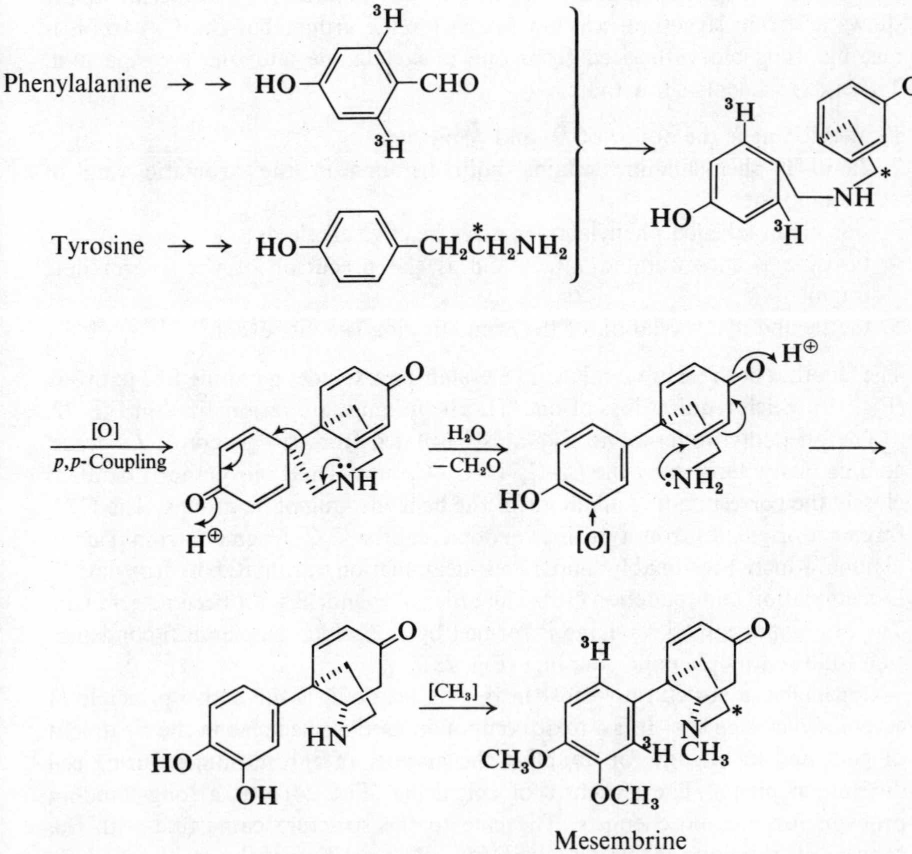

Fig. 22 Biosynthesis of mesembrine

H, COOH, H_S, NH_2, H_5C_6, H_R $\xrightarrow{NH_3\text{-Lyase}}$ H_R, COOH, H $\xrightarrow{1.\ [O]\quad 2.\ [CH_3]\quad 3.\ [H]}$

HO, COOH, CH_3O, OCH_3 (10)

The C_6–C_1–N–C_2C_6 unit is uncommon outside *Amaryllidaceae*. Cryptostyline,[32] which belongs to a small group of alkaloids found in *Orchidaceae* is a 1-phenylisochinoline derivative formed analogously to the 1-benzylisoquinolines. Mesembrine (Fig. 22) found in *Sceletium strictum* of the *Aizoaceae* family shows a strong structural kinship to the crinine group, but the C_1 carbon is missing. It is biosynthesized from one phenylalanine and one tyrosine unit. Tracer experiments show that:

1. methionine is the donor of *O*- and *N*-methyls;
2. 2,6-di-^{3}H-phenylalanine retains both tritiums in the aromatic ring of mesembrine;
3. side chain labelled phenylalanine gives inactive alkaloids;
4. tyrosine is incorporated intact and is the precursor of the hydroindole fragment;
5. the second hydroxylation of the aromatic ring is a late event.[33]

The fact that both ^{3}H are retained in mesembrine excludes a crinine-like pathway (Fig. 19) which requires loss of one ^{3}H. The findings are rationalized in Fig. 22.

The phenethylisoquinoline alkaloids[34] isolated from six genera of *Liliaceae*, contain characteristically the C_6–C_3–N–C_2–C_6 unit. Their biosynthesis parallels closely the corresponding alkaloids of the benzylisoquinoline groups. The C_6C_2 fragment originates from tyrosine via dopa and the C_6C_3 fragment from phenylalanine which presumably undergoes deamination, aromatic hydroxylation, *O*-methylation and reduction (10). The order of events has not been determined. The homoaporphine kreysigine is formed by a Bischler–Napieralski condensation followed by phenolic coupling (Fig. 23).

Colchicine, a structurally most intriguing alkaloid, is the active principle of several *Colchicum* spp. It is a toxic compound used in medicine in the treatment of gout and in biology for doubling the number of chromosomes during cell division in plants. The structure of colchicine (Fig. 24) was a long-standing problem for organic chemists. The clue to the structure came first with the concept of tropolone aromaticity by Dewar. The position of the carbonyl group was finally determined by an X-ray diffraction study. Tracer studies in *Colchicum*

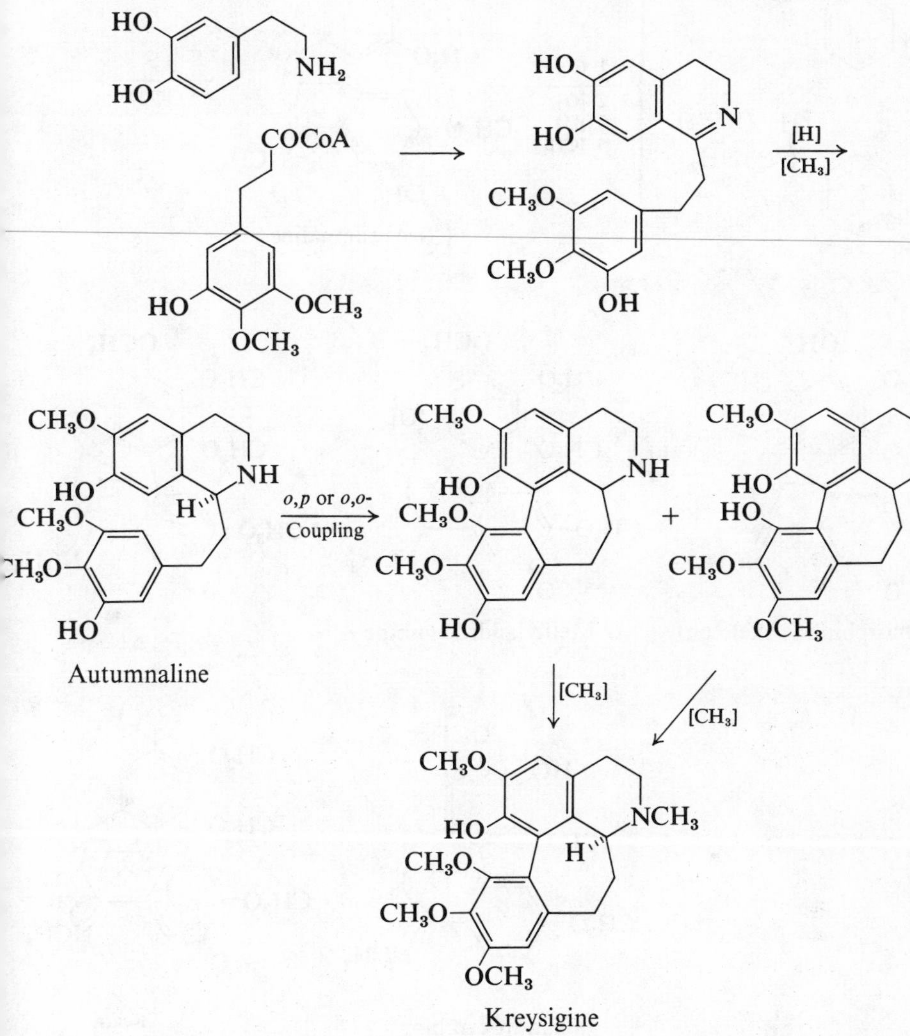

HO
HO
NH_2
COCoA
HO
OCH$_3$
OCH$_3$
HO
HO
N
[H]
[CH_3]
CH_3O
CH_3O
OH

CH_3O
HO
NH
H
CH_3O
CH_3O
HO
Autumnaline
o,p or *o,o*-
Coupling
CH_3O
HO
NH
CH_3O
CH_3O
HO
+
CH_3O
HO
NH
HO
CH_3O
OCH$_3$
[CH_3]
[CH_3]
CH_3O
HO
NCH_3
H
CH_3O
CH_3O
OCH$_3$
Kreysigine

Fig. 23 Biosynthesis of kreysigine, a homoaporphine alkaloid

spp. revealed that intact phenylalanine provided the C_6–C_3 fragment and the aromatic nucleus of tyrosine plus C^3 provided the seven membered tropolone system by ring enlargement. The nitrogen originates from tyrosine which was proved by a double labelling technique. The $^{14}C/^{15}N$ ratio of autumnaline was identical to the $^{14}C/N^{15}N$ ratio in colchicine. Thus, it cannot, for example, be introduced by some late transamination reaction occurring in the condensed 6,7,7-membered tricyclic structure.

The lyase mediated elimination of ammonia involves loss of the pro-Hs at C^3

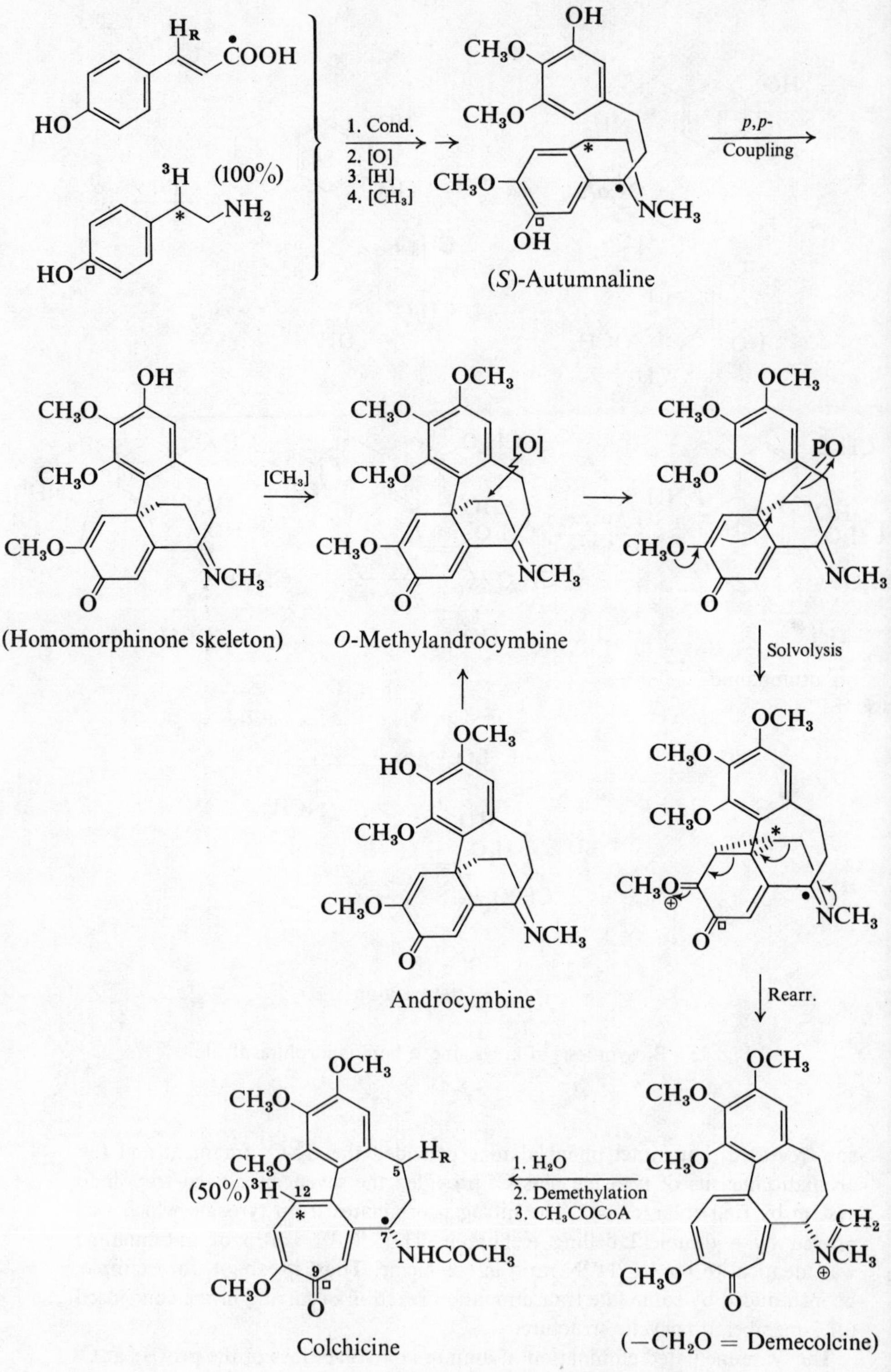

H_R
COOH
HO
³H (100%)
NH₂
HO
1. Cond.
2. [O]
3. [H]
4. [CH₃]
OH
CH₃O
CH₃O
CH₃O
NCH₃
OH
(S)-Autumnaline
p,p-
Coupling
(Homomorphinone skeleton)
[CH₃]
O-Methylandrocymbine
[O]
PO
Solvolysis
Androcymbine
Rearr.
1. H₂O
2. Demethylation
3. CH₃COCoA
(50%)³H
NHCOCH₃
Colchicine
CH₂
NCH₃
(−CH₂O = Demecolcine)

Fig. 24 Biosynthesis of colchicine

of phenylalanine (10), consequently stereospecifically labelled H_R should be retained at C^5 in colchicine which was shown to be the case.[35] 1-^{14}C-Labelled phenylalanine, 3-^{14}C- and 4′-^{14}C-labelled tyrosine appear at C^7, C^9 and C^{12}, respectively. The specific labelling of colchicine proves that the metabolic paths of phenylalanine and tyrosine follow separate lines in the plant. Little by little a fascinating biosynthetic scheme emerged out of the results of numerous feeding experiments (Fig. 24). The two amino acids give first a phenethylisoquinoline derivative which undergoes a directed *p,p*-phenol coupling to give the homomorphinone skeleton. The presumed relay alkaloid autumnaline proved, in fact, to be efficiently incorporated, whereas phenethylisoquinolines with other methoxylation patterns were incorporated poorly. It so happened that androcymbine, an isomer of the presumed *p,p*-phenol coupling product from autumnaline, was isolated from *Androcymbium melanthioides*, a relation of *Colchicum autumnale*, and its *O*-methyl derivative was found to be incorporated remarkably well.[35,36] Therefore, the next step is a methylation of the homomorphinone to *O*-methylandrocymbine. The insertion of C^{12} is best explained by enzymatic controlled hydroxylation because half of the tritium was removed from 3-^{3}H-labelled autumnaline. Non-specific hydroxylation would result in an isotope effect and oxidation to the carbonyl state with a complete loss of tritium. Solvolysis of the phosphate group supported by electron donation by the methoxy group leads to the presumed cyclopropane derivative which eventually rearranges to the tropolone nucleus. C^2 of tyrosine is believed to be lost as formaldehyde. The intermediate *N*-methyl derivative, demecolcine is efficiently incorporated into colchicine. The ring expansion has, in fact, a laboratory analogy in reaction (11) which further strengthens the suggested biosynthetic route.

H_2C—OH ; CH_3O ; OCH_3 $\xrightarrow{TsCl}$ H_2C—OTs ; CH_3O ; OCH_3 $\xrightarrow{\Delta}$

H ; C—H ; $CH_3\overset{\oplus}{O}$; OCH_3 $\longrightarrow$ OCH_3 ; OCH_3

(11)

7.4 Alkaloids derived from tryptophan. The indole alkaloids

Apart from some simple derivatives of tryptophan, e.g. indolylalkylamines, physostigmines, and β-carbolines (Fig. 25), the formation of which have ample analogies in the tyrosine series, the overwhelming majority of the indole alkaloids

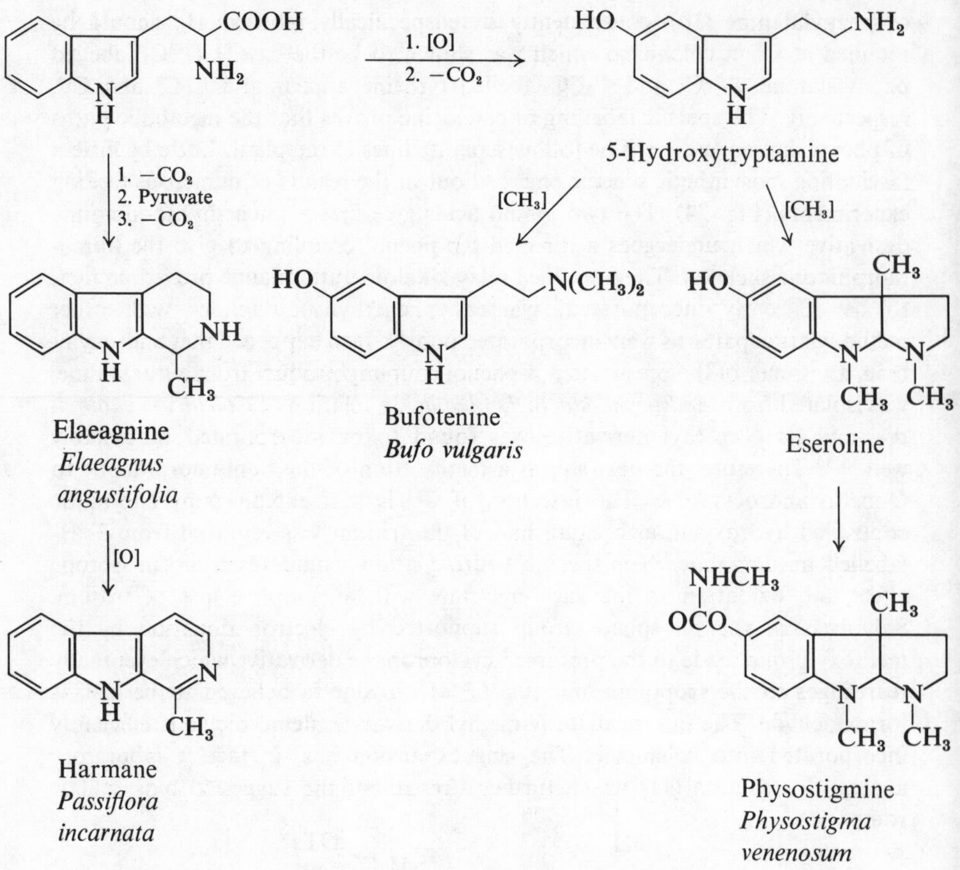

1. [O]
2. $-CO_2$

5-Hydroxytryptamine

1. $-CO_2$
2. Pyruvate
3. $-CO_2$

[CH_3]

[CH_3]

Elaeagnine
Elaeagnus angustifolia

Bufotenine
Bufo vulgaris

Eseroline

[O]

Harmane
Passiflora incarnata

Physostigmine
Physostigma venenosum

Fig. 25 Biosynthesis of some simple tryptophan alkaloids

is distinguished by a formidable structural variation and complexity. The tryptamine is visibly an invariant feature, but the C_9 or C_{10} aliphatic unit concealed for a long time its true origin. The indole alkaloids, which primarily are confined to three plant families, *Apocynaceae*, *Loganiaceae* and *Rubiaceae*, attracted interest, partly because of their significant neurophysiological action, but not least because their structural elucidation, syntheses and biogenesis presented to the organic chemist a challenge on the highest intellectual level. In this area biosynthesis has celebrated triumphs in bringing sense and order to a bewildering variety of structures.

In a few instances the indole nucleus is modified to an isoquinoline nucleus, e.g. in the calycanthine and cinchonine alkaloids.

There were wild speculations on the origin of the aliphatic part until the monoterpene hypothesis was presented.[37,38] Subsequently it was shown to be correct by the incorporation of mevalonate or more efficiently, geranyl phos-

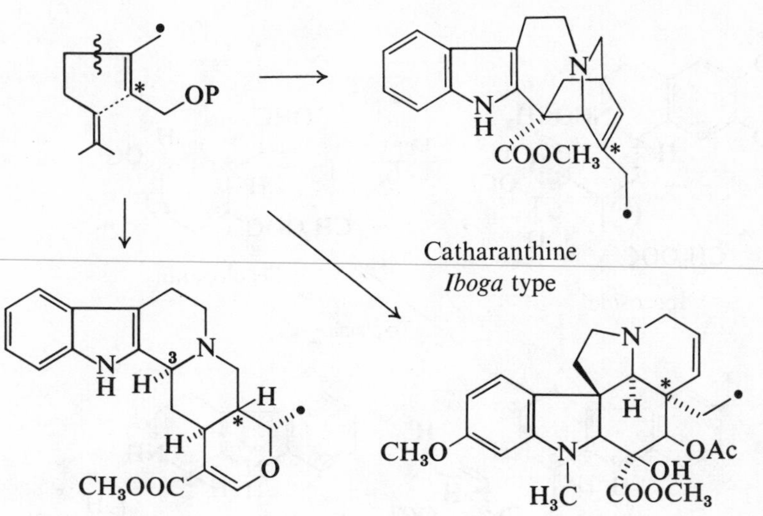

Catharanthine
Iboga type

Ajmalicine
Corynanthe type

Vindoline
Aspidosperma type

Fig. 26 Incorporation of 2- and 4-^{14}C-geraniol into indole alkaloids

phate.[39] Specific incorporation was ensured by administering synthetic 2- and 4-^{14}C-geranyl phosphate to *Catharanthus roseus* (*Vinca rosea*) which is known to produce a variety of indole alkaloids of widely different structures (Fig. 26). Degradation showed that the label was entirely located at the marked positions. Thus, being confident of the origin of the non-tryptamine part, the next step was to identify the intermediates en route from geraniol. The monoterpene hypothesis suggested cyclopentene derivatives as likely candidates. Aid came from structural work on ipecoside (section 7.3) that was carried out at the same time.[40] It contains a monoterpenoid part which could be traced back to loganin as an attractive precursor which by cleavage of the five-membered ring gives secologanin (section 5.3). Condensation of secologanin with dopa gives *N*-deacetyl ipecoside. Labelled loganin as well as secologanin were subsequently shown to be efficiently incorporated into a number of indole alkaloids, and moreover, loganin was shown to be present in *Catharanthus roseus*. Hence, it is quite plausible that tryptamine and secologanin in the plant give rise to an analogous indole glycoside located somewhere on the path between loganin and the indole alkaloids (Fig. 27). Direct *in vitro* condensation of secologanin with tryptamine afforded two isomeric glycosides, vincoside, and isovincoside which by dilution analysis were proved to be present in *C. roseus*.[41] This technique is often used in cases where the presumed intermediate occurs in amounts so small that it cannot be isolated. The intermediate, the carrier, is therefore synthesized and added together with a radioactive precursor to the plant. After a suitable time it is again recovered and purified. The compound appears on the pathway if it has acquired radioactivity.

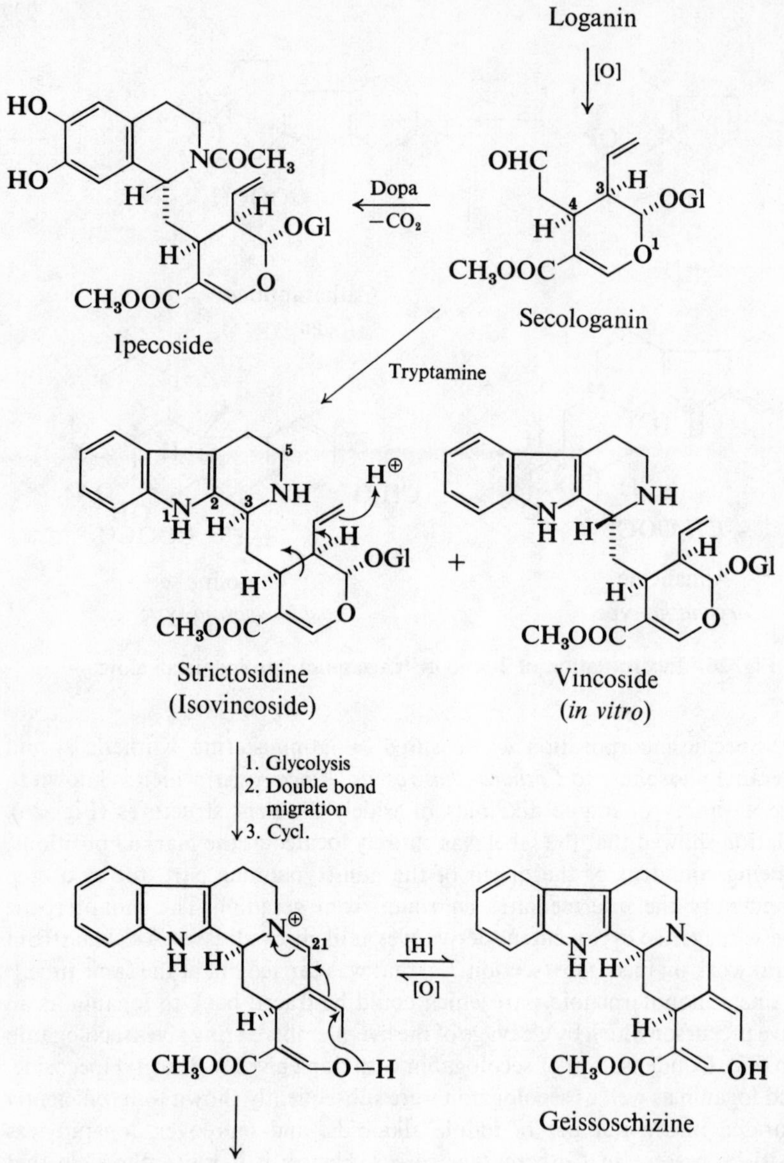

Loganin

[O]

Secologanin

Dopa
$-CO_2$

Ipecoside

Tryptamine

Strictosidine
(Isovincoside)

+

Vincoside
(*in vitro*)

1. Glycolysis
2. Double bond migration
3. Cycl.

[H]
[O]

Geissoschizine

NADPH

Cathenamine

Ajmalicine

Fig. 27 Biosynthesis of ajmalicine

Strictosidine, and indole glycoside of the same gross structure, was simultaneously isolated from another plant *Rhazya stricta*,[42] and later proved to be identical with isovincoside.[43] Only this isomer with (*S*)-configuration at C^3 is incorporated into ajmalicine, vindoline, catharanthine, and other indole alkaloids. The biosynthesis of ajmalicine is illustrated in Fig. 27. If the scheme is correct, the C^3–H of secologanin is lost, but the C^4–H is retained during the transformations.

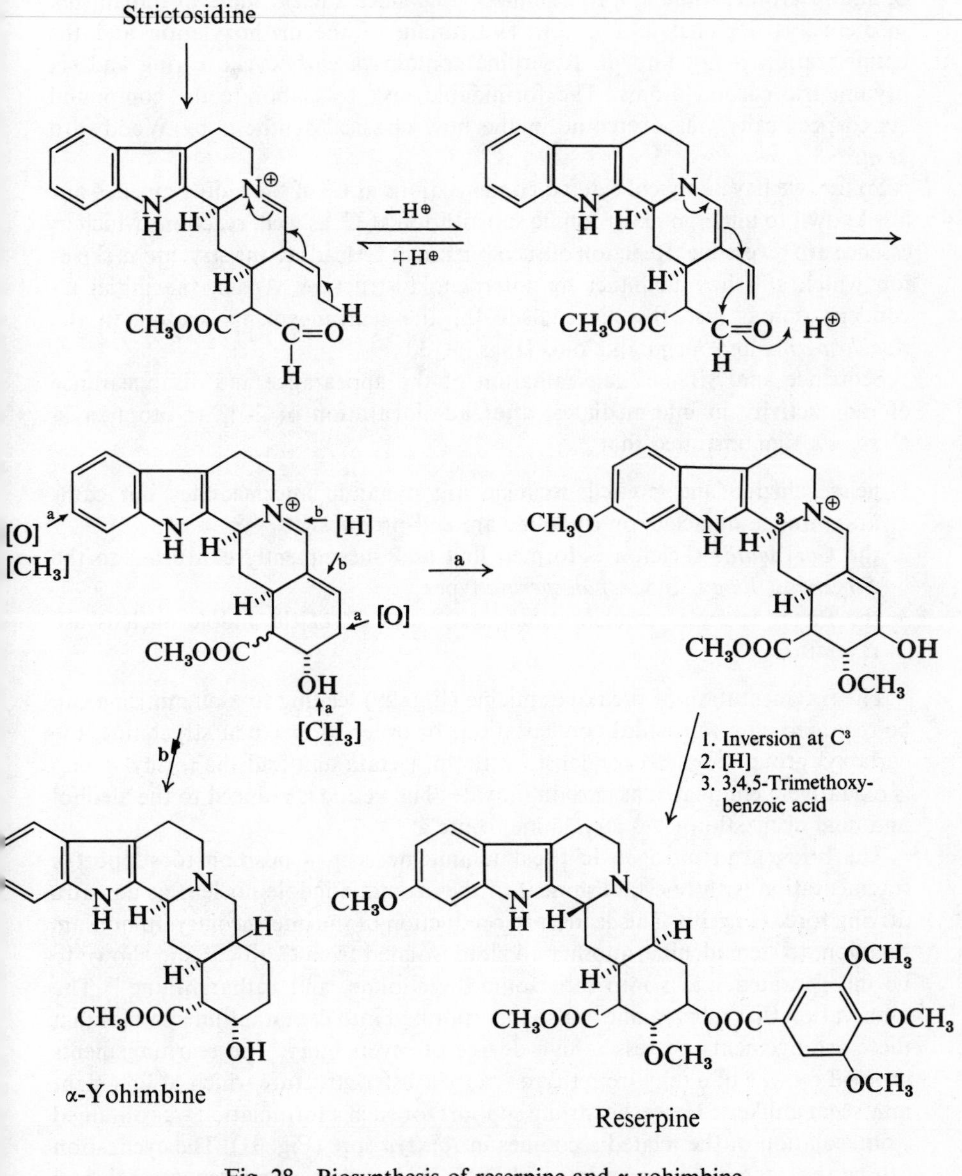

Strictosidine

$-H^{\oplus}$

$+H^{\oplus}$

CH_3OOC C=O H

CH_3OOC C=O $H^{\oplus}$

[O] [CH$_3$] [H] [O] [CH$_3$] OH CH_3OOC

a

CH_3O CH_3OOC OH OCH_3

1. Inversion at C^3
2. [H]
3. 3,4,5-Trimethoxy-benzoic acid

CH_3OOC OH

α-Yohimbine

CH_3O CH_3OOC OCH_3 OOC OCH_3 OCH_3 OCH_3

Reserpine

Fig. 28 Biosynthesis of reserpine and α-yohimbine

This was also borne out in practice. Geissoschizine was proved by double labelling to be incorporated intact into ajmalicine and several other indole alkaloids. The intermediacy of cathenamine was proved by experiments in cell free cultures. In the absence of NADPH the synthesis stopped at this stage but proceeded enzymatically to ajmalicine on addition of NADPH.

Strictosidine appears to be a universal intermediate in the biosynthesis of indole alkaloids.[44] The biosynthesis of the *Rauwolfia* alkaloids yohimbine, route b, and reserpine, route a, a tranquillizer used since ancient times in Indian folk medicine, is depicted in Fig. 28. The timing of the methoxylation and the epimerization is not known. Reserpine contains a carbocyclic E-ring and six asymmetric carbon atoms. The formidable task to elaborate the compound stereospecifically was overcome in the now classical synthesis by Woodward *et al.*[45]

So far, we have only considered condensations at C^2 of the indole nucleus but it is known to undergo electrophilic substitution at C^3 as well, reactions which in essence are reversible. Reaction of secologanin at C^3 leads to the strychnine skeleton which still has an intact monoterpenoid structure. Again, the initial C^3 condensation is also the prerequisite for the rearrangements leading to the *Aspidosperma* and *Iboga* alkaloids (Figs 29, 30).

Sequence analysis, i.e. determination of the appearance and disappearance of radioactivity in intermediates, after administration of 2-^{14}C-tryptophan in *C. roseus*, demonstrated that

1. geissoschizine and preakuammicine are dynamic intermediates but catharanthine, ajmalicine, or vindoline are end-products;
2. the *Corynanthe* skeleton is formed first and subsequently converted to the *Strychnos*, *Iboga*, and *Aspidosperma* types;
3. the Pictet–Spengler condensations at C^2 and C^3 of the indole nucleus are reversible.

The fragmentation of preakuammicine (Fig. 29) leading to akuammicine can be regarded as a retroaldol condensation. In order to arrive at strychnine, the carboxyl group (Fig. 29) condenses with an acetate unit and the formyl group is oxidatively eliminated as carbon dioxide. The ketide is reduced to the alcohol and final cyclization gives strychnine, route a.

The bridgehead nitrogen in preakuammicine is in a position to support a fragmentation with re-establishment of the aromatic indole nucleus as an extra driving force (Fig. 30). This leads, after reduction of the intermediary immonium function, to stemadenine, another alkaloid isolated from *C. roseus* and shown to be incorporated intact into tabersonine, vindoline, and catharanthine.[46] The observation that tabersonine is also incorporated into catharanthine implies that these arrangements possess a high degree of reversibility. The rearrangements proceed by way of a dihydropyridine—acrylic ester structure which at first sight may seem unlikely. However, strong support for such a formulation was obtained from isolation of the related secodines in *Rhazya* spp. (Fig. 31). The cyclization can be regarded either as an internal Diels–Alder reaction or stepwise Michael

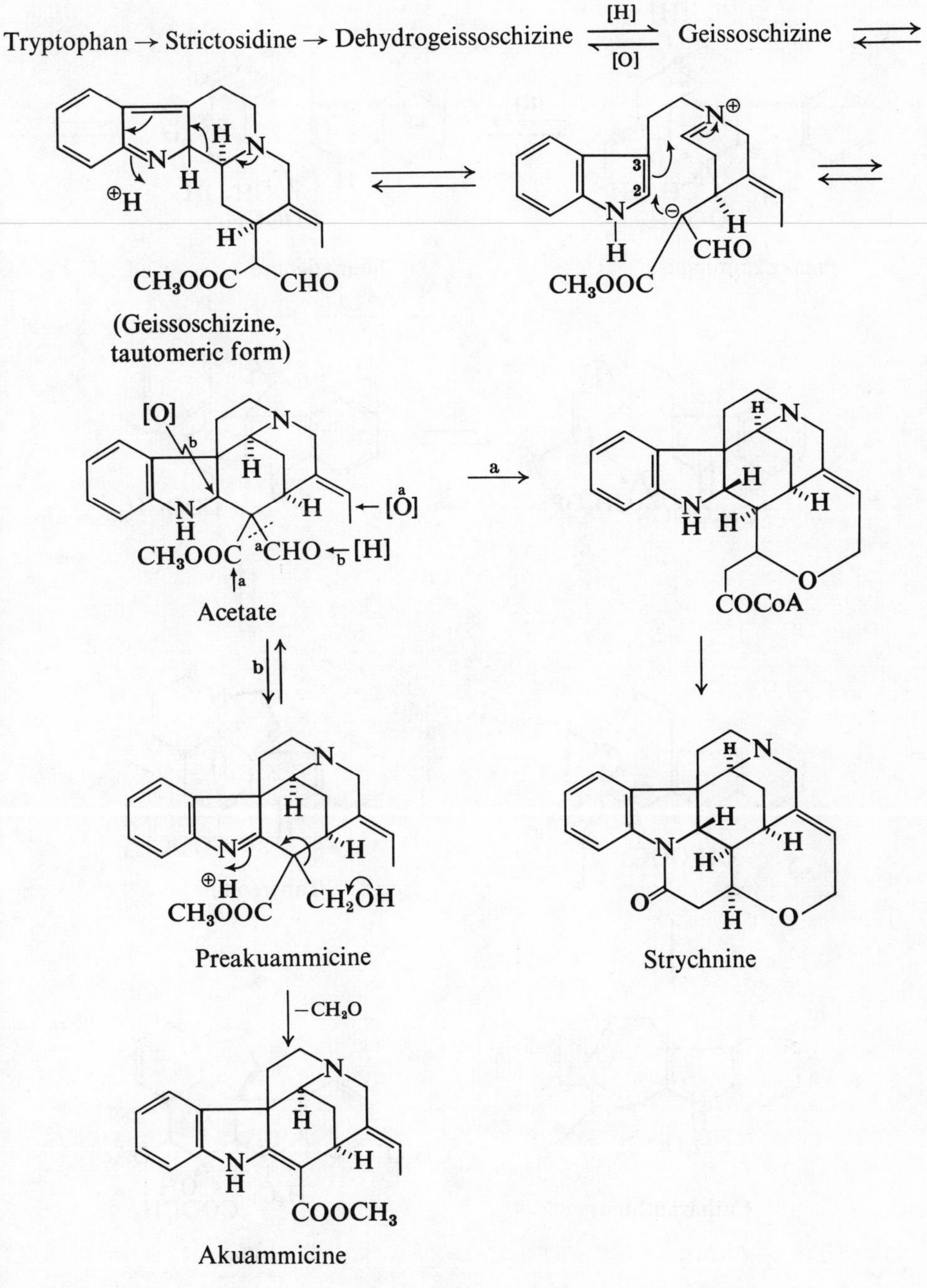

Tryptophan → Strictosidine → Dehydrogeissoschizine $\underset{[O]}{\overset{[H]}{\rightleftharpoons}}$ Geissoschizine ⇌

(Geissoschizine, tautomeric form) ⇌ ⇌

Acetate

a →

b ⇅

Preakuammicine

$-CH_2O$

Akuammicine

Strychnine

Fig. 29 Biosynthesis of akuammicine and strychnine

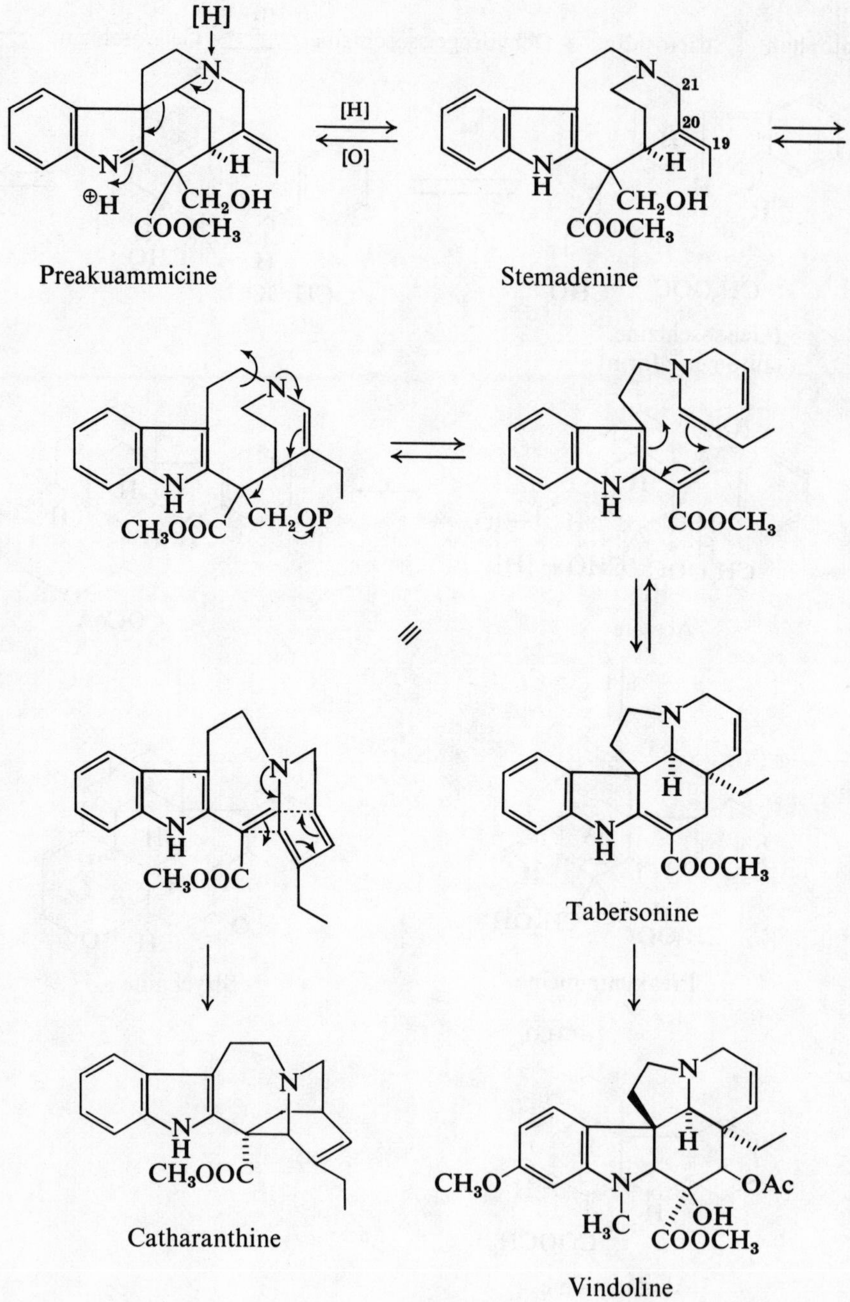

Fig. 30 Biosynthesis of *Aspidosperma* and *Iboga* alkaloids

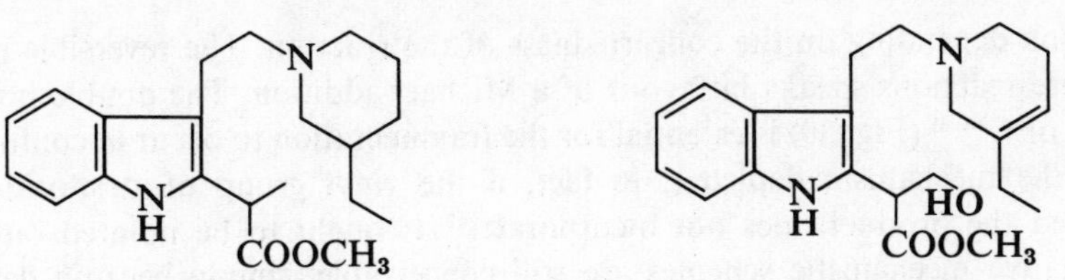

Tetrahydrosecodine Dihydrosecodin-17-ol

Fig. 31 Structures of secodines

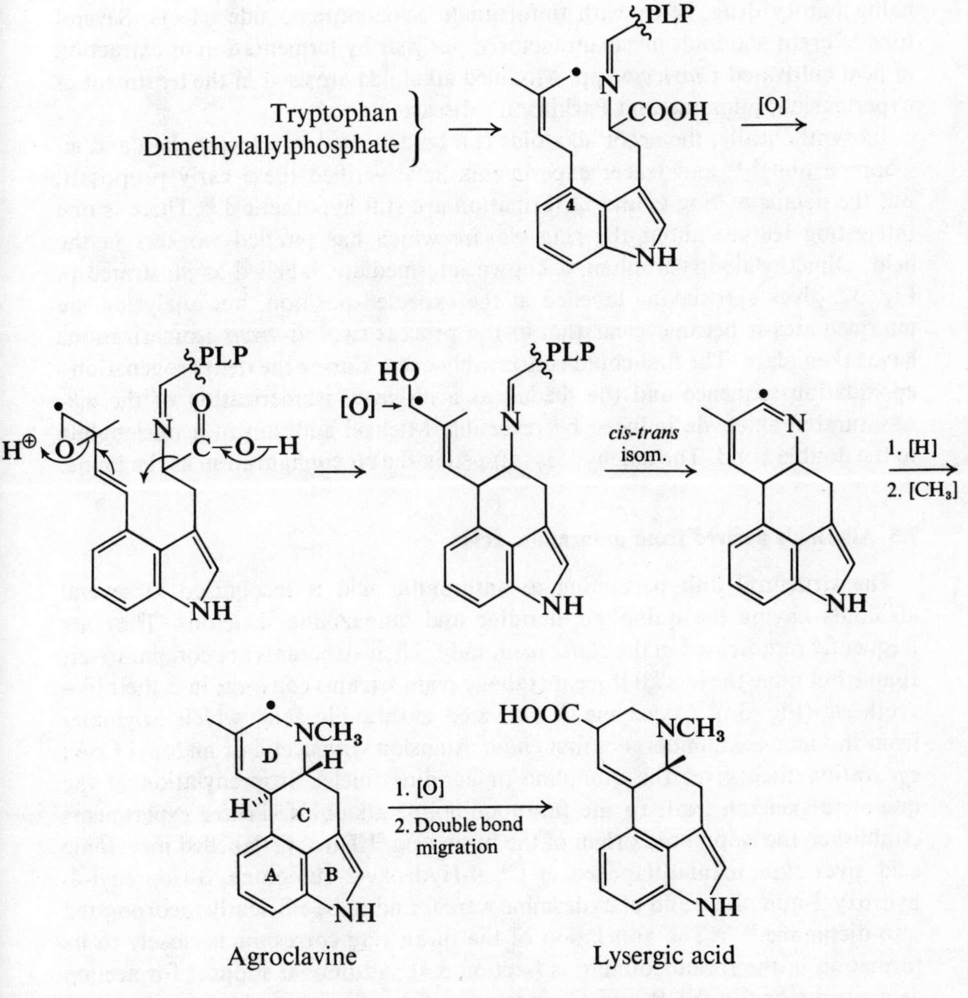
Tryptophan
Dimethylallylphosphate
[O]
cis-trans
isom.
1. [H]
2. [CH3]
1. [O]
2. Double bond
migration

Agroclavine Lysergic acid

Fig. 32 Biosynthesis of ergot alkaloids

addition depending on the concertedness of the reaction. The reversible nature of these reactions speaks in favour of a Michael addition. The double bond at $C^{19,20}$ or $C^{20,21}$ (Fig. 30) is essential for the fragmentation to occur in conformity with the mechanism depicted; in fact, if the vinyl group of strictosidine is reduced the product does not incorporate.[47] It ought to be pointed out that alternative mechanistic schemes are still conceivable, simply because detailed experimentation is still lacking.

The ergot alkaloids are produced by the fungus *Claviceps purpurea* which is parasitic on rye and certain grasses. The metabolites attracted the interest of chemists because of their strange and dramatic action on the human mind, not least after Hofmann's heroic experiments on himself in the 1940s. One of the most potent derivatives, LSD or lysergic acid diethylamide, is misused as a hallucinatory drug, often with unfortunate schizophrenic side effects. Several tons of ergot alkaloids are manufactured per year by fermentation or extraction of field cultivated *Claviceps* spp. Modified alkaloids are used in the treatment of hypertension, migraine and Parkinson's disease.

Biosynthetically, the ergot alkaloids can be dissected into tryptophan and an isoprene unit[48,49] and tracer experiments have verified these early proposals, but the details of ring C and D formation are still hypothetical.[50] There is one interesting feature about the ring closure which has puzzled workers in the field. Dimethylallyltryptophan, a known intermediate, labelled as illustrated in Fig. 32, gives agroclavine labelled at the expected position, but analysing the intermediates it became clear that in the process two *cis–trans* isomerizations have taken place. The first could conceivably occur during the dehydrogenation–epoxidation sequence and the second as a *cis–trans* isomerization of the α,β-unsaturated aldehyde induced by reversible Michael addition of a nucleophile to the double bond. The aldehyde is trapped in the *cis* configuration as the imine.

7.5 Alkaloids derived from anthranilic acid

The structural unit pertaining to anthranilic acid is recognized in several alkaloids having the quinoline, acridine and quinazoline skeletons. They are frequently represented in the *Rutaceae* family. Their structures are comparatively simple but none the less all three metabolic main streams converge into their biosynthesis (Fig. 33). Coenzyme A activated anthranilic acid, which originates from shikimic acid, undergoes first chain extension with acetyl or malonyl CoA; cyclization then gives the quinoline or acridine nuclei. Isoprenylation of the quinoline skeleton leads to the furanoquinoline alkaloids. Tracer experiments established the isoprenoid origin of the furan ring.[51] Thus, C^5-labelled mevalonic acid gives skimmianine labelled at C^3. 4-Hydroxy-2-quinolone, 3-isoprenyl-4-hydroxy-2-quinolone and platydesmine were found to be efficiently incorporated into dictamine.[52,53] The annelation of the furan ring corresponds closely to its formation in the furanocoumarins (section 3.4). Additional support for acetate as a precursor for the B and C rings came from tracer work on a quinoline alkaloid produced by *Pseudomonas aeruginosa*[54] (Fig. 34). The oxygenation

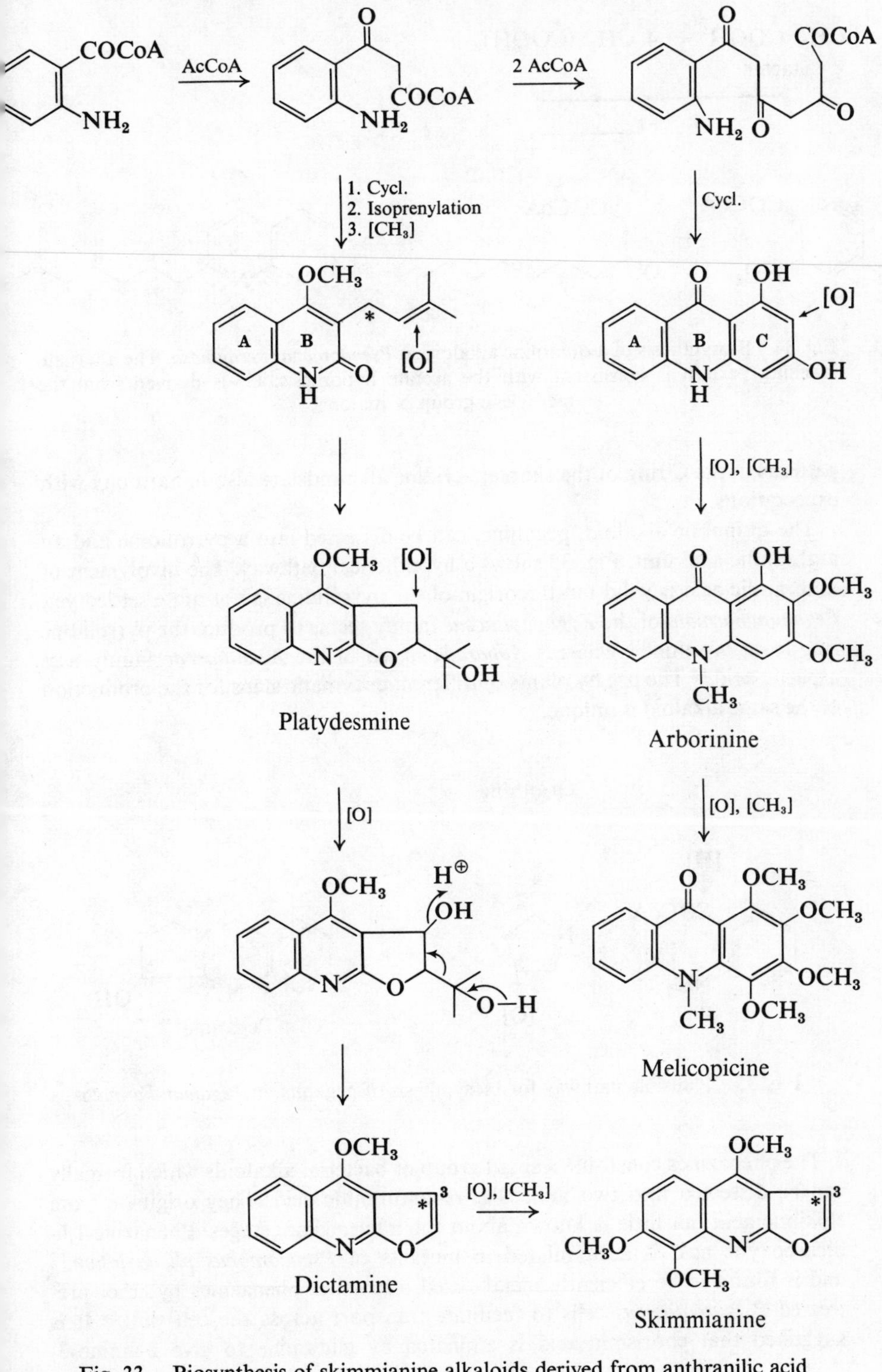
COCoA
NH2
AcCoA
O
COCoA
NH2
2 AcCoA
O
COCoA
NH2 O
O
1. Cycl.
2. Isoprenylation
3. [CH3]
Cycl.
OCH3
A
B
N
H
O
[O]
O
OH
[O]
A
B
C
N
H
OH
[O], [CH3]
OCH3 [O]
N
O
OH
Platydesmine
O
OH
OCH3
OCH3
N
CH3
Arborinine
[O]
[O], [CH3]
OCH3
H⊕
OH
N
O
O—H
O
OCH3
OCH3
OCH3
OCH3
N
CH3
Melicopicine
OCH3
3
N
O
Dictamine
[O], [CH3]
OCH3
3
CH3O
N
O
OCH3
Skimmianine

Fig. 33 Biosynthesis of skimmianine alkaloids derived from anthranilic acid

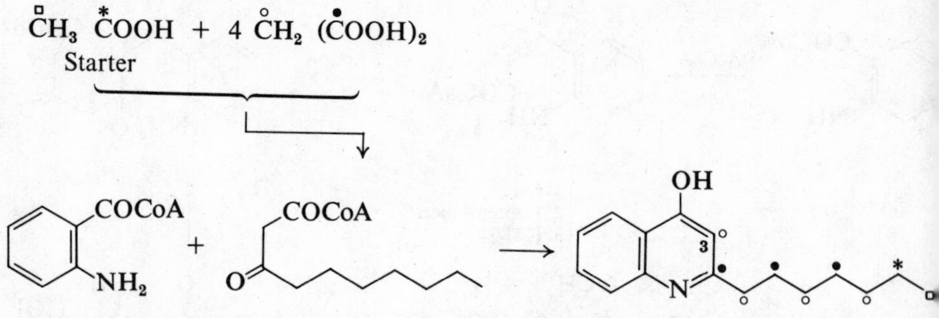

$\overset{\square}{C}H_3\ \overset{*}{C}OOH + 4\ \overset{\circ}{C}H_2\ (\overset{\bullet}{C}OOH)_2$
Starter

COCoA NH$_2$ + COCoA O → OH 3 N

Fig. 34 Biosynthesis of a quinoline alkaloid in *Pseudomonas aeruginosa*. The alternate labelling pattern is consistent with the acetate hypothesis. C^3 is derived from the methylene group of malonate

patterns in the C ring of the simpler acridine alkaloids are also in harmony with expectations.

The quinoline alkaloid, peganine, can be dissected into a pyrrolidine and an anthranilic acid unit. Fig. 35 shows a hypothetical pathway. The involvment of anthranilic acid is valid but the origin of the pyrrolidine is not quite settled yet. *Peganum harmala* of the *Zygophyllaceae* family seems to produce the pyrrolidine ring from ornithine[55] whereas *Adhatoda vasica* of the *Acanthaceae* family uses aspartic acid.[56] The use by plants of different enzymatic steps for the production of the same alkaloid is unique.

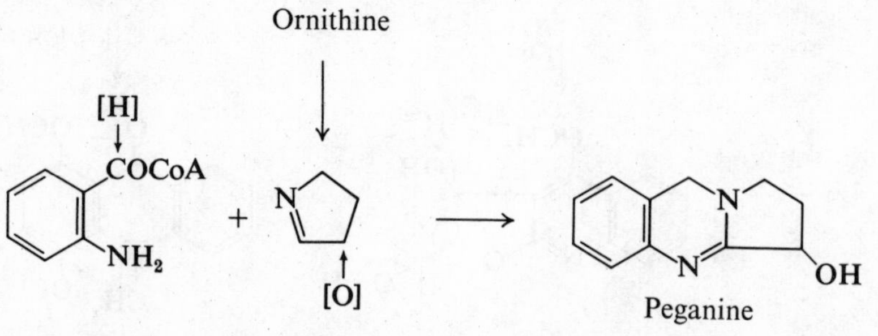

Ornithine

[H] COCoA NH$_2$ + N [O] → N N OH

Peganine

Fig. 35 Plausible pathway for biosynthesis of peganine in *Peganum harmala*

The phenazines constitute a small group of bacterial alkaloids which formally can be dissected into two molecules of anthranilic acid. They originate from shikimic acid but little is known about the intermediate stages. Phenazine-1,6-dicarboxylic acid is accumulated in mutants of *Pseudomonas phenazinium*,[57] and is found to be efficiently metabolized into other phenazines by ether pre-treated *P. aureofaciens* cells to facilitate transport across the cell wall.[58] It is suggested that chorismic acid is aminated by glutamine to give 2-amino-3-

hydroxycyclohexan-4,6-dienoic acid, which is further oxidized and dimerized to phenazine-1,6-dicarboxylic acid (Fig. 36). Shikimic acid labelled at C^6 is found to give $C^{5a,10a}$-labelled iodinin.[59]

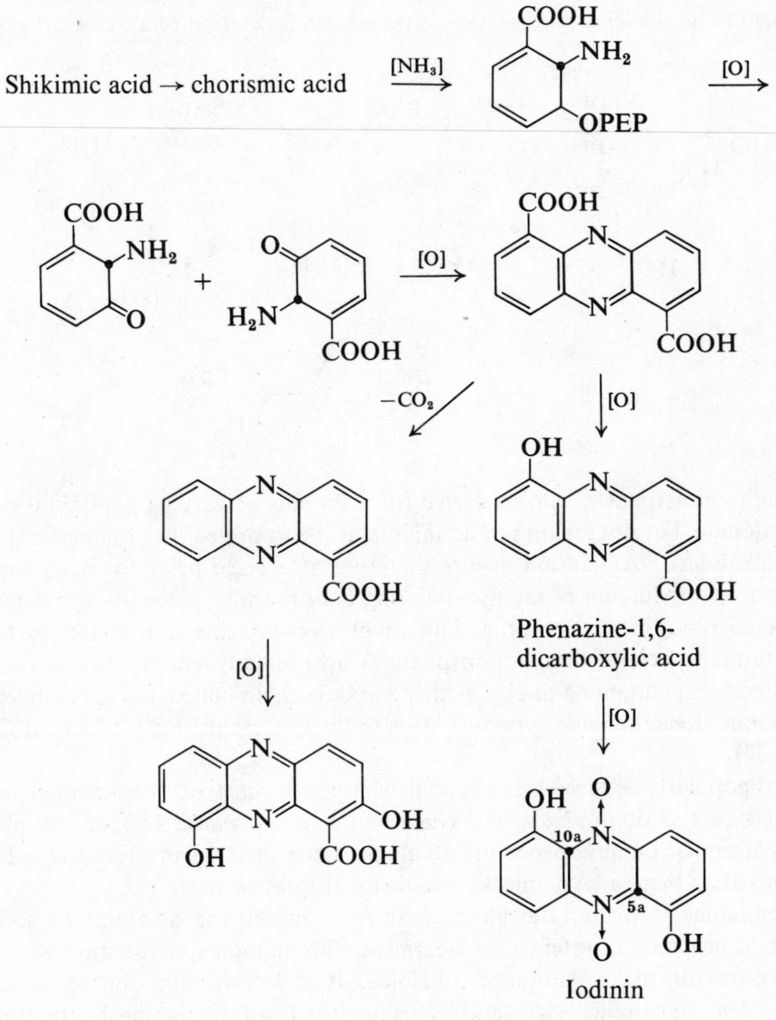

Shikimic acid → chorismic acid $\xrightarrow{[NH_3]}$ [COOH, NH_2, OPEP] $\xrightarrow{[O]}$

[COOH, NH_2, O] + [O, H_2N, COOH] $\xrightarrow{[O]}$ [COOH, N, N, COOH] Phenazine-1,6-dicarboxylic acid

$-CO_2$ ↙ [N, N, COOH] → [O] → [N, N, OH, OH, COOH]

[O] ↓ [OH, N, N, COOH] → [O] → [OH, O, 10a, N, N, 5a, O, OH] Iodinin

Fig. 36 Plausible route to phenazine alkaloids

7.6 Alkaloids derived by amination of terpenes

The biosynthesis of terpene alkaloids is largely a question of the biosynthesis of the parent compounds. Therefore they could equally well be categorized with them as functionalized derivatives in the same way as amino sugars are classified as carbohydrates. Diversification or structural modifications occur on the terpenoid level, i.e. amination is a late event. In this presentation we have chosen to let the basic character of the compound rule the systemization. The amination can take place according to a redox process (12), a substitution (13) or an addition (14). The substitution is expected to proceed with inversion. Loss of label at the α-position implies that the reaction proceeds via formation of a carbonyl group.

$$\mathrm{HO{-}CH} \underset{\mathrm{NADPH}}{\overset{\mathrm{NADP}^{\oplus}}{\rightleftharpoons}} \mathrm{O{=}C} \xrightarrow{[\mathrm{NH_3}]} \mathrm{HN{=}C} \underset{\mathrm{NADP}^{\oplus}}{\overset{\mathrm{NADPH}}{\longrightarrow}} \mathrm{H_2N{-}CH} \qquad (12)$$

$$\mathrm{HO{-}CH} \xrightarrow{\mathrm{HOP}} \mathrm{PO{-}CH} \;\; [\mathrm{NH_3}] \longrightarrow \mathrm{H{-}C{-}NH_2} \qquad (13)$$

$$\mathrm{C{=}C} \xrightarrow[\mathrm{X}^{\oplus}]{[\mathrm{NH_3}]} \mathrm{X{-}C{-}C{-}NH_2} \qquad (14)$$

The monoterpene alkaloids derive from iridoids of varying oxidation levels. Mevalonate, but not loganin or actinidine, is incorporated into the skytanthine-like alkaloids.[60] Amination occurs therefore at a stage prior to formation of loganin and reduction of the pyridine ring does not take place. As suggested in section 5.3, ring closure is initiated by solvolysis of C^9 phosphate giving rise to an unsaturated cyclopentanoid phosphate which directly reacts with ammonia released from glutamine or else is further oxidized, aminated and aromatized to actinidine. Reaction with tyramine leads to the quaternary *Valeriana* alkaloids[61] (Fig. 37).

It is popularly believed that the somewhat musty smell of *Valeriana officinalis* attracts cats as do *Actinidia polygama* and *Nepeta cataria*. In fact, the plants contain similar monoterpenes and alkaloids. Actinidine has an effect on the EEG of the cat so there are chemical grounds for the old sayings.

Gentianine is by far the most common monoterpene alkaloid. In several plants it becomes an artefact on treatment with ammonia, a common work-up procedure for the isolation of alkaloids. It is known that the secoirodoid glycosides, genetiopicroside, and swertiamarin form gentianine by treatment with ammonia *in vitro* (15). *Enicostema littorale*, known to contain swertiamarin, gave 0.18 per cent of gentianine in the presence of ammonia, whereas when processed without ammonia none was found.[62] However, gentianine has been isolated from several plants without use of ammonia in the work-up procedure.[63]

Sesquiterpene alkaloids are found in *Nymphaeaceae*, e.g. in the romantic water lily, *Nuphar* spp, and in *Orchidaceae*, e.g. in *Dendrobium nobile*. Castoramine[64] was isolated from the scent glands of the Canadian beaver, and it is most likely that it is not endogenous, but accumulated and modified by the beaver feeding on *Nuphar* roots. The C^{15} *Nuphar* alkaloids can visually be

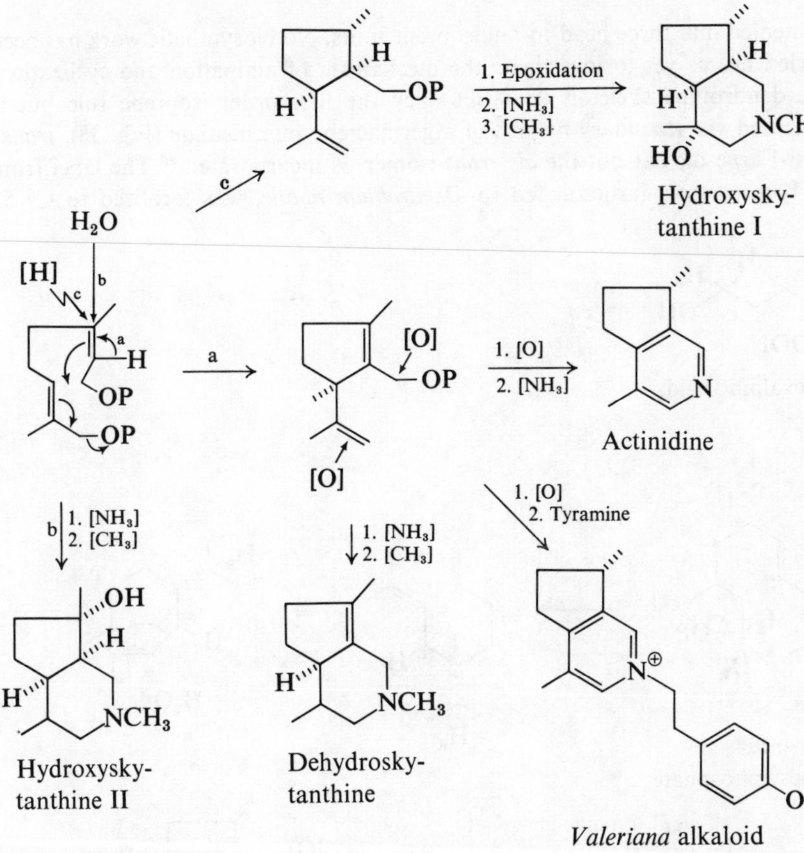

H
OP
H
1. Epoxidation
2. [NH3]
3. [CH3]
H
H
NCH3
HO
Hydroxysky-
tanthine I
c
H2O
[H]
c
b
a
H
OP
OP
a
[O]
OP
[O]
1. [O]
2. [NH3]
N
Actinidine
1. [O]
2. Tyramine
b
1. [NH3]
2. [CH3]
1. [NH3]
2. [CH3]
OH
H
H
NCH3
Hydroxysky-
tanthine II
H
NCH3
Dehydrosky-
tanthine
N
OH
Valeriana alkaloid

Fig. 37 Hypothetical routes to monoterpene alkaloids

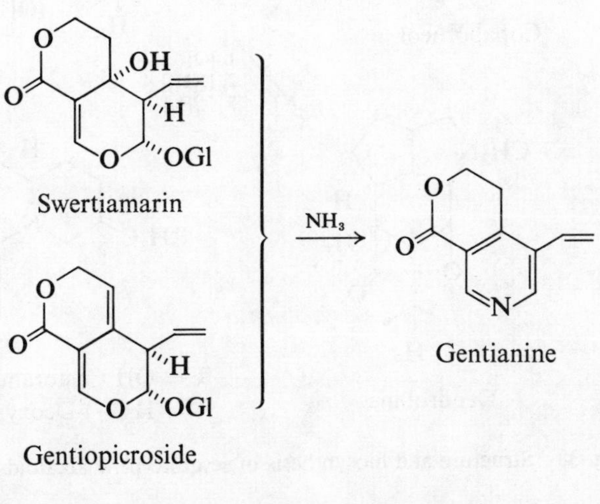

O
OH
O
H
O
OGl
Swertiamarin
O
O
H
O
OGl
Gentiopicroside
NH3
O
O
N
Gentianine

(15)

segmented into three head-to-tail isoprene units. No biosynthetic work has been carried out as yet to investigate the mechanism of amination and cyclization. The dendrobine skeleton does not obey the first order isoprene rule but is generated via secondary fissions of a germacrene intermediate (Fig. 38). *trans–trans*-Farnesol, but not the *cis–trans*-isomer, is incorporated.[65] The label from 1-$^3H_2$-*trans–trans*-farnesol, fed to *Dendrobium nobile*, was localized to C^5 52

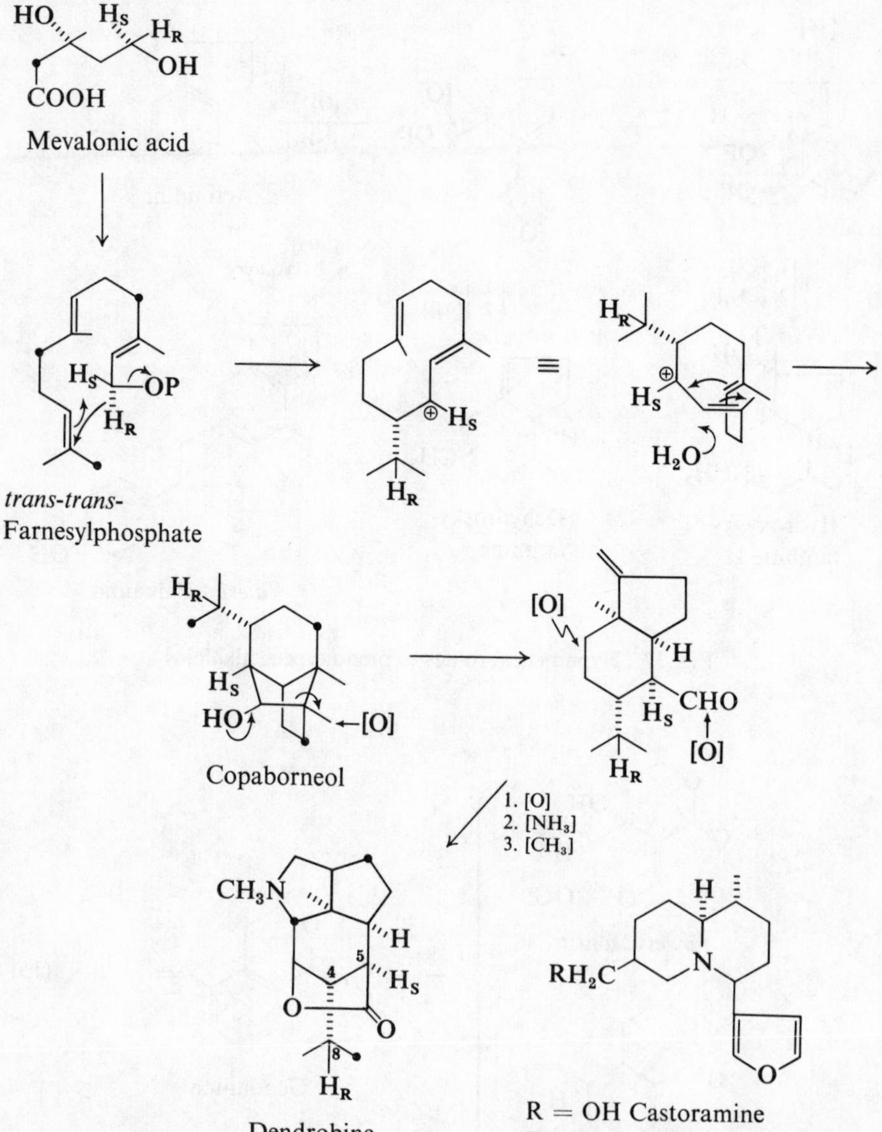

Mevalonic acid

trans-trans-Farnesylphosphate

Copaborneol

1. [O]
2. [NH_3]
3. [CH_3]

Dendrobine

R = OH Castoramine
R = H (−)-Deoxynupharidine

Fig. 38 Structure and biosynthesis of sesquiterpene alkaloids

per cent and C^8 47 per cent, and when 1-^{3}H(*S*)-*trans–trans*-farnesol was administered, 85 per cent of the label was recovered at C^5, i.e. the 1,3-H_R shift is a key step and it has a high degree of stereospecificity. A proposal involving two consecutive 1,2 shifts is thus eliminated. In this case one tritium would be located at C^4.

The diterpene alkaloids occur in the genera *Aconitum* and *Delphinium* of the *Ranunculaceae* family and in *Garrya* spp. of the *Cornaceae* family. *Aconitum* spp. are ubiquitously distributed plants cultivated in gardens and also to be found growing tall in the fertile valleys of Sarek, the last European wilderness north of the arctic circle in Sweden. They contain a group of extremely poisonous alkaloids. Poisoning is diagnosed by a tingling sensation in the whole body followed by numbness. The skeletal elucidation turned out to be difficult with classical methods, and as in several other cases the key breakthrough came with an X-ray study which finally established the structure of lycoctonine[66] (Fig. 39). There is an obvious structural resemblance between garryfoline and the diterpene (—)kaur-16-en-15β-ol. Characteristic for the diterpene alkaloids are the *N*-ethyl or *N*-β-hydroxyethyl bridge across ring A.

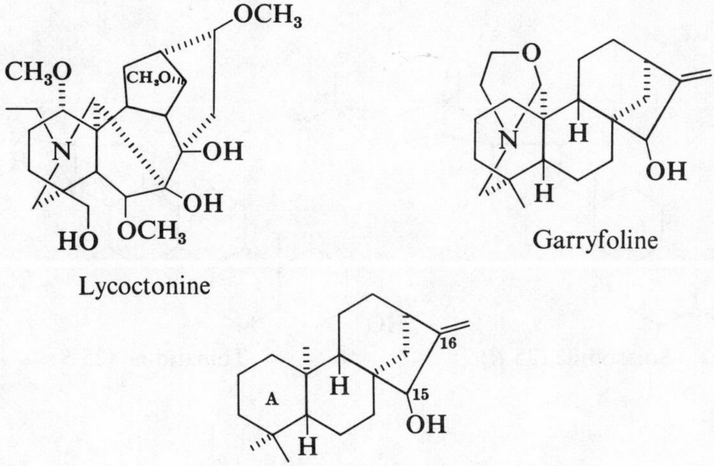

OCH$_3$ CH$_3$O CH$_3$O N OH OH HO OCH$_3$

Lycoctonine

O N H H OH

Garryfoline

A H H 16 15 OH

(—) Kaur-16-en-15β-ol

Fig. 39 Structure of diterpene alkaloids

The steroid alkaloids are characteristic metabolites of the *Solanaceae*, *Liliaceae*, and *Buxaceae* families. *Zygadenus* spp. containing zygadenine (Fig. 40) are poisonous plants causing losses among livestocks in Northern America. The *C*-nor-D-homo-steroid nucleus is formed via C^{12} hydroxylation which initiates the rearrangement (16).

Cholesterol, though incorporated with low efficiency into some plants, is thought to be a precursor for the steroid alkaloids. Solasodine and tomatidine

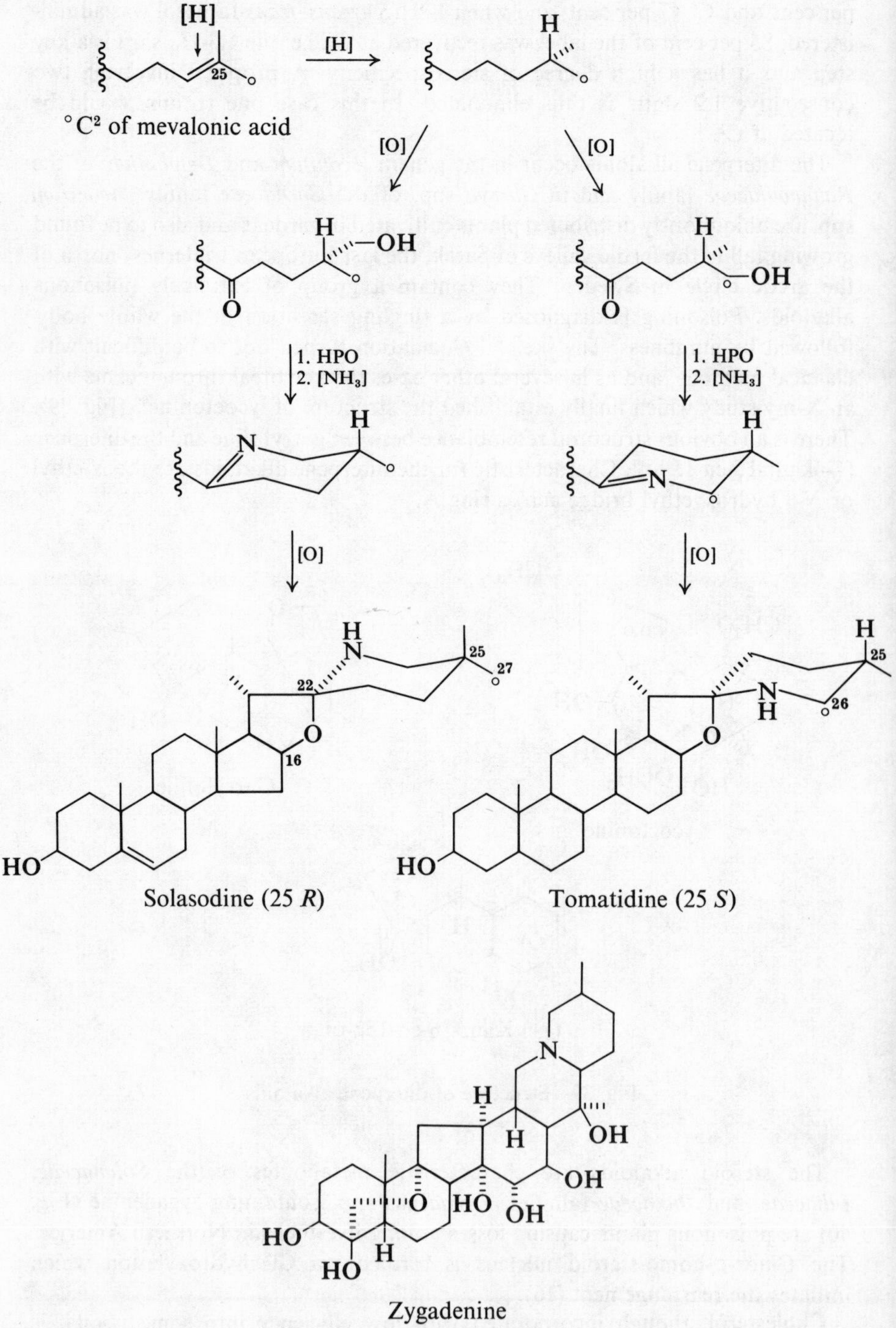

[H]

25

[H]

H

°C^2 of mevalonic acid

[O]

[O]

H —OH

O

H

OH

O

1. HPO
2. $[NH_3]$

1. HPO
2. $[NH_3]$

H

N

H

N

[O]

[O]

H N 25 27 22 O 16

HO

Solasodine (25 *R*)

H 25 N H 26 O

HO

Tomatidine (25 *S*)

N

H H OH

OH

O HO OH

HO H

HO

Zygadenine

Fig. 40 Steroid alkaloids

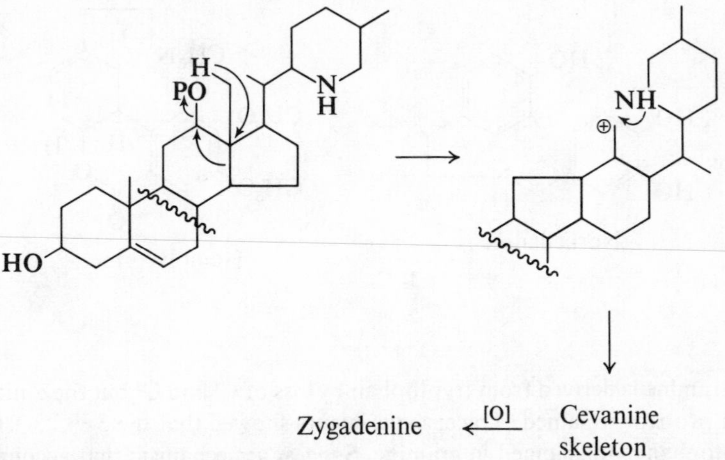

H
PO
N
H
NH
⊕
HO
(16)

Zygadenine $\xleftarrow{[O]}$ Cevanine skeleton

have an intact steroid skeleton. As pointed out earlier, amino functionalization is a late event. Cholesterol is specifically hydroxylated at C^{26} or C^{27}, aminated and cyclized to the piperidine derivative (Fig. 40).[67–69] In tomatidine the C^{26} and in solasodine the C^{27} is derived from mevalonic acid. The amination proceeds without loss of ^3H label at $C^{26/27}$ consistent with direct amination without intervention of a carbonyl group. On the other hand, formation of the piperidine ring proceeds via an imine. Introduction of a hydroxyl at C^{16} and formation of the furan ring complete the biosynthesis.[70]

7.7 Problems

7.1 (+)-Isothebaine is generated from (—)-orientaline. Suggest a mechanism for this transformation that accounts for the unusual oxygenation pattern.

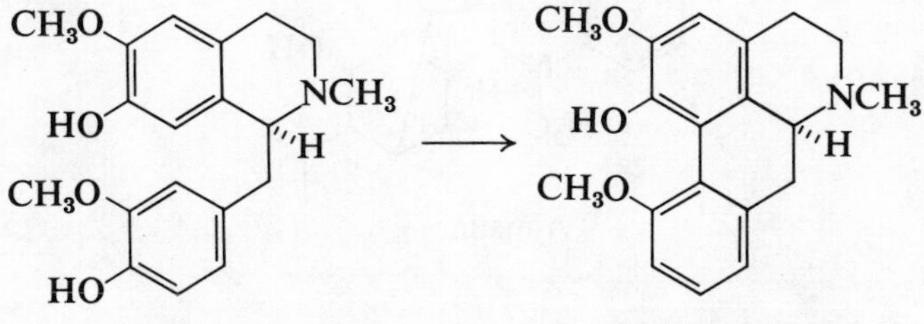

CH_3O HO NCH_3 H CH_3O HO

(—)-Orientaline

CH_3O HO NCH_3 H CH_3O

(+)-Isothebaine

7.2 The formation of the *Amaryllis* alkaloid homolycorine can be explained by a sequence of reactions starting from norbelladine which undergoes phenolic coupling, hydroxylation, ring scission, methylations and finally ring closure. Formulate the intermediate stages.

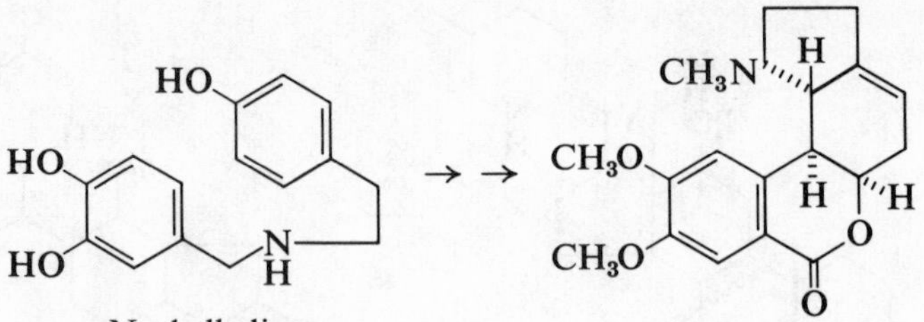

Norbelladine → → Homolycorine

7.3 Gramine is derived from tryptophan by loss of C^1 and C^2 but the α-nitrogen is most probably retained. Tracer experiments showed that the 3-^{14}C:3-^{3}H ratio of tryptophan is unchanged in gramine. Suggest a mechanism that accounts for these facts. (Gross, D., Nemeckova, A. and Schütte, H. R. *Z. Pflanzenphysiol.* **57** (1967) 60).

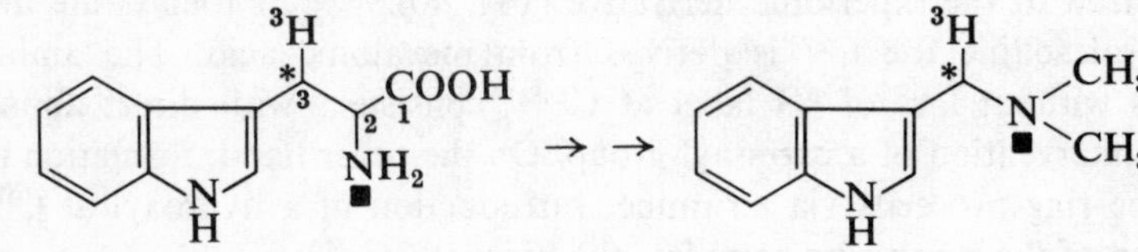

7.4 Suggest a plausible biosynthesis of ajmaline from strictosidine.

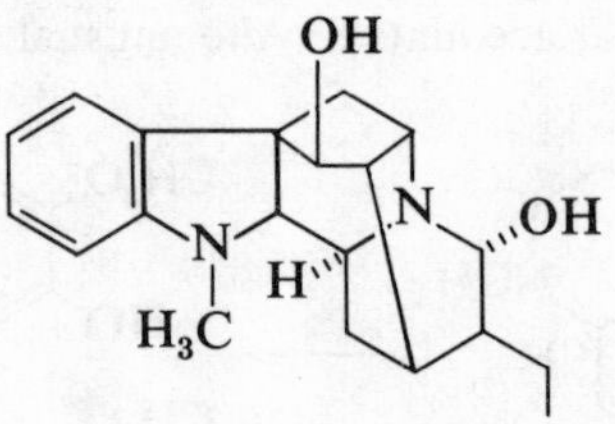

Ajmaline

7.5 Goniomine, isolated from *Goniomina malagasy*, *Apocynaceae*, has an unusual ring structure that can be related to the monoterpenoid indole alkaloids. Suggest a reasonable biosynthetic pathway from tryptamine and secologanin. The hydrolysed secologanin is supposed to be oriented as illustrated for the Pictet–Spengler condensation. (Chiaroni, A., Randriambola, L., Riche, C. and Husson, H.-P. *J. Am. Chem. Soc.* **102** (1980) 5921).

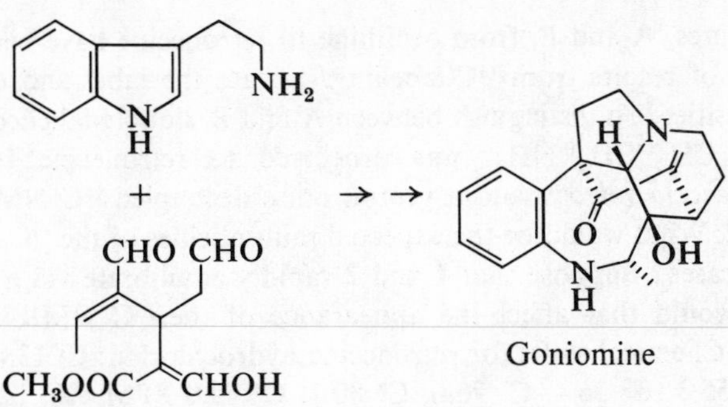

Goniomine

7.6 Suggest a biosynthetic pathway for quinine from methoxygeissoschizine. The label from 2-^{14}C-tryptophan is located at the α-carbon of the quinoline ring in quinine.

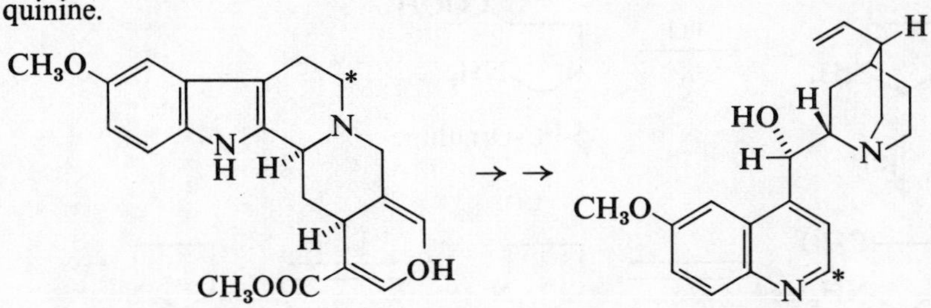

Methoxygeissoschizine → → Quinine

7.7 When 1-$^{3}H_2$-geraniol, 2-^{14}C-geraniol, and 5-^{3}H-loganin are fed to *Catharanthus roseus*, radioactive tabersonine (Fig. 30) and catharanthine (Fig. 26) can be isolated. Where are the alkaloids labelled?

7.8 Tylophorinine has its genesis in tyrosine, phenylalanine and ornithine. Phenacylpyrrolidine and its mono- and dihydroxylated derivatives are incorporated intact. Phenol oxidation occurs at one stage of the biosynthesis. Discuss a plausible pathway. (Herbert, R. B., Jackson, F. B. and Nicolson, I. T. *J. Chem. Soc. Chem. Commun.* **1976**, 450).

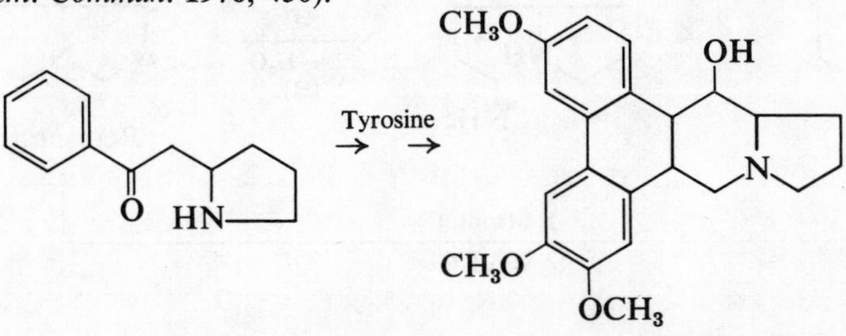

Tyrosine → →

Tylophorinine

7.9 Two routes, A and B, from ornithine to retronecine have been suggested on the basis of results from ^{14}C-labelling. Locate the label and calculate the relative intensities. To distinguish between A and B, double-labelled putrescine, $NH_2CH_2CH_2CH_2{}^{13}CH_2{}^{15}NH_2$, was processed to retronecine by the wick method in *Senecio vulgaris* and a proton noise decoupled ^{13}C NMR spectrum was recorded. What would be the expected multiplicities of the ^{13}C NMR peaks for the two cases? Suppose that **1** and **2** rapidly equilibrate via a sigmatropic shift. How would that affect the appearance of the ^{13}C NMR spectrum of retronecine? Chemical shifts for retronecine hydrochloride: C^1 138.0, C^2 122.6, C^3 62.7, C^5 55.3, C^6 36.4, C^7 70.6, C^8 80.1, C^9 59.0 PPM, $^1J_{^{13}C^{13}N}$ is *ca* 5 Hz (Grue-Soerensen, G. and Spenser, I. D. *J. Am. Chem. Soc.* **103** (1981) 3208).

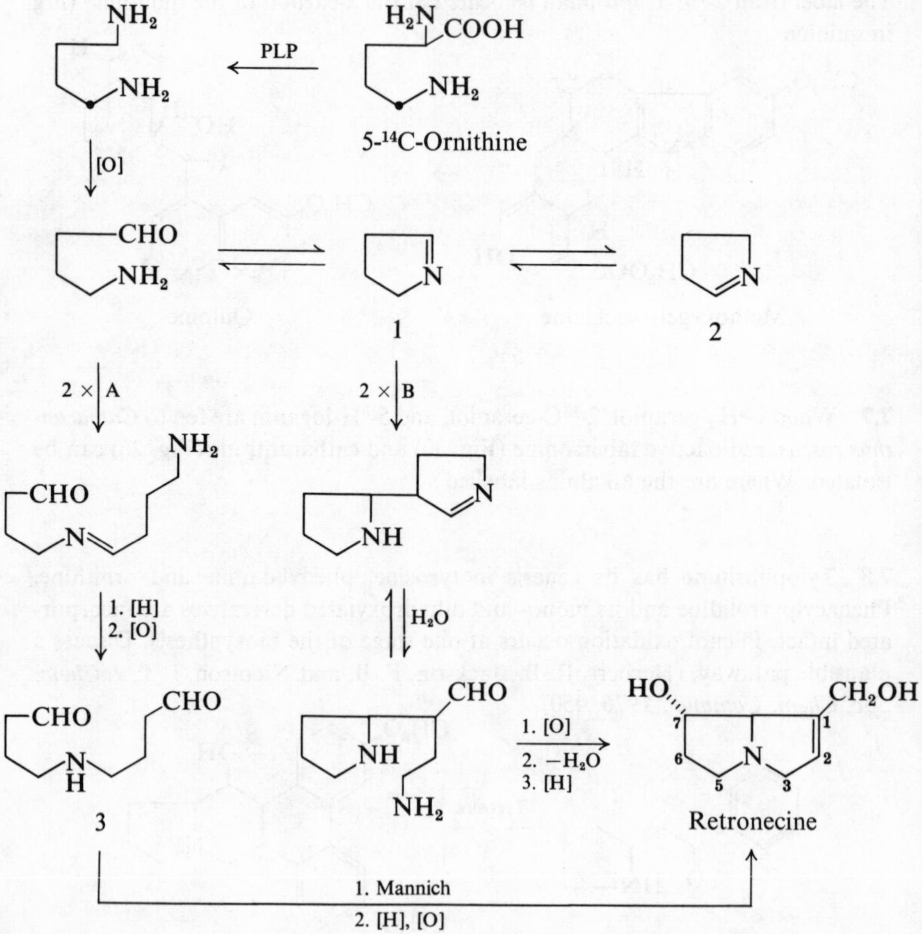

NH_2 / NH_2 ← PLP — H_2N COOH / NH_2

5-^{14}C-Ornithine

[O]

CHO / NH_2 ⇌ N (**1**) ⇌ N (**2**)

2 × A — 2 × B

NH_2 / CHO / N — N / NH

1. [H]
2. [O]

H_2O

CHO / CHO / N H (**3**) — CHO / NH / NH_2

1. [O]
2. $-H_2O$
3. [H]

HO / CH_2OH / 7 / 1 / 6 / N / 2 / 5 / 3

Retronecine

1. Mannich
2. [H], [O]

Bibliography

1. Seigler, D. S. in *The Alkaloids*, Vol. XVI, p. 1, Manske, R. H. F. (Ed.), Academic Press, New York, 1977.
2. Ahmad A. and Leete, E. *Phytochemistry* **9** (1970) 2345.
3. Liebisch, H. W., Radwan, A. S. and Schütte, H. R. *Liebigs Ann.* **721** (1969) 163.
4. Leete, E. and Yu, M. L. *Phytochemistry* **19** (1980) 1093.
5. Bottomley W. and Geissman, T. A. *Phytochemistry* **3** (1964) 357.
6. Gupta, R. N. and Spenser, I. D. *Phytochemistry* **9** (1970) 2329.
7. Keogh, M. F., and O'Donovan, D. G. *J. Chem. Soc. (C)* **1970**, 1792, 2470.
8. Leistner, E. and Spenser, I. D. *J. Am. Chem, Soc.* **95** (1973) 4715.
9. Schütte, H. R., Hindorf, H., Mothes, K. and Hübner, G. *Liebigs Ann.* **680** (1964) 93.
10. Golebiewski, W. M. and Spenser, I. D. *J. Am. Chem. Soc.* **98** (1976) 6729.
11. Leete, E. and Olsen, J. O. *J. Am. Chem. Soc.* **94** (1972) 5472.
12. Roberts, M. F. *Phytochemistry* **14** (1975) 2393.
13. Leete, E., Leichleiter, J. C. and Carver, R. A. *Tetrahedron Lett.* **1975,** 3779.
14. Dawson, R. F., Christman, D. R., D'Adamo, A. F., Solt, M. L. and Wolf, A. P. *J. Am. Chem. Soc.* **82** (1960) 2628.
15. Solt, M. L., Dawson, R. F. and Christman, D. R. *Plant. Physiol.* **35** (1960) 887.
16. Leete, E. and Slattery, S. A. *J. Am. Chem. Soc.* **98** (1976) 6326.
17. (a) Yamamura, S. and Hirata, Y. in *The Alkaloids*, Vol. XV, p. 41, Manske, R. H. F. (Ed.), Academic Press, New York, 1975. (b) Snieckus, V. in *The Alkaloids*, Vol. XIV, p. 425, Manske, R. H. F. (Ed.), Academic Press, New York, 1973. (c) MacLean, D. B. in *The Alkaloids*, Vol. XIV, p. 347, Manske, R. H. F. (Ed.), Academic Press, New York, 1973.
18. O'Donovan, D. G. and Barry, E. *J. Chem. Soc. Perkin I* **1974,** 2528.
19. Schütte, H. R. and Seelig, G. *Liebigs Ann.* **730** (1970) 186.
20. Bhakuni, D. S., Singh, A. N., Tewari, S. and Kapil, R. S. *J. Chem. Soc. Perkin I* **1977,** 1662.
21. Wilson, M. L. and Coscia, C. J. *J. Am. Chem. Soc.* **97** (1975) 431.
22. Tewari, S., Bhakuni, D. S. and Kapil, R. S. *J. Chem. Soc. Chem. Commun.* **1975,** 554.
23. Battersby, A. R., Sheldrake, P. W., Staunton, J. and Summers, M. C. *Bioorg. Chem.* **6** (1977) 43.
24. Battersby, A. R. and Parry, R. J. *J. Chem. Soc. Chem. Commun.* **1971,** 901.
25. Battersby, A. R., McHugh, J. L., Staunton, J. and Todd, M. *J. Chem. Soc. Chem. Commun.* **1971,** 985.
26. Barton, D. H. R., Bhakuni, D. S., Chapman, G. M., Kirby, G. W., Haynes, L. J. and Stuart, K. L. *J. Chem. Soc. (C)* **1967**, 1295.
27. Tewari, S., Bhakuni, D. S. and Kapil, R. S. *J. Chem. Soc. Chem. Commun.* **1974,** 940.
28. Barton, D. H. R. and Cohen, T. *Festschrift A. Stoll*, Birkhäuser, Basel, 1957, p. 117.
29. Bhakuni, D. S. and Singh, A. N. *J. Chem. Soc. Perkin I* **1978,** 618.
30. Bowman, W. R., Bruce, I. T. and Kirby, G. W. *Chem. Commun.* **1969,** 1075.
31. Fuganti, C. in *The Alkaloids*, Vol. XV, p. 83, Manske, R. H. F. (Ed.), Academic Press, New York, 1975.
32. Leander, K., Lüning, B. and Ruusa, E. *Acta Chem. Scand.* **23** (1969) 244.
33. Jeffs, P. W., Karle, J. M. and Martin, N. H. *Phytochemistry* **17** (1978) 719.
34. Kametani, T. and Koizumi, M. in *The Alkaloids*, Vol. XIV, p. 265, Manske, R. H. F. (Ed.), Academic Press, New York, 1973.
35. Battersby, A. R., Herbert, R. B., Pijewska, L., Šantavý, F. and Sedmera, P. *J. Chem. Soc. Perkin I* **1972,** 1736.

36. Battersby, A. R., Herbert, R. B., McDonald, E., Ramage, R. and Clements, J. H. *J. Chem. Soc. Perkin I* **1972,** 1741.
37. Thomas, R. *Tetrahderon Lett.* **1961,** 544.
38. Wenkert, E. *J. Am. Chem. Soc.* **84** (1962) 98.
39. Battersby, A. R. in *Pure and Applied Chemistry* **14** (1967) 117, Butterworths, London, and in *The Alkaloids* **1** (1971) 31, Specialist Periodical Reports, The Chemical Society, Saxton, J. E. (Ed.). Scott, A. I. in *MTP Int. Rev. Sci.* **9** (1973) 105, Hey, D. H. and Wiesner, K. F. (Eds.), Butterworths, London.
40. Battersby, A. R., Gregory, B., Spenser, H., Turner, J. C., Janot, M. M., Potier, P., François, P. and Levisalles, J. *Chem. Commun.* **1967,** 219.
41. Battersby, A. R., Burnett, A. R. and Parsons, P. G. *Chem. Commun.* **1968,** 128.
42. Smith, G. N. *Chem. Commun.* **1968,** 912.
43. Stöckigt, J. and Zenk, M. H. *J. Chem. Soc. Chem. Commun.* **1977,** 646.
44. Rueffer, M., Nagakara, N. and Zenk, M. H. *Tetrahedron Lett.* **1978,** 1593.
45. Woodward, R. B., Bader, F. E., Bickel, H., Frey, A. J. and Kierstad, R. W. *Tetrahedron* **2** (1958) 1.
46. Qureshi, A. A. and Scott, A. I. *Chem Commun.* **1968,** 948.
47. Brown, R. T., Smith, G. F., Stapleford, K. S. J. and Taylor, D. A. *Chem. Commun.* **1970,** 190.
48. Mothes, K., Weygand, F., Gröger, D. and Griesebach, H. *Z. Naturforsch.* **13b** (1958) 41.
49. Birch, A. J. in *Ciba Foundation Symposium on Amino Acids and Peptides with Antimetabolic Activity*, Wolstenholme, G. E. W. and O'Connor, C. M. (Eds.), Churchill, London, 1958, p. 247.
50. Floss, H. G. *Tetrahedron* **32** (1976) 873.
51. Colonna, A. O. and Gros, E. G. *Chem. Commun.* **1970,** 674.
52. Grundon, M. F., Harrison, D. M. and Spyropoulos, C. G. *J. Chem. Soc. Chem. Commun.* **1974,** 51.
53. Collins, J. F., Donelly, W. J., Grundon, M. F. and James, K. J. *J. Chem. Soc. Perkin I* **1974,** 2177.
54. Ritter, C. and Luckner, M. *European J. Biochem.* **18** (1971) 391.
55. Liljegren, D. R. *Phytochemistry* **10** (1971) 2661.
56. Johne, S., Waiblinger, K. and Gröger, D. *Pharmazie* **28** (1973) 403.
57. Byng, G. S. and Turner, J. M. *J. Gen. Microbiol.* **97** (1976) 57.
58. Buckland, P. R., Herbert, R. B. and Holliman, F. G. *Tetrahedron Lett.* **1981,** 595.
59. Hollstein, U., Mock, D. L., Sibbit, R. R., Roisch, U. and Lingens, F. *Tetrahedron Lett.* **1978,** 2987.
60. Gross, D., Berg, W. and Schütte, H. R. *Biochem. Physiol Pflanz.* **163** (1972) 576.
61. Torssell, K. and Wahlberg, K. *Acta Chem. Scand.* **21** (1967) 53.
62. Govindachari, T. R., Sathe, S. S. and Viswanathan, N. *Indian J. Chem.* **4** (1966) 201.
63. Cordell, G. A. in *The Alkaloids* Vol. XVI, p. 431, Manske, R. H. F. (Ed.) Academic Press, New York, 1977
64. Valenta, Z. and Khaleque, A. *Tetrahedron Lett.* **12** (1959) 1.
65. Corbella, A., Gariboldi, P., Jommi, G. and Sisti, M. *J. Chem. Soc. Chem. Commun.* **1975,** 288.
66. Przybylska, M. and Marion, L. *Can. J. Chem.* **34** (1956) 185; Pelletier, S. W., Mody, N. V., Varughese, K. I., Maddry, J. A. and Desai, H. K. *J. Am. Chem. Soc.* **103** (1981) 6536.
67. Tschesche, R. and Hulpke, H. Z. *Naturforsch,* **21b** (1966) 893.
68. Guseva, A. R. and Paseshnichenko, V. A. *Biochemistry* (USSR) **27** (1962) 721.
69. Ronchetti, F. and Russo, G. *J. Chem. Soc. Chem. Commun.* **1974,** 785.
70. Kaneko, K., Seto, H., Motoki, C. and Mitsuhashi, H. *Phytochemistry* **14,** (1975) 1295.

Chapter 8

N-Heteroaromatics

8.1 Introduction

An account of secondary metabolism cannot be considered to be complete if one omits heteroaromatics like pyrimidines, purines, porphyrins, etc., even though they are considered first order constituents participating in the fundamental chemistry of life and replication. However, this class of compounds does contain several secondary metabolites. Furthermore, their mode of formation is of general chemical interest. Several *N*-heteroaromatics are vitamins and coenzymes catalysing the most intriguing reactions—vitamin B_{12} in particular. We have already met representatives of the pteridine nucleus in the monoxygenase cofactors riboflavin, vitamin B_2, and the thiazole and pyrimidine nuclei in the cofactor thiamine, vitamin B_1. These and other nuclei will be discussed in this chapter. The redox reaction of the riboflavins are treated in section 3.3 in connection with biological hydroxylation.

Before we discuss the contemporary production of *N*-heterocycles, we shall once more go back a few billion years to the time of the prebiotic brew. We now know that the mechanisms of replication are intimately connected with the action of genes, which can be chemically defined as self-replicating pieces of nucleic acids, and which possess the ability to direct polypeptide or enzyme synthesis in a defined manner. This process is outlined in textbooks of biochemistry. The building blocks of these nucleic acids are ribose, pyrimidines or purines linked together with phosphoric acid (Fig. 1). Consequently these compounds required also to be synthesized by chance, which although it may seem highly unlikely, turns out to be quite possible. Electric discharges in the presumably reducing atmosphere containing methane, carbon dioxide, carbon monoxide, ammonia, water, and nitrogen can give rise to hydrocyanic acid, cyanoacetylene, urea, and formaldehyde in reactions catalysed by various metals and oxides (Al_2O_3, Fe, Ni, etc.). Oligomerization of hydrocyanic acid can eventually lead to adenine (1)[1] and from cyanoacetylene and urea cytosine can be formed (2).[2] Formaldehyde can, in principle, condense to ribose[3] which then condenses with the nucleic acid bases and polyphosphate to nucleotides in a magnesium ion catalysed reaction.[4]

Fig. 1 Section of a ribonucleic acid (RNA)

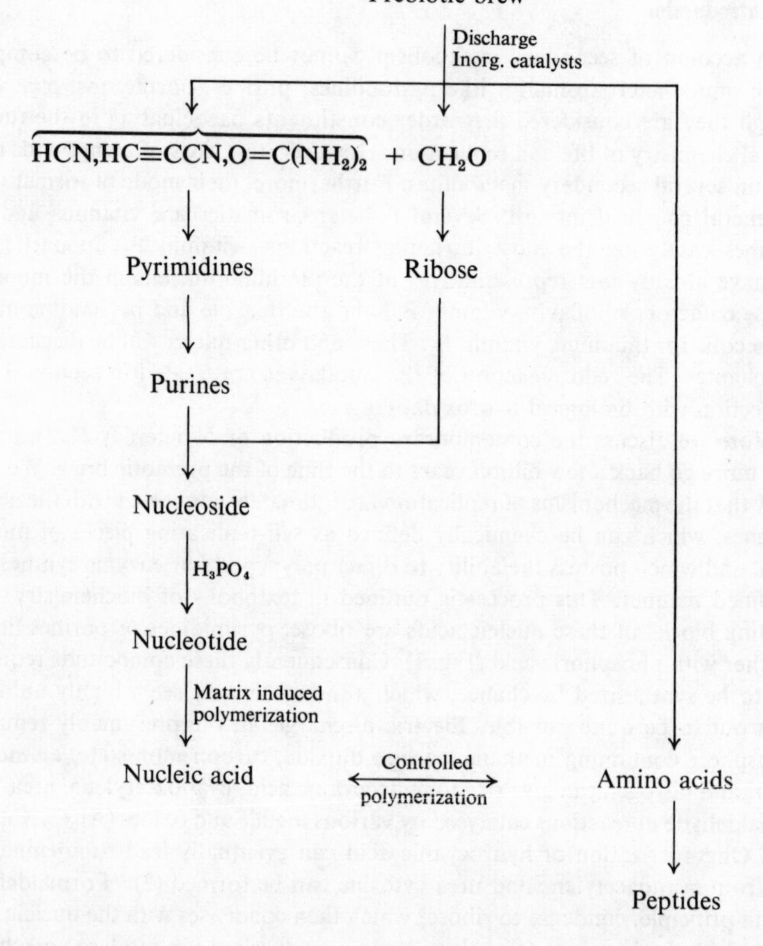

Fig. 2 Suggested prebiotic pathway to nucleic acids and peptides

Eventually, a condensing agent, such as a water-soluble carbodiimide, effects polymerization and selected condensation could have an influence on the order of units in the nucleic acid formed. The prebiotic steps are summarized in Fig. 2. Finally, biomimetic studies show that the complicated porphyrin nucleus of the cytochromes is formed in an astonishingly simple manner by acid catalyzed tetramerization and cyclization of porphobilinogen. Furthermore, it is conceivable that porphobilinogen can be formed in a primitive process from glycine and succinate as building blocks, as depicted in section 8.3. Complexation of porphyrin with the ferrous ion gives us the first organic catalyst of our planet. A primitive stage was set for the chemical evolution.

$$\text{HCN} \;(\times 3,\ \text{H}^{\oplus}) \longrightarrow \text{4-amino-5-cyanoimidazole} \xrightarrow{\text{HCN}} \text{Adenine} \qquad (1)$$

$$\text{HC}\equiv\text{C}-\text{C}\equiv\text{N} + \text{H}_2\text{N}-\text{CO}-\text{NH}_2 \;(\text{H}^{\oplus}) \longrightarrow \text{Cytosine} \qquad (2)$$

8.2 Pyrimidines, purines and pteridines

The major bases in RNA, ribonucleic acid, are adenine, guanine, cytosine and uracil. In DNA, deoxyribonucleic acid, uracil is replaced by thymine (5-methyluracil). Like their ribosides the free bases are present in trace amounts in the cell as catabolic products from the nucleic acids or as intermediates in the biosynthesis of nucleic acids. In some fungi and sponges, nucleosides are produced *per se* in larger quantities having antibiotic properties, e.g. nebularine from *Clitocybe nebularis*[5] (Fig. 3). The N^6-prenylpurines and their ribosides, the so-called cytokinins, are growth hormones strongly promoting cell division, enlargement, and differentiation in plants.[6] These prenylated purine derivatives are also found as minor constituents in transfer RNA.[7] A hydrogenated purine skeleton is contained in the fatally poisonous saxitoxin which is produced by the marine dinoflagellate *Gonyaulax catanella* and also by the blue-green alga *Aphanizomenon flos-aque*, and occasionally accumulated in clams and mussels feeding on the dinoflagellates during blooms (red tide). When the toxin reaches man in the feeding chain, it acts by blocking the sodium ion channel and causes what is known as paralytic shellfish poisoning. The structure of saxitoxin is unusual and posed serious problems to the organic chemist. It was eventually

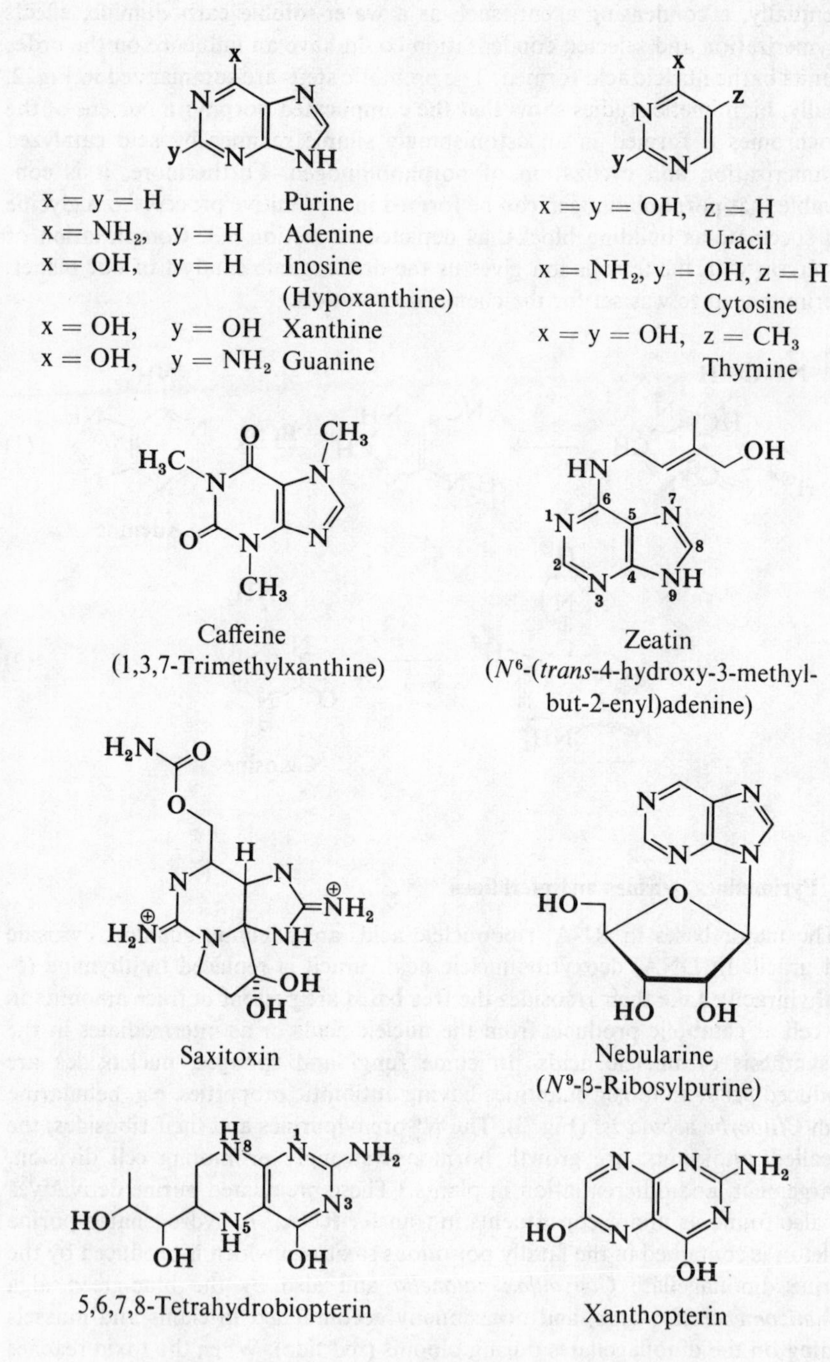

Fig. 3 Structure of purine and pteridine derivatives

solved by X-ray crystallography.[8] The purine derivatives caffeine and theobromine (3,7-dimethylxanthine) are widely used as stimulants. Caffeine is found in *Coffea* spp. (*Rubiaceae*), tea, *Thea sinensis* (*Theaceae*), in yerba maté from the evergreen shrub *Ilex paraguariensis* (*Aquifoliaceae*) and in the bark of sapindaceous jungle lianes in the Amazon, *Paullinia* spp. The fatty cacao seeds from *Theobroma cacao* (*Sterculiaceae*) contains *ca.* 1 per cent theobromine. The old Aztecs called the tree chocoatl and not very much more remains etymologically in our modern culture from this once so powerful people. The British are the largest consumers of tea per head. Coffee became a popular beverage rather late and was once treated with suspicion. In Sweden in the eighteenth century it was forbidden by law to drink coffee; now the Swedes are among the largest coffee consumers in the world, whatever conclusion one can draw from that. Apart from functioning as a building block in nucleic acids, pyrimidines are rare molecules in nature.

The bicyclic pteridine (from Gr. *pteron*, wing) skeleton formed by a pyrimidine ring fused to a pyrazine ring is contained in the biologically important coenzymes tetrahydrofolic acid, biopterine and riboflavin. The pleasing, colourful and fluorescent pigments found in the wings of butterflies and in the skin of amphibians are hydroxy- and amino-substituted pteridines which on oxidation can form an extended quinonic system. These pigments play an important role in visual chemical communication in the terrestrial and marine environments.

Visual inspection of the structure is a misleading guide for chemical intuition concerning the biosynthesis of purines. Contrary to expectations the nucleoside is not formed by ribosidation of a preformed purine derivative. Instead ribose is attached already at the start of the assembly of the purine ring system. The condensed heteroaromatic can be dissected into a pyrimidine ring, an imidazole ring and urea but no such derivatives are incorporated (Fig. 4, A). Feeding experiments with pigeons and analysis of uric acid (2,6,8-trihydroxypurine) excreted, indicated that carbon dioxide, ammonia, formate and intact glycine are built into the molecule. The biosynthetic scheme which emerged, is as follows (Fig. 5).[9] C^1 of ribose-5-phosphate is activated by phosphorylation with ATP. The pyrophosphate group is displaced with 'active' ammonia, generated from glutamine, with inversion of configuration at C^1 giving the β-N^9 atom of the final

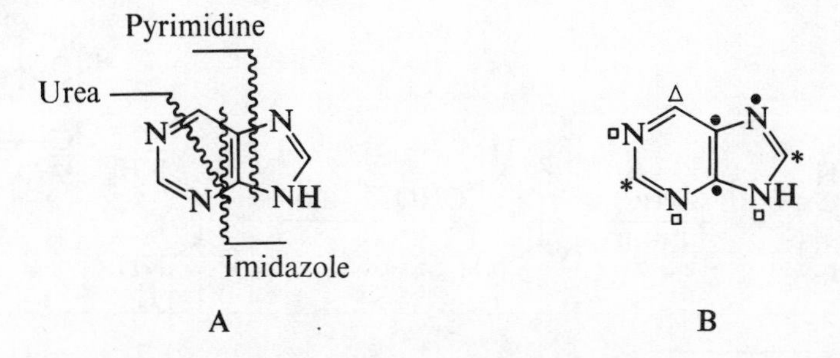

Fig. 4 A, Imaginable precursors of the purine ring; B, precursors of the purine ring. ● Glycine, * Formate, □ $[NH_3]$, △ CO_2

purine nucleotide. Glycine is activated by ATP and introduced intact to form an amide. In the following steps the atoms are introduced one by one. The free amino group of the glycine amide is formylated with N^5,N^{10}-formamidinyl-tetrahydrofolate, thus accomplishing the introduction of C^8. The next step before cyclization to the imidazole involves an ATP assisted amidination of the glycine amide by ammonia, generated from glutamine. The purine C^6 atom is introduced as carbon dioxide in an electrophilic substitution at C^4 of the imidazole nucleus. Nature uses yet another technique to introduce the N^1 atom. The carboxyl group of the imidazole is phosphorylated with ATP and reacted with aspartic acid. β-Elimination gives the carboxamide and fumaric acid. C^2 is finally introduced as formate in a second folate-assisted reaction and cyclization gives the nucleotide inosinic acid (IMP) which is a key intermediate for other purine nucleotides. The conversion of IMP to adenylic acid (AMP) is accomplished by amination of C^6 via the aspartic acid route (above) by assistance of GTP. The biosynthesis of guanylic acid (GMP) is accomplished by oxidation of C^2 with NAD⊕ to xanthylic acid (XMP) followed by amination of C^2.

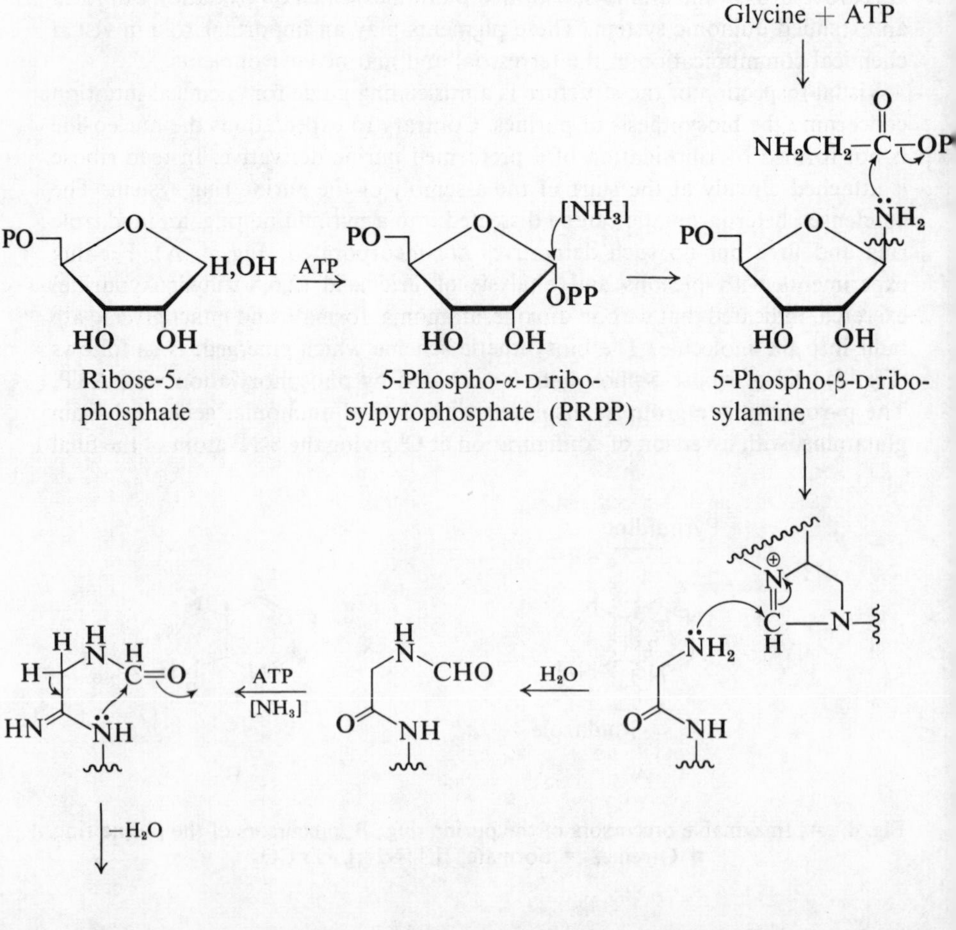

Glycine + ATP

$NH_2CH_2-C(=O)-OP$

Ribose-5-phosphate —ATP→ 5-Phospho-α-D-ribosylpyrophosphate (PRPP) —$[NH_3]$→ 5-Phospho-β-D-ribosylamine

—H_2O→ ←ATP, $[NH_3]$—

—H_2O→

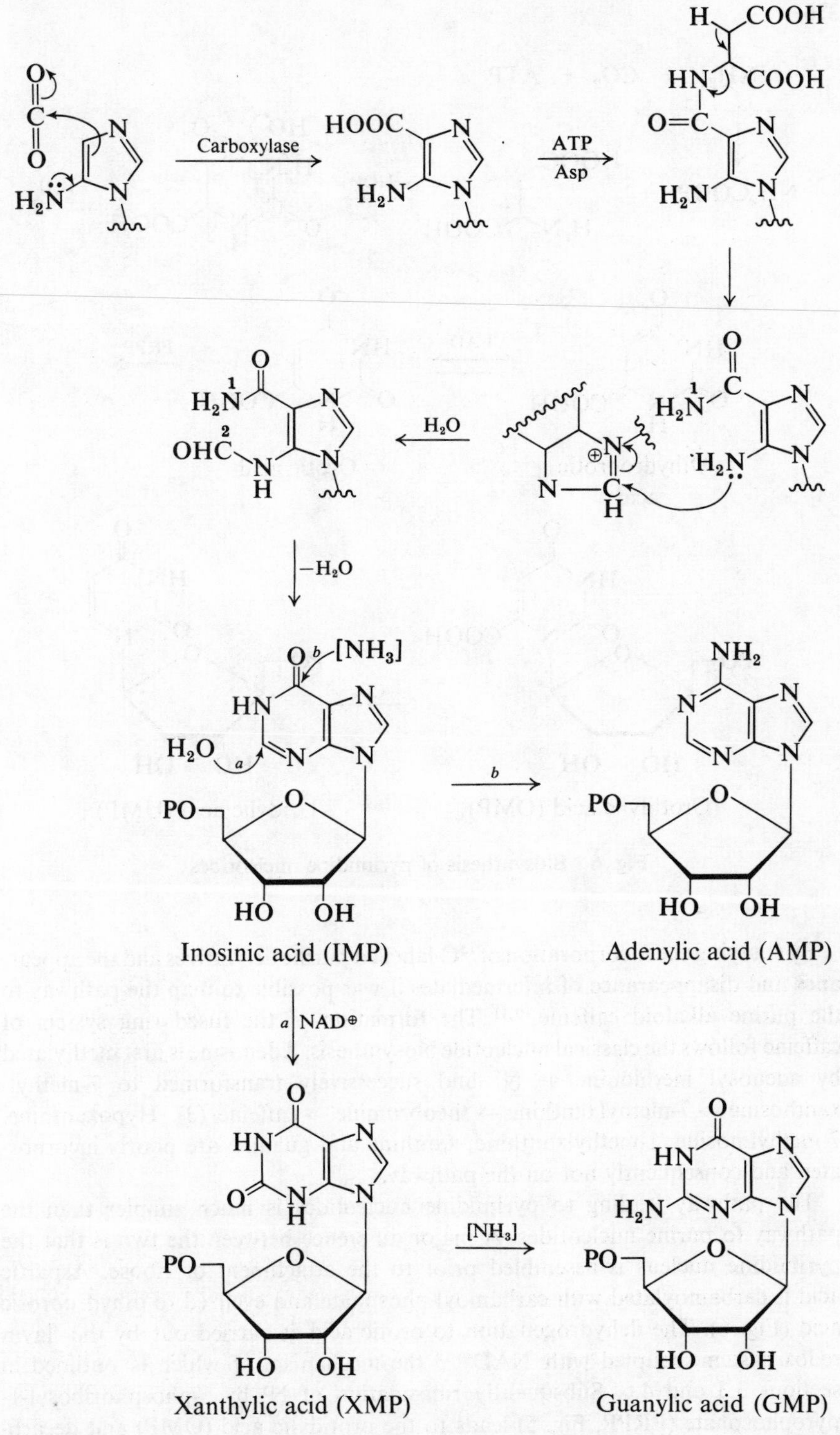

Carboxylase

ATP
Asp

H_2O

$-H_2O$

$[NH_3]$

b

a NAD⊕

$[NH_3]$

Inosinic acid (IMP)

Adenylic acid (AMP)

Xanthylic acid (XMP)

Guanylic acid (GMP)

Fig. 5 Biosynthesis of purine nucleotides. The 5-phosphoribosyl moiety is omitted in some of the intermediates

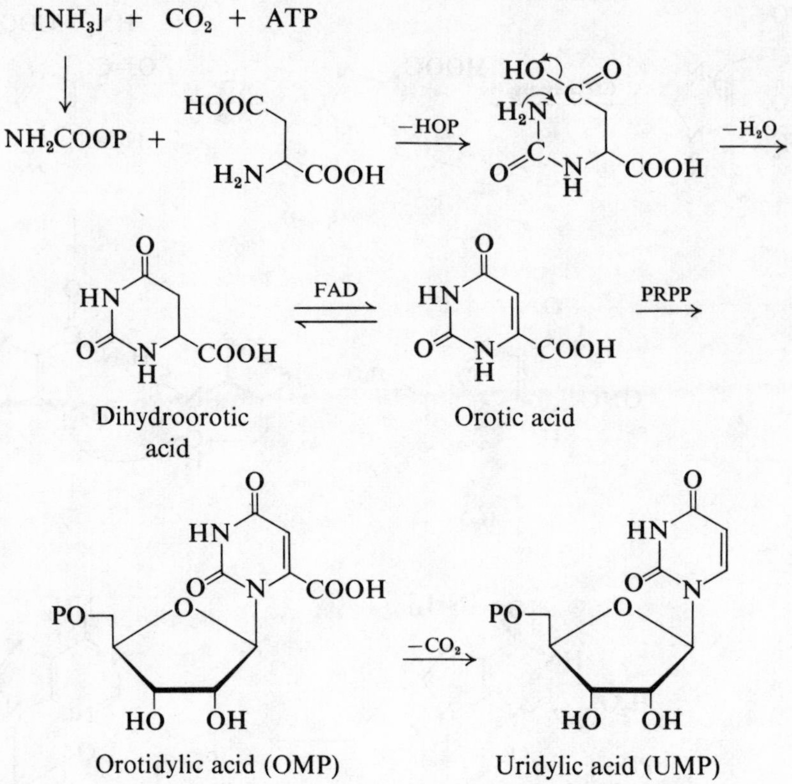

$[NH_3] + CO_2 + ATP$

$\downarrow$

NH_2COOP + HOOC, H_2N, COOH $\xrightarrow{-HOP}$ HO, O, C, H_2N, C, O, N, H, COOH $\xrightarrow{-H_2O}$

O, HN, O, N, H, COOH

Dihydroorotic acid

$\underset{}{\overset{FAD}{\rightleftharpoons}}$

O, HN, O, N, H, COOH

Orotic acid

$\xrightarrow{PRPP}$

O, HN, O, N, COOH, PO, O, HO, OH

Orotidylic acid (OMP)

$\xrightarrow{-CO_2}$

O, HN, O, N, PO, O, HO, OH

Uridylic acid (UMP)

Fig. 6 Biosynthesis of pyrimidine nucleotides

By studying the incorporation of ^{14}C-labelled purine derivatives and the appearance and disappearance of intermediates it was possible to map the pathway to the purine alkaloid caffeine.[10,11] The formation of the fused ring system of caffeine follows the classical nucleotide biosynthesis. Adenosine is first methylated by adenosyl methionine at N^7 and successively transformed to 7-methylxanthosine → 7-methylxanthine → theobromine → caffeine (3). Hypoxanthine, 7-methylinosine, 1-methylxanthine, xanthine and guanine are poorly incorporated and consequently not on the pathway.

The pathway leading to pyrimidine nucleotides is much simpler than the pathway to purine nucleotides. A major difference between the two is that the pyrimidine nucleus is assembled prior to the attachment of ribose. Aspartic acid is carbamoylated with carbamoyl phosphate and cyclized to dihydroorotic acid (Fig. 6). The dehydrogenation to orotic acid is carried out by the flavin redox system, coupled with $NAD^{\oplus}$,[12] the mechanism of which is outlined in sections 3.3 and 4.6. Subsquently ribosidation of N^1 by 5-phosphoribosyl-1-pyrophosphate (PRPP, Fig. 5) leads to the orotidylic acid (OMP) and decarboxylation gives uridylic acid (UMP).

Adenosine → $[CH_3]$ → 7-Methyladenosine → 7-Methylxanthosine → 7-Methylxanthine → 3,7-Dimethylxanthine (Theobromine) → 1,3,7-Trimethylxanthine (Caffeine)

(3)

The biosynthesis of thymine (5-methyluracil) is of mechanistic interest. It represents one of a few instances where methylation is carried out by 5,6,7,8-tetrahydrofolate (THF) rather than by methionine.[13] Uridylic acid (UMP) is the substrate but for the sake of clarity only the pyrimidine moiety is shown in Fig. 7. Formaldehyde is trapped by THF and forms the $N^{5,10}$-methylene bridge. THF represents the reduced partner of a quinonoid redox couple, the oxidized form of which can form two tautomers of dihydrofolic acid (DHF) (4). The

5,6,7,8-Tetrahydrofolate (THF, partial structure) ⇌ [O] ⇌ 7,8-Dihydrofolate (DHF)

CH_2O ↓ ; $H^{\oplus}$ ⇌

(4)

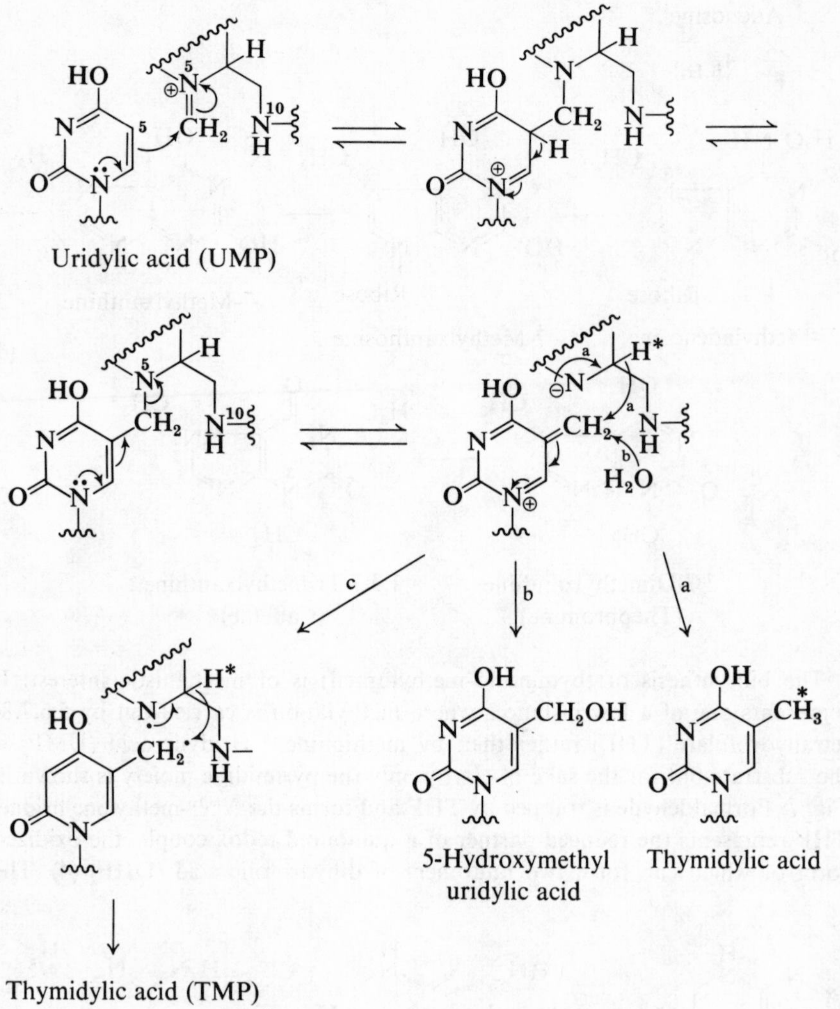

Fig. 7 Mechanism for methylation and hydroxymethylation of uridylic acid (UMP). Partial structure; 5-phosphoribase moiety is omitted

7,8-dihydrofolic acid is the most stable tautomer. It is reduced back to THF by NADH. The C^6–H bond is consequently rather weak. Uridine undergoes electrophilic substitution at C^5 and rearomatization occurs by elimination of C^5–H, whereby the methylene bridge now moves over into the uridyl nucleus. The N^5-nitrogen serves as a leaving group assisted by the lone pair of uridine-N^3. Cleavage leads to an intermediate quinonoid uridyl moiety which picks up the C^6–H as a hydride (route a) or reacts with water to the hydroxymethyl group (route b). Route a can equally well be formulated as a radical reaction. The cleavage occurs homolytically or alternatively the quinonoid uridyl moiety oxidizes the THF ion which gives rise to two semiquinonoid radicals and the uridyl radical abstracts finally the C^6–H atom (route c). Both the hydride shift and the radical abstraction are consistent with the finding that thymidine is produced with no loss of activity from 6-^{3}H-THF. A $^3H/^1H$ isotope effect of 5.2 has been observed indicating that the hydrogen transfer is taking part in the rate determining step. The preceding steps are probably equilibria.[14]

It is inferred that the enzyme which participates in the methylation of cytidine acts by adding across the 5,6 double bond thereby initiating the reaction with $N^{5,10}$-methylene-THF (5)[15] as shown by experiments with 5-fluorodeoxyuri-

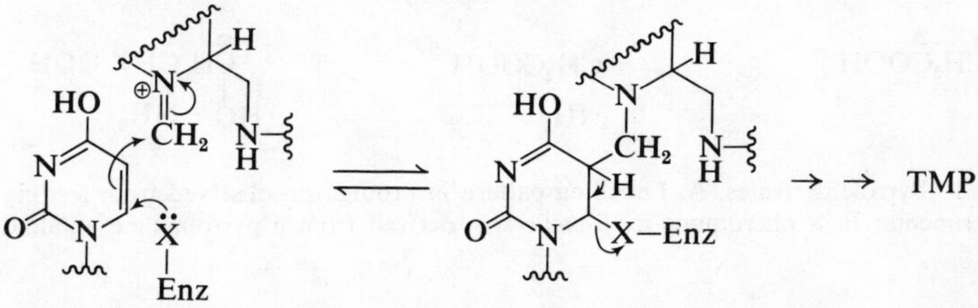

→ → TMP (5)

dylate, F-dUMP, (6). It was found that the fluorine derivative causes rapid inactivation of the enzyme and that it loses UV absorbance due to loss of the pyrimidine chromophore on incubation with enzyme and $N^{5,10}$-methylene-THF in consistency with the formulation in (6), where fluorine now blocks the elimination of the enzyme. A slow exchange of 5-C-^{3}H in uracil was also noted which could be referred to as a reversible addition of enzyme across the 5,6 double bond. However, these observations can also be accommodated in the scheme in Fig. 7, if we assume that addition to the enzyme is a secondary process which takes place, when rearomatization upon alkylation of cytidine cannot occur.

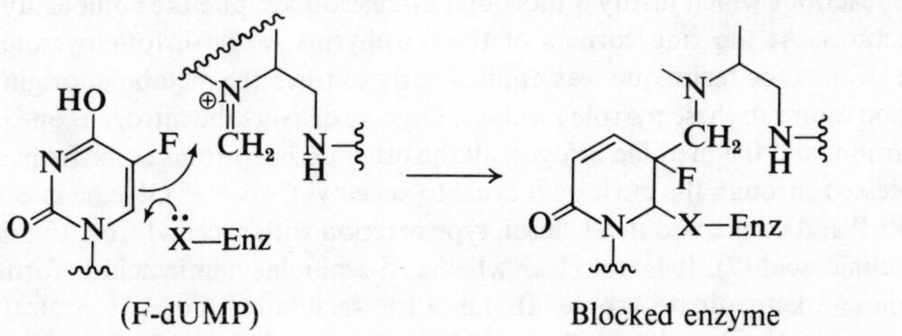

(6)

(F-dUMP) Blocked enzyme

8.3 Pyrroles and porphyrins

Apart from being a building block in the porphyrins, corrins, and some gall pigments of common biogenesis, pyrrole derivatives are rare in nature. A few pyrrole derivatives (Fig. 8) are produced by microorganisms, sponges, higher plants and animals. The intensely red prodigiosins in *Serratia marcescens* are

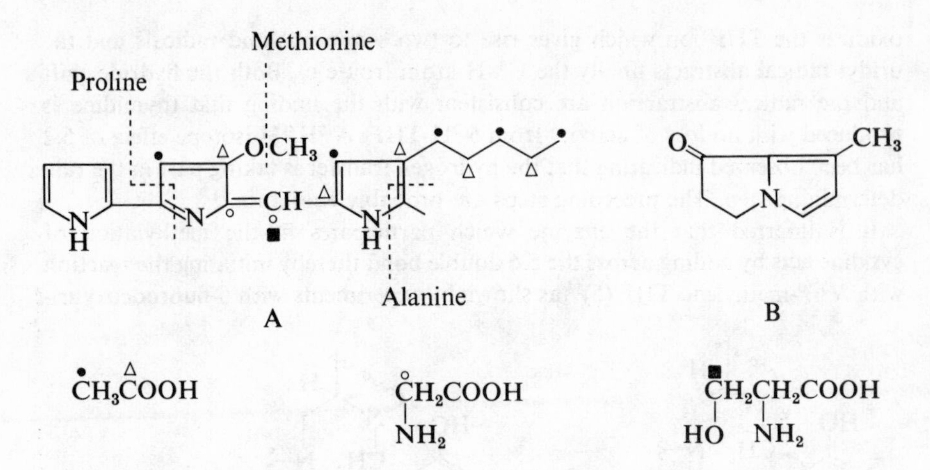

Fig. 8 Pyrrole derivates. A, Labelling pattern in prodigiosin observed from feeding experiments; B, a pheromone in *Danaus* spp. derived from a pyrrolizidine alkaloid

derived from acetate and amino acids as established by MS and ^{13}C NMR.[16] However, this biosynthetic pathway is quite different from that of the porphyrins. The pheromone from the hair pencils of *Danaus* spp. originates from certain pyrrolizidine alkaloids produced by plants on which the butterfly thrive and it is further metabolized by the butterfly.[17] Simple amino acids were not incorporated when fed to the insect.

In view of their immense biological importance the macrocyclic pigments of life have attracted special interest among scientists. By the combined efforts of workers in all fields of chemistry a detailed and coherent picture of the biosynthesis starts to emerge. The biosynthesis provides several instructive and remarkable reactions which justify a thorough discussion, despite the complexity of the structures. At the ring corners of the porphyrins we have four pyrrole units. The ^{14}C isotope technique was applied early to trace the metabolic origin of the carbon atoms in these pyrrole residues. Glycine delivers the nitrogen, one pyrrole α-carbon, and the methine bridges, all the other carbon atoms come from acetate processed through the citric acid cycle to succinyl CoA.[18,19] Glycine is activated by PLP and condensed in a Claisen type reaction with succinyl CoA to 5-aminolaevulinic acid (7). It is not clear whether 5-aminolaevulinic acid is formed via 2-amino-3-ketoadipate (route a), i.e. PLP facilitates and α-H ionization of glycine, or by decarboxylative alkylation (route b), i.e. PLP catalyses the decarboxylation as in decarboxylation of amino acids. In any event, 2-amino-3-ketoadipate is found to decarboxylate spontaneously in less than one minute at pH 7[20] and its PLP derivative, i.e. the intermediate in route a, will most likely decarboxylate immediately. When 5-aminolaevulinic acid was added to enzyme preparations from duck blood it served as an *in vitro* precursor of haem.[21] Two molecules of 5-aminolaevulinic acid dimerize in a Knorr-type reaction to the pyrrole derivative porphobilinogen (PBG,8) which also incorporates well.

$$HOOCCH_2CH_2\overset{O}{\overset{\|}{C}}-CoA \quad \ominus CHCOOH-N{=}CH-PLP \xrightarrow{a} HOOCCH_2CH_2\overset{O}{\overset{\|}{C}}-CH(N{=}CH\sim)\overset{O}{\overset{\|}{C}}-O-H \xrightarrow{H_2O}$$

$$HOOCCH_2CH_2\overset{O}{\overset{\|}{C}}-CoA \quad H_2C(N{=}CH\sim)-\overset{O}{\overset{\|}{C}}-O-H \xrightarrow[H_2O]{b} HOOCCH_2CH_2\overset{O}{\overset{\|}{C}}-CH_2NH_2 \qquad (7)$$

5-Aminolaevulinic acid

Tetramerization of PBG head to tail, catalysed by deaminase, gives symmetrical and colourless uroporpyhrinogen I with alternating arrangement of acetate and propionate groups which on oxidation, e.g. by O_2, easily forms the strongly coloured conjugated aromatic 18-π-system, uroporphyrin I (Fig. 9). The mode of condensation was followed by ^{13}C NMR of 5-amino-5-^{13}C-laevulinic acid, enriched at 90 atom per cent, which gives rise to 2,11-labelled PBG and symmetrically labelled uroporphyrin I.[22] The ^{13}C signal from the equivalent *meso* carbons, appears characteristically as a double doublet, rel. intensity *ca*. 90 per cent, $^1J_{^{13}C-^{13}C}$ 72 Hz, $^3J_{^{13}CNC^{13}C}$ 5 Hz centred on a broad singlet, rel. intensity *ca*. 10 per cent at δ 97.5. The four pyrrole α-carbon atoms give a similar pattern at δ 143.5. This is in agreement with a straightforward cyclization.

In the biologically active porphyrins, A–C have alternating arrangements of the acetic and propionic acid groups, as in uroporphyrinogen I but ring D is 'turned around' so that C and D now have adjacent propionic acid groups as in

$$\text{2 ALA (HOOC, COOH, } O{=}, \ominus, H_2N, H{-}NH) \xrightarrow{-2\,H_2O} \text{PBG (COOH, COOH, 2, 11, N-H, } NH_2) \qquad (8)$$

Porphobilinogen (PBG)

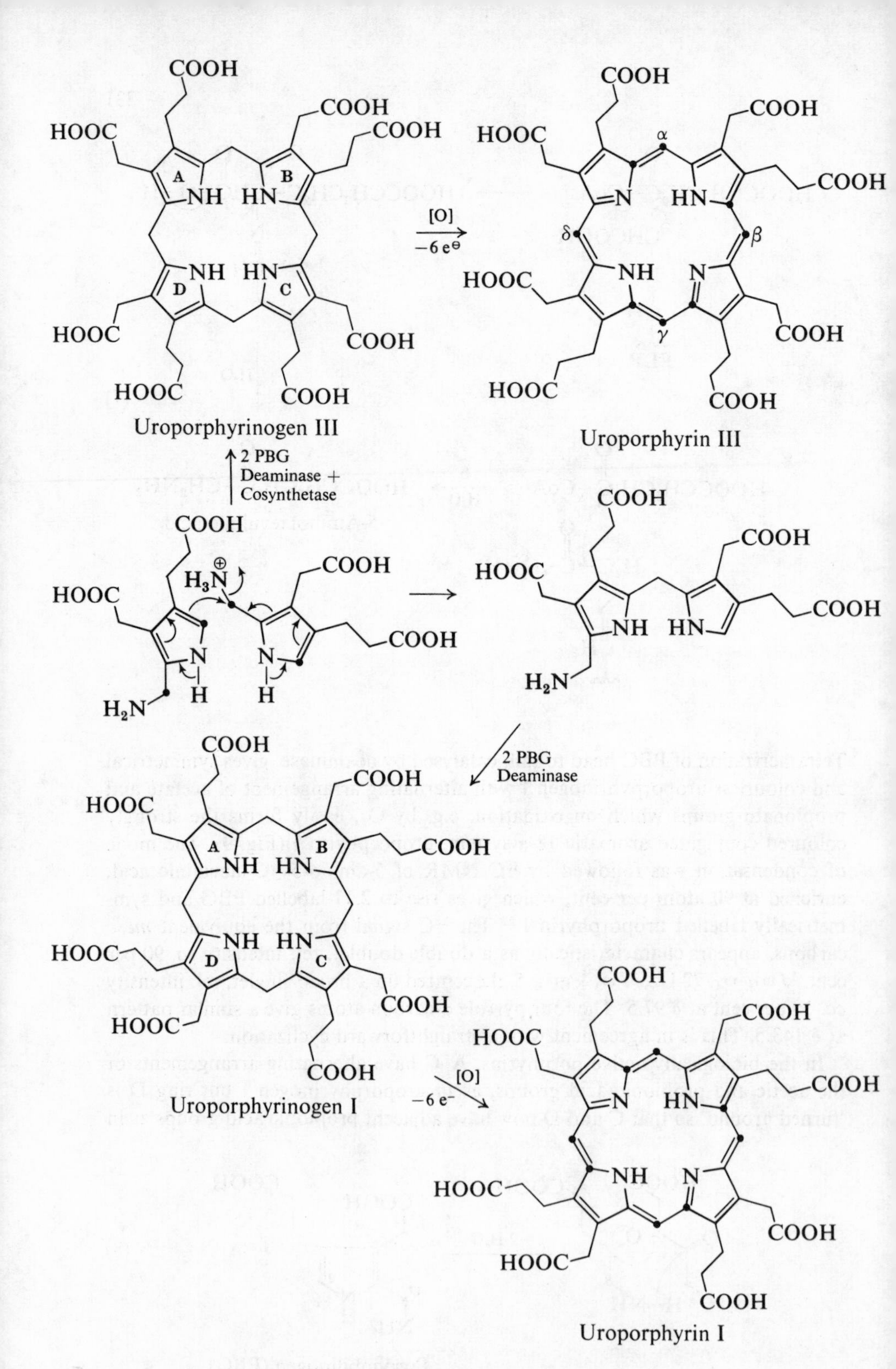

Fig. 9 Biosynthesis of uroporphyrins

uroporphyrinogen III, a central intermediate on the road to chlorophylls, corrins and haems. It is also synthesized from ALA via PBG but needs for cyclization a second enzyme, the so-called cosynthetase which is only active on PBG in the presence of the deaminase.[23,24] The enzymatic studies indicate that deaminase starts the tetramerization and at some stage cosynthetase takes over, picks up an intermediate released, and diverts it to the rearranged uroporphyrinogen III. Cosynthetase is unable to convert uroporphyrinogen I into isomer III. The ^{13}C NMR spectrum of uroporphyrine III, produced by the combined enzyme complex, obtained either from avian blood or from the alga *Euglena gracilis*, was in agreement with the labelling pattern in Fig. 9. By dilution of the double-labelled (90 per cent) PBG with four parts of unlabelled PBG it is arranged so that the majority of the pyrrole units have only two interacting ^{13}C atoms which simplifies the fine structure of the proton decoupled ^{13}C NMR spectrum. The α-, β- and δ-carbons show primarily a fine doublet, $^3J_{^{13}CNC^{13}C}$ 5 Hz, proving that the A, B and C units are incorporated intact. The γ carbon signal is a doublet, $^1J_{^{13}C^{13}C}$ 72 Hz, indicating an intramolecular rearrangement of the side chain from one α-carbon to the other in ring D.[22]

It was possible to separate cleanly the two enzymes. By quenching the action of the aminase it was possible to identify an open chain tetramer which slowly *chemically* cyclized to uroporphyrinogen I. The tetramer gave rapidly the rearranged uroporphyrinogen III by treatment with cosynthetase.[25] It was also shown that the assembly of the four rings started with ring A followed by sequential addition of B, C and D. A mechanism for the rearrangement is depicted in (9).[25,26] The open chain head-to-tail tetramer holds a conformation on the cosynthetase which makes it possible for the δ-*meso* carbon (dotted carbon atom) to attack at the substituted α-carbon of ring D forming a spiro intermediate. This is an example of a biosynthetic *ipso*-substitution. Cleavage of the γ-*meso*-α-pyrrole carbon bond, rotation of the D unit and recyclization at the other α-pyrrole carbon complete the rearrangement.

Uroporphyrinogen III is converted to coproporphyrinogen III by stepwise decarboxylation of the four acetic acid side chains to methyl groups (Fig. 10). The reaction proceeds stereospecifically with retention of configuration (10).[27] This was demonstrated by incubation of haemolysed avian erythrocytes with chiral succinate, isolation of the biosynthesized haem and analysis of the acetic acid formed by oxidative degradation of the haem by the malate-fumarate procedure (section 5.2). It turns out that 2-*R*-succinic acid gives 2-*S*-acetic acid. The oxidative decarboxylation of the propionic acid side chains is also sterically controlled. The decarboxylation takes place first at ring A and subsequently at ring B as proved by the inability of the enzyme to metabolize a synthetically prepared derivative with a propionic group on A and a vinyl group on B.[28] The steric course of the decarboxylation of the propionic acid side chain was followed by investigation of the fate of ALA specifically deuterated in the side chain (11). The ^1H NMR signal of H_R in the vinyl group of protoporphyrin IX appears as a doublet, J_{H-H_R} 18 Hz, characteristic for a *trans* coupling. Earlier work with specifically labelled succinate has shown that only the pro-*S*-hydrogen

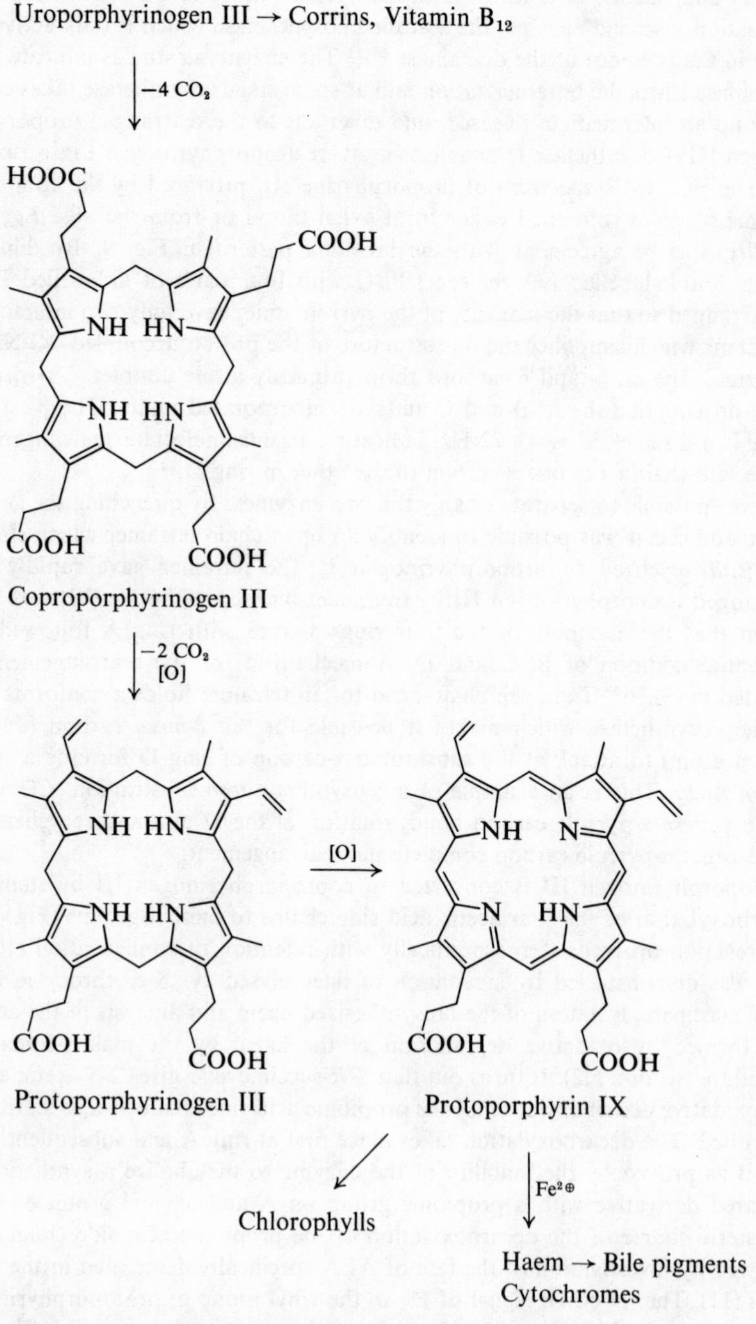

Uroporphyrinogen III → Corrins, Vitamin B_{12}

$-4\ CO_2$

HOOC

COOH

NH HN

NH HN

COOH

COOH

Coproporphyrinogen III

$-2\ CO_2$

[O]

NH HN

NH HN

[O]

NH N

N HN

COOH

COOH

COOH

COOH

Protoporphyrinogen III

Protoporphyrin IX

Chlorophylls

$Fe^{2\oplus}$

Haem → Bile pigments

Cytochromes

Fig. 10 Biosynthesis of the corrins, chlorophylls and haems

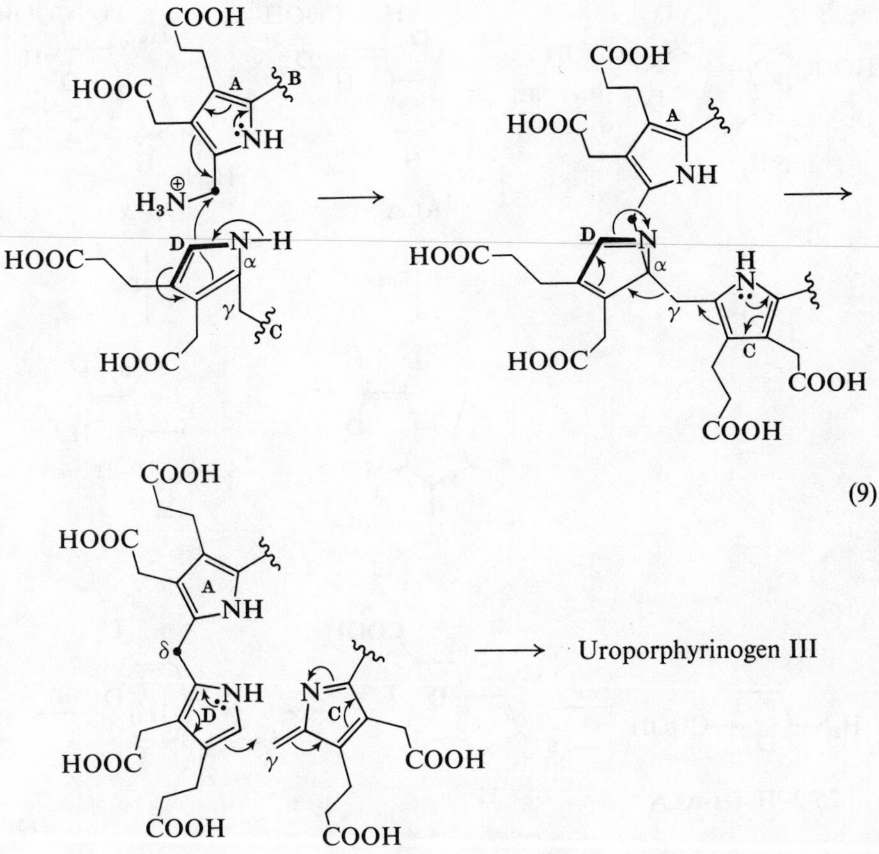

COOH HOOC A B NH $H_3\overset{\oplus}{N}$ D N—H α γ C HOOC HOOC

→ COOH HOOC A NH D N α H N γ C HOOC HOOC COOH COOH →

COOH HOOC A NH δ NH D N C HOOC γ COOH COOH COOH

→ Uroporphyrinogen III

(9)

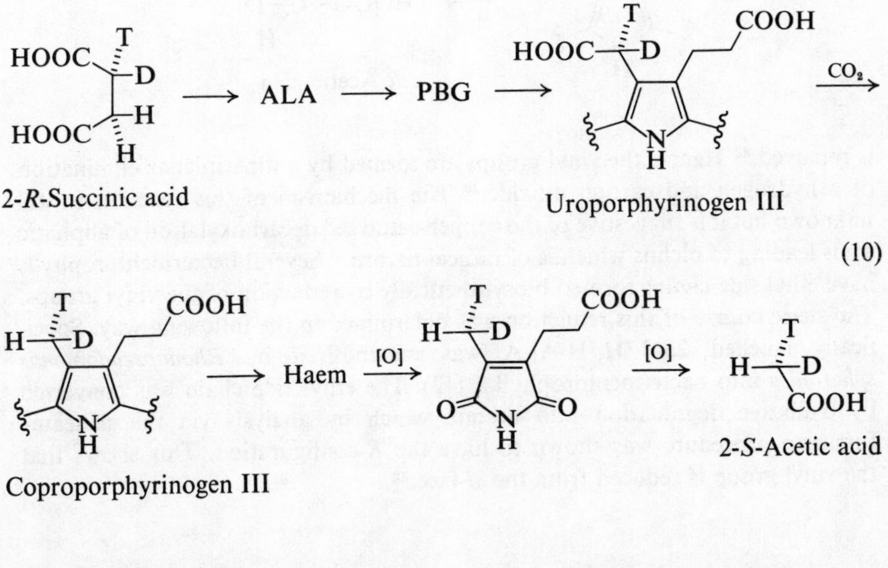

HOOC T D H HOOC H → ALA → PBG → HOOC—T D COOH N H → CO_2

2-*R*-Succinic acid

Uroporphyrinogen III

(10)

T H—D COOH N H → Haem →[O] T H—D COOH O N H O →[O] T H—D COOH

Coproporphyrinogen III

2-*S*-Acetic acid

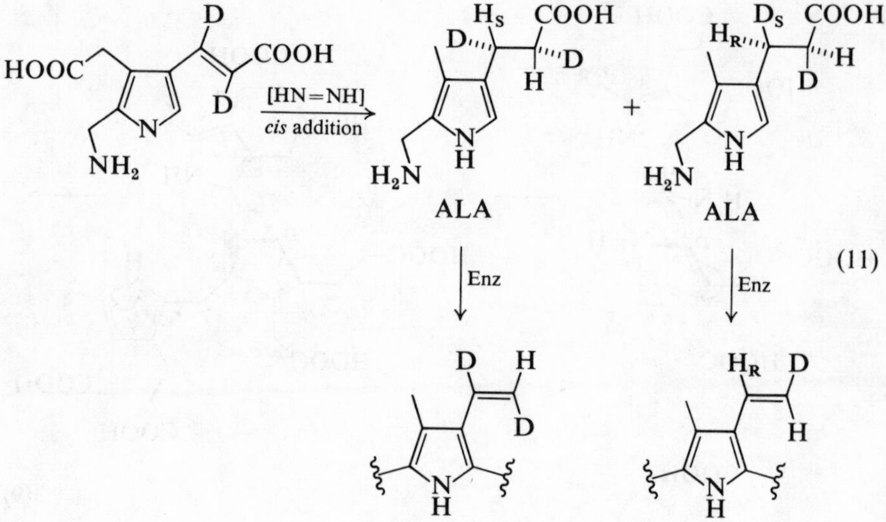

(11)

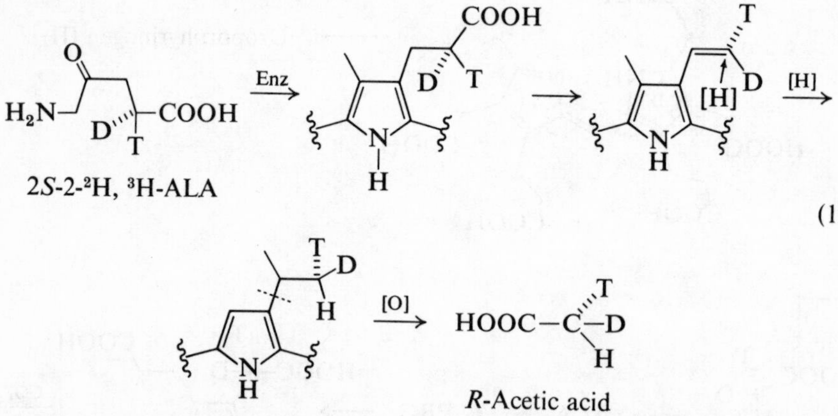

(12)

is removed.[29] Hence, the vinyl groups are formed by antiperiplanar elimination of a hydrogen and carbon dioxide.[30] The mechanism of this reaction is still unknown but it is suggestive of the copper-catalysed decarboxylation of aliphatic acids leading to olefins which is of radical nature.[31] Several bacteriochlorophylls have ethyl side chains formed biosynthetically by reduction of the vinyl groups. The steric course of this reduction was determined in the following way. Specifically labelled 2*S*-2-^{2}H,^{3}H-ALA was metabolized by *Rhodopseudomonas spheroides* into bacteriochlorophyll a (12). The ethyl side chain was converted by oxidative degradation into acetate which by analysis via the maleate–fumarate procedure was shown to have the *R*-configuration. This shows that the vinyl group is reduced from the *si*-face.[32]

8.4 Biosynthesis of the corrin skeleton

The isolation, structure and action of vitamin B_{12} is an exciting story starting with the discovery by Minot and Murphy in 1926 that liver extracts cure pernicious anaemia. The active factor, called vitamin B_{12}, was, after painstaking efforts, isolated in crystalline form as a cyanide twenty years later.[33,34] The compound is present in minute quantities in liver and a test on patients suffering from pernicious anaemia was the only assay method accessible. The structural elucidation proved to be an extremely difficult problem which ultimately was solved by X-ray crystallography.[35] A year later a considerable stir was caused when it was found that the cyanide, now called cyanocobalamine, $[Co^{3\oplus}]$-CN was an artefact produced in the work-up.[36]

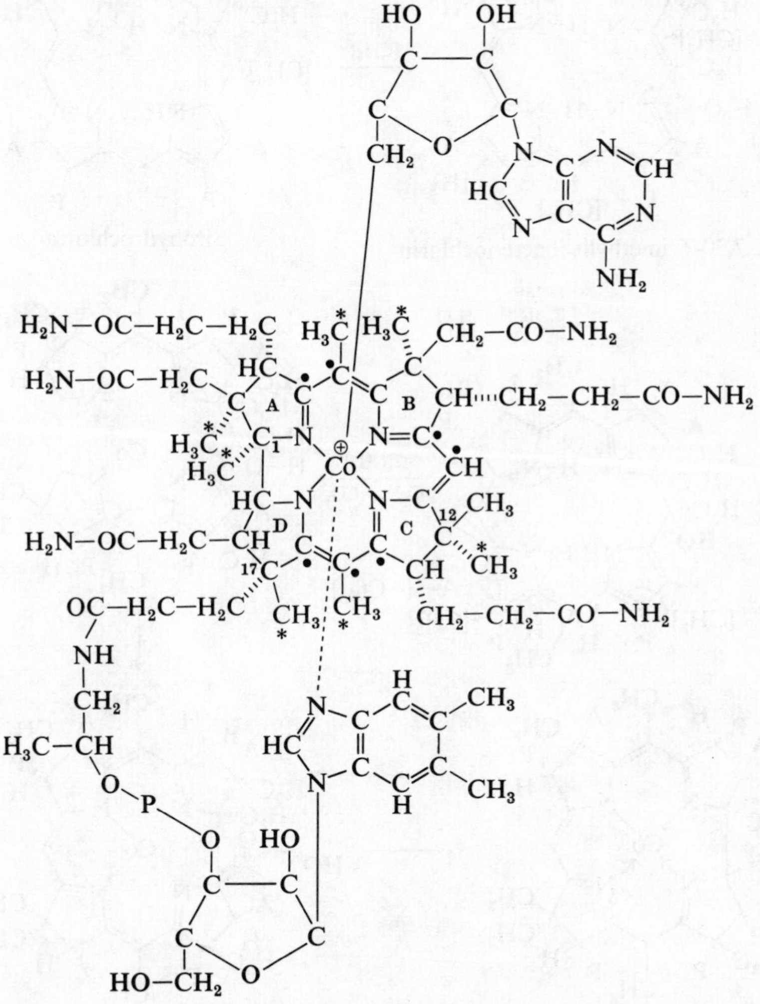

Fig. 11 Structure of vitamin B_{12}

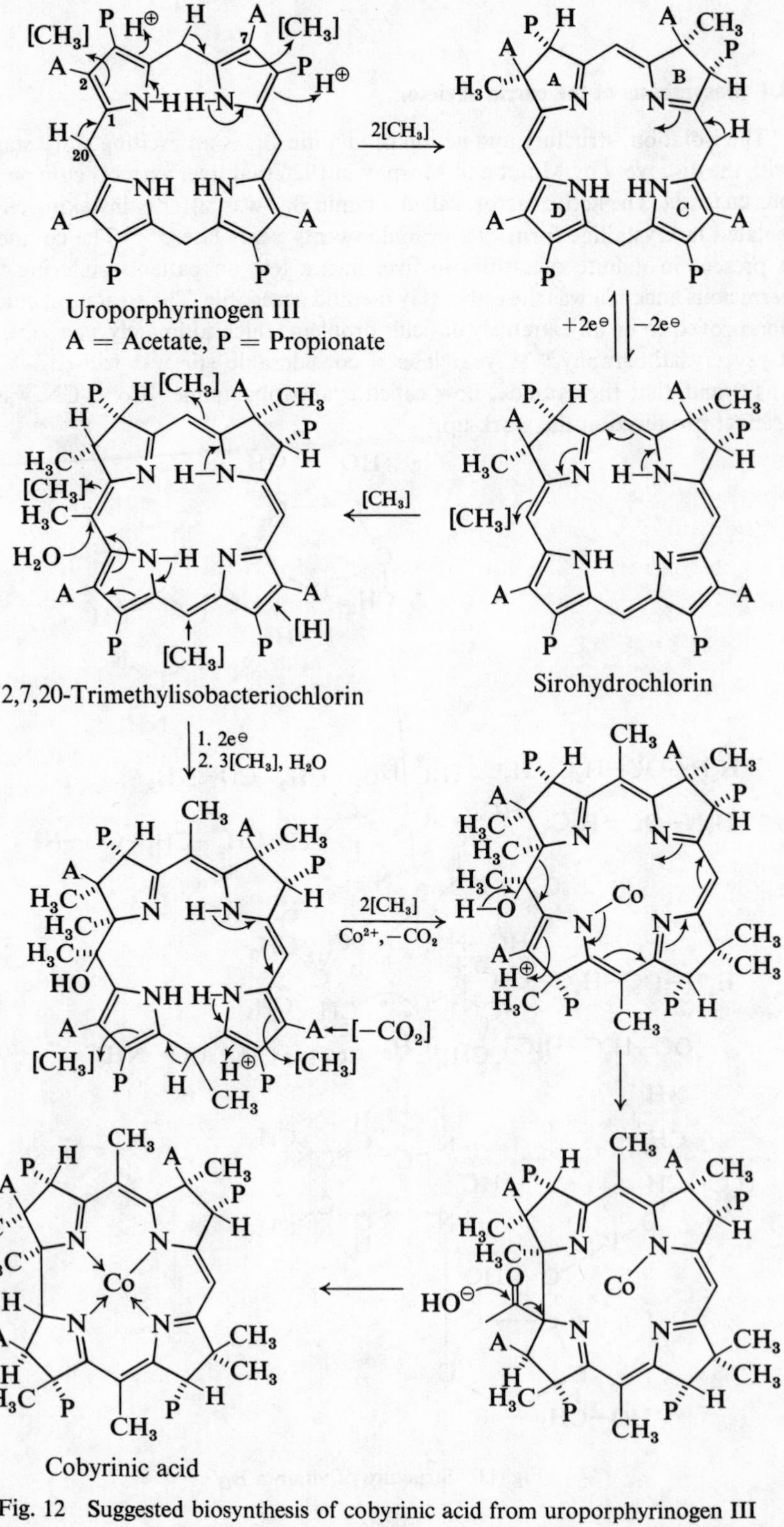

Uroporphyrinogen III
A = Acetate, P = Propionate
2[CH3]
+2e⊖
−2e⊖
Sirohydrochlorin
[CH3]
2,7,20-Trimethylisobacteriochlorin
1. 2e⊖
2. 3[CH3], H2O
2[CH3]
Co2+, −CO2
Cobyrinic acid

Fig. 12 Suggested biosynthesis of cobyrinic acid from uroporphyrinogen III

The active natural derivative has a 5′-deoxyadenosyl group as ligand (Fig. 11). Thus, cyanide displaces the adenosyl during the work-up but if cyanocobalamine is administered to the organism, the adenosyl derivative is formed again in the cell. Its mode of action as cofactor is a riddle but a problem of considerable importance. The problems arise partly because we have as yet no suitable descriptive formalism to handle reactions of transition metals in organic chemistry. As a rule these organo-transition-metallics have no σ-C-metal bonds and are unstable unless their d-orbitals are tied up by ligands.

The biosynthesis of the corrins presents additional problems with reference to the origin and mode of addition of the methyl groups and the mechanism of the ring contraction. Cultures of *Propionibacterium shermanii* incorporate 5-amino-5-^{13}C-laevulinic acid into the dotted positions in agreement with the substitution pattern in haem. One of the labels, the γ-*meso* carbon, is lost (Fig. 11) in the ring contraction. Seven of the eight methyls come from methionine,[37,38] the eighth at C^{12} arises by decarboxylation of the acetic acid side chain.[39] Taking advantage of the different shielding effects[40] from *cis* and *trans* vicinal alkyl groups, it was demonstrated by ^{13}C NMR that the geminal C^{12}–α–methyl originates from methionine.[41] All methyl groups but C^{17}–CH_3 are introduced from the α-side. It is worth noting that the carbon skeleton of vitamin B_{12} and its immediate precursor cobyrinic acid are on the same oxidation level as uroporphyrinogen III from which they originate. Furthermore, it is found that the methylated intermediates sirohydrochlorin and 2,7,20-trimethylisobacteriochlorin are on the pathway (Fig. 12). One surprising feature is that nature first methylates C^{20} and then extrudes the CH_3–C^{20} fragment as acetic acid in the ring contraction step as shown by specific labelling of the two carbon atoms.[42] Biomimetic model experiments indicate that the ring contraction can be referred to as an α-hydroxyketone rearrangement.[43] The electron shuffling technique, brought to its extreme, is used in the mechanistic rationalization of all the known facts of the sequence from uroporphyrin III to cobyrinic acid (Fig. 12). The exact order of events following the C^{20} methylation is not known yet. It is hypothesized in the scheme that C^1 methylation follows but methylation at *e.g.* C^{17} works mechanistically equally well."

8.5 Reactions of vitamin B_{12}

Vitamin B^{12}, $[Co^{3\oplus}]$-R, is reduced under controlled potential to $[Co^{2\oplus}]$ and by $NaBH_4$ to $[Co^{\oplus}]$ (13), formally corresponding to homolytic and heterolytic cleavages of the Co–C bond (14, 15). The monovalent complex $[Co^{\oplus}]$ is an efficient nucleophile and reductant, reminiscent of magnesium in its properties.

$$[Co^{3\oplus}]\text{-R} \overset{e^{\ominus}}{\rightleftharpoons} [Co^{2\oplus}] \overset{e^{\ominus}}{\rightleftharpoons} [Co^{\oplus}] \qquad (13)$$

$$[Co^{3\oplus}]\text{-R} \rightleftharpoons [Co^{2\oplus}] + R^{\cdot} \qquad (14)$$

$$[Co^{3\oplus}]\text{-R} \leftrightharpoons [Co^{\oplus}] + R^{\oplus} \qquad (15)$$

It reacts with alkyl halides or phosphates with formation of alkylcobalamines, thus providing a facile route to organocobalt compounds (16). Vitamin B_{12} is

$$[Co^{\oplus}] + RX \rightleftharpoons [Co^{3\oplus}]\text{-}R + X^{\ominus} \qquad (16)$$

analogously formed from $[Co^{\oplus}]$ and ATP by expulsion of triphosphate. The alkylcobalamines undergo electrophilic as well as nucleophilic substitution (17, 18) but the homolytic equilibrium (14) seems to be the best base of support

$$[Co^{3\oplus}]\text{-}CH_3 + Hg^{2\oplus} \rightarrow [Co^{3\oplus}] + CH_3Hg^{\oplus} \qquad (17)$$

$$RS^{\ominus} + [Co^{3\oplus}]\text{-}CH_3 \rightarrow RSCH_3 + [Co^{\oplus}] \qquad (18)$$

for exploration and explanation of vitamin B_{12} mediated reactions, categorized as 1,2 shifts (19), where X = alkyl carboxyl, hydroxyl and amino, here exemplified by the propandiol dehydrase reaction (20).[44] In the biomethylation of the

$$-\overset{X}{\underset{|}{\overset{|}{C}}}-\overset{H}{\underset{|}{\overset{|}{C}}}- \xrightarrow{B_{12}} -\overset{H}{\underset{|}{\overset{|}{C}}}-\overset{X}{\underset{|}{\overset{|}{C}}}- \qquad (19)$$

$$\underset{\text{1R, 2R}}{H_3C(H)(HO)C\text{-}C(D)(H)\,{}^{18}OH} \longrightarrow \underset{\text{2S}}{H_3C(H)(D)C\text{-}C(H)({}^{18}OH)(OH)} \xrightarrow{-H_2{}^{18}O} H_3C(D)(H)C\text{—}CHO \qquad (20)$$

mercuric ion—and other metal salts (17)—a transient methyl carbanion is formed, whereas in *S*-methylation the methyl group is transferred as a carbonium ion equivalent (18), actually a reversal of (16). The overall propan-1,2-diol dehydrase reaction is similar to the elimination–addition of water in the citrate–isocitrate rearrangement in the citric acid cycle or the 1,2-diol–one rearrangement mediated by sugar dehydrase (section 2.5). These dehydrase reactions are characterized by an initial carbanion formation and subsequent elimination of the adjacent hydroxyl group (21). Contrary to expectations, no isotope effect is observed for the citrate-isocitrate, and maleate–fumarate interconversions. It is thought that the dissociation of the product from the enzyme complex might be the rate determining step. Furthermore, these reactions are not vitamin B_{12} dependent and extensive oxygen and proton incorporation from water is observed. On the other hand, propandiol contains no such activated proton which could initiate the rearrangement, nor has the propanal formed exchanged protons with the solvent.

$$\text{HO}\!\!-\!\!\text{CH}(\text{CH}_3)\!-\!\text{C(OH)(H)}\!-\!\text{C(=O)CH}_3 \;\; \ddot{\text{B}}\!-\!\text{Enz} \longrightarrow \text{CH}_3\text{CH}\!=\!\text{C(OH)}\!-\!\text{C(=O)CH}_3 \;\; \text{HB}^{\oplus}\!-\!\text{Enz} \longrightarrow \text{CH}_3\text{CH}_2\text{C(=O)C(=O)CH}_3 \qquad (21)$$

The acid catalysed diol–one rearrangement (22) represents another conceivable model reaction but it fails to account for the unexpected findings that the tritium transfer is intermolecular in (23) and that the coenzyme also becomes tritiated implicating that vitamin B_{12} serves as a hydrogen carrier.[45]

$$\text{CH}_3\text{C(OH)(H)}\!-\!\text{C(OH)(H)CH}_3 \xrightarrow{\text{H}^{\oplus}} \text{CH}_3\text{C(=O)}\!-\!\text{CH}_2\text{CH}_3 \qquad (22)$$

$$\underset{\text{OH}\quad\text{OH}}{\text{CH}_2\!-\!\text{CH}_2} + \text{CH}_3\underset{\text{OH}}{\overset{\text{H}}{\text{C}}}\!-\!\underset{\text{OH}}{\overset{^3\text{H}}{\text{C}}}\!-\!\text{H} \xrightarrow{\text{Enz-B}_{12}} {}^3\text{HCH}_2\!-\!\text{CHO} + \text{CH}_3\underset{^3\text{H}}{\text{CH}}\!-\!\text{CHO} \qquad (23)$$

The following experimental results have to be considered in any mechanistic interpretation of the propandiol–propanal conversion (20):

1. no incorporation of protons from the solvent;
2. C^1–H of propandiol is recovered both at C^2 of propanal and at $C^{5'}$ of the deoxyadenosyl group of B_{12}. 5′-CH_2-Labelled coenzyme exchanges hydrogen atoms with C^2–H of propanal;
3. the choice as to which hydrogen atom migrates from C^1 to C^2 and loss of C^1–^{18}O label depends upon the chirality at C^2;
4. substitution at C^2 occurs with inversion of configuration;
5. an isotope effect, k_H/k_D *ca.* 10, was observed for *R,S* 1-$^2H_2$-propandiol;
6. oxygen acts as an inhibitor.

A radical mechanism (24) has been advanced featuring initial homolysis of the Co–C bond (14) and an astonishingly selective hydrogen abstraction of the C^1–H by the deoxyadenosyl radical. The substrate radical rearranges within the ligand sphere of $[Co^{2\oplus}]$ to the product radical which finally abstracts a hydrogen atom from the methyl group of deoxyadenosine thus regenerating the coenzyme for a new cycle. This mechanism excludes hydrogen exchange with the solvent, accounts for the isotope effect and the facts that both the coenzyme and C^2 of propanal become labelled by C^1–H-labelled propandiol and explains the cross-over of label according to (23). The product radical does not necessarily abstract the same labelled hydrogen from the centrosymmetric methyl group of 5′-deoxyadenosine. This nucleoside has in fact been isolated from the reaction mixture but it has not been possible to achieve incorporation of added 5-deoxy-adenosine probably because it is strongly bound to the enzyme. According to expectations 5′-C-labelled coenzyme did give C^2-labelled propanal. The inhibitory effect of oxygen can be explained by its nearly diffusion controlled interceptive reaction with the C-centred radicals.

$$[Co^{3\oplus}]\text{—R} \rightleftharpoons [Co^{2\oplus}] + R^{\bullet}\ \ \text{H—}\underset{\text{HO}}{\text{CH}}\underset{\text{OH}}{\text{CH}}\text{CH}_3 \longrightarrow \text{RH} + {}^{\bullet}\underset{\text{HO}}{\text{CH}}\underset{\text{OH}}{\text{CH}}\text{CH}_3 \rightleftharpoons \tag{24}$$

$$[Co^{3\oplus}]\text{—}\underset{\text{HO}}{\text{CH}}\underset{\text{OH}}{\text{CH}}\text{CH}_3 \rightleftharpoons (\text{OH})_2\dot{\text{C}}\text{HCH}_3 + \text{HR} + [Co^{2\oplus}] \rightleftharpoons$$

$$\text{OHCCH}_2\text{CH}_3 + [Co^{3\oplus}]\text{—R}$$

Examination of the stereospecificity of the 1,2 shift shows that 1-pro-*R*–H and 1-pro-*S*–H migrate selectively in 2*R* and 2*S* propandiol with inversion at C^2 (20). Another interesting feature of the reaction is that 2*R* and 2*S* propandiol labelled with ^{18}O at C^1 retain 8 per cent and 88 per cent of the label in propanal, respectively. This shows that the rearrangement stereospecifically passes via a 1,1-geminal diol (20), the hydrolysis of which is enzymatically controlled.[46]

The hydroxyl migration in the radical is unique. It has one analogue in the radical abstraction of a proton from glycol by the hydroxy radical; acetaldehyde radicals were observed by ESR (25).[47] The last radical hydrogen abstraction

$$\text{HO}^{\bullet} + \underset{\text{HO}}{\text{CH}_2}\underset{\text{OH}}{\text{CH}_2} \longrightarrow \text{H}_2\text{O} + {}^{\bullet}\underset{\text{H—O}}{\text{CH}}\text{—}\underset{\text{OH}}{\text{CH}_2} \longrightarrow \underset{\text{O}}{\overset{\|}{\text{CH}}}\dot{\text{C}}\text{H}_2 \quad (+\ \text{H}^{\oplus}) \tag{25}$$

step in (24) is endothermic by at least 5 kcal and therefore very slow. It may be explicable in terms of fixation of reactants on the enzyme and fast diffusion of the product from the site of reaction but otherwise unlikely on thermodynamic grounds.

A similar mechanism is formulated for e.g. the 3,6-diaminohexanoate to 3,5-diaminohexanoate rearrangement[48] (26) and for the carbon skeletal rearrangement of methylmalonate to succinate (27), reactions which are still less well understood.

NH_2 COOH NH_2 $\xrightarrow{B_{12}}$ NH_2 NH_2 COOH (26)

COOH COCoA $\xrightarrow{B_{12}}$ CoAOC COOH (27)

In reaction (27) an unactivated methyl group is inserted into the carbon chain; in this case as well as in (20) and (26) an exchange of hydrogens between substrate and coenzyme is observed. Although cation, anion and carbene mechanisms have been claimed for the vitamin B_{12} mediated isomerization only the radical mechanism has found support from biomimetic studies[49,50] (28, 29).

$H_5C_2{-}SC(=O)$ $COOC_2H_5$ $[Co^{3\oplus}]$ $\xrightarrow[\text{dark}]{\text{24 h}}$ $H_5C_2SC(=O)$ $COOC_2H_5$ (28)

O Ph $[Co^{3\oplus}]$ $\xrightarrow[a]{h\nu}$ O Ph

O Ph Br $\xrightarrow[b]{Ag^{\oplus}}$ O Ph (29)

O Ph Br $\xrightarrow[c]{Bu_3SnH,\ \Delta}$ O Ph

The radical intermediate gives acyl migration (29a, c) and the cationic intermediate gives phenyl migration (29b). An anionic intermediate ought to incorporate protons from the solvent which in fact is not observed.

8.6 Retrospect and prospect

We expect a student in physics to have a knowledge of the theoretical background for a mathematical formula describing a physical phenomenon. The same should be applied to the study of natural products. The knowledge of structures for a number of compounds is not sufficient; more essential is the knowledge of how they are formed. It is evident that natural product chemistry made the great leap forward with the elucidation of the biosynthetic mainstreams which brought system to a seemingly chaotic network of processes. Considering the diversity and complexity of naturally occurring compounds one is amazed by the relatively small number of reaction types and agents nature utilizes for the synthesis of its products. In the laboratory we are aided by at least a hundred reducing or oxidizing agents to perform various redox reactions, while nature accomplishes its task with just a few. Nature has already made its entry into the laboratory in more than one way. Microorganisms and purified enzymes are routinely used in industry to perform stereospecific transformations in practically all classes of compounds, natural as well as synthetic. Redox reactions, epoxidations, hydroxylations, dehydrations, condensations (formations of C–C bonds), cleavages, rearrangements, etc., are now performed by the aid of certain strains of microorganisms as single steps in a synthetic sequence and of special advantage is the introduction of chirality in such a process.[51,52] This is an area where a joint venture of microbiologists and chemists can lead to significant scientific and technological progress.

Another astonishing feature is that, essentially, biosynthesis is identical in the simplest prokaryote, in the highest plant, and in the human being, implying that the refined synthetic machinery arose very early in the biological evolution.

The main biosynthetic pathways are well known today but still many connecting paths have to be explored. However, the enzymatic reaction mechanisms are still imperfectly known.

Much is to be learned from the synthetic skill of nature. Thus nature has in photosynthesis found an efficient method to convert the energy of light into chemical energy, thereby producing reactive ATP and splitting water into oxygen and the equivalent of hydrogen, NADPH. Man has not yet been successful in efficiently solving this long-standing significant problem for utilizing the energy of the sun.

As pointed out in the introduction our knowledge of the biological and ecological functions of secondary metabolites is embryonic and a large field of important research is ahead of us. This will give us a deeper understanding of the conditions for coexistence of species on earth and will teach human beings to treat nature kindly. Up until now we can say that man's activities have only been a threat to nature and our own existence.

The plant kingdom is our major renewable resource both for nutrients, fibres, various chemicals such as drugs, essential oils, resins, and for raw material for various chemical processes. The importance of drugs in human affairs appears from the fact that in 1967 *ca.* 23 per cent of all prescription (hormones, cardiovascular agents, analgesics, etc.) in the United States contained secondary natural products of plant origin, vitamin pills excluded. Another 20 per cent are of microbial (antibiotics) and animal origin.[53] Considering that the majority of plant species, especially in the tropics, have not been investigated yet, the prospects for the development of natural product chemistry are good. Structural elucidation will always be a necessary prerequisite for research in this field but is no longer considered prestigious as in the earlier days. Traditionally chemists have turned to terrestrial flora and fauna. Technological advances now make it easier to collect marine organisms at various depths and recent research on marine natural products has indeed given interesting results, widening our perspectives at the interface between chemistry and biology. The aquatic environment has an influence on the structural pattern of marine metabolites. A number of unusual structures have been solved spreading enthusiasm amongst workers in the field.

Secondary metabolites play a distinguished role in all manifestations of human behaviour. Utilization of narcotics and stimulants is pervading all cultures. Poisons have always been feared and used for various purposes. Exotic perfumes and essential oils are more harmless but none the less of pleasant significance, whenever people gather. Nor should we forget the exciting spices stimulating our appetite and giving new dimensions to the pleasures of eating. And there is no prospect of substituting these subtle products with synthetic analogues. Evidently, natural product chemistry will continue to be part of the age in which we live.

Bibliography

1. Oro, J. *Nature* **191** (1961) 1193.
2. Ferris, J. P., Sanchez, R. A. and Orgel, L. E. *J. Mol. Biol.* **33** (1968) 693.
3. Gabel, N. W. and Ponnamperuma, C. *Nature* **216** (1967) 453.
4. Orgel, L. E. and Lohrmann, R. *Acct. Chem. Res.* **7** (1974) 368.
5. Löfgren, N. and Lüning, B. *Acta Chem. Scand.* **7** (1953) 225.
6. Miller, C. O. *Science* **157** (1967) 1055.
7. Burrows, W. J., Armstrong, D. J., Kaminek, M., Skoog, F., Bock, R. M., Hecht, S. M., Dammann, L. G., Leonard, N. J. and Occolowitz, J. *Biochemistry* **9** (1970) 1867.
8. Bordner, J., Thiessen, W. E., Bates, H. A. and Rapoport, H. *J. Am. Chem. Soc.* **97** (1975) 6008.
9. Buchanan, J. M. and Hartman, S. C. *Adv. Enzymol.* **21** (1959) 199.
10. Looser, E., Bauman, T. W. and Wanner, H. *Phytochemistry* **13** (1974) 2518.
11. Suzuki, T. and Takahashi, E. *Phytochemistry* **15** (1976) 1235.
12. Kondo, H., Friedmann, H. C. and Vennesland, B. *J. Biol. Chem.* **235** (1960) 1533.
13. Friedkin, M. in *Adv. Enzymology* **38** (1973) 235, Meister, A. (Ed.), J. Wiley, New York.
14. Sigman, D. and Mooser, G. *Ann Revs. Biochem.* **44** (1975) 895.

15. Santi, D. and McHenry, C. *Proc. Natl. Acad. Sci. USA* **69** (1977) 1855.
16. Wasserman, H. H., Sykes, R. J., Peverada, P., Shaw, C. K., Cushley, R. J. and Lipsky, S. R. *J. Am. Chem. Soc.* **95** (1973) 6874.
17. Schneider, D., Boppre, M., Schneider, H., Thompson, W. R., Boriack, C. J., Petty, R. L. and Meinwald, J. *J. Comp. Physiol.* **97** (1975) 245.
18. Shemin, D. and Rittenberg, D. *J. Biol. Chem.* **166** (1946) 621, 637.
19. Shemin, D. and Wittenberg, J. *J. Biol. Chem.* **192** (1951) 315.
20. Lawer, W. G., Neuberger, A. and Scott, J. J. *J. Chem. Soc.* **1959,** 1474.
21. Shemin, D. and Russel, D. S. *J. Am. Chem. Soc.* **75** (1953) 4873.
22. Battersby, A. R., Hunt, E. and McDonald, E. *J. Chem. Soc. Chem. Commun.* **1973,** 442.
23. Bogorad, L. and Granick, S. *Proc. Natl. Acad. Sci. USA* **39** (1953) 1176.
24. Bogorad, L. *J. Biol. Chem.* **233** (1958) 501, 510.
25. Battersby A. R., Fookes, C. J. R., Matcham, G. W. J. and McDonald, E. *Nature* **285** (1980) 17; Battersby, A. R. and McDonald, E. *Acct. Chem. Res.* **12** (1979) 14.
26. Mathewson, J. A. and Corwin, A. H. *J. Am. Chem. Soc.* **83** (1961) 135.
27. Barnard, G. F. and Akhtar, M. *J. Chem. Soc. Chem. Commun.* **1975,** 494.
28. Cavaliero, J. A. S., Kenner, G. W. and Smith, K. M. *J. Chem. Soc. Chem. Commun.* **1973,** 183.
29. Zaman, Z., Abboud, M. M. and Akhtar, M. *J. Chem. Soc. Chem. Commun.* **1972,** 1263.
30. Battersby, A. R., McDonald, E., Wurziger, H. K. W. and James, K. J. *J. Chem. Soc. Chem. Commun.* **1975,** 493.
31. Sheldon, R. A. and Kochi, J. K. *Organic Reactions* **19** (1972) 279.
32. Battersby, A. R., Gutman, A. L., Fookes, C. J. R., Günther, H. and Simon, H. *J. Chem. Soc. Chem. Commun.* **1981,** 645.
33. Rickes, E. L., Brink, N. G., Koniuszy, F. R., Wood, T. R. and Folkers, K. *Science* **107** (1948) 396.
34. Smith, E. L. and Parker, L. E. J. *Biochem. J.* 43 (1948) viii.
35. Hodgkin, D. C., Kamper, J., Lindsey, J., MacKay, M., Pickworth, J., Robertson J. H., Shoemaker, C. B., White, J. G., Prosen, R. J. and Trueblood, K. N. *Proc. Roy. Soc.* **A242** (1957) 228.
36. Barker, H. A., Weissbach, H. and Smyth, R. D. *Proc. Natl. Acad. Sci. USA* **44** (1958) 1093.
37. Brown, C. E., Katz, J. J. and Shemin, D. *Proc. Natl. Acad. Sci. USA* **69** (1972) 2585.
38. Scott, A. I., Townsend, C. A., Okada, K., Kajiwara, M., Whitman, P. J. and Cushley, R. J. *J. Am. Chem. Soc.* **94** (1972) 8267, 8269.
39. Bray, R. C. and Shemin, D. *J. Biol. Chem.* **238** (1963) 1501.
40. Dalling, D. K. and Grant, D. M. *J. Am. Chem. Soc.* **94** (1972) 5318.
41. Scott, A. I., Townsend, C. A. and Cushley, R. J. *J. Am. Chem. Soc.* **95** (1973) 5759.
42. Nussbaumer, C., Infeld, M., Wörner, G., Müller G. and Arigoni, D. *Proc. Natl. Acad. Sci. USA* **78** (1981) 9. Mombelli, L., Nussbaumer, C., Weber, H., Müller, G. and Arigoni, D. *ibid.* **78** (1981) 11. Battersby, A. R., Bushell, M. J., Jones, C., Lewis, N. G. and Pfenninger, A. *ibid.* **78** (1981) 13.
43. Rasetti. U., Pfaltz, A., Kratky, C. and Eschenmoser, A., *Proc. Natl. Acad. Sci., USA* **78** (1981) 16.
44. Abeles, R. H. and Dolphin, D. *Acct. Chem. Res.* **9** (1976) 114.
45. Frey, P. A., Eisenberg, M. K. and Abeles, R. H. *J. Biol. Chem.* **242** (1967) 5369.
46. Rétey, J., Umani-Ronchi, A., and Arigoni, D. *Experientia* **22** (1966) 502.
47. Buley, A. L., Norman, R. O. C. and Pritchett, R. J. *J. Chem. Soc.* (*B*) **1966,** 849.
48. Rétey, J., Kunz, F., Arigoni, D. and Stadtman, T. C. *Helv. Chim. Acta* **61** (1978) 2989.
49. Scott, A. I. and Kang, K. *J. Am. Chem. Soc.* **99** (1977) 1997.

50. Tada, M., Miura, K., Okabe, M., Seki, S. and Mizukami, H. *Chem. Letters*, *Japan* **1981,** 33.
51. Kieslich, K. *Microbiol. Transformation.* G. Thieme, Stuttgart, 1976.
52. Industrial Microbiology, *Scientific American*, Sept. 1981.
53. Farnsworth, N. R. in *Phytochemistry III*, Miller, L. P., (Ed.) Van Nostrand Reinhold Co., New York, 1973, p. 351.

Answers to problems

Chapter 2

2.1

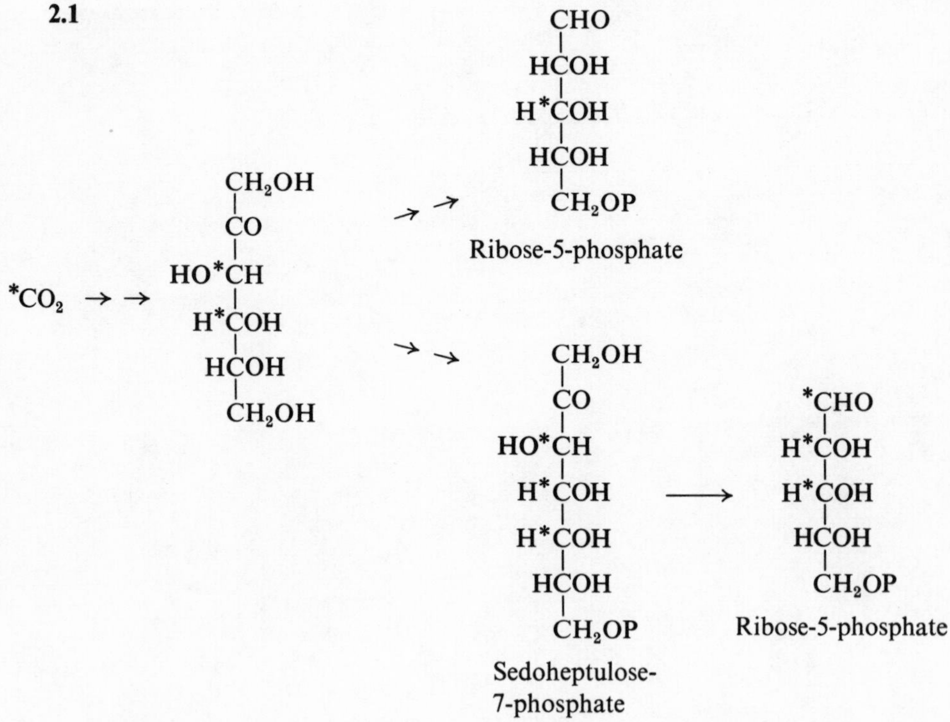

$*CO_2 \rightarrow \rightarrow$

```
 CH2OH
 |
 CO
 |
HO*CH
 |
H*COH
 |
HCOH
 |
 CH2OH
```

$\rightarrow \rightarrow$

```
 CHO
 |
HCOH
 |
H*COH
 |
HCOH
 |
 CH2OP
```

Ribose-5-phosphate

$\rightarrow \rightarrow$

```
 CH2OH
 |
 CO
 |
HO*CH
 |
H*COH
 |
H*COH
 |
HCOH
 |
 CH2OP
```

Sedoheptulose-7-phosphate

$\longrightarrow$

```
 *CHO
 |
H*COH
 |
H*COH
 |
HCOH
 |
 CH2OP
```

Ribose-5-phosphate

2.2

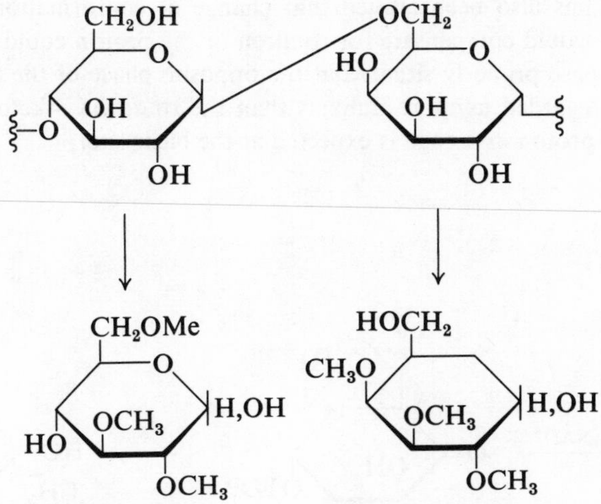

CH₂OH, OCH₂, HO, OH, OH, OH, OH

2,3,6-tri-*O*-methylglucose

2,3,4-tri-*O*-methylgalactose

The chain contains *ca.* 2.6×10^4 residues

2.3

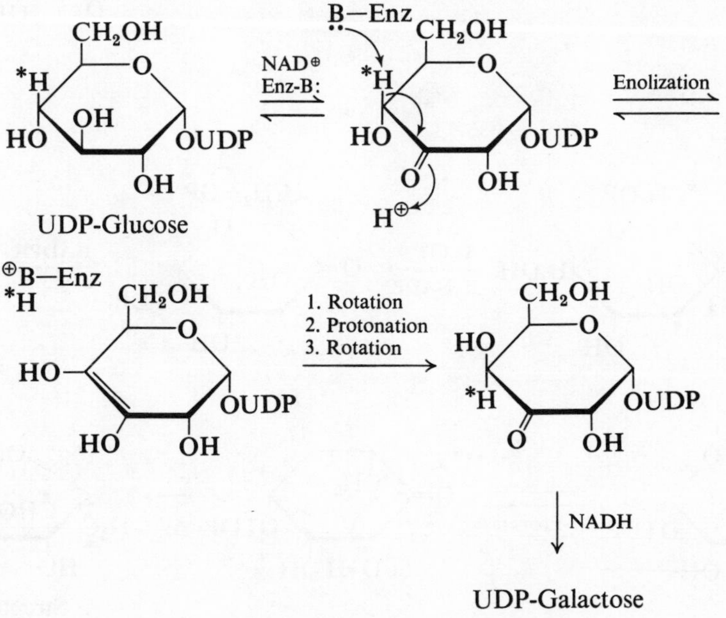

UDP-Glucose

1. Rotation
2. Protonation
3. Rotation

NADH

UDP-Galactose

The sugar must rotate twice around the C^1–OP axis to account for the steric outcome. It has also been argued that change of conformation of the sugar (chair–boat) would compensate for rotation or the proton could be transferred to a second base properly situated at the opposite phase of the molecule. This sequence is regarded as more unlikely than the route via 4-keto-UDP-glucose because fast proton exchange is expected at the basic site.

2.4

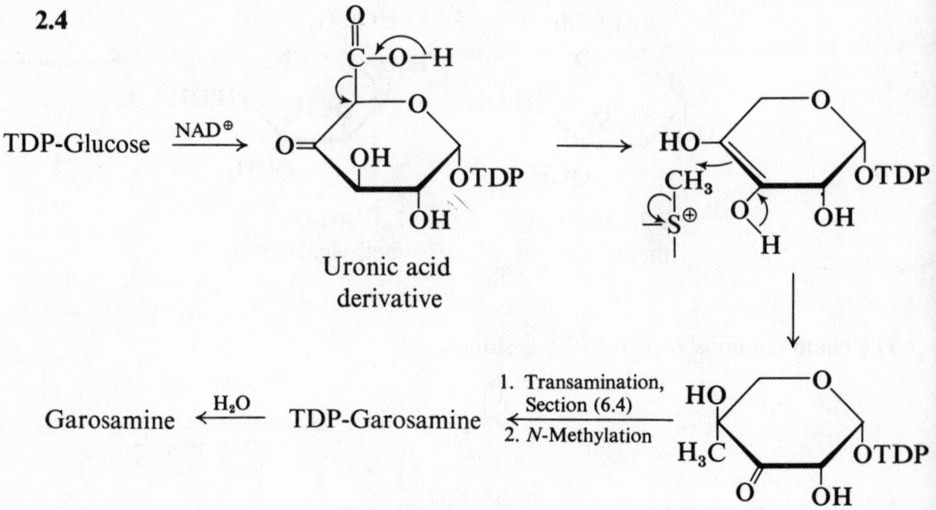

TDP-Glucose $\xrightarrow{NAD^{\oplus}}$ Uronic acid derivative $\longrightarrow$ ... $\longrightarrow$ TDP-Garosamine $\xleftarrow{\text{1. Transamination, Section (6.4); 2. } N\text{-Methylation}}$

Garosamine $\xleftarrow{H_2O}$ TDP-Garosamine

2.5

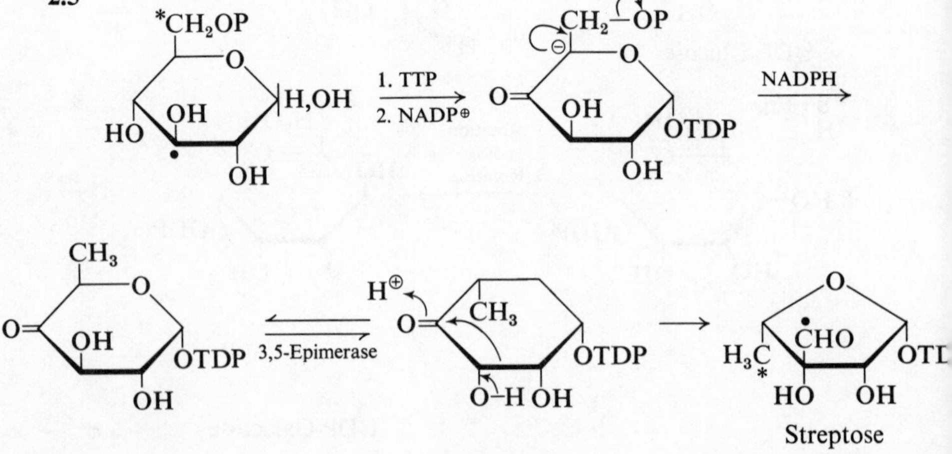

$\xrightarrow{\text{1. TTP; 2. } NADP^{\oplus}}$ $\xrightarrow{NADPH}$

$\rightleftharpoons$ 3,5-Epimerase $\longrightarrow$ Streptose

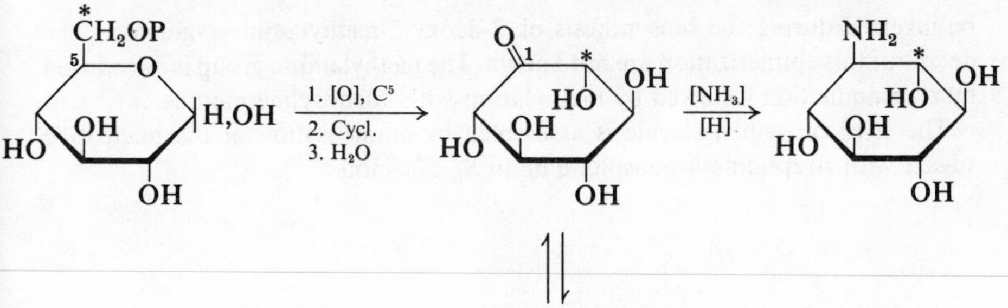

CH_2OP (5) O, H,OH, OH, HO, OH — 1. [O], C^5; 2. Cycl.; 3. H_2O → O (1), HO, OH, OH, HO, OH — $[NH_3]$, [H] → NH_2, HO, OH, OH, HO, OH

myo-Inositol

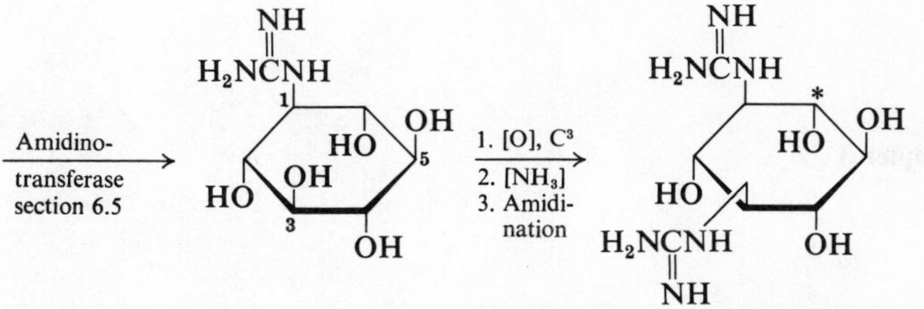

Amidino-
transferase
section 6.5

NH, $H_2N\overset{\|}{C}NH$ (1), HO, OH (5), OH, HO (3), OH

1. [O], C^3
2. $[NH_3]$
3. Amidi-
nation

NH, H_2NCNH, HO, OH, HO, H_2NCNH, NH, OH

Streptidine

The intermediate monoaminocyclitol has a plane of symmetry. The chiral enzyme oxidizes only C^3 but not C^5 as indicated by the absorption at 72.4 PPM. The peaks at 13.4 PPM and 61.2 PPM originate from the CH_3 and CH_2OH groups, respectively.

Since D-glucose (but not L-glucose) is incorporated, all chiral centres have to

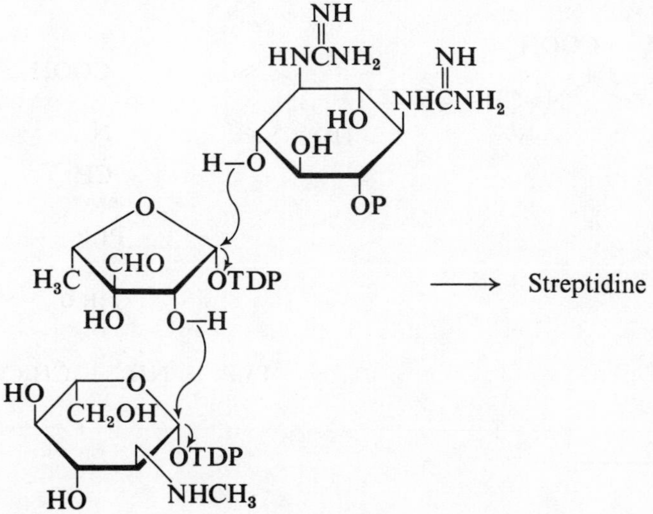

NH, $HNCNH_2$, NH, $NHCNH_2$, HO, OH, H—O, OP

O, H_3C, CHO, OTDP, HO, O—H

HO, O, CH_2OH, OTDP, HO, $NHCH_3$

→ Streptidine

be inverted during the biosynthesis of 2-deoxy-2-methylamino-L-glucose. The details of this epimerization are not known. The methylamino group is introduced by transamination followed by methylation with adenosylmethionine.

The streptomycin molecule is assembled by condensation of the nucleotide sugars with streptidine-6-phosphate in an S_N2 fashion.

Chapter 3

3.1

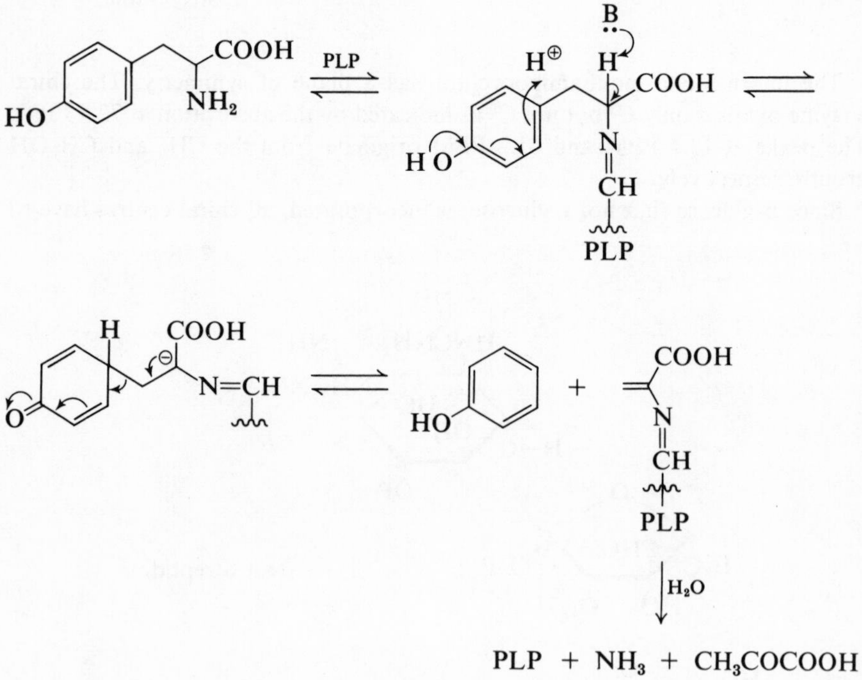

COOH NH₂ HO PLP ⇌ H⊕ B H COOH HO N CH PLP ⇌

H COOH ⊖ N=CH O ⇌ HO + COOH N CH PLP

H_2O

PLP + NH_3 + $CH_3COCOOH$

3.2

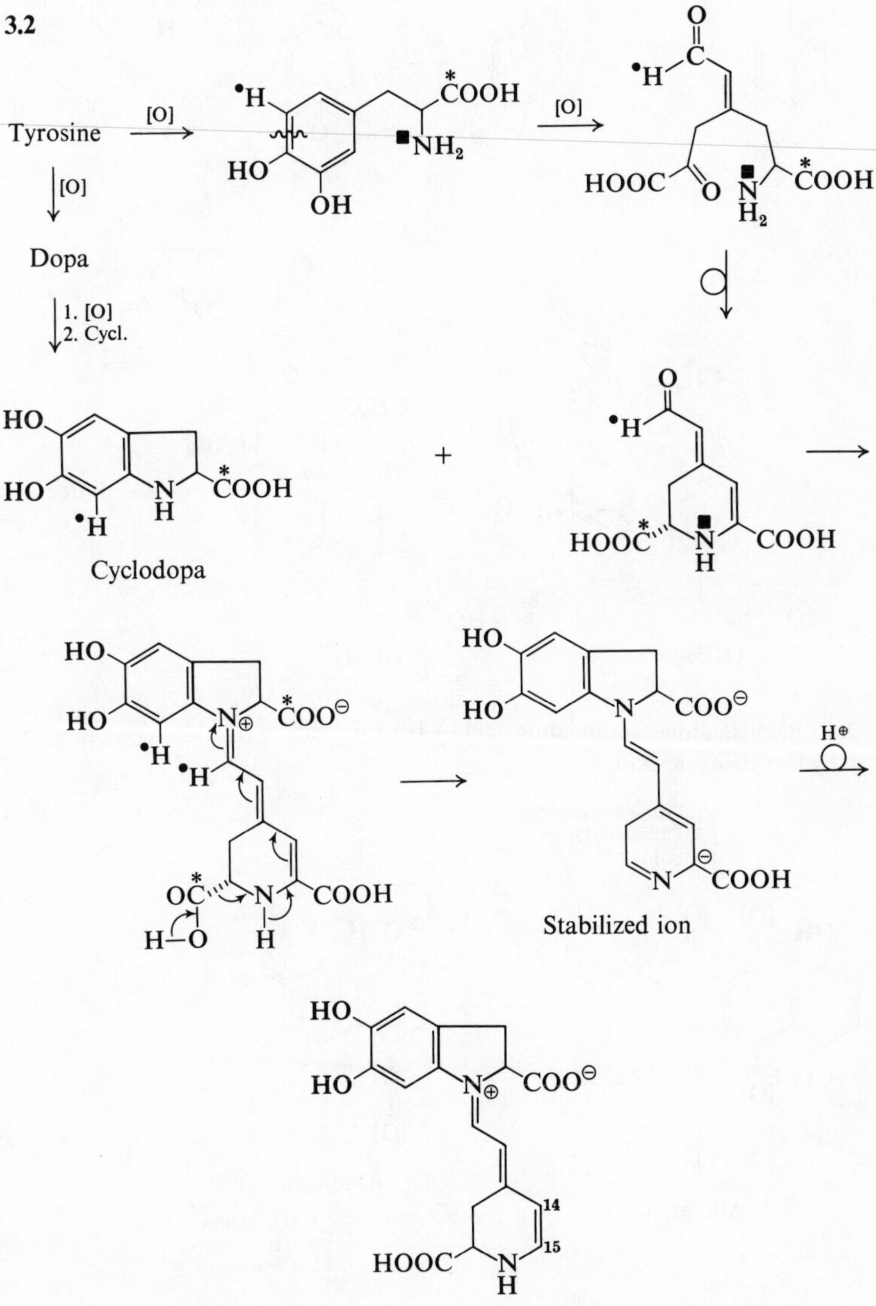

Tyrosine —[O]→ •H ... COOH, ■NH$_2$, HO, OH —[O]→ •H, O, C, HOOC, O, N■H$_2$, *COOH

Tyrosine ↓[O] Dopa ↓ 1. [O] 2. Cycl.

HO, HO, N H, *COOH, •H

Cyclodopa

\+ O, •H, HOOC*, N■ H, COOH ⟶

HO, HO, N⊕, *COO⊖, •H, •H, OC*, H—O, N H, COOH ⟶

HO, HO, N, COO⊖, ⊖ N, COOH

Stabilized ion

H⊕

HO, HO, N⊕, COO⊖, 14, 15, HOOC, N H

3.3

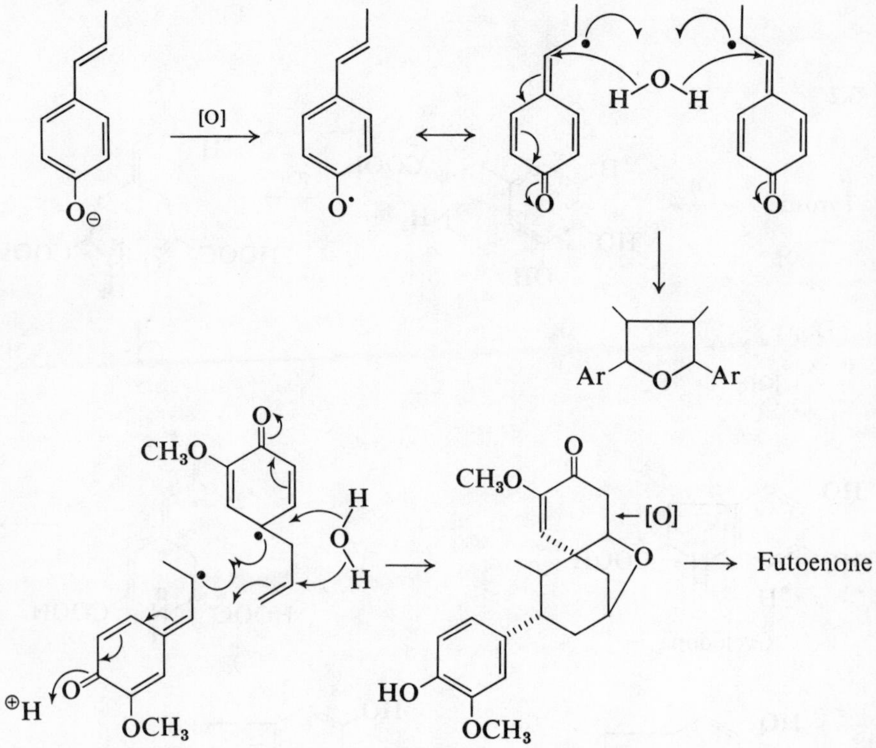

[O]

Ar O Ar

CH$_3$O

H O H

$^{\oplus}$H O

OCH$_3$

CH$_3$O ← [O]

O → Futoenone

HO

OCH$_3$

3.4 Phenylalanine → cinnamic acid → benzoic acid → *p*-hydroxybenzoic acid

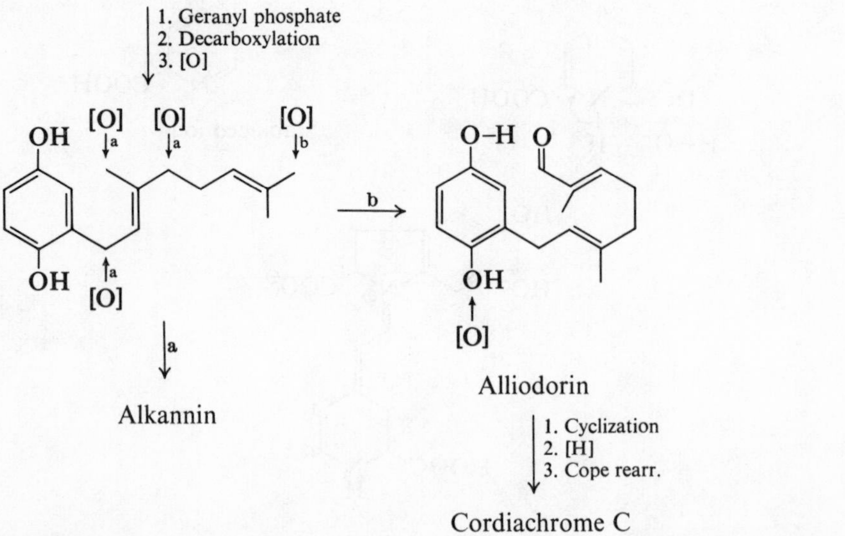

1. Geranyl phosphate
2. Decarboxylation
3. [O]

OH [O] a [O] a [O] b

OH [O] a

b →

O–H O

OH

[O]

a

Alkannin

Alliodorin

1. Cyclization
2. [H]
3. Cope rearr.

Cordiachrome C

The cordiachrome pathway is supported by isolation of aldehyde alliodorin.

3.5

(a)

$\overset{\oplus}{S}$ — $^*CH_3$; PLP-bound phenylalanine (N=CH—PLP), C(=O)—O—H ⟶ [O] → $C_6H_5CH_2CH(NH_2)\overset{*}{C}H_3$ ⟶

HO— (4-allyl-2-methoxyphenol, * at terminal =CH$_2$) OCH_3

Eugenol

(b)

H—O— (coniferyl alcohol) OCH_3, OH; $^*CH_3$—$\overset{\oplus}{S}$ ⟶ O= (quinone methide) OCH_3, $^*CH_3$, O—H $\xrightarrow{-CH_2O}$

HO— (isoeugenol) OCH_3, $\overset{*}{C}H_3$ ⟶ Eugenol

3.6

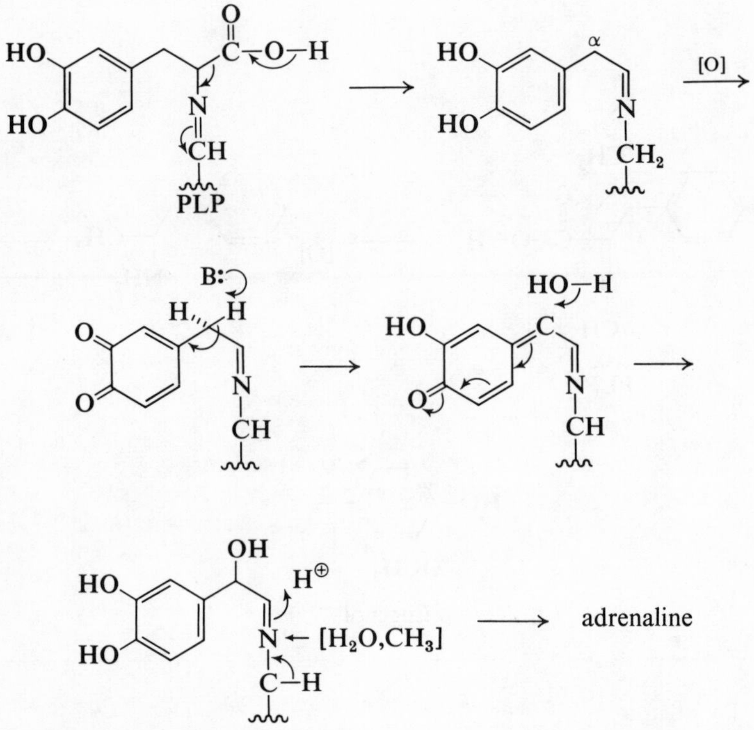

HO, HO, O, C, O—H, N, CH, PLP → HO, HO, α, N, CH_2 $\xrightarrow{[O]}$

B:, H, H, O, O, N, CH → HO—H, HO, C, O, N, CH →

OH, HO, HO, $H^{\oplus}$, N ← $[H_2O,CH_3]$, C—H ⟶ adrenaline

PLP catalysis in conjunction with oxidation of dopamine to the *o*-quinone activates the α-protons for abstraction by a base. Michael addition of water to the quinoid structure leads to α-hydroxylation.

3.7

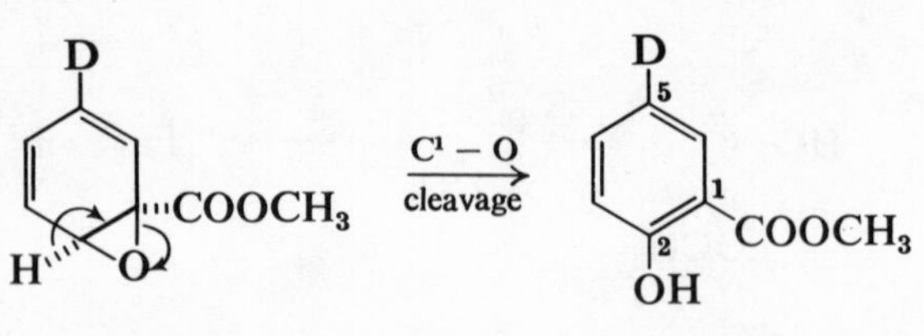

D, $COOCH_3$, H, O $\xrightarrow[\text{cleavage}]{C^1-O}$ D, 5, 1, $COOCH_3$, 2, OH

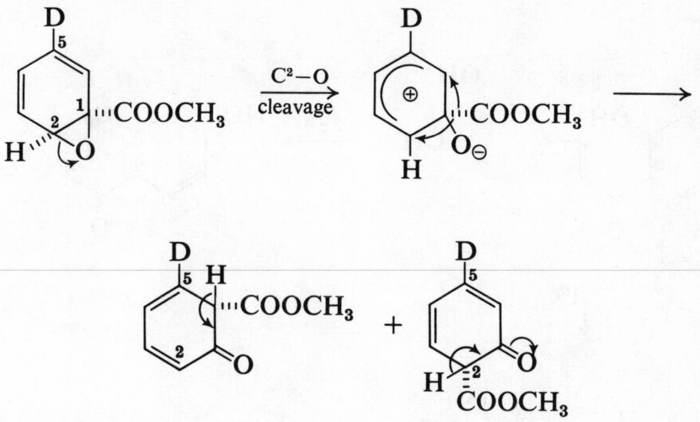

D, 5, 1, 2, H, O, COOCH$_3$ $\xrightarrow[\text{cleavage}]{C^2-O}$ D, ⊕, COOCH$_3$, O⊖, H ⟶

D, 5, H, COOCH$_3$, 2, O + D, 5, O, H, 2, COOCH$_3$

^{2}H-Labelling at C^5 gives three different products.

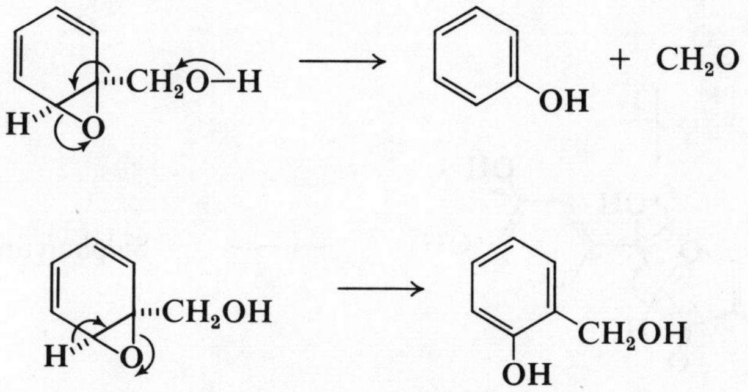

H, O, CH$_2$O—H ⟶ OH + CH$_2$O

H, O, CH$_2$OH ⟶ CH$_2$OH, OH

Chapter 4

4.1

4 Acetate → → OH, COCoA $\xrightarrow{-H_2O}$

3-OH-Octanoyl CoA

COOH $\xrightarrow{\text{3 Acetate}}$ 14: 1 (9c)

8: 1 (3c)

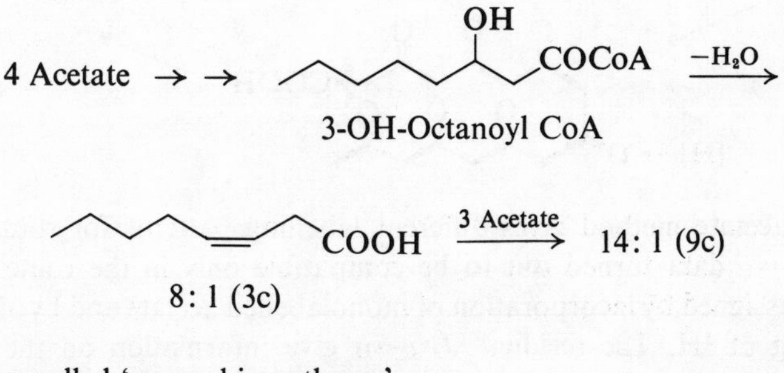

This is the so-called 'anaerobic pathway'.

4.2

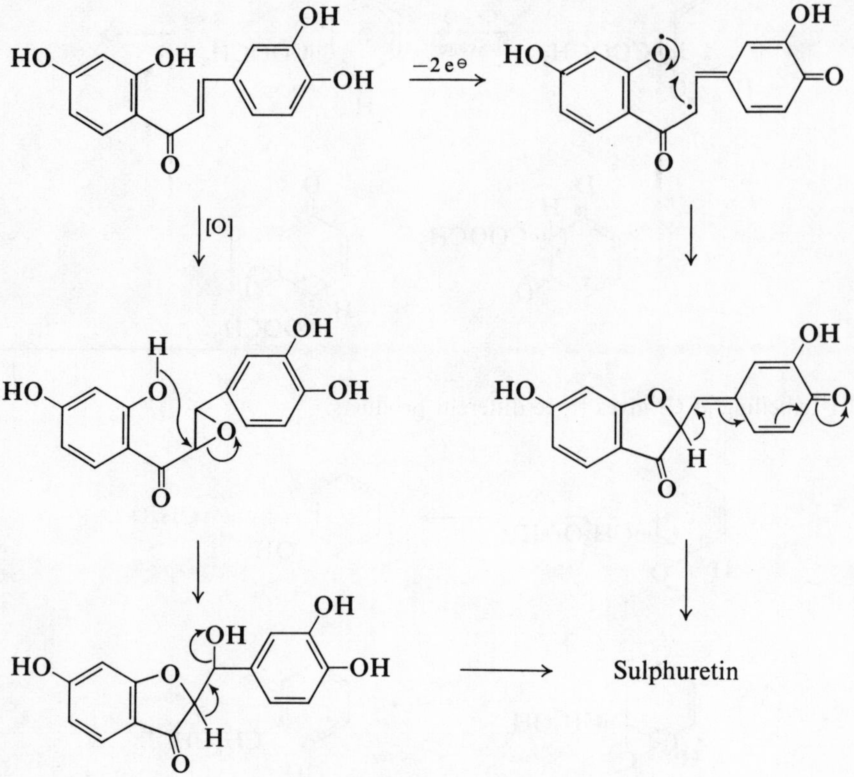

HO OH OH OH $\xrightarrow{-2e^{\ominus}}$ HO O OH O

O O

[O]

H OH OH HO O OH O

HO O OH O H O

OH OH HO O OH H O

Sulphuretin

4.3

O O O O O O O COOH

a

8 $-\overset{*}{C}OOH$

b

O O O COOH O O O [H] → O

OH O OH O OH

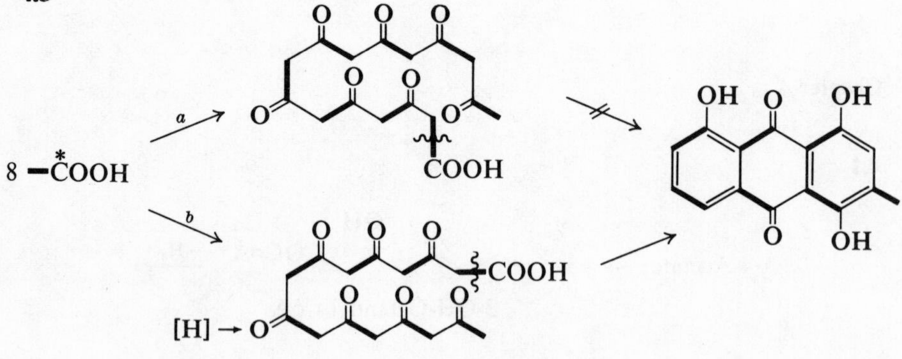

The $^{13}C_2$-acetate method gives different labelling patterns for routes a and b. The $J_{^{13}C-^{13}C}$ data turned out to be compatible only in the route b. The ^{13}C shifts are assigned by incorporation of monolabelled acetate and by off-resonance decoupling of ^1H. The residual $^1J_{^{13}C-^1H}$ give information on the number of hydrogens attached to a certain carbon atom.

4.4

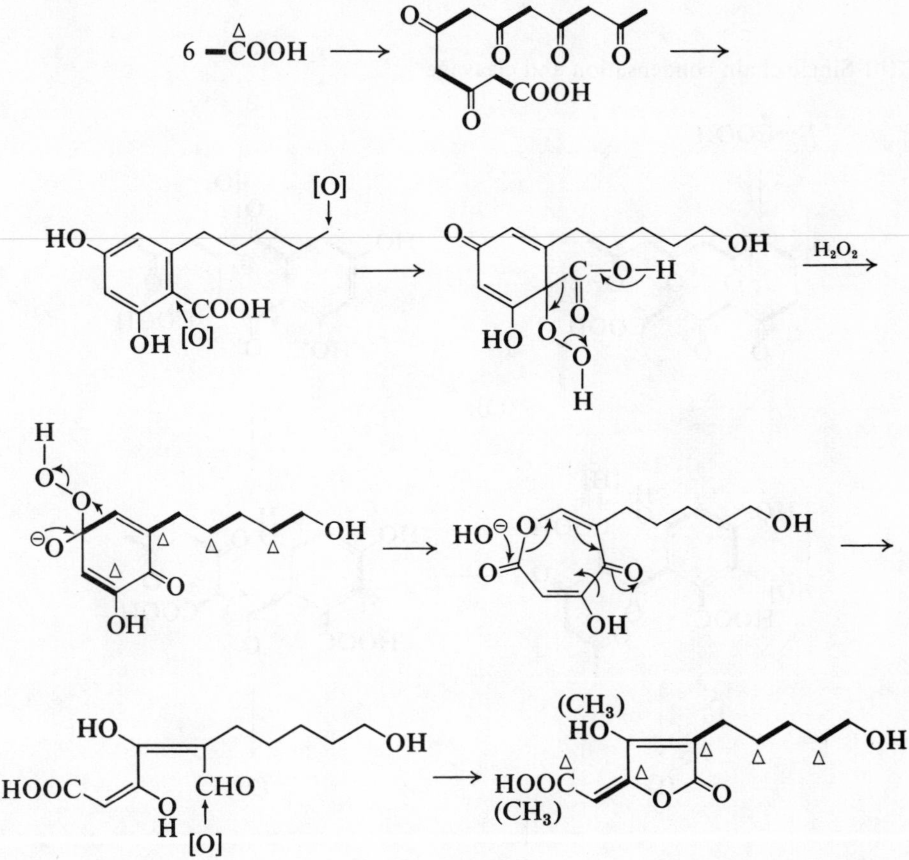

6 ▬$\overset{\Delta}{C}$OOH → [polyketide chain] COOH →

[O] ↓ HO, OH, COOH, [O] → O, OH, C–O–H, O, HO, O, O, H $\xrightarrow{H_2O_2}$

H, O, O, ⊖O, Δ, Δ, Δ, Δ, OH, O, OH → HO⊖, O, O, O, OH, OH →

HO, OH, HOOC, O, H, CHO, [O] → (CH$_3$), HO, Δ, Δ, Δ, OH, HOOC, Δ, (CH$_3$), O, O

4.5

(a) Separate chain condensation

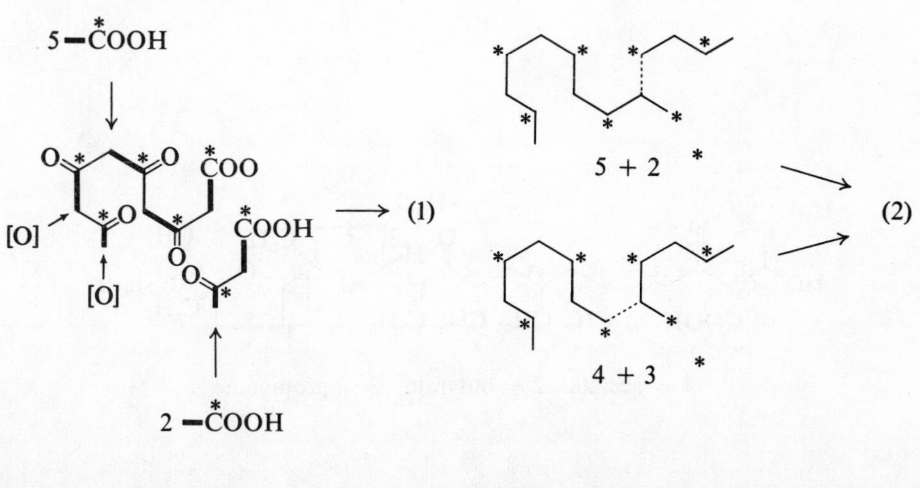

5 ▬$\overset{*}{C}$OOH ↓ O, O, O, O, $\overset{*}{C}$OO, $\overset{*}{C}$OOH, [O], [O], O ↑ 2 ▬$\overset{*}{C}$OOH → (1)

5 + 2 → (2)

4 + 3 → (2)

(b) Single chain condensation and cleavage

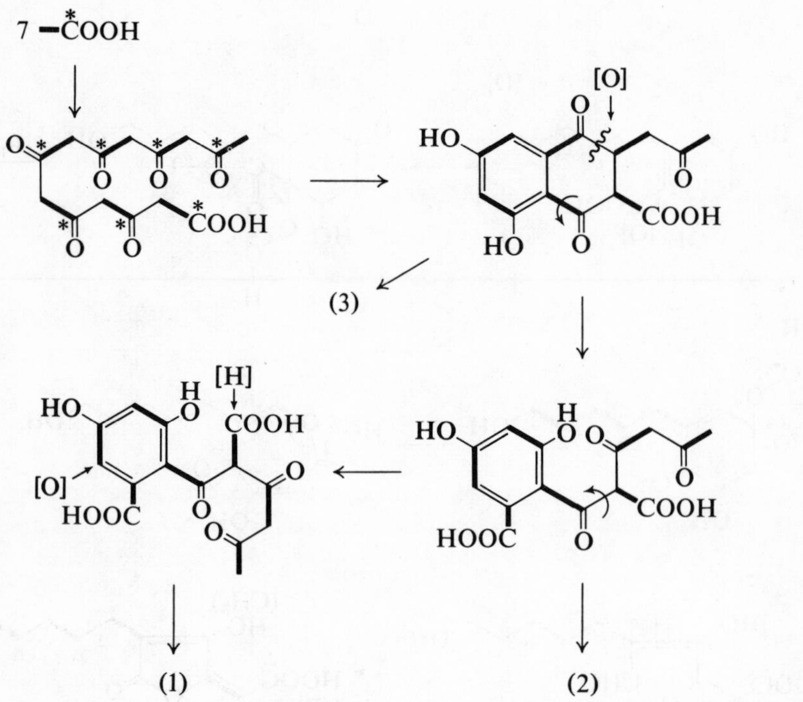

7 —$\overset{*}{C}OOH$

↓

O, O, O, O, O, O, *COOH → HO, O, [O], O, COOH, HO, O

(3)

↓

HO, H O, [H], COOH, [O], HOOC, O, O, O ← HO, H O, O, O, COOH, HOOC, O

↓ (1) ↓ (2)

The oxidative cleavage of the fusarubin precursor must occur between two acetate units. Rotation and recyclization give (1) and (2). The nearly equal enrichment of label indicates that one single chain precursor is more likely. A two chain assembly should conceivably give rise to higher enrichment in the starter units.

4.6

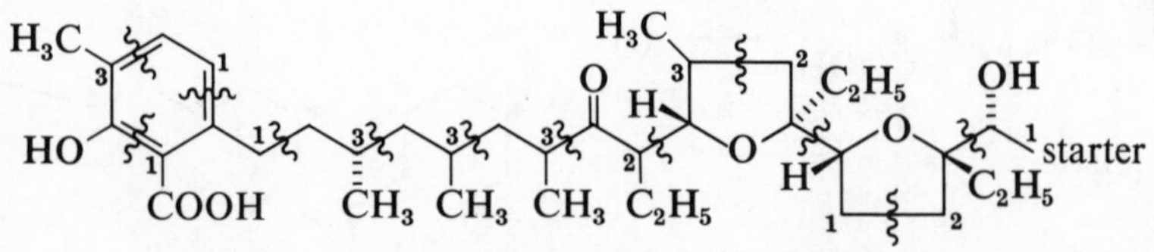

H_3C, HO, COOH, CH_3 CH_3 CH_3 C_2H_5, O, H_3C, H, O, C_2H_5, H, O, OH, starter, C_2H_5

1 = acetate 2 = butyrate 3 = propionate

Formation of the benzene ring:

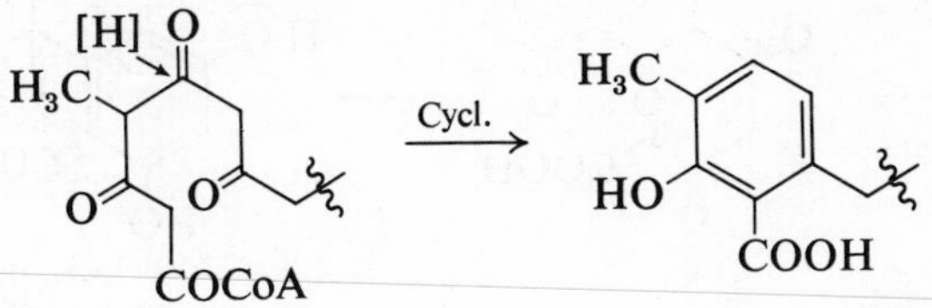

Formation of the tetrahydrofuran ring:

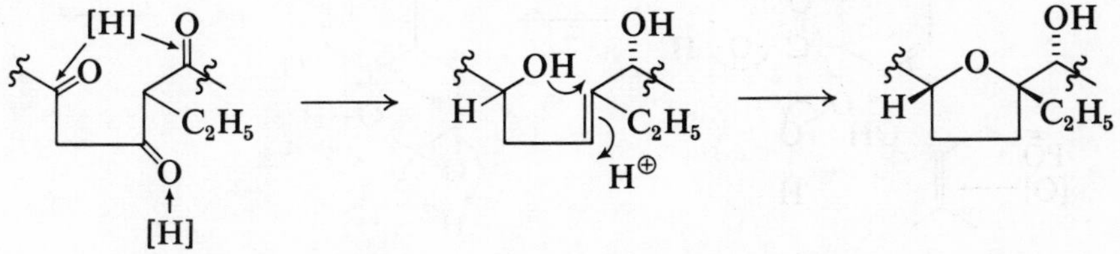

Another possibility: (*cf.* Cane, D. E., Liang, T.-C. and Hasler, H. *J. Am. Chem. Soc.* **103** (1981) 5962; Hutchinson, C. R., Sherman, M. M., McInnes, A. G., Walter, J. A. and Vederas, J. C. *J. Am. Chem. Soc.* **103** (1981) 5956), the vicinal hydroxy groups are derived from molecular oxygen via the olefins and epoxides.

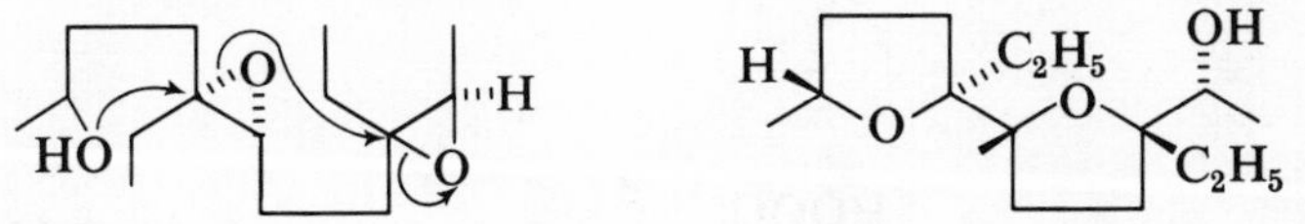

The methyl groups could conceivably come from methionine, but labelling experiments show that this is not the case.

4.7

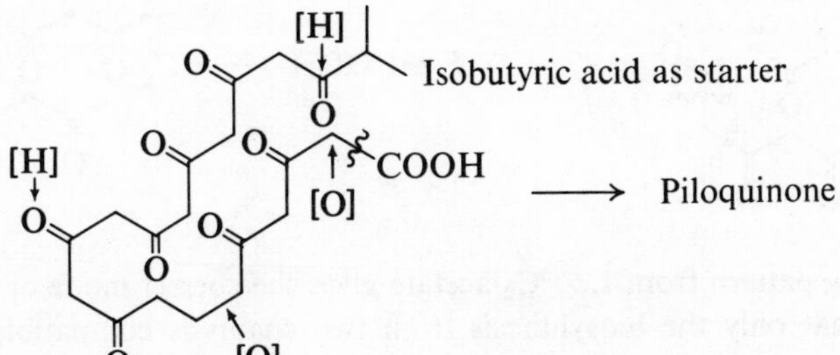

Isobutyric acid as starter

→ Piloquinone

4.8

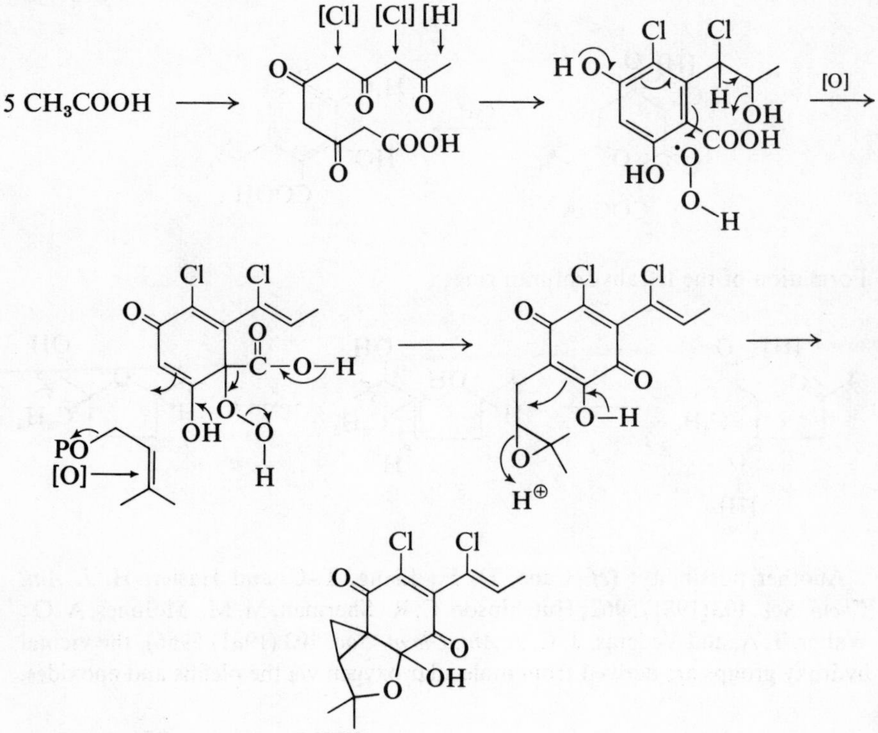

5 CH_3COOH → [Cl] [Cl] [H] → O, COOH → HO, Cl, Cl, H, OH, COOH, HO, O–O–H → [O] →

Cl, Cl, O, C–O–H, OH, O, O, H, PO, [O] → → O, Cl, Cl, O, O–H, O, $H^{\oplus}$ →

Cl, Cl, O, O, OH, O

4.9

6 $\overset{\bullet}{C}H_3\overset{*}{C}OOH$ → HOOC, C, O, O, C, O, O, O → HO, O, O

C, O, O, O, HOOC, C, O, O

5 ▬COOH → COOH, O, O, OH, O, O

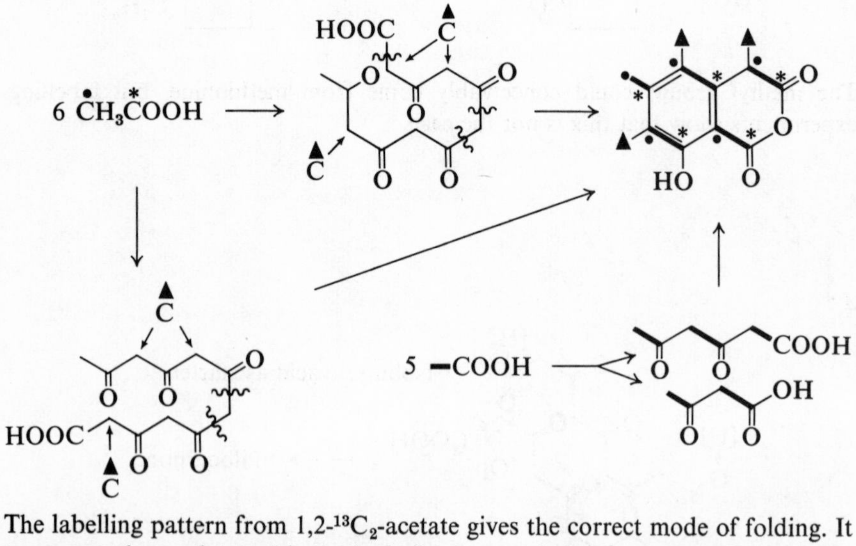

The labelling pattern from 1,2-$^{13}C_2$-acetate gives the correct mode of folding. It turns out that only the biosynthesis from two chains is compatible with the ^{13}C NMR spectroscopic results ($^2J_{^{13}C^{13}C}$).

Chapter 5

5.1 The key step is the 1,3 shift. C^5–$^2H_2$-labelled mevalonic acid is expected to label artemisia ketone at C^1 and at C^9 or C^{10}.

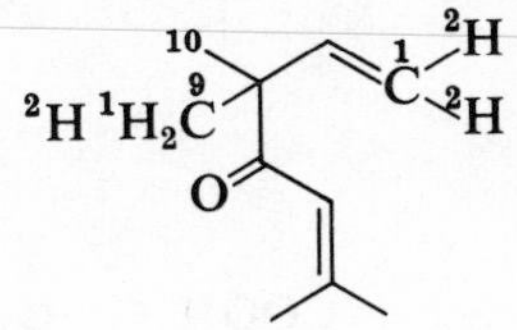

^1H NMR should show a triplet (1: 1: 1) for the methylene protons of C^9 Artemisia alcohol should be a precursor for the ketone. Electrophilic aliphatic C^4–C^2 condensation leads to the chrysanthemyl skeleton (section 5.3).

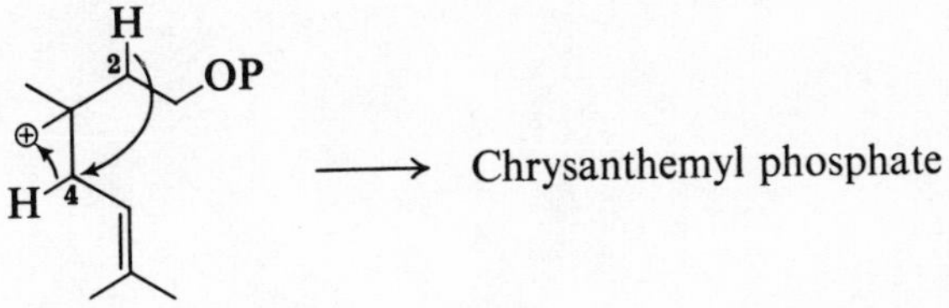

→ Chrysanthemyl phosphate

5.2

→ Eremophilone

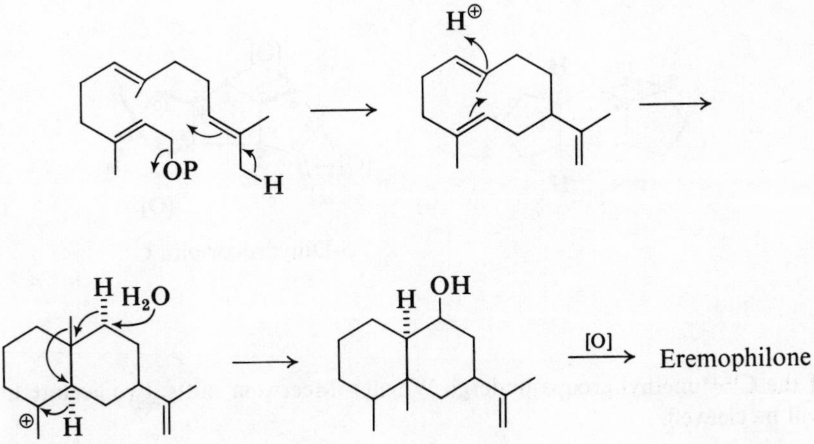

5.3

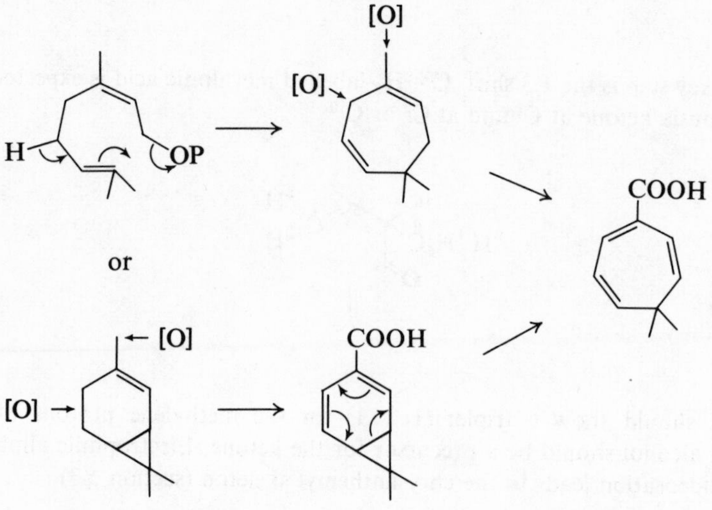

[O]

[O]

H

OP

COOH

or

[O]

[O]

COOH

4-Carene

5.4

$H_3C-COOH$

H

OP

12

H

13

H

H

[O]

7

8

6

15

9

10

5

11

3

[O]

4

2

1

14

13

12

[O]

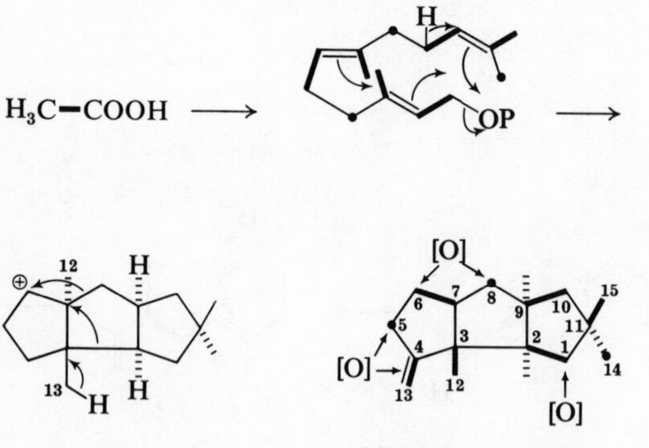

5-Dihydrocoriolin C

If the $C^{12,13}$ methyl groups undergo Wagner–Meerwein shifts, two acetate units will be cleaved.

5.5

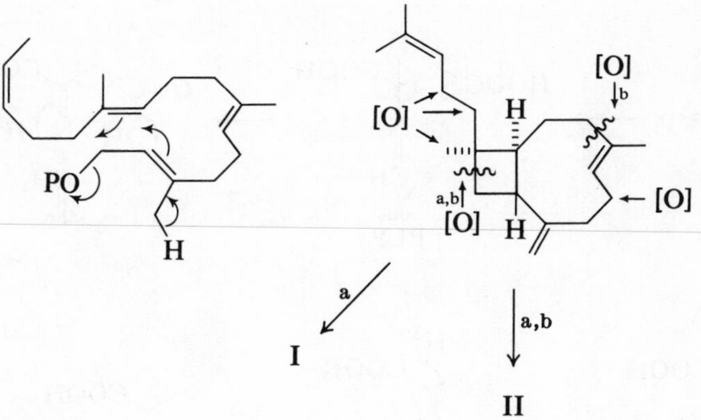

I

II

5.6 There are several ways to coil the C_{20} geranylgeranyl phosphate chain to obtain this cembrene derived diterpene, e.g.

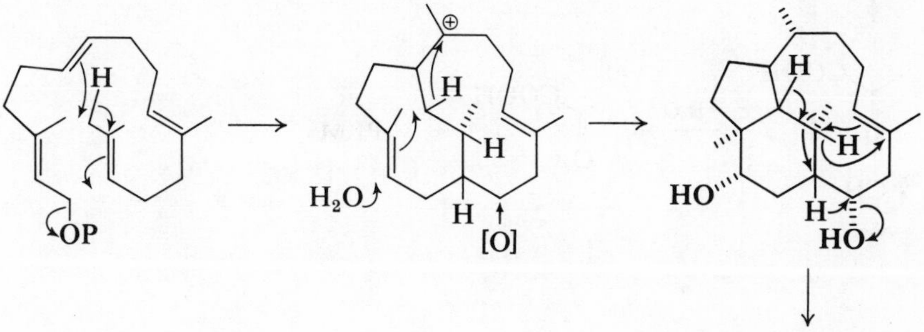

3α-Hydroxy-15-rippertene

This sequence starts from all-*trans*-geranylgeranyl phosphate which undergoes two 1,3 H shifts and one 1,2 methyl shift.

Chapter 6

6.1 PLP gives a Schiff base with aspartic acid and rearranges to the ketimine. This structure stabilizes a negative charge on C^3, i.e. it facilitates the cleavage of the C^3–COOH bond. Protonation of C^3 and C^2 gives alanine after hydrolysis. Protonation of C^3 and the pyridoxamine moiety followed by hydrolysis gives pyruvic acid and PMP.

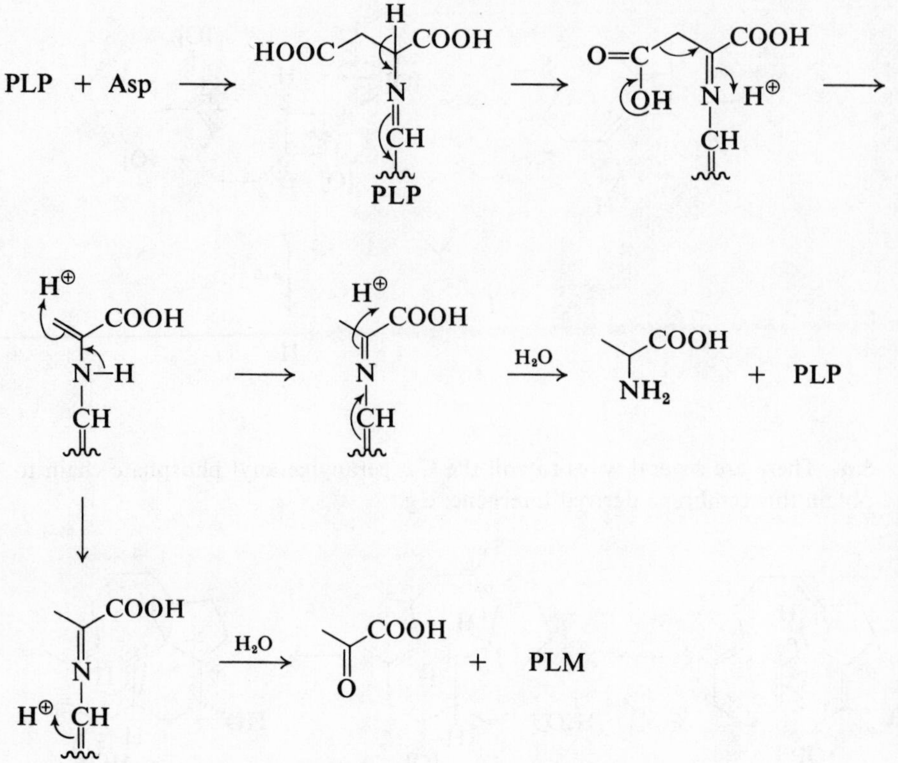

$$\text{PLP} + \text{Asp} \longrightarrow \text{HOOC-CH}_2\text{-C(H)(COOH)-N=CH-PLP} \longrightarrow \text{O=C(OH)-CH}_2\text{-C(COOH)=N(H}^{\oplus}\text{)-CH=} \longrightarrow$$

$$\text{CH}_2\text{=C(COOH)-N(H)-CH=} \longrightarrow \text{CH}_3\text{-C(COOH)=N-CH=} \xrightarrow{H_2O} \text{CH}_3\text{CH(NH}_2\text{)COOH} + \text{PLP}$$

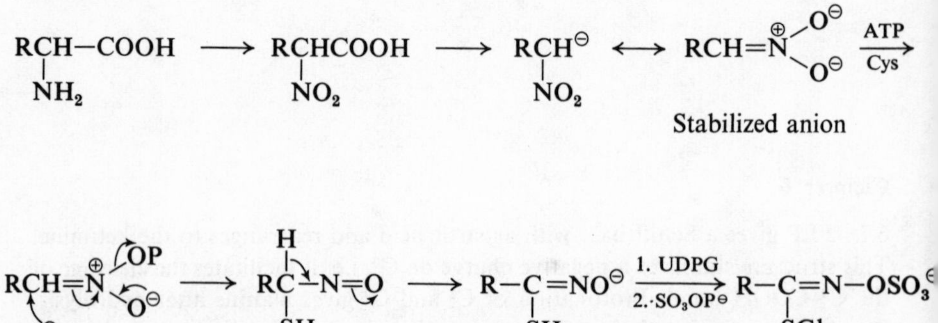

$$\downarrow$$

$$\text{CH}_3\text{-C(COOH)=N-CH=} \xrightarrow{H_2O} \text{CH}_3\text{C(=O)COOH} + \text{PLM}$$

6.2

$$\text{RCH(NH}_2\text{)-COOH} \longrightarrow \text{RCH(NO}_2\text{)COOH} \longrightarrow \text{RCH}^{\ominus}\text{(NO}_2\text{)} \longleftrightarrow \text{RCH=N}^{\oplus}\text{(O}^{\ominus}\text{)}_2 \xrightarrow[\text{Cys}]{\text{ATP}}$$

Stabilized anion

$$\text{RCH=N}^{\oplus}\text{(OP)(O}^{\ominus}\text{)} + \text{HS}^{\ominus} \longrightarrow \text{RC(H)(SH)-N=O} \longrightarrow \text{R-C(SH)=NO}^{\ominus} \xrightarrow[2.\cdot SO_3OP^{\ominus}]{1.\ UDPG} \text{R-C(SGl)=N-OSO}_3^{\ominus}$$

6.3

Gabaculine 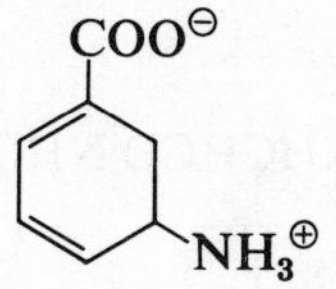 competes with $COO^{\ominus}$ ... $NH_3^{\oplus}$

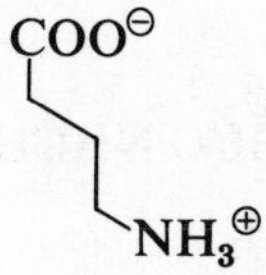

at the active site of the transaminase forming

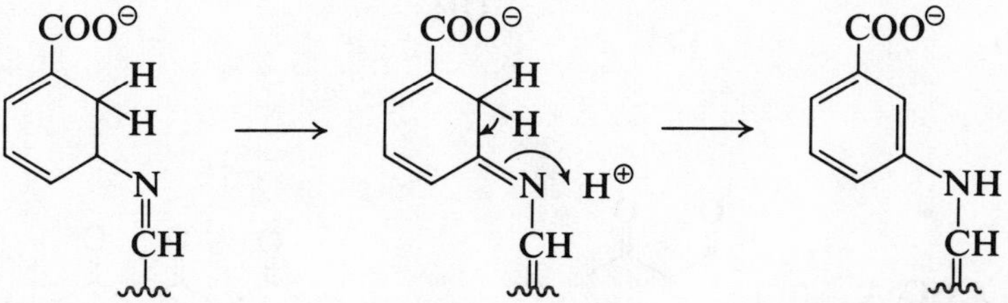

PLP-aldimine → PLP-ketimine → Stable aromatic and blocked PLP-derivative

PLP-aldimine rearranges to the ketimine and aromatizes to a non-hydrolysable PLP derivative blocking its further reactions as coenzyme.

6.4

Tryptophan $\xrightarrow[\text{Decarboxylation}]{\text{PLP}}$ Tryptamine $\xrightarrow[\text{2. Br}^{\ominus}\text{/peroxidase}]{\text{1. [CH}_3\text{]}}$

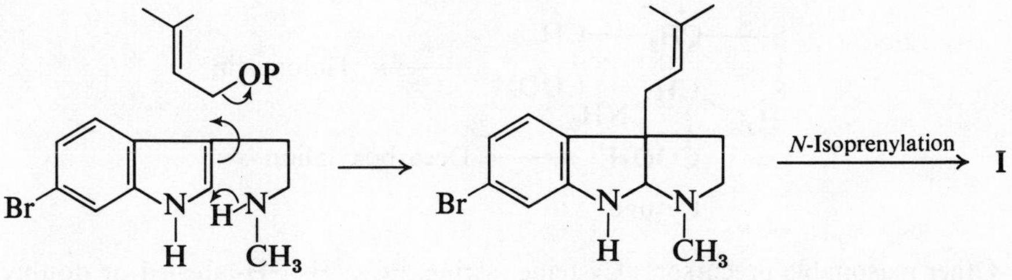

$\xrightarrow{N\text{-Isoprenylation}}$ I

The order of events is not known. The inverted isoprene unit at C^3 of II is either introduced by an S_N2' reaction or an *N*-isoprenylation followed by a Claisen rearrangement and cyclization

6.5

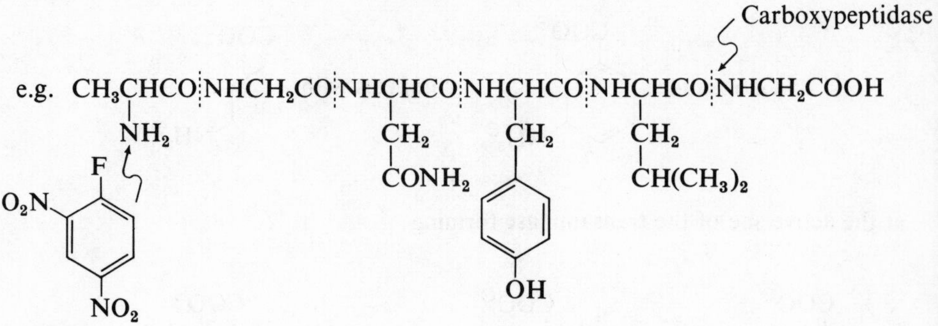

Carboxypeptidase

e.g. $CH_3CHCO \vdots NHCH_2CO \vdots NHCHCO \vdots NHCHCO \vdots NHCHCO \vdots NHCH_2COOH$

NH_2 ; CH_2CONH_2 ; CH_2–C$_6$H$_4$–OH ; $CH_2CH(CH_3)_2$

F, O_2N, NO_2

6.6

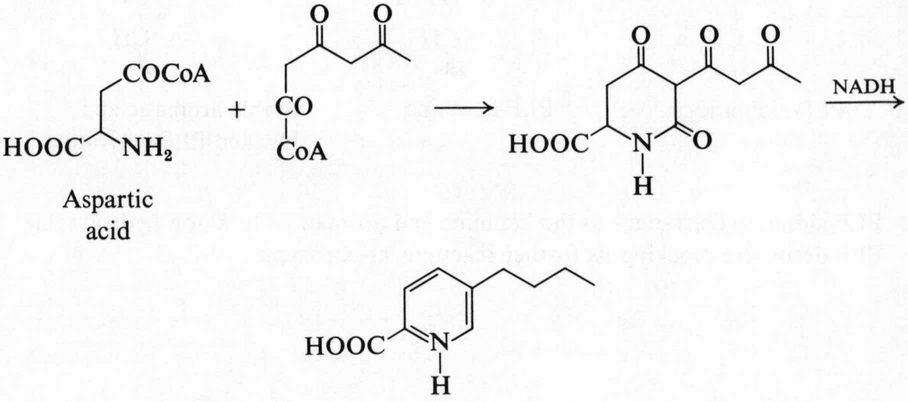

COCoA, HOOC, NH_2 + O O, CO, CoA ⟶ O O O, HOOC, N, O, H ⟶ NADH

Aspartic acid

HOOC, N, H

6.7

A working hypothesis:

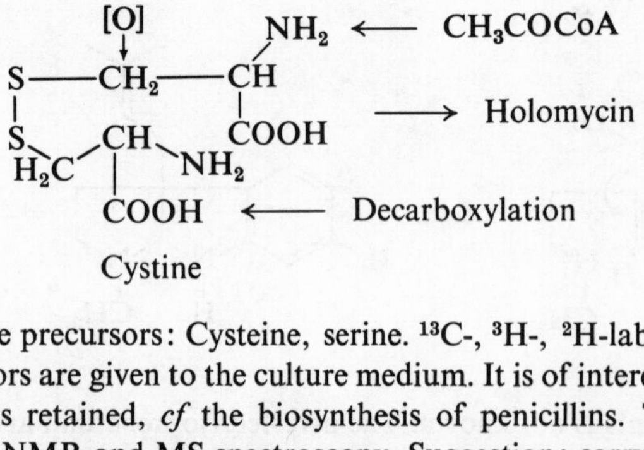

[O] ↓ ; NH_2 ⟵ CH_3COCoA

S—CH_2—CH ; COOH ⟶ Holomycin

S, H_2C, CH, NH_2

COOH ⟵ Decarboxylation

Cystine

Other reasonable precursors: Cysteine, serine. ^{13}C-, ^{3}H-, ^{2}H-labelled or doubly labelled precursors are given to the culture medium. It is of interest to investigate whether C_α–H is retained, *cf* the biosynthesis of penicillins. The metabolites are analysed by NMR and MS spectroscopy. Suggestion: carry out a literature search to see if the problem is solved.

6.8 Anthramycin can by visual inspection be dissected into an anthranilic acid derivative and a substituted proline derivative, the latter originating from tyrosine by ring cleavage.

[O] [O] [C*H3] H N NH2 COOH → OH H3C* NH2 COOH + HOOC• NH2 3H OH 3H [O] →

HO H3C* H N O NH O 3H [C*H3] [O] HO OH [O] → → Anthramycin

The order of events is still unknown but methylation of the proline part occurs late since one ^3H is retained.

6.9 The compounds can be dissected into two identical amino acids, e.g. homoserine or homocysteine which by a PLP-catalysed process combine head to tail or head to head. Aromatization gives the metabolites. A simple biomimetic synthesis along these lines is published (Brown, M. and Büchi, G. *J. Am. Chem. Soc.* **98** (1976) 3049).

COOH H2N– –NH2 HOOC or H2N– –NH2 HOOC COOH $\xrightarrow{[O]}$ Zoanthoxanthins

Carbamoyl phosphate + NH_3

7.1

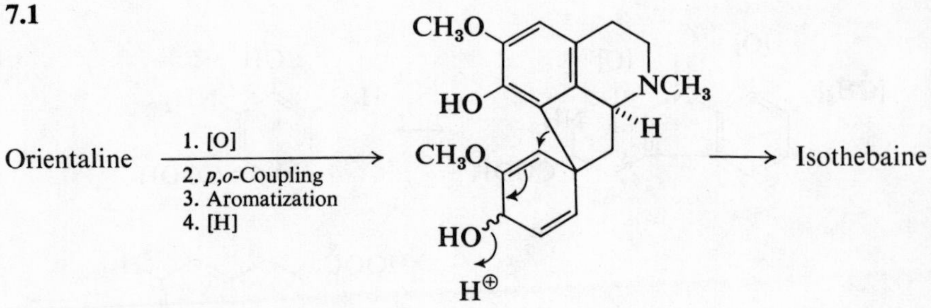

Orientaline → (1. [O] 2. *p,o*-Coupling 3. Aromatization 4. [H]) → Isothebaine

CH_3O HO NCH_3 H CH_3O HO $H^{\oplus}$

7.2

Norbelladine → (1. Methylation 2. *p,o*-Coupling 3. Michael addition) →

[H] O H CH_3O H N HO $[CH_3]$

HO H CH_3O H CH_3O N [O]

Pluviine

Ring-fission →

$[CH_3]$ → HN H CH_3O H OH CH_3O CHO [O]

Lactonization

Homolycorine

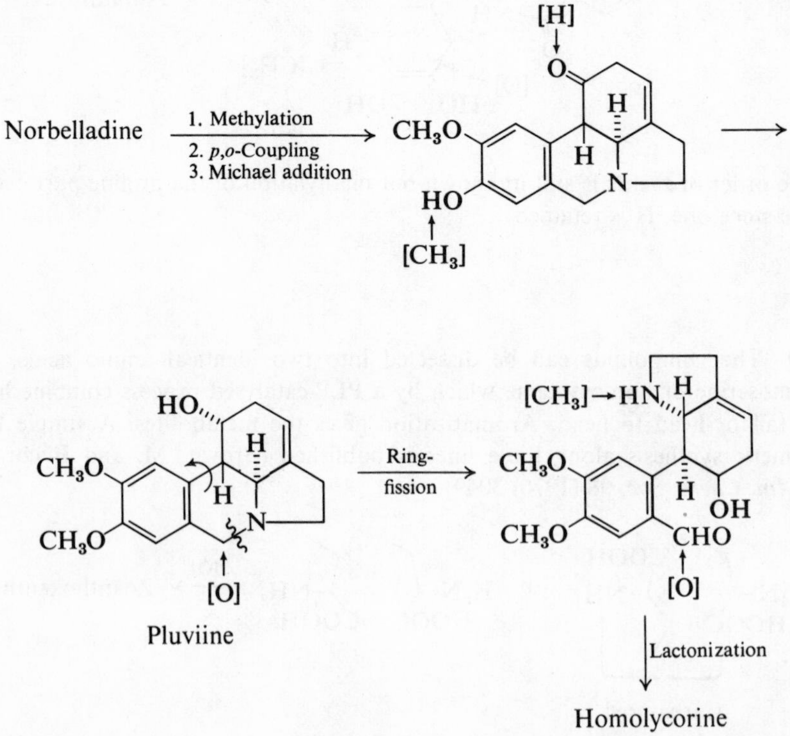

7.3

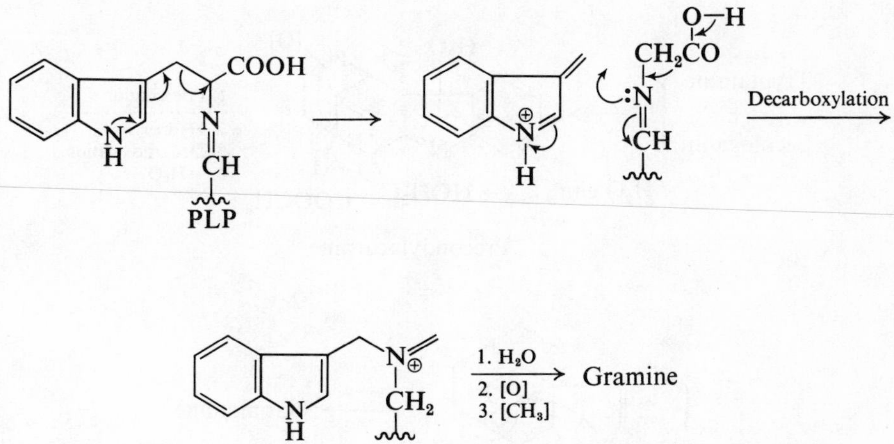

COOH
N
H
N
CH
PLP

N
H
O—H
CH_2CO
:N
CH
Decarboxylation

N
H
N
CH$_2$
1. H_2O
2. [O]
3. [CH_3]
Gramine

7.4

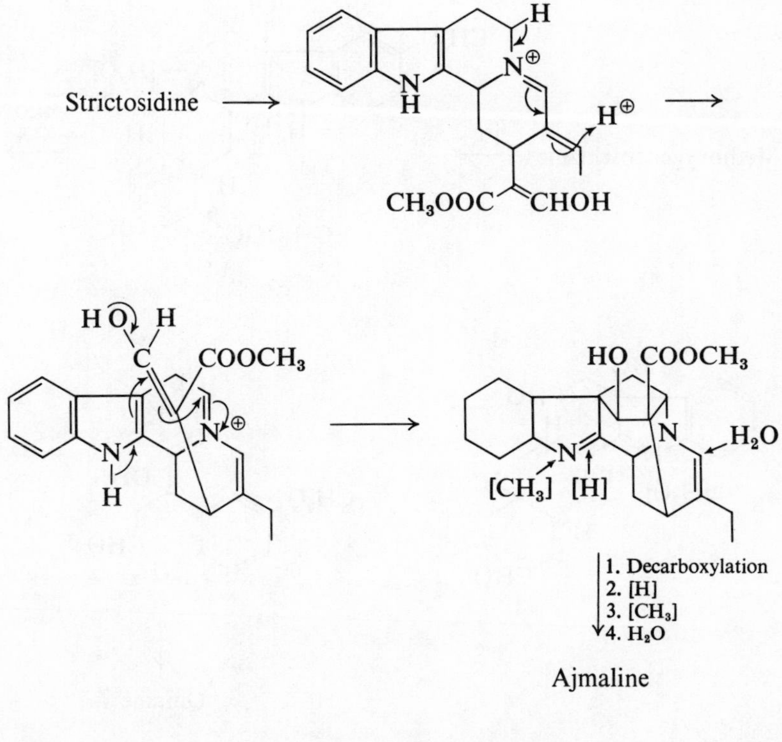

Strictosidine
H
N
H
N
$H^{\oplus}$
CH_3OOC
CHOH

H O H
C
$COOCH_3$
N
H
N

HO $COOCH_3$
N
N
H_2O
[CH_3] [H]
1. Decarboxylation
2. [H]
3. [CH_3]
4. H_2O

Ajmaline

7.5

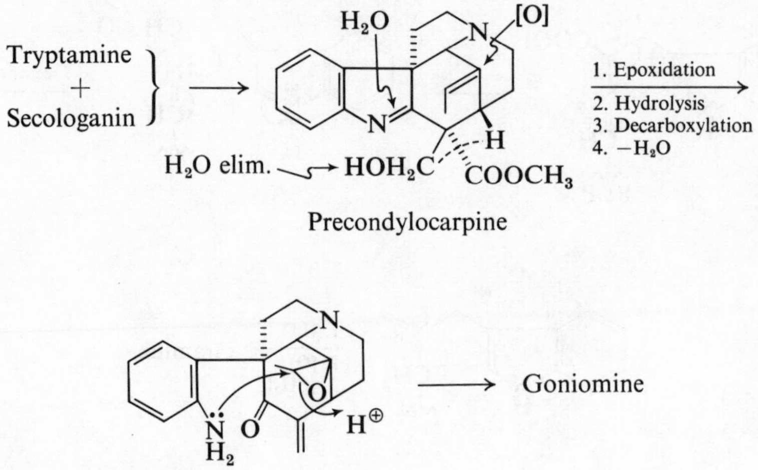

Tryptamine + Secologanin → Precondylocarpine

H_2O, [O], H_2O elim. → HOH_2C, H, $COOCH_3$

Precondylocarpine

1. Epoxidation
2. Hydrolysis
3. Decarboxylation
4. $-H_2O$

N, O, $\ddot{N}H_2$, $H^{\oplus}$ → Goniomine

7.6

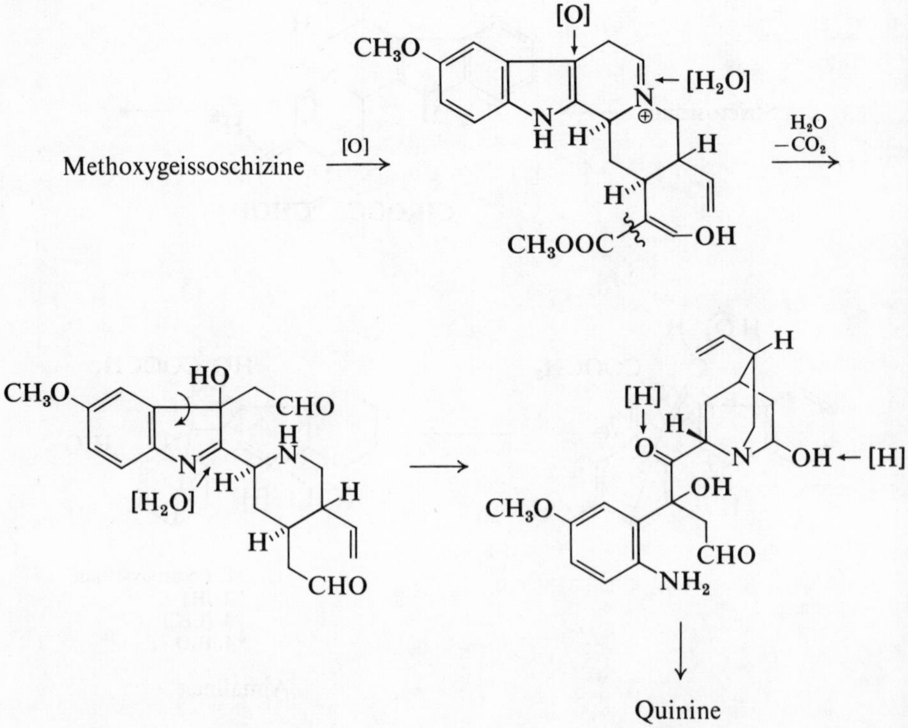

Methoxygeissoschizine $\xrightarrow{[O]}$

CH_3O, [O], $[H_2O]$, N, H, H, H, H, CH_3OOC, OH

$\xrightarrow[-CO_2]{H_2O}$

CH_3O, HO, CHO, H, N, N, H, $[H_2O]$, H, H, CHO →

H, [H], H, O, N, OH ← [H], OH, CH_3O, CHO, NH_2

↓

Quinine

7.7

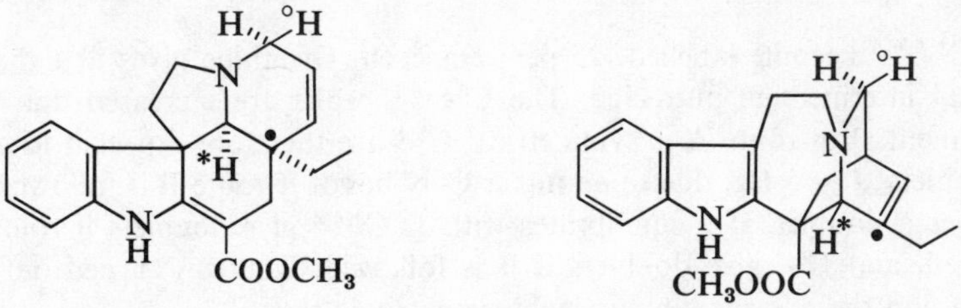

H,,, °H
N
*H
N
H
$COOCH_3$

Tabersonine

H,,, °H
N
N
H
H*
CH_3OOC

Catharanthine

* 1-^3H_2-Geraniol
• $2\text{-}^{14}C$-Geraniol
∘ 5-^3H-Loganin

7.8

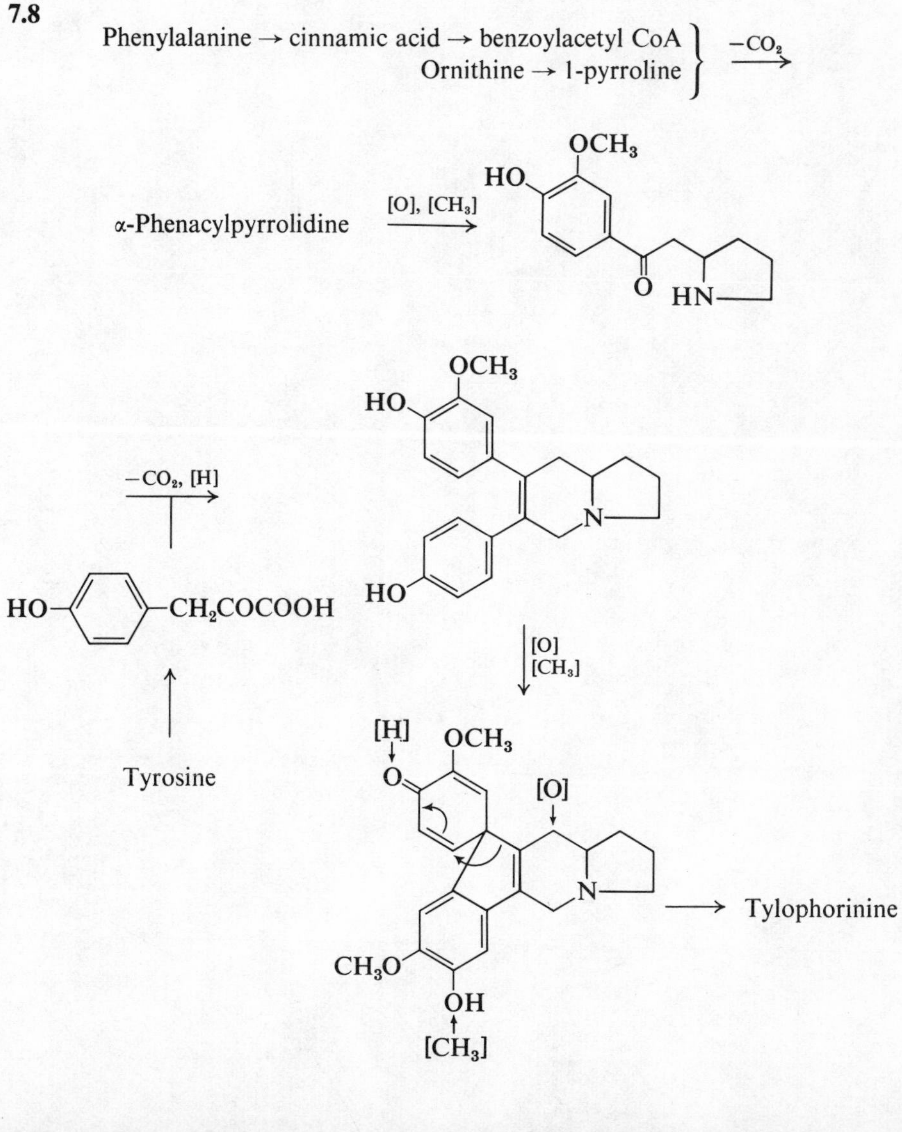

Phenylalanine → cinnamic acid → benzoylacetyl CoA
Ornithine → 1-pyrroline
$\xrightarrow{-CO_2}$

α-Phenacylpyrrolidine $\xrightarrow{[O], [CH_3]}$

OCH_3
HO
O
HN

$\xrightarrow{-CO_2, [H]}$

OCH_3
HO
N
HO

HO—C_6H_4—$CH_2COCOOH$

↑

Tyrosine

↓ [O] $[CH_3]$

[H]
↓
O
OCH_3
[O]
↓
N
CH_3O
OH
↑
$[CH_3]$

→ Tylophorinine

7.9 $C^{3,5,8,9}$ become labelled, 25 per cent each. Ornithine gives first the symmetrical intermediate putrecine. The $C^{3,5,8,9}$ peaks are increased due to ^{13}C enrichment. 3 in route A is symmetrical $C^{3,5}$ are therefore expected to appear as doublets, $J \sim 5$ Hz, due to an intact C–N bond. If route B is followed, only C^{5} gives a doublet. If 2 equilibrates with 1, $C^{3,5,8}$ give doublets if route A is followed, and $C^{5,8}$ give doublets if B is followed. Route A turned out to be correct and there is actually no rapid sigmatropic shift.

Name Index

Subject Index